THE UNITED STATES ENERGY ATLAS

THE UNITED STATES

ENERGY ATLAS

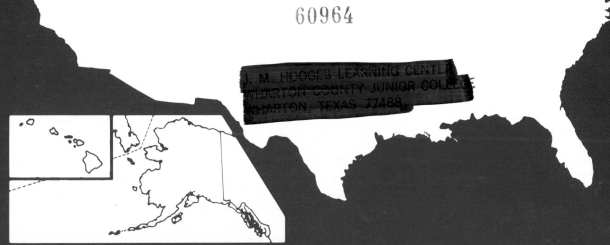

David J. Cuff / William J. Young

THE FREE PRESS   A Division of Macmillan Publishing Co., Inc.   New York

COLLIER MACMILLAN PUBLISHERS   London

To Ethel M. Cuff, whose personal energy and love of language have been an inspiration to her son.

D.J.C.

To my wife Edna for her unceasing encouragement and to Beth, Dona, Jo Ann, and John, in the hope that today's energy decisions will assure them a healthy environment and productive lives.

W.J.Y.

Copyright © 1980 by The Free Press
A Division of Macmillan Publishing Co., Inc.

The Free Press
A Division of Macmillan Publishing Co., Inc.
866 Third Avenue, New York, N.Y. 10022

Collier Macmillan Canada, Ltd.

Library of Congress Catalog Card Number: 80-24942

Printed in the United States of America

printing number

1   2   3   4   5   6   7   8   9   10

Library of Congress Cataloging in Publication Data

Cuff, David.
   The United States energy atlas.

   Includes bibliographies.
   1. Power resources—United States. I. Young,
William, 1935-   joint author. II. Title.
TJ163.25.U6C83   333.79'0973   80-24942
ISBN 0-02-691250-3

Contents

Acknowledgments

Obtaining all the information for this book would have been impossible without those who responded so readily to our requests by telephone, by mail, and in person. We would like to thank the following people; and we apologize to those whose names have been omitted. Of course, any errors of fact or interpretation must be blamed on the authors alone.

Most important were the cooperation and patience of George Brightbill and Sandy Thompson in the documents room of Temple's Paley Library. They found many valuable items for us, and were more interested in the documents being useful than in having them returned promptly.

Four energy transportation flow maps in the atlas are part of a remarkable series produced under the direction of John Jimison of the Congressional Research Service at the Library of Congress. With his cooperation and the aid of Jack Wittman of the Special Mapping Division of the U.S. Geological Survey, we were able to reproduce those maps.

Information on coal resources was provided generously by Donald Ralston and Zane Murphy of the U.S. Bureau of Mines, and by George Nielsen, Editor in Chief of Keystone Coal Industry Manual.

Wind data and advice on its interpretation were obtained from Lowell Kravitz, meteorologist with General Electric at King of Prussia, Pennsylvania. Louis Divone, of the U.S. Department of Energy (then ERDA) helped us obtain contractors' research reports. Jack Reed, of Sandia Laboratories in Albuquerque, New Mexico was generous with his time and with the results of his research.

Obtaining oil shale information in the field was eased by James Hager and Peter Rutledge of the U.S. Geological Survey in Grand Junction, Colorado. Trips to commercial operations in the area were made possible by L. Ludlum of the Colony Development Corporation, W. Herget of the Rio Blanco Oil Shale Project, Martin Redding of C-b Shale Oil Project, and R. Thomason of Occidental Oil Shale Incorporated.

Latest data on hydroelectric potential was obtained with cooperation of Donald Gunn, Randy Hanchey, and Shap Zanganeh of the U.S. Army Corps of Engineers Institute for Water Resources, as well as Neal Jennings and Eugene Jarecki of the Federal Regulatory Commission.

Guidance on biomass data was provided by Pete Schauffler of the Bio-Energy Council in Washington, Dan Fink of the U.S. Department of Agriculture, Dan Hathaway in the Fuels for Biomass Branch of the U.S. Department of Energy, and John Zerbe of the University of Wisconsin Forest Products Laboratory.

Information on ocean thermal gradients was found

through Robert Cohen, and Lloyd Lewis of the Division of Solar Technology in the U.S. Department of Energy, and from J. P. Walsh, OTEC Project Manager of Value Engineering, Alexandria, Virginia.

John Duffie and William Beckman of the Solar Energy Laboratory at the University of Wisconsin freely shared their unpublished research results and were most patient with our questions. Paul Curto and George Bennington of the Mitre Corporation, Maclean, Virginia, directed us to reports on the potential of solar applications. Richard Meyer of the U.S. Geological Survey in Reston, Virginia, gave early encouragement as well as advice on interpreting data on undiscovered crude oil and natural gas. Donald White of the U.S. Geological Survey in Menlo Park, California provided perspective on the state of the information on the country's geothermal resources.

Ed Barry, Senior Vice President of Macmillan Publishing Co., Inc., Charles Smith, Vice President of Macmillan Publishing Co., Inc. and Kitty Moore, Editor, all lent their enthusiasm to the project at a time when others saw only its difficulties. Kitty Moore's insistence on clarity and completeness substantially improved the text and made the book more useful as a reference work.

We offer special thanks to Mark Mattson, of *Cartografik* in Philadelphia, for designing the three-dimensional symbols used for resource amounts.

THE UNITED STATES ENERGY ATLAS

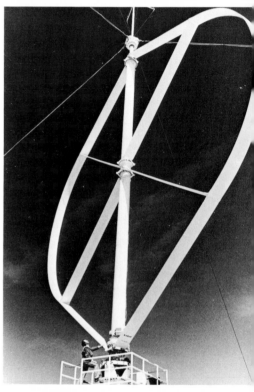

Introduction

- Organization

- Methods of Presentation

- Nonrenewable Resources

- Renewable Resources

Since 1973, when OPEC nations asserted their control of world oil supplies and prices, developing nations and industrial powers alike have faced ever increasing costs of crude oil and the very real possibility of supplies being interrupted. Now, the continuing political instability in the Middle East further compromises future oil supplies.

Despite the uncertainty of supply and the certainty of even higher prices, the United States has done little to move away from crude oil as a major energy source. Currently, the country obtains roughly 50 percent of its raw energy from crude oil—almost half of which is imported. This continued dependence on expensive foreign crude has not only made the country vulnerable, but has contributed heavily to inflation and the erosion of the value of the dollar.

It is imperative that energy supplies *within the United States* be clearly understood: first, to demonstrate the inevitability of the domestic petroleum shortage; and second, to delineate alternatives to crude oil and natural gas. An understanding of the broad spectrum of possible energy sources is not easily attained because resource information for mineral fuels and for renewable energy sources resides in a variety of government documents, trade journals, scientific journals, monographs, and reports to the government by research contractors. Furthermore, each document has its own form of presentation, and employs units that are unique to the resource being studied. It is rare to find the amounts of different resources all expressed in common units which make comparisons easy. While a great deal has been written about energy policies and new technologies, very little has been written to promote an understanding of the magnitudes and locations of energy resources within the country.

This atlas strives to present a complete review of both renewable and non-renewable energy resources. It will serve as a reference for those who need detailed information on a specific resource and for those who need an overview of the various possibilities. Specifically, the atlas provides the following.

- A detailed analysis of the amounts of mineral fuels and their locations within the country.
- Clarification of the uncertainties that cloud the amounts of mineral fuels.
- Highlights of the production and transportation of fuels.
- An assessment of the renewable energy sources, with emphasis on their geographic aspects.
- A regional and national overview of the energy attainable from both renewable and non-renewable sources, with resource amounts expressed in energy units to facilitate comparisons.

◀

A composite photograph of the different energy resources used in the United States.

- A profile of each of the nation's ten economic regions that captures their energy resource character for planning purposes.
- An introduction to selected energy futures, and an estimate of the impact of future demands upon the existing resources.

Technical and policy matters that bear on the use of fuels or the transition to renewable energy sources are not included, except to the extent that they illuminate some resource amounts or the possibility of their extraction. The opposite emphasis is seen in a number of works that study energy policy and exclude any consideration of resource amounts (see, for instance, Lovins, 1975). This atlas will be complementary to such studies, and also to those recent publications that do compare energy resources but do not provide full information on resource amounts or their locations in the country (see Schurr, 1979, and National Academy of Sciences, 1980).

Figure 1 shows that a great deal of the energy that flows into and through the United States economy is not used, but is rejected. Thus, *primary energy* is roughly twice the nation's *end-use energy*. Of paramount importance are strategies that will re-use energy that is rejected from power plants and engines of all sorts, and strategies that will diminish demand for the end-use energy. A barrel of oil that is gained through conservation is just as real as one gained through production in a frontier area or by some synthetic fuel process—and it is less costly both in dollar and energy expenditure. Recognizing the urgency of conservation, this atlas nevertheless is concerned

largely with the left side of Figure 1: the sources of primary energy. It also deals not only with the traditional fuels that now account for roughly 95 percent of the nation's primary energy, but also with renewable and semi-renewable sources that may soon displace much of the mineral fuel and provide energy that is more healthful and more sustainable.

Organization

The atlas has three parts and an Appendix. The first deals with nonrenewable (mineral) sources of energy. The second deals with renewable sources. The third part is a summary and overview, and is followed by the Appendix.

ONE. NONRENEWABLE SOURCES

One chapter is devoted to each of the following nonrenewable resources: coal, crude oil and natural gas, shale oil and tar sands, nuclear fuels, and geothermal heat. Geothermal heat is grouped with nonrenewable sources because the accessible occurrences of usable high temperatures can be exhausted by development.

TWO. RENEWABLE SOURCES

Solar radiation plays a major role as a primary energy source. It is discussed first, and is followed by chapters on those energy sources that depend upon solar radiation: wind, hydropower, ocean thermal gradients, and biomass. Tidal power, despite its being independent of

Fig. 1 Primary energy inputs, and end-use sectors in the United States (1976).

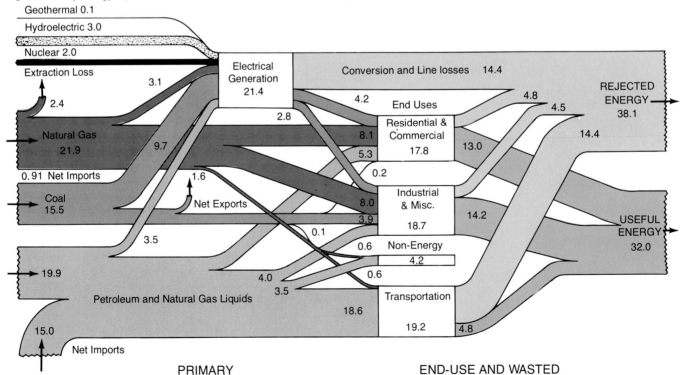

solar radiation, is discussed briefly in a section at the end of the hydropower chapter.

THREE. CONCLUSION

The overview and comparison of renewable and non-renewable energy sources maps energy amounts thought to be recoverable, and shows how the energy amounts from different sources coincide in certain states and regions. Energy profiles for economic regions are presented here, as are a few projections of energy needs for the future.

APPENDIX

The appendix provides a glossary of terms, a geologic time scale, and a table of factors for converting energy units and other measurements. Suggested readings are located at the end of the book, as are the *References* which list the source documents used by the authors.

Methods of Presentation

Resource amounts and their locations are expressed graphically in a variety of maps and graphs. The most prevalent types are previewed in Fig. 2.

SEGMENTED CIRCLES FOR RELATIVE AMOUNTS

The traditional scale circle with segments (pie chart style) is an invaluable tool for showing amounts of an energy source in different locations as well as the relative size of the components that make up the total at any one location. Circle areas are made proportional to the amounts represented. A guide to the amounts is provided in a legend of nested circles as shown in Fig. 2. The key to circle segments is the divided bar that accompanies each map of this type and shows how the components apply as a national average.

DISCRETE SYMBOLS FOR ABSOLUTE AMOUNTS

It is difficult to judge actual amounts from circles. For this reason, important quantities are mapped by a collection of symbols, each of which represents a certain tonnage or volume. Tons, whether coal or nuclear fuel, are shown by blocks; volumes of oil are shown by barrels; and volumes of gas are shown by cylinders. A quick impression can be gained from the size of the blockpile or barrel pile and a fairly accurate estimate can be obtained by counting symbols in the accumulation (Fig. 2B). Important amounts are tabulated for the reader who needs precise data.

Fig. 2 Symbolization used in the atlas: segmented circles for relative amounts; and five symbols used for absolute amounts of resources or their energy equivalents.

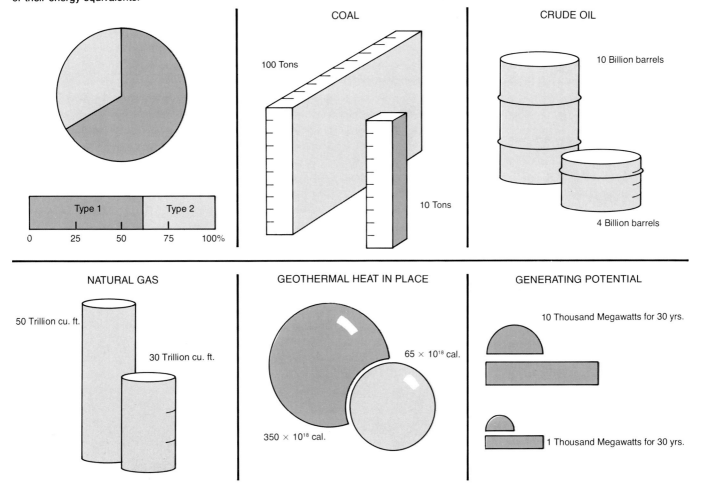

BAR GRAPHS FOR RELATIVE AND ABSOLUTE AMOUNTS

In a bar graph, amounts can be ranked and compared compactly. Often the quantities shown this way in the atlas have been mapped in an earlier illustration and are presented again in bar graph form to emphasize comparisons. With the exception of one graph in the coal chapter, all bar lengths are scaled in *linear* fashion (the length of 200 units is twice that for 100 units) so that relative magnitudes are clear.

Part One: Nonrenewable Resources

The energy sources surveyed here, with the exception of geothermal heat, are mineral fuels: coal, crude oil and natural gas, and the fissionable metals, uranium and thorium. The term *nonrenewable* is applied broadly to all mineral substances; but it applies in a most absolute way to the mineral fuels. A metal such as copper is re-usable, and may be recycled in order to reduce the need to mine the virgin ore. The mineral fuels, however, are irrevocably lost as their chemical or physical energy is converted to heat.

The mineral fuels have two important traits in common. First, their occurrence is controlled by the types of rock that occur in different areas, making an understanding of major rock types very useful. Second, the mineral fuels occur in limited amounts which are not all of the same degree of certainty or availability. This matter must be clarified by careful use of resource terminology.

ROCK TYPES AND OCCURRENCE OF MINERAL FUELS

Accumulations of crude oil, natural gas, and coal, like concentrations of metallic minerals in ores, do not occur randomly in the earth's crust but show a strong tendency to associate with rocks of certain kinds. Some metallic ores, especially those of abundant metals such as iron and aluminum, occur in such a variety of rocks that generalization is difficult. For fossil fuels, however, the relationship is simple.

IGNEOUS ROCKS   Igneous rocks form from the cooling of molten rock, which is called magma if it cools within the earth, and lava if it cools at the surface. For example, deeply buried magma of a certain chemical composition will become a mass of granite, and will have the large crystals and coarse texture permitted by slow cooling at depth. Molten rock reaching the surface as lava in a volcanic flow or eruption cools quickly to form a finer-texture rock, such as *basalt* which now covers much of the states of Washington and Oregon. A great number of rock textures and chemical compositions exist in rocks classified as igneous, but all originate in high temperature magma or lava.

SEDIMENTARY ROCKS   Unlike igneous rocks, sedimentary rocks form entirely at or near the earth's surface. Practically all are due to materials settling and collecting quietly in oceans, lakes, or river beds. A sandstone, for example, is composed of sand grains compressed and cemented together. Shale is composed of compacted clay particles. Limestones are made of shell fragments in some cases, or of chemically-precipitated lime in other cases. The materials deposited become rock as they are subjected to pressure of overlying materials, which are usually younger sediments.

METAMORPHIC ROCKS   As the name implies, metamorphic rocks are the result of some kind of alteration of an existing rock. The sedimentary rock sandstone, for instance, may be altered to a dense hard siliceous rock called quartzite. Similarly, a shale can be altered to slate; and heat and pressure can turn a limestone into marble. Although the most familiar examples of metamorphics were originally sedimentary, igneous and metamorphic rocks are also altered. Frequently the process of metamorphism occurs at great depth and under high temperatures and pressures that accompany the compressive forces of mountain building. As a result, the earlier rock may be wholly or partially melted and allowed to crystallize.

CRYSTALLINE ROCKS VERSUS SEDIMENTARY, AND THE OCCURRENCE OF FOSSIL FUELS   For simplicity, igneous and metamorphic rocks may be considered together and called, for lack of a better term, *crystalline rocks.* The crystalline texture in igneous rocks is due to formation of crystals directly from magma or lava: in metamorphics it is due to crystals forming from re-melted materials or from percolating fluids rich in minerals.[1] All igneous and most metamorphic rocks have a history of high temperatures, pressures, and often violent deformation. Since hydrocarbon fuels are obviously susceptible to destruction at high temperatures, they would not form in igneous rocks and would not survive in most metamorphics.

Sedimentary rocks are the logical hosts for fossil fuels because they have a history that is relatively quiet, without excessive temperature or pressure. Equally important, however, is the environment in which both fossil fuels and sedimentary rocks are born. Coal is formed from plant materials that have fallen into swampy areas near an ocean shore: as sea level changes, silts and clay are deposited and eventually enclose the coal in a sedimentary sandwich. Natural gas and crude oil are thought to form from the alteration of marine plant and animal organisms whose organic remains collect on the ocean floor, along with the sediments that become rocks. Rocks of sedimentary type, then, are formed in the same environments as coal and petroleum, and have a history consistent with the preservation of the delicate hydrocarbons.

The locations of sedimentary versus crystalline rocks can be understood with the aid of a cross-section (Fig. 3). This figure shows crystalline rocks to be on the continent's basement, exposed at the surface—to the exclusion of sedimentary—in two kinds of areas. One is broad low 'shield' areas, such as the Canadian Shield north of the Great Lakes; and the other is where basement rocks poke through sedimentary rock in mountain ranges such as the Southern Rockies in Colorado, the Sierra Nevada's, and Ozark Dome in Missouri, and the Black Hills of South Dakota.

Areas not identified as crystalline (Fig. 3) may be assumed to have some, and possibly very thick, sedimentary rocks; those areas, therefore, are where fossil fuels may be expected. In the following pages, the amounts of coal or petroleum mapped will be influenced by the presence or absence of sedimentary rocks in the regions or states for which resource data is gathered.

ROCK TYPES AND OCCURRENCE OF NUCLEAR FUELS Although uranium and thorium are not abundant metals, some of their ores occur in sedimentary rocks, and quite different ores occur in igneous and metamorphic rocks. There is no simple relationship between areas of occurrence and the patterns of Fig. 3. The various types of deposits and their locations will be discussed in a later chapter.

RESOURCE TERMINOLOGY, AND THE LIFE OF RESOURCES

HOW MUCH DO WE HAVE LEFT? Despite frequent references to mineral resources being *finite* amounts, the answer to the question of how much remains is quite *indefinite* because of two uncertainties. The first is due to incomplete exploration. There is, and always will be, *some* possibility of undiscovered uranium oxide deposits, coal beds, or petroleum accumulations. The amounts remaining can never be predicted accurately. The second is due to changing economic factors and technological advance that both can alter the definition of what constitutes a usable resource. Both these variables must be kept in mind in any review of mineral resources, whether fuels, metallics, or other types.

The following scheme, used by the U.S. Geological Survey and initiated by its recent director, Vincent E. McKelvey (McKelvey, V. E., and F. H. Wang, 1969), is used in this book to organize resource amounts for coal, crude oil and natural gas, nuclear fuels, and geothermal heat.

Fig. 3 U.S. areas that are dominantly crystalline rocks, and therefore *unfavorable* to fossil fuels. Included is a cross-sectional view showing how crystalline rocks are overlain in some areas by sedimentary which are favorable to fossil fuels.

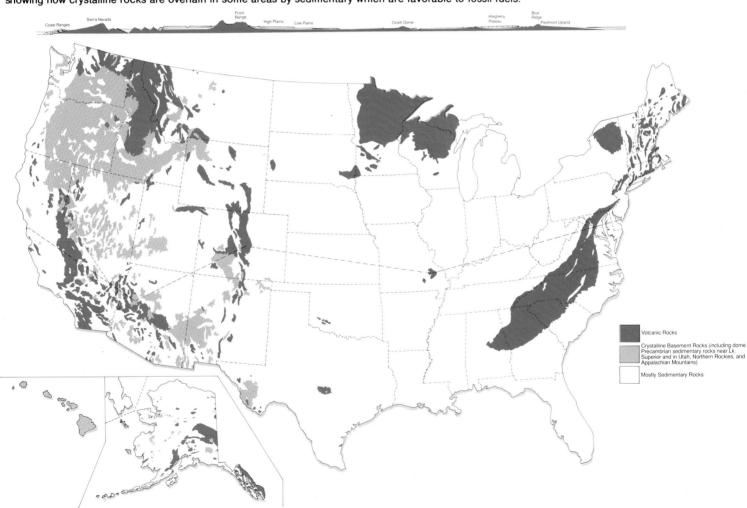

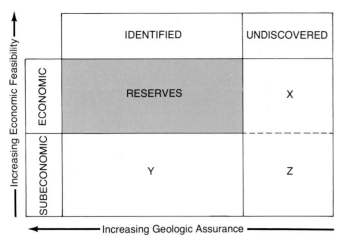

Fig. 4 Scheme for organizing mineral resources of differing degrees of certainty and economic feasibility.

The rectangular diagram in Fig. 4 comprehends *all* of a particular resource that exists within a given area, such as a country or continent. Some of that resource quantity is better known than other parts of it. To reflect that, *identified* amounts are assigned to the left side while *undiscovered* amounts occupy the right side of the diagram. Some of the resources, whether *identified* or *undiscovered,* are more easily recovered from the earth because they are rich concentrations, near the surface, easily treated after mining, or have some combination of these characteristics. Such resources are located near the top of the diagram, while the resources less easily recovered are placed toward the bottom. The upward direction, then, is one of increasing economic feasibility, and leftward is the direction of increasing geologic certainty.

Because the diagram takes into consideration both geological knowledge and economic factors, it is an indispensable aid to classifying resources and assigning them unambiguous terms. *Reserves,* shown by the shaded area, are those amounts that are well-defined through geologic exploration and development, and at the same time are *recoverable* at current prices and with current technology. Other resources of the same substance may be disqualified as reserves for one of two major reasons: they may be attractive economically but as yet undiscovered (position X); or they may be well-defined but not easily recovered because of low grade or difficult access (position Y). Some undiscovered resources are not now attractive economically (position Z) and are thus the most remote and the least likely to be used.

Although the definition of reserves is firm, the actual magnitude of reserves changes with time. Obviously, production (consumption) will diminish reserves; but there are also two other general processes that tend to augment reserves. First is the *discovery* of resources which allow the reserve category to expand to the right. Equally important is the influence that economic and

technological changes can have upon the lower limit of the reserves category. For example, when copper ores were more abundant in 1920 the average ores mined and considered economic contained 1.5 percent copper. In the 1970's, with usable ores more scarce, and with bigger machinery available for handling large quantities of rock, the ores mined averaged around 0.6 percent copper. Rock that was *not* reserve in 1920 *became* reserve as the boundary between economic and non-economic shifted downward in the diagram. Rising prices for mineral fuels can have a similar effect, as will be shown in connection with crude oil.

HOW LONG WILL IT LAST? Raising the question of the *life* of a resource is disturbing, because of the implicit assumption that we plan to use all of the resource in question. Nevertheless, the exercise is useful because comparing resource amounts with rates of use is one of the best ways to bring meaning to large and unfamiliar numbers such as billions of tons of coal or trillions of cubic feet of natural gas.

There is, of course, no way of predicting resource life with confidence. As the foregoing section pointed out, there is uncertainty about resource amounts remaining to be used. In addition, there is uncertainty about future rates of use. One device used in this atlas is the *static reserve index* which is simply current reserves divided by current annual rate of use (i.e., rate of production or mining of the raw resource). A more thorough approach is to project both resource amounts and use rates into the future. Resource amounts may be assumed to include all resources that may be useful in the future, including those amounts now undiscovered and those identified but non-economic (Fig. 4). The use rate may be assumed to remain constant, or to grow at a certain rate: in either case, the accumulated production through a period of years can be estimated. This accumulated production can be compared with total resources remaining to show the expected status at the end of the period chosen.

If rates of use increase continually at a fixed percentage rate, the impact upon resource amounts is staggering. For instance, if production increases steadily at a rate of 6 percent per year, then the cumulative amount used will *double in 12 years.* At such a rate, as one geologist put it, the world's petroleum will be half gone by the year 2003, and all gone by the year 2015 (Skinner, 1976). Or, to put in another way, if the world's petroleum resource is actually twice what we thought it to be, the reprieve is only twelve years!

Apparently, ever-increasing annual rates of resource use tend to make less significant the question of how much of a resource remains, because vast amounts can be consumed in a very few years. Happily, the increasing rates of energy use can be avoided in the future, and can be replaced by annual rates that level off and fall. Some projections of energy futures, both expansionist and more conservative, are referred to in Part Three of this atlas.

Part Two:
Renewable Resources

Renewable sources of energy are either the primary energy flows that reach the earth's surface, or the various forms of energy that depend upon a primary flow and can be utilized year after year if properly managed.

Figure 5 represents schematically the three fundamental sources from which energy flows continuously to the earth's surface: geothermal heat, gravitational forces, and solar radiation.

Geothermal heat originates in the hot semi-molten mantle rock at the base of the crust and in radioactive decay of elements within the crust. The supply of heat, especially from the deep source, is practically unlimited; but the portions that are in high-temperature occurrences accessible for exploitation are not, and for this reason geothermal heat is treated here as a non-renewable resource. Recent work by the U.S. Geological Survey has defined and measured the resource so thoroughly that a substantial chapter can be devoted to interpretation of geothermal heat in the country.

Gravitational forces that lead to rise and fall of tides are due to influence of the moon and the sun, and are completely divorced from solar radiation. Usable mostly for electrical generation, the resource does not appear to be large, either on world or national scale. The potential is dealt with in a brief section at the end of the hydroelectric chapter.

SOLAR RADIATION

Radiation can be employed for its thermal and its photovoltaic effects. In addition, a number of familiar physical and biological processes may be exploited—all of which are dependent upon solar radiation.

PHYSICAL EFFECTS   Unequal heating of the earth's surface leads to winds, which can be harnessed for power and which, themselves, contribute to ocean currents and wave pulsations. Heating by the sun also causes surface waters of tropical and subtropical oceans to be warmer than deep waters, making a thermal gradient or contrast that can be used for power generation through ocean thermal energy conversion (OTEC). Solar radiation provides the energy by which water is evaporated from the oceans and dropped on highlands as water that can be used for hydroelectric power generation. Separate chapters are devoted to solar radiation, wind, hydroelectric power, and ocean thermal gradients. No material is presented on either ocean currents or wave potential.

BIOLOGICAL EFFECTS   Because photosynthesis is essential to plant life, it is clear that all natural and domestic food chains are based on solar radiation. The energy afforded by harnessing draft animals such as horses is plant energy converted. The energy in coal and petroleum represents vestiges of the very large amounts of plant life that has existed on the planet throughout its history. Further opportunities exist today in the various organic materials, both plant and animals, that can be produced continuously through photosynthesis. These materials, broadly considered *biomass*, include wastes and crops grown specifically for their energy content. They are dealt with in a separate chapter.

THE QUESTION OF RENEWABLE ENERGY AMOUNTS

For mineral fuels, and even for some dimensions of geothermal energy, finite amounts of energy in place and

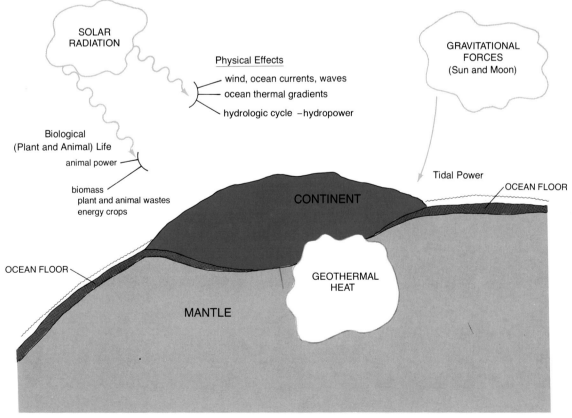

Fig. 5 Renewable energy flows from three sources: geothermal heat, gravitational forces, and solar radiation.

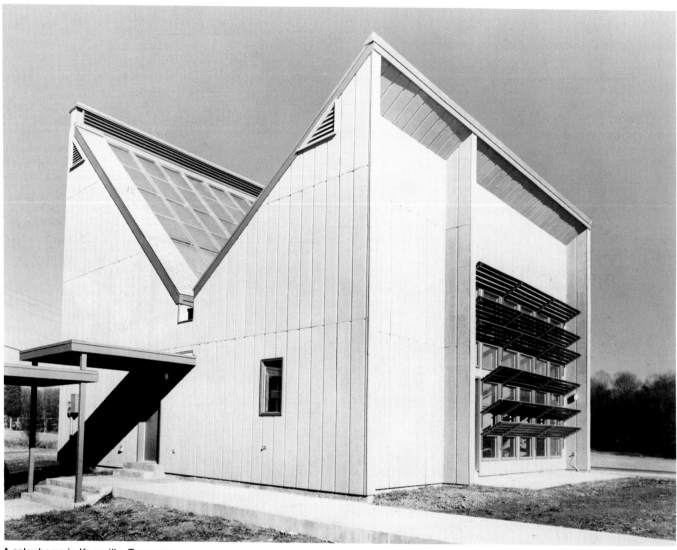

A solar home in Knoxville, Tennessee.

An electrical generating station in Homer City, Pennsylvania. This is one of several giant mine-mouth power plants which consume coal from nearby mines to make electricity for markets many miles away. *Courtesy New York State Electric and Gas Corporation.*

energy recoverable can be estimated—though with some nagging uncertainties about definition of "usable resource." For renewable sources the definition of total potential is less clear, although it can be approached. The total potential of hydroelectric is the easiest to define, at least conceptually, because river systems embrace only so much hydrostatic head, and only so many installations can be tolerated if certain conditions are specified. Total wind power can be defined if certain areas are delineated as suitable, and if those areas are then theoretically 'saturated' with wind machines of a specified type and generating capacity. Nonetheless, new technologies, especially those that would capture light winds or allow more densely packed installations, can make earlier estimates obsolete.

The total potential of ocean thermal gradients can be approached in a similar fashion, by defining what would constitute saturation of suitable ocean areas. Most elusive is the potential of biomass, because in *energy crop* endeavors, as opposed to utilizing wastes, the energy use will compete with other uses of the land or of the crop. Future prices of various commodities, not only competing fuels, can therefore influence how much biomass energy can be produced in a given region and in the nation as a whole.

It is important to recognize that *energy delivered* through the use of radiation, wind, or falling water, leads to savings in mineral fuels that are larger than may appear at first glance. Ten thousand British Thermal Units (BTUs) of energy delivered through a solar heating system does not simply avert the burning of equivalent amounts of fuel oil or natural gas. If the furnace is 80 percent efficient, then the fuel needed to supply the 10,000 BTUs is 12,500 BTUs. In the case of electrical power generation the savings are more dramatic, because only 35 to 40 percent of fuel energy is converted to electrical in a conventional power plant. *Fuel savings* is the term used to denote the amounts of fuel energy (or actual amounts of fuel) that would be displaced by the use of some renewable energy supply.

NOTES

[1]In fact, some sedimentary rocks, especially the carbonates, have crystalline texture, too; so the term is not ideal.

PART ONE

Nonrenewable Resources

Chapters 1 through 5 deal with nonrenewable resources: coal, crude oil and natural gas, oil shales and tar sands, nuclear fuels, and geothermal heat. With the exception of geothermal heat, these nonrenewable energy sources are mineral fuels. Their resource amounts are therefore subject to the uncertainties of limited geologic knowledge and changing economic factors.

For each of the five resources in Part One, careful distinction must be made between *in place* amounts and amounts that are thought to be *recoverable.* The chapters vary in that respect, because the information available on coal resources is for tonnage in place, whereas nuclear fuel resources are represented in official documents as tonnages recoverable under certain assumptions.

In each of the five chapters the units employed are those that are peculiar to the resource, such as tons for coal, barrels for crude oil, and cubic feet for natural gas. Translation of these amounts into energy units, for comparisons among the five sources, is presented in the *Overview,* Part Three.

Coal

Although coal is thought to have been used as an energy source in China as long ago as 1000 B.C., it was not until the thirteenth century that notable coal mining operations appeared in England, Scotland, and on the continent of Europe. During this period, coal was not regarded as an important energy resource because of both the noxious fumes it emitted on combustion and the abundant supply of timber available at the time in Europe. With the development of coal-fired brick kilns in the fifteenth and sixteenth centuries, demand for coal began to increase. The establishment of iron smelting works and other engineering and metallurgical developments in England during the seventeenth and eighteenth centuries brought about a dramatic increase in demand for coal and ushered in an era in which coal was to be the dominant source of energy.

In the United States, coal appears to have been first mined in the early 1700s along the James River in Virginia. Almost 150 years elapsed before the discovery of the rich and expansive Appalachian bituminous coal field which was to become the focus of coal commercialization in the United States. In these early years, movement of coal overland was difficult, but the advent of the steam locomotive solved the land transport problem and opened huge new markets. In 1900, coal fulfilled about 90 percent of the national energy needs despite the development of oil and gas reserves some years earlier. From 1900 to 1920, however, oil began to compete with coal as an important source of energy (Fig. 1–1). From 1920 to the mid-1970s, coal's share of the United States' energy market declined rather steadily in face of competition from the cleaner-burning oil and gas. By 1976, coal's share of the nation's energy market had declined to less than 20 percent of the total (Executive Office of the President, 1977). Despite its diminished role in the United States, the volume of coal production has significantly increased in the past two decades due to a large increase in the country's demand (Fig. 1–2).

With the alarming and rapid depletion of domestic reserves of oil and natural gas and foreign imports constituting almost half of the domestic oil consumption in 1977, there has been a call to significantly increase the use of coal for the next two to three decades (Executive Office of the President, 1977). In the near term, most coal will continue to be burned directly in electric generating plants with the use of devices to control stack emissions of sulfur oxides and particulates. If coal usage is to increase at the rate projected, more effective and economical methods of cleaning coal before combustion must be developed. In addition, present emission standards will have to be abided by and perhaps made even

← Miner testing for undesirable gas levels. *Courtesy Bethlehem Steel Corporation.*

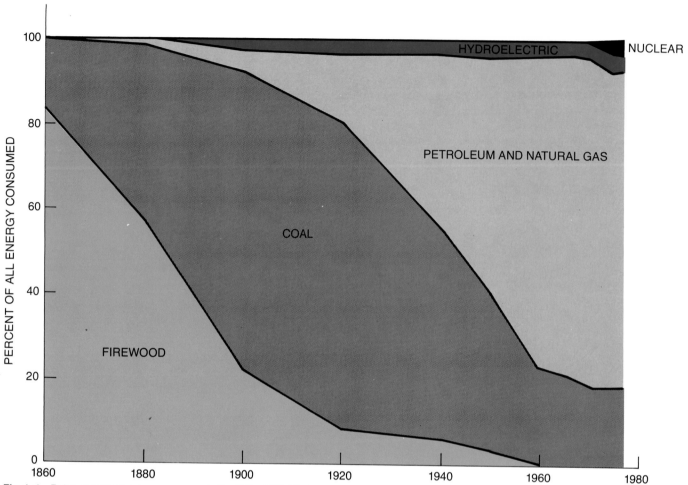

Fig. 1–1 Roles of various energy sources in the U.S.: 1860-1976.

Fig. 1–2 U. S. production of all coal: 1950-1978.

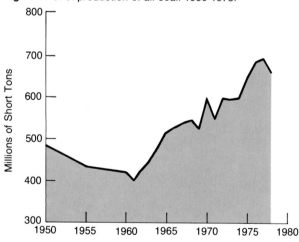

more rigorous. Long-term projections indicate that conversion of coal to synthetic crude and synthetic natural gas could provide important substitutes for conventional oil and gas, and would help to greatly reduce air pollution from coal burning. Unfortunately, greatly increased coal mining could be accompanied by the environmental damage caused by strip mining. Surface mining should be avoided if possible, and requirements for land restoration after surface mining must not be relaxed. As shown in this chapter, very substantial amounts of desirable low-sulfur coal are available through underground mining methods. Economics aside, this coal should be preferred to that obtainable by surface mining.

Origin and Character of Coal

Coal was formed from vegetal matter that once thrived in some areas of the earth. The accumulation of coal-forming plant debris occurred millions of years ago, mainly during the Carboniferous period, and to a lesser degree during the Cretaceous and Tertiary periods. The earth's climate was particularly favorable for extravagant plant growth during this time. In addition, areas of level swamp land where plant life thrived were in abundance.

In the swamps, materials were preserved by virtue of the low-oxygen stagnant waters which arrested the bacterial action causing plant decomposition. Vegetal matter, accumulated in the shallow swamp waters, formed peat. As the land subsided or the sea rose (or both), the plant debris was covered by clays, sands and lime muds, which were the basis of the shales, sandstones, and limestones found on top of the coal seams today. Over thousands of years the peat was compacted, became more dense, and was gradually transformed into coal.

Coal deposits are distributed worldwide although they are much more common in the Northern Hemisphere. Within the United States, coal-bearing strata underlie about 13 percent of the land area and are present in varying degrees in thirty-seven states (Averitt, 1973). In many areas of the United States, coal seams continue over great distances and large areas as, for instance, the Pittsburgh Seam which underlies some 15,000 square miles of the Appalachian Highlands. Coal seams in the United States range in thickness from a thin film to 100 feet or more, but most are between 2 and 10 feet thick (Bateman, 1951).

Rank and Sulfur Content

Coal is classified on the basis of the amount of fixed carbon, moisture, and volatile matter present. Fixed carbon is the percentage of stable carbon found in coal. Volatile matter is the proportion of organic gases, such as oxygen and hydrogen. The rank of coal is generally based on the percentage of carbon, which increases through the ranks from lignite to subbituminous to bituminous to anthracite (Fig. 1–3). Figure 1–3 also shows that the highest heat content (energy content) of all four ranks is found in bituminous coal which has relatively few volatiles and low moisture content. The heat or energy content is lowest in lignite which has a high moisture content.

The changes in fixed carbon, moisture, and volatile matter content through the ranks are an expression of the progressive alteration of original peat materials and depend upon depth and heat of burial, compaction, time and structural deformation (Averitt, 1973). Energy content by rank ranges from 14×10^6 BTUs per short ton for lignite to 26.4×10^6 BTUs for anthracite (Department of Interior, 1976).

Figure 1–4 shows the distribution of United States coal deposits by rank. Lignite deposits, excepting those in Alabama, are found exclusively in young rocks west of the Mississippi River, the largest one being located in North Dakota. Subbituminous coals are found only in the western part of the United States, particularly in Montana, Wyoming, and New Mexico. Bituminous coals are found throughout the United States, but most are confined to Paleozoic rocks in the Appalachian and

Fig. 1–3 Constituents and energy content of different ranks of coal.

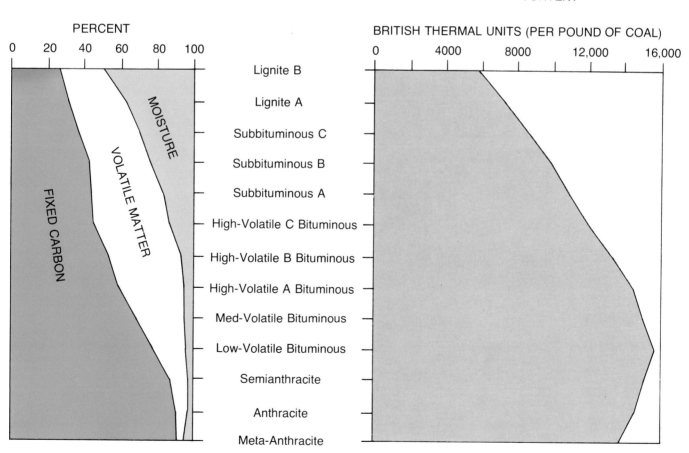

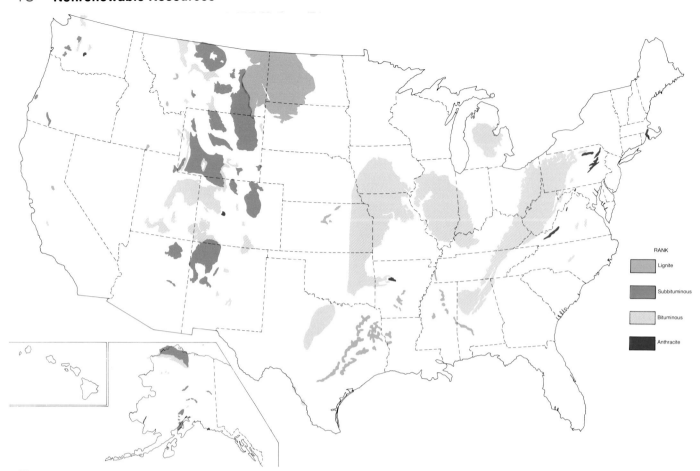

Fig. 1–4   Occurrence of coal of four ranks.

RANK

Lignite

Subbituminous

Bituminous

Anthracite

Mid-Continent regions. Anthracite deposits are scarce and are found only in Pennsylvania, Virginia, Arkansas, Colorado, Washington, and Alaska.

The quality or grade of coal is determined by the relative amounts of ash, sulfur, and certain other noxious constituents. Thus far, coal resources of the United States have been classified only according to sulfur content (Bureau of Mines, 1975). Sulfur is a particularly undesirable constituent of coal because it lowers its quality, contributes to corrosion in boilers, creates air and water pollution problems, and inhibits plant growth in spoil banks. Although the amounts vary in both chemical composition and weight percentage, all coals contain some sulfur. According to Averitt, the sulfur content of coal in the United States ranges from 0.2 percent to about 7 percent, but the average in all coal is 1 to 2 percent. The bituminous coals of the Pennsylvanian Age in the Appalachian and Interior coal basins have the highest sulfur content, with over 40 percent of the identified resources containing at least 3 percent. The subbituminous coal and lignite of the Rocky Mountains and Northern Great Plains, however, generally contain less than 1 percent sulfur (Averitt, 1973).

In order to control sulfur emissions, the Environmental Protection Agency has established national stack emission limits for new stationary sources with a capacity of greater than 250 million BTUs per hour heat input

(Bureau of Mines, 1975). The EPA regulates stack emissions, not sulfur content of coal burned, the limit being 1.2 pounds of sulfur dioxide per million BTUs of heat input. Because coals of different ranks have different heat values, the sulfur contents commensurate with allowable emissions will vary (Fig. 1–5). For example, a lignite with heat value of 7,000 BTUs per pound could hold no more than 0.4 percent sulfur, while bituminous coal of 13,000 BTUs per pound could hold up to 0.78 percent sulfur.[1]

Mining Methods and Recovery Rates

There are two basic methods used for extracting coal from the seam: underground and surface. In underground methods, coal is extracted without removal of overburden, that is, the soil deposits found on top of the coal. Strip mining involves removal of the overburden and extracting the coal from the exposed seam. Underground mining is used when the coal is buried too deeply in the ground to make surface mining feasible or possible. There are three different types of underground mines, as illustrated in Fig. 1–6: shaft, drift, and slope types. The type of mining method used to recover coal depends upon the total area of coal available, the thickness and inclination of the seam, the thickness of the overburden, the value of the surface land, as well as other economic fac-

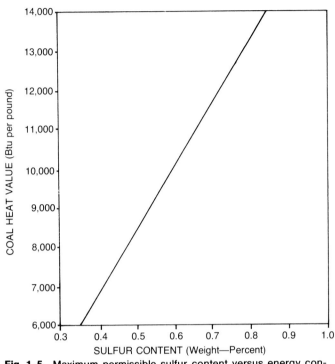

Surface Mining in Kentucky. The large bucket in foreground is used to remove overburden material and expose the coal which is in the floor of the area shown. The smaller power shovel in the background loads coal into large trucks with capacity of 135 tons. *Courtesy National Coal Association.*

Fig. 1–5 Maximum permissible sulfur content versus energy content of coal commensurate with EPA standard for Air Quality.

Fig. 1–6 Types of mines include the shaft, drift and slope types, used to obtain coal that lies in seams underground, and the strip type, used to remove coal that lies very close to the surface. Although not all four types would be found on one hill, this diagram shows that the type of mine chosen depends on the depth of the seam, the surface contours, and on whether an outcrop occurs.

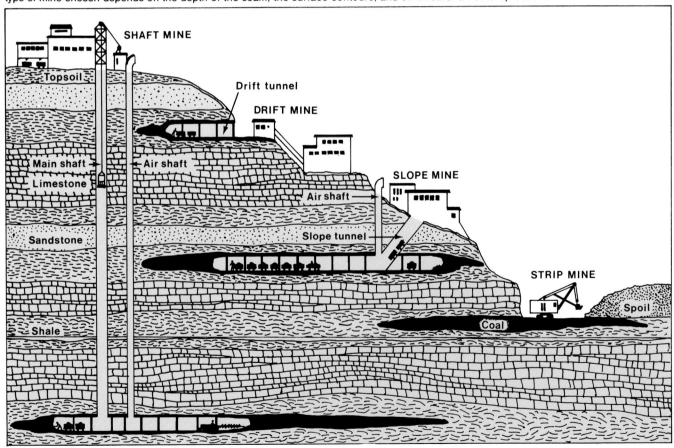

tors. In 1978, almost all of the coal produced in the United States was recovered by either underground (37 percent) or strip mining methods (63 percent) (Department of Energy, March, 1979).

There are two basic types of underground mining methods: room and pillar and longwall. In room and pillar mines, coal is removed by cutting rooms, or large tunnels, in the solid coal leaving pillars of coal for roof support. Longwall mining takes successive slices over the entire length of a long working face (Fig. 1–7). In the United States, almost all of the coal recovered by underground mining is by room and pillar method. In Great Britain, however, most of the coal is mined by longwall methods. Longwalling is advantageous because it can be carried out at depths greater than those at which the room and pillar system is used. This is simply because at great depths the pillars might break under the weight of overburden.

Up to now, longwalling has not been used extensively in the United States because most of the coal has been extracted from the thin and relatively shallow seams of the eastern part of the country. One would expect longwall methods to be used more as the demand increases for the thick and deep coal deposits of the western states.

The percentage of coal recovered from a mineable seam depends on a number of factors. These include: seam thickness, number of pillars needed to safely support the roof, and the degree to which the surface land is protected. Figure 1–8 compares the amount of coal recovered by room and pillar versus longwall recovery for beds of various thickness. Longwall mining can recover

Fig. 1–7 Underground mining methods: room and pillar versus longwall.

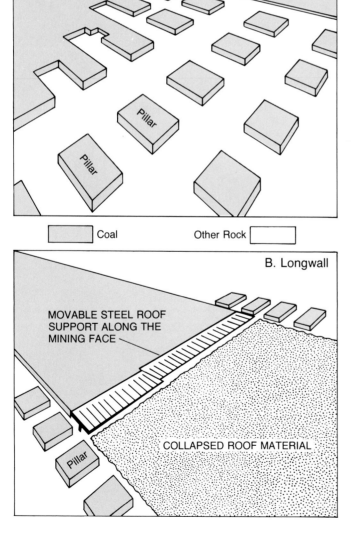

Fig. 1–8 Recovery ratios: room and pillar versus longwall methods, for beds of various thickness.

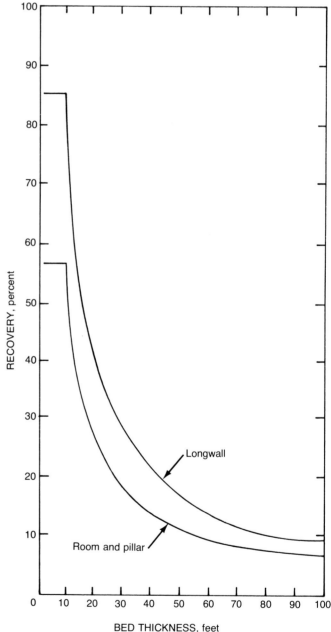

85 percent of the coal in beds less than 10 feet thick whereas the recovery rate in these same beds using room and pillar techniques is only 57 percent. The recovery factor for both methods drops off very rapidly for beds over 10 feet thick.

As improved earth-moving equipment became available, strip mining developed very rapidly in the United States. Strip mining consists of removing the soil and rock above a coal seam and then removing the exposed coal. Modern strip mines employ equipment capable of removing overburden of more than 200 feet in thickness. Strip mining has surpassed underground techniques in the United States as the favored recovery method because it is cheaper and allows a higher recovery rate (85 to 90 percent). Strip mining, however, has been extremely destructive to the land surface, as it leaves huge scars and piles of broken rock, denudes the area of vegetation, and greatly promotes water pollution in adjacent streams. To counteract this destruction of the land, the state and federal governments now require the coal mining industry to restore stripped land to conditions favorable to future productivity. It is seldom possible, however, to restore a stripped area to its precise original condition.

Resource Terminology

The U.S. Geological Survey and the Bureau of Mines have classified coal resources to depths of 6000 feet (Averitt, 1975 and Bureau of Mines, 1975). As indicated in Fig. 1–9, coal resources are designated as either *Identified* or *Hypothetical*. Identified resources, assessed to a depth of 3000 feet, consist of mineral-bearing rock whose existence and location are known. The quantity of these deposits that meet specific depth and thickness criteria is termed the *Reserve Base*. The criteria used by the Bureau of Mines stipulate a thickness of 28 inches or

A giant dragline removing overburden from coal in an Ohio strip mining operation. Each bite of the bucket holds 220 cubic yards, or roughly 325 tons of material. Notice the size of the two men standing to the right of the power cable spool in the foreground. *Courtesy Bucyrus-Erie Co.*

Glossary of Coal Resource Terms

Resources Total quantity of coal in the ground within specified limits of bed thickness and over-burden thickness. Comprises Identified and Hypothetical resources.

Identified Resources Coal-bearing rock whose location and existence is known.

Hypothetical Resources Estimated tonnage of coal in the ground in unmapped and unexplored parts of known coal basins to an overburden of 6,000 feet; determined by extrapolation from nearest areas of Identified resources.

Measured Resources Tonnage of coal in the ground based on assured coal bed correlations and on closely spaced observations about one-half mile apart. Computed tonnage judged to be accurate within 20 percent of the true tonnage.

Indicated Resources Tonnage of coal in the ground based partly on specific observations and partly on reasonable geologic projections. The points of observation and measurement are about one mile apart from beds of known continuity.

Inferred Resources Tonnage of coal in the ground based on assumed continuity of coal beds adjacent to areas containing Measured and Indicated resources.

Demonstrated Reserve Base A selected portion of coal in the ground in the Measured and Indicated category. Restricted primarily to coal in thick and intermediate beds less than 1,000 feet below the surface and deemed economically and legally available for mining at the time of the determination.

Recoverability Factor The percentage of coal in the Reserve Base that can be recovered by established mining practices.

Reserve Tonnage that can be *recovered* from the Reserve Base by application of the recoverability factor.

Source: *U.S. Geological Survey Bulletin 1412*, 1975.

Fig. 1–9 Coal resource categories, combining concepts of U. S. Geological Survey and U. S. Bureau of Mines.

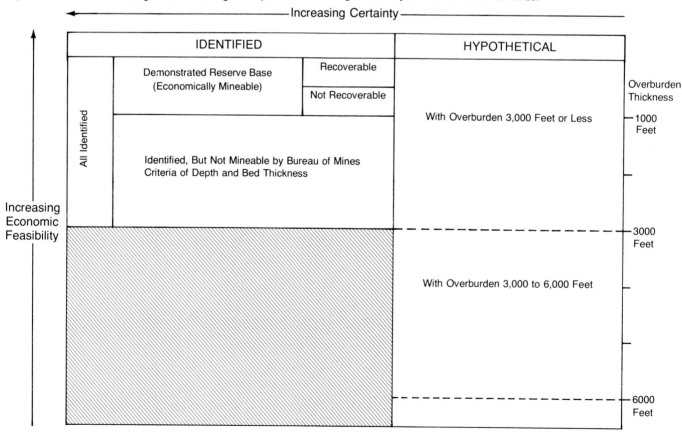

Conventional, or Room and Pillar Mining. A five-foot seam is slotted by this cutting machine in a Kentucky mine. The six-inch slot will allow the coal to expand and fall when it is blasted. Not shown are pillars of coal that are left to support the roof. *Courtesy Westinghouse Air Brake.*

Longwall mining. Whirling cutters on the mining machine shave coal from the face on the left. The exposed coal face and the work area may be several hundred feet across with no coal pillars to support the roof. Support is provided by the steel roof and hydraulic jacks shown here. As the coal face retreats, the roof and supports are moved to the left. And the roof behind (to the right) is allowed to fall. Because no coal is left as pillars, the recovery rate is much higher than in room and pillar mining. *National Coal Association.*

more for bituminous and anthracite coals, and 60 inches for subbituminous and lignite. The depth criterion for all ranks, except lignite, is 1000 feet (Bureau of Mines, 1975). In the case of lignite, only beds that can be mined by surface methods are included in the Reserve Base. These generally lie no more than 120 feet below the surface. In some instances, however, coal deposits which do not conform to the depth criteria are included in the Reserve Base if they are either presently being mined or could be mined commercially. The term "Demonstrated" refers to both Measured and Indicated categories of coal as defined by the Bureau of Mines and the U.S. Geological Survey (Bureau of Mines, 1975). *In summary, the Demonstrated Reserve Base refers to in-place coal, both Measured and Indicated, that is technically and economically mineable.*

The part of the Reserve Base that can be recovered is termed the *Reserve.* Whether or not the coal will be recovered depends on a number of factors: the characteristics of the coal bed; mining methods used; and any legal constraints placed upon mining a deposit because of natural and cultural features. Although the recovery ratio may vary from 40 to 90 percent, it is thought that at least half of all in-place coals in the country can be recovered.

Hypothetical resources can be estimated on the basis of broad geologic knowledge and theory. They are defined as undiscovered deposits, whether of recoverable or subeconomic grade, whose existence in certain districts is geologically predictable (Averitt, 1975).

World Coal Resources

Coal resources of the world have been estimated at about 10.8 trillion short tons (*World Coal*, November 1975). Measured reserves account for approximately 12 percent of the total, and Indicated and Inferred resources the remaining 88 percent.[2] Of the total coal resources identified in Fig. 1–10 and Table 1–3, the U.S.S.R. holds 53 percent, North and Central America 28 percent, and Asia 10 percent. Most of the remainder is located in Europe.

The reserve portion of world coal resources amounts to 1,297 billion short tons, 90 percent of which is found in four regions: European and Asian U.S.S.R., North and Central America, Europe, and Asia. North and Central America (primarily the United States) and the U.S.S.R. have 29 and 21 percent of the reserves, respectively, followed by Europe and Asia with 23 and 10 percent.

Despite the phenomenal growth in the use of oil and gas in the past three decades, world coal production has almost doubled (Figure 1–11). In 1978, three countries, the U.S.S.R., the United States, and China, accounted

Fig. 1–10   World coal resources divided by region.

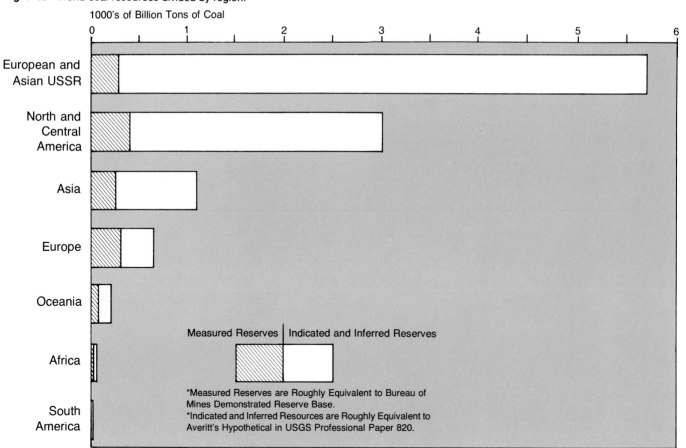

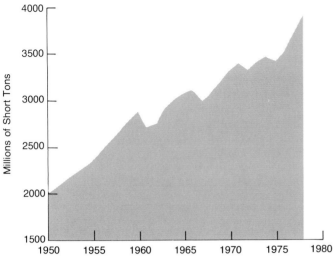

Fig. 1–11 World coal production: 1950-1978.

for almost three-fourths of the world's production. The U.S.S.R. produced 27 percent of the world's bituminous and lignite coals, and the United States and China each produced 23 percent of the total (*Coal International*, May 1979). At the same time, over 50 percent of all energy consumed in the U.S.S.R. and China was derived from coal, whereas in the United States less than 20 percent of total energy consumed came from coal.

United States Coal Resources

IDENTIFIED AND HYPOTHETICAL RESOURCES

As of January 1974, the estimated remaining coal resources of the United States were 3.9 trillion short tons (Table 1–1). Of the total, 48 percent was classified as Identified and the remaining 52 percent as Hypothetical. Table 1–2 presents recent revisions of Averitt's coal resources estimates for six states. According to information published by the Department of Energy in 1978, total Identified and Hypothetical coal resources of the United States should be increased by 2,450 billion tons (Department of Energy, February, 1978). Alaska accounts for 72 percent of this increase. All of the additional resources in Alaska are Hypothetical. They are believed to be located in the Northern Alaska coal resource region. Montana's and Wyoming's total coal resources have been upwardly revised based on new estimates of Hypothetical coals in the Powder River Basin area.

Almost 90 percent of the Identified coal is found at depths of 1000 feet or less, and most is in thick beds. Therefore, of the total Identified coal, about 400 billion tons are in thick seams with no more than 1000 feet of overburden. This is the portion of the coal resources that is either being mined at the present or is of current economic interest.

The Hypothetical amounts are estimates based on extrapolations of Identified resources in adjacent areas. Although Hypothetical resources in total are slightly larger than the Identified resources, they are considered inaccessible for mining at the present. About 20 percent of these resources are thought to lie at depths of 3000 to 6000 feet. Less than 200 billion tons of the Hypothetical resources are in seams at depths of less than 1000 feet.

Figures 1–12 and 1–13 show Identified coal as a proportion of all coals, and as amounts in each state. Although coal deposits of the United States are widely distributed (being found in thirty-six states), five western states, Wyoming, North Dakota, Montana, Colorado, and Alaska, account for almost 67 percent of total estimated resources, 60 percent of the Identified resources, and 73 percent of the Hypothetical.[3] Montana and North Dakota alone house over 37 percent of the total Identified coal and 20 percent of the Hypothetical.[4]

In the eastern part of the country, Illinois, Kentucky, West Virginia, Pennsylvania, and Indiana house the largest share of the total remaining coal. These states account for over 17 percent of the total, 25 percent of the Identified, and 10 percent of the Hypothetical coal. Within these five states, 71 percent of the remaining coal deposits are classified as Identified and 29 percent Hypothetical.

Figure 1–14 shows the differing amounts of Identified coals by rank. In the United States, 28 percent are lignites, 28 percent subbituminous, 43 percent bituminous, and 1 percent is anthracite. Bituminous coal predominates in the eastern United States whereas subbituminous coals are found exclusively in the West. Over 80 percent of all Identified subbituminous coals are located in Alaska, Montana, and Wyoming. North Dakota contains only lignite deposits accounting for 70 percent of the total lignite deposits in the United States. Ninety-five percent of the country's anthracite deposits are located in Pennsylvania.

As ranks of coal differ in their BTU values, the tonnage does not represent a clear indication of how much usable energy is contained in the coal. Bituminous coals, due to their high BTU values, account for 53 percent of the total energy contained in Identified coals, but only 43 percent of the tonnage. On the other hand, lignites account for only 19 percent of the total energy content, but 28 percent of the tonnage. In practical terms, this means that the energy content of each ton of bituminous coal is equivalent to that of 1.3 tons of subbituminous coal and 1.9 tons of lignite. The difference in the energy content of the various ranks of coal prejudices the uses of the abundant but low-energy western coals because approximately one-third more western subbituminous coal than eastern bituminous coal would have to be transported to meet the same energy need.

DEMONSTRATED RESERVE BASE

The Demonstrated Reserve Base refers to in-place coal deposits that are presently mineable both technically and economically (Bureau of Mines, 1974). Essentially, they represent the mineable portion of Averitt's Identified resources. Although the Reserve Base is referred

Table 1-1.   All U.S. Coal Resources, Identified and Hypothetical, to Depths of 3,000 and 6,000 Feet as of January 1, 1974, Coal in Place (in millions of short tons)

| | OVERBURDEN 0-3,000 FEET | | | | | | | OVER-BURDEN 3,000-6,000 FEET | OVER-BURDEN 0-6,000 FEET |
| | Remaining Identified Resources, January 1, 1974 | | | | | Estimated Hypothetical Resources in Unmapped and Unexplored Areas | Estimated Total Identified and Hypothetical Resources | Estimated Additional Hypothetical Resources in Deeper Structural Basins | Estimated Total Identified and Hypothetical Resources |
STATE	Anthracite and Semi-Anthracite[1]	Bituminous Coal[1]	Subbituminous Coal[2]	Lignite	TOTAL				
Alabama	0	13,262	0	2,000	15,262	20,000	35,262	6,000	41,262
Alaska	0	19,413	110,666	0	130,079	130,000	260,079	5,000	265,079
Arizona	0	21,234	0	0	21,234	0	21,234	0	21,231
Arkansas	428	1,638	0	250	2,416	4,000	6,416	0	6,416
Colorado	78	109,117	19,733	20	128,948	161,272	290,220	143,991	434,211
Georgia	0	24	0	0	24	60	81	0	84
Illinois	0	146,001	0	0	146,001	100,000	246,001	0	246,001
Indiana	0	32,868	0	0	32,868	22,000	54,868	0	54,868
Iowa	0	6,505	0	0	6,505	14,000	20,505	0	20,505
Kansas	0	18,668	0	0	18,668	4,000	22,668	0	22,668
Kentucky									
Eastern	0	28,226	0	0	28,226	24,000	52,226	0	52,226
Western	0	36,120	0	0	36,120	28,000	64,120	0	64,120
Maryland	0	1,152	0	0	1,152	400	1,552	0	1,552
Michigan	0	205	0	0	205	500	705	0	705
Missouri	0	34,184	0	0	31,184	17,489	18,673	0	48,673
Montana	0	2,299	176,819	112,521	291,639	180,000	171,639	0	171,639
New Mexico	4	10,748	50,639	0	61,391	65,556	126,947	74,000	200,947
North Carolina	0	110	0	0	110	20	130	5	135
North Dakota	0	0	0	350,602	350,602	180,000	530,602	0	530,602
Ohio	0	41,166	0	0	41,166	6,152	17,318	0	17,318
Oklahoma	0	7,117	0	0	7,117	15,000	22,117	5,000	27,117
Oregon	0	50	284	0	334	100	434	0	434
Pennsylvania	18,812	63,940	0	0	82,752	4,000	86,752	3,600	90,352
South Dakota	0	0	0	2,185	2,185	1,000	3,185	0	2,185
Tennessee	0	2,530	0	0	2,530	2,000	4,530	0	4,530
Texas	0	6,048	0	10,293	16,341	112,100	128,441	0	128,411
Utah	0	23,186	173	0	23,359	22,000	45,359	35,000	80,359
Virginia	335	9,216	0	0	9,551	5,000	14,551	100	14,651
Washington	5	1,867	4,180	117	6,169	30,000	36,169	15,000	51,169
West Virginia	0	100,150	0	0	100,150	0	100,150	0	100,150
Wyoming	0	12,703	123,240	0	135,943	700,000	835,943	100,000	935,943
Other States	0	610	32	46	688	1,000	1,688	0	1,688
TOTAL	19,662	747,357	485,766	478,134	1,730,919	1,849,649	3,580,649	387,696	3,968,264

Source: Averitt, Paul. *Coal Resources of the United States, January 1, 1974,* U.S. Geological Survey Bulletin 1412, (Washington: Government Printing Office, 1975).

[1]for bituminous and anthracite coals the estimates include beds 14 inches or more in thickness.

[2]for subbituminous and lignite coals the estimates include beds of 2½ feet or more.

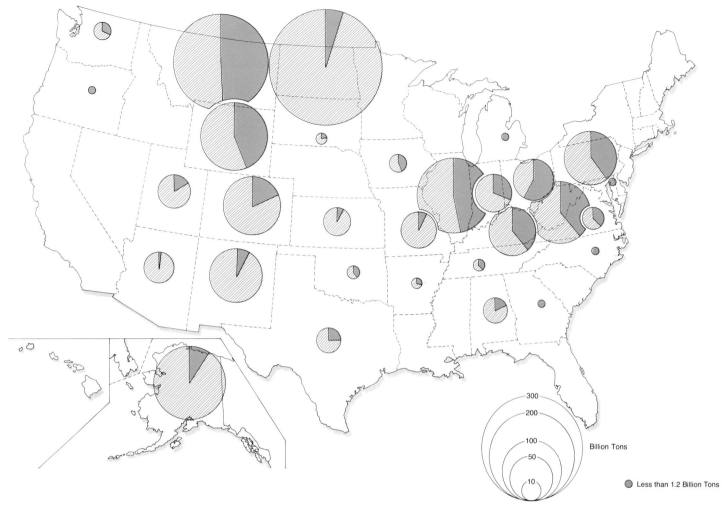

Fig. 1–12 Total U. S. coal resources (1974) showing proportions identified and hypothetical.

Table 1–2 Recent Revisions to Averitt's Total United States Coal Resources, 1978 (in billions of tons)

STATE	Averitt's Estimate	TOTAL IDENTIFIED AND HYPOTHETICAL (Overburden to 6,000 Feet) D.O.E. Revision	Change
Alaska	265.10	2,029.00	+ 1,763.90
Illinois	246.00	331.60	+ 85.60
Montana	471.64	670.00	+ 198.36
New Mexico	200.95	283.00	+ 82.05
West Virginia	100.15	116.60	+ 16.45
Wyoming	935.94	1,240.00	+ 304.06
TOTAL	2,219.78	4,670.20	+2,450.42

Source: U.S.G.S. *Bulletin 1412,* and Department of Energy, *Underground Coal Conversion Program,* Volume III, *Resources,* February 1978.
Note: In Alaska, most of the additional coal recognized is classified as Hypothetical. In western states, some is Identified, and in the Powder River Basin (Wyoming and Montana) some is at depths less than 3,000 feet.

to as mineable throughout the following pages, it is mineable only on the basis of geologic criteria. Other factors, such as competing land use or legal and social considerations could make it unfeasible to recover the coal.

Of the total Identified coal in the United States, over 27 percent of the tonnage and 30 percent of the energy content is considered Reserve Base. As shown in Fig. 1–15, the mineable portion of Identified coal resources varies widely from state to state. For example, 58 percent of the in-place coal in Ohio is considered mineable as compared to only 2 percent in Arizona. In general, since the coals in the eastern United States are more accessible, a larger percentage is mineable. The depth and thickness of seams in western states account for the considerably smaller portions of the in-place coals that are considered mineable. The most extreme example is found in North Dakota where only 5 percent of the more than 350 billion tons of Identified lignite can be mined by present methods.

RANK   The total Demonstrated Reserve Base of coal in the United States amounts to almost 437 billion short tons. Roughly 54 percent of the Reserve Base is located in states west of the Mississippi River (Table 1–3). Figure

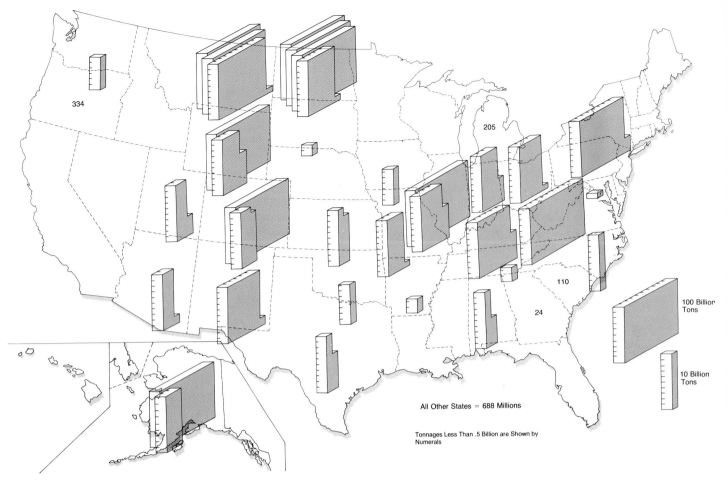

334

205

110

24

100 Billion
Tons

10 Billion
Tons

All Other States = 688 Millions

Tonnages Less Than .5 Billion are Shown by
Numerals

Fig. 1–13 Identified coal amounts to 3,000 feet (1974).

Fig. 1–14 All identified coal (1974) divided by rank.

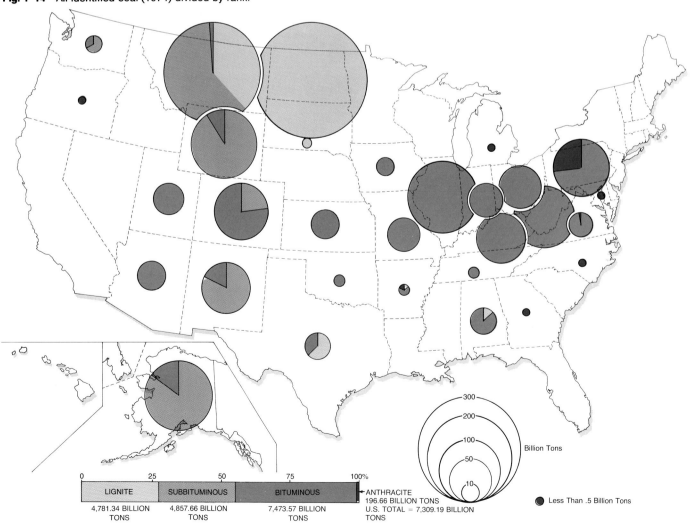

0	25	50	75	100%

LIGNITE	SUBBITUMINOUS	BITUMINOUS	←ANTHRACITE
4,781.34 BILLION TONS	4,857.66 BILLION TONS	7,473.57 BILLION TONS	196.66 BILLION TONS U.S. TOTAL = 7,309.19 BILLION TONS

300
200
100
50
10

Billion Tons

● Less Than .5 Billion Tons

DEMONSTRATED RESERVE BASE
436.7 BILLION TONS

OTHER IDENTIFIED
1,294.2 BILLION TONS

U.S. TOTAL- 1,730.9 BILLION TONS

0   25   50   75   100%

300
200
100
50
Billion Tons
10

⬤ Less than 1.2 Billion Tons

Fig. 1–15 Mineable coal (Demonstrated Reserve Base) as proportion of Identified coal resources (1974).

1–16 shows the division of the Reserve Base by rank for each state. When measured by tonnage, the breakdown of the Reserve Base for the entire country by rank is: 6 percent for lignite; 38 percent for subbituminous; 53 percent for bituminous; and 2 percent for anthracite. If the energy content by rank is considered, the percentage of the Reserve Base for lignite and subbituminous is slightly smaller and slightly larger for bituminous. The distribution of coal deposits by rank again reveals the dominance of bituminous coal in the eastern states and subbituminous and lignite in the West.

MINING METHODS The mining method used to extract coal determines, in part, the amount of different ranks of coal that are usable. The coal Reserve Base is presented according to mining method in Table 1–4 and Fig. 1–17. Some coals can be mined *only* by one of the two methods; other coals can best be mined by one or the other methods. About 69 percent of the Reserve Base is extracted by underground mining, while 31 percent can be mined by surface methods. This ratio varies greatly from state to state and region to region. It differs radically in eastern and western parts of the country. In states east of the Mississippi most of the coal (84 percent) is under-

ground mineable, while only 17 percent can be mined by surface methods. In contrast, 56 percent of the western coals are mineable by underground methods and 44 percent by surface methods. In states east of the Mississippi River with large Reserve Bases, such as Illinois, West Virginia, and Pennsylvania, the portion that can be mined by underground methods typically exceeds 80 percent. Wyoming is typical of western states, with 55 percent of the Reserve Base underground mineable.

MINING METHOD AND RANK The recovery of coal by surface mining depends primarily on the ratio of the thickness of the overburden to that of the coal bed. Basically, a limit of 15 feet of overburden per foot of coal thickness is used to calculate the surface mineable Reserve Base. Available machinery currently limits surface mining to depths less than 180 feet (Bureau of Mines, *Information Circular 8680*, 1975).

Figures 1–18 through 1–21 show the amounts of coal mineable by surface and underground methods for each of the four ranks. The *lignites* located in Montana, North Dakota, Texas, and Alabama are all surface mineable. For the more deeply buried *subbituminous* coals of the western states only 40 percent are surface mineable.

Nevertheless, there is a considerable difference from state to state in the amounts of subbituminous coals that can be mined by surface methods. For example, 77 percent of these coals in New Mexico are surface mineable, but none of those in Colorado. Approximately 18 percent of the *bituminous* coals are surface mineable, a percentage that roughly holds true for states both east and west of the Mississippi River. Major exceptions are found in Kansas and Alaska where bituminous coals are not deeply buried. Only 1 percent of the significant amounts of *anthracite* coal, found almost exclusively in Pennsylvania, is amenable to surface mining. Figure 1–22

provides a summary of both the tonnage and energy content of surface and underground mineable coals for each of the four ranks.

SULFUR CONTENT Stimulated by the need to identify clean-burning coals, the Bureau of Mines has delineated the coal Reserve Base on the basis of sulfur content, using four major categories: (1) *low-sulfur*, less than or equal to 1.0 percent sulfur; (2) *medium-sulfur*, 1.1–3.0 percent sulfur; (3) *high-sulfur*, greater than 3.0 percent sulfur; and (4) unknown sulfur content. As was discussed before in the section on rank and sulfur content, sulfur is an undesirable constituent of coal as it lowers the quality

Table 1–3 Mineable Amounts of Coal (Demonstrated Reserve Base) According to Rank, 1974 (in millions of tons)

STATE	Anthracite	Bituminous	Subbituminous	Lignite	TOTAL
Alabama		1,954		1,027	2,981
Alaska		1,021	10,146	296	11,643
Arizona			350		350
Arkansas	96	536		32	664
Colorado	28	10,095	4,746		14,869
Illinois		65,665			65,665
Indiana		10,623			10,623
Iowa		2,883			2,883
Kansas		1,388			1,388
Kentucky		25,533			25,533
Maryland		1,048			1,048
Michigan		118			118
Missouri		9,488			9,488
Montana		1,384	99,950	7,097	108,431
New Mexico	2	1,777	2,642		4,421
North Dakota				15,998	15,998
Ohio		21,081			21,081
Oklahoma		1,293			1,293
Oregon			1		1
Pennsylvania	7,120	23,866			30,986
South Dakota				428	428
Tennessee		983			983
Texas				3,272	3,272
Utah		4,043			4,043
Virginia	137	3,510			3,647
Washington		250	1,694		1,944
West Virginia		39,574			39,574
Wyoming		4,524	48,807		53,331
West of Mississippi	126	38,864	168,336	27,123	234,450
East of Mississippi	7,257	193,955		1,027	202,238
U.S. TOTAL	7,383	232,819	168,336	28,150	436,688

Source: Bureau of Mines, *Information Circulars #8678* and *#8655*.

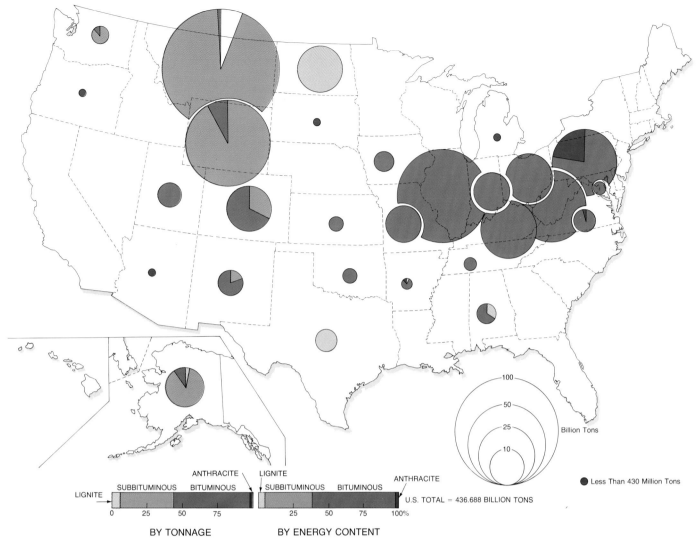

Fig. 1–16 Mineable coals (Demonstrated Reserve Base) divided by rank.

LIGNITE   SUBBITUMINOUS   BITUMINOUS   ANTHRACITE

0   25   50   75

BY TONNAGE

LIGNITE   SUBBITUMINOUS   BITUMINOUS   ANTHRACITE

25   50   75   100%

BY ENERGY CONTENT

U.S. TOTAL = 436.688 BILLION TONS

100
50
25
10
Billion Tons

● Less Than 430 Million Tons

Fig. 1–17 Total tonnage mineable by surface and underground methods.

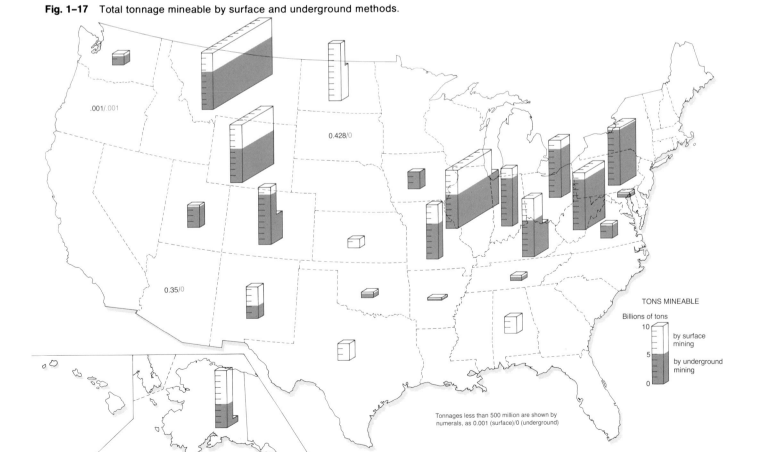

.001/.001

0.428/0

0.35/0

TONS MINEABLE

Billions of tons

10
5
0

by surface mining
by underground mining

Tonnages less than 500 million are shown by numerals, as 0.001 (surface)/0 (underground)

Billion Tons

10

5   (all by surface mining)

0

0.032/0

0.296/0

ALASKA

Tonnages less than 500 million are shown by numerals, as 0.032 (surface)/0 (underground)

Fig. 1–18 Lignite tonnage mineable by surface and underground methods.

Fig. 1–19 Subbituminous tonnage mineable by surface and underground methods.

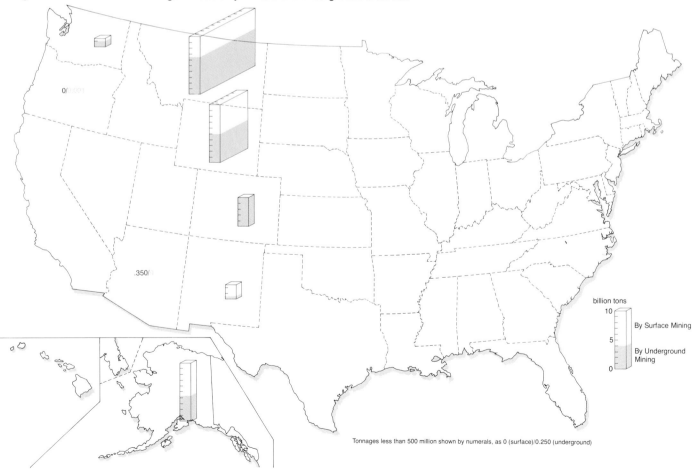

0/0.001

.350/

billion tons

10

5   By Surface Mining

     By Underground
     Mining

0

Tonnages less than 500 million shown by numerals, as 0 (surface)/0.250 (underground)

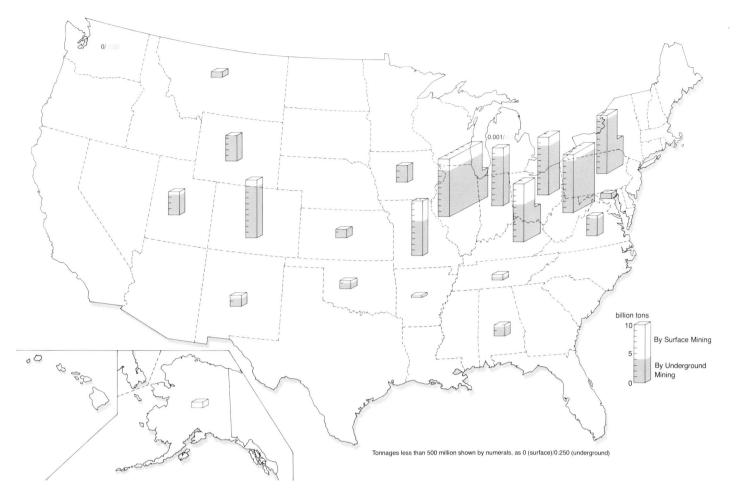

0/.250

0.001/

billion tons

10

5

0

By Surface Mining

By Underground Mining

Tonnages less than 500 million shown by numerals, as 0 (surface)/0.250 (underground)

Fig. 1–20 Bituminous tonnage mineable by surface and underground methods.

Fig. 1–21 Anthracite tonnage mineable by surface and underground methods.

0/27

0/2

0/96

0/137

Billion Tons

10

5

0

By surface mining

By Underground Mining

Tonnages Less Than 500 Million (0.5 Billion) Are Shown By Numerals In Millions of Tons.

Table 1–4 Mineable Amounts of Coal (Demonstrated Reserve Base) of All Ranks by Underground and Surface Methods, 1974 (in millions of tons)

STATE	By Surface	By Underground	TOTAL MINE-ABLE
Alabama	1,183	1,797	2,981
Alaska	7,399	4,244	11,643
Arizona	350		350
Arkansas	263	402	665
Colorado	870	13,998	14,868
Illinois	12,223	53,442	65,665
Indiana	1,674	8,948	10,622
Iowa		2,883	2,883
Kansas	1,388		1,388
Kentucky	7,352	18,181	25,533
Maryland	146	902	1,048
Michigan	1	117	118
Missouri	3,414	6,075	9,489
Montana	42,571	65,860	108,431
New Mexico	2,285	2,137	4,422
North Dakota	15,998		15,998
Ohio	3,653	17,429	21,081
Oklahoma	434	860	1,294
Oregon	1	1	2
Pennsylvania	1,178	29,808	30,986
South Dakota	428		428
Tennessee	318	665	983
Texas	3,272		3,272
Utah	262	3,781	4,043
Virginia	678	2,969	3,647
Washington	500	1,445	1,945
West Virginia	5,208	34,366	39,574
Wyoming	23,846	29,485	53,331
West of Mississippi	103,281	131,169	234,450
East of Mississippi	33,614	168,624	202,238
UNITED STATES TOTAL	136,895	299,793	436,688

Source: Bureau of Mines, *Information Circulars #8678* and *#8655*.

of coal, corrodes boilers, and creates pollution problems.

Of the total United States coal Reserve Base, 46 percent is low-sulfur, 21 percent medium-sulfur, 22 percent high-sulfur, and 11 percent unknown (Figs. 1–23 and Table 1–5). As with the location of different ranks of coal in eastern and western states, the amount of sulfur in the east and west region varies widely. In the western states, where 54 percent of the coal Reserve Base is located, over 70 percent of the proportion of coals is low-sulfur, with 87 percent having less than 3 percent sulfur. By contrast, only 15 percent of the coal Reserve Base of the eastern states is low-sulfur and 43 percent has a sulfur content of more than 3 percent.

Montana and Illinois demonstrate the variations in sulfur content between eastern and western coal Reserves. In Illinois, with about 15 percent of all the mineable coals in the United States, few low-sulfur coals are found and almost 70 percent is classified as high-sulfur. On the other hand, in Montana, which has almost 25 percent of the Nation's Reserve Base of coal, about 94 percent is low-sulfur. In the East, West Virginia has the largest amount of low-sulfur coal, over 14 billion tons.

SULFUR CONTENT AND RANK To a large extent sulfur content varies with the rank of coal. For example, low-sulfur coal is most often subbituminous, and high-sulfur is usually bituminous. Most anthracites, however, are low in sulfur so there is no reliable relationship between sulfur content and rank. Table 1–6 shows the amount of *low-sulfur* coal by rank. Of the more than 198 billion tons of low-sulfur coal, 5 percent is lignite, 74 percent subbituminous, 18 percent bituminous and 3 percent anthracite. All of the low-sulfur lignite and subbituminous coal is located in the western states. The states with the largest amounts of low-sulfur subbituminous coal are Montana and Wyoming, with a combined total of more than 130 billion tons. Montana and North Dakota house most of the 10.2 billion tons of low-sulfur lignite. Slightly more than 60 percent of the 35 billion tons of low-sulfur bituminous coal is located in the eastern states, with more than 70 billion tons in West Virginia and Kentucky. In the West, Colorado, Utah, New Mexico, Wyoming, and Alaska have over 90 percent of the low-sulfur bituminous. More than 6.3 billion tons of low-sulfur anthracite coal are found in five states: Pennsylvania, Virginia, Colorado, Arizona, and New Mexico, but Pennsylvania houses all but 88 million tons of it. It is interesting to note that 85 percent of all low-sulfur coal in Pennsylvania is anthracite.

As Table 1–7 shows, a total of 91.9 billion tons of medium-sulfur coal Reserves exists in the United States. By rank, the division is: 15 percent lignite, 18 percent subbituminous, 66 percent bituminous, and 1 percent anthracite. Medium-sulfur lignite and subbituminous coals are again confined to the western states. Wyoming houses over 12 billion tons of medium-sulfur subbituminous, and North Dakota approximately 10 billion tons of medium-sulfur lignite. Of the roughly 60 billion tons of

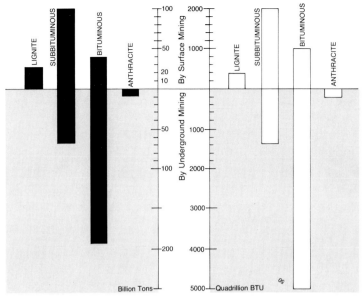

Fig. 1–22 Summary of surface and underground mineable coals of four ranks, showing tonnage and energy content.

Category	Sulfur Content
Low-sulfur	Less than or equal to 1.0 percent
Medium-sulfur	1.1 to 3.0 percent
High-sulfur	Greater than 3.0 percent
Unknown sulfur	Unknown percent

Fig. 1–23 Mineable coals (Demonstrated Reserve Base) divided by sulfur content.

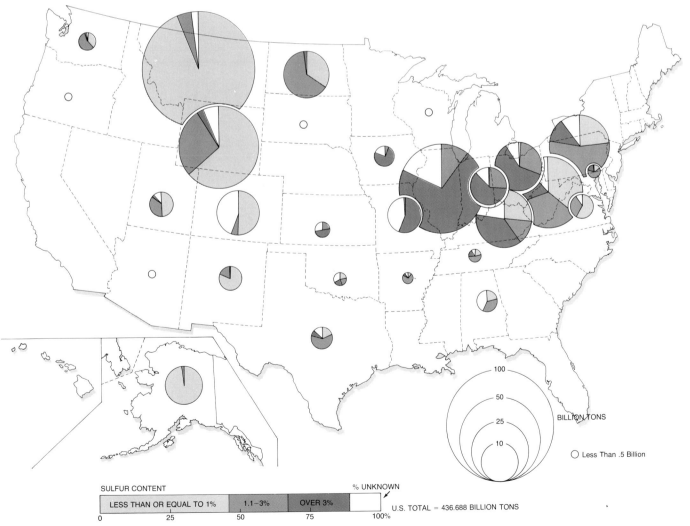

medium-sulfur bituminous coal, almost 90 percent is located in the eastern United States. Most of these eastern Reserves are found in West Virginia, Pennsylvania, Illinois, Ohio, and Kentucky. Four western states, Montana, Wyoming, Utah, and Colorado, have a combined total of 4.5 billion tons of medium-sulfur bituminous coal. In addition, there are small amounts of medium-sulfur anthracite coal in Pennsylvania and Arkansas.

High-sulfur coals account for more than 97 billion tons of the coal Reserve Base. Over 98 percent of all of the high-sulfur coal is bituminous (Table 1–8). Only 1.3 billion tons of the lignite and subbituminous coal of the western United States has a high sulfur content. On the other hand, approximately 90 percent of the high-sulfur bituminous coal occurs in the eastern United States. Over one-half of it is found in Illinois. None of the Pennsylvania anthracite is classified as high-sulfur.

A total of 48.5 billion tons of the coal Reserve Base is of *unknown sulfur* content because the sulfur has not yet been measured (Table 1–9). Over 80 percent of these coals are bituminous, while subbituminous and lignite account for 11 and 6 percent respectively (anthracite is too small to measure). Almost three-fourths of the bituminous coals of unknown sulfur content are located in the eastern United States, Illinois having the largest amount, approximately 12 billion tons. Figure 1–24 presents a national summary of mineable coals showing which ranks account for the low-, medium-, and high-sulfur coals.

Table 1–5 Mineable Amounts of Coal Showing Four Sulfur Categories, 1974 (in millions of tons)

STATE	DEMONSTRATED RESERVE BASE	AMOUNTS BY SULFUR CATEGORY			
		Up to 1%	1.1–3.0%	> 3.1%	Unknown
Alabama	2,981	625	1,100	16	1,239
Alaska	11,643	11,459	184		
Arizona	350	173	177		
Arkansas	664	81	463	46	74
Colorado	14,869	7,488	786	47	6,547
Illinois	65,665	123	6,970	46,372	12,200
Indiana	10,623	98	2,609	6,626	1,290
Iowa	2,883	1	227	2,106	549
Kansas	1,388	0	309	696	383
Kentucky	25,533	6,559	3,886	9,543	5,545
Maryland	1,048	135	691	187	35
Michigan	118	5	85	21	7
Missouri	9,488		182	5,226	4,080
Montana	108,431	101,646	4,115	503	2,167
New Mexico	4,421	3,575	818	1	28
North Dakota	15,998	5,389	10,325	269	15
Ohio	21,081	134	6,441	12,634	1,872
Oklahoma	1,293	275	327	241	450
Pennsylvania	30,986	7,318	16,914	3,800	2,954
South Dakota	428	103	288	36	1
Tennessee	983	205	533	157	88
Texas	3,272	660	1,885	284	444
Utah	4,043	1,969	1,547	49	478
Virginia	3,647	2,140	1,163	14	330
Washington	1,944	597	1,265	39	43
West Virginia	39,574	14,092	14,006	6,823	4,653
Wyoming	53,331	33,912	14,657	1,701	3,061
UNITED STATES TOTAL	436,687	198,763	91,953	97,437	48,534

Source: Bureau of Mines, *Information Circulars 8680* and *8693*, 1975.

SULFUR CONTENT, RANK, AND MINING METHOD   In order to assess the overall quality and ease of recovery of the coal Reserve Base, it is helpful to look at the rank, sulfur content and mining method together.

Tables 1–6 through 1–9 reveal, for the four sulfur levels in turn, the amounts of coal of each rank that are considered surface mineable and underground mineable.

For *low-sulfur coals*, 37 percent is surface mineable and 63 percent underground mineable. Over 75 percent of the coal mineable by surface methods is subbituminous, 14 percent lignite, 10 percent is bituminous, and a very small proportion is anthracite. For the portion mineable by underground methods, 72 percent is subbituminous, 23 percent bituminous, and 5 percent is anthracite. Lignite is found at shallow depths and is therefore mined by surface methods. Over half of the low-sulfur coal is in Montana.

Most of it is *subbituminous* and underground mineable. In the East, West Virginia, Pennsylvania, and Kentucky house approximately 14 percent of the low-sulfur coal. Excepting the more than 6 billion tons of low-sulfur anthracite, low-sulfur coal in the East is bituminous and over 80 percent is recoverable by underground mining.

Predominantly bituminous, *medium-sulfur coals* are usually recoverable by underground mining. Surface mineable medium-sulfur coals account for 36 percent of the total, with 44 percent being lignite, 35 percent subbituminous, and 21 percent bituminous. Underground mineable medium-sulfur coals amount to more than 59 billion tons, roughly 90 percent of which are bituminous. Three states, Pennsylvania, Wyoming, and West Virginia, contain almost one-half of the total medium-sulfur coal. In Pennsylvania, more than 95 percent of the

Table 1–6   Amounts of Low-Sulfur Coal by Rank and Mining Method, 1974 (in millions of short tons)

STATE	SURFACE MINEABLE				UNDERGROUND MINEABLE				TOTAL MINE- ABLE
	Lignite	Sub- bituminous	Bituminous	Anthracite	Lignite	Sub- bituminous	Bituminous	Anthracite	
Alabama	0	0	35				589		624
Alaska	275	5,902	1,201			4,081			11,459
Arizona	0	173							173
Arkansas	0		38				35	8	81
Colorado	0		724			3,030	3,706	27	7,487
Illinois	0		4				119		123
Indiana	0		17				80		97
Iowa	0						1		1
Kentucky	0		1,516				5,043		6,559
Maryland	0		29				106		135
Michigan	0						5		5
Montana	3,795	34,387				63,306	158		101,646
New Mexico	0	1,452	229			430	1,463	1	3,575
North Dakota	5,389								5,389
Ohio	0		19				115		134
Oklahoma	0		120				154		274
Oregon	0					1			1
Pennsylvania	0		55	83			981	6,199	7,318
South Dakota	103								103
Tennessee	0		66				139		205
Texas	660								660
Utah	0		52				1,916		1,968
Virginia	0		412				1,676	52	2,140
Washington	0	167				252	178		597
West Virginia	0		3,005				11,087		14,092
Wyoming	0	13,192				19,487	1,232		33,911
TOTAL	10,222	55,273	7,522	83		90,587	28,783	6,287	198,757

Sources: Bureau of Mines, *Circulars 8680* and *8693*, 1975.

medium-sulfur coal is underground mineable bituminous. Two-thirds of the Wyoming medium-sulfur coal is subbituminous and recoverable by surface methods.

High-sulfur coals, characteristically bituminous, are recovered by underground mining. Illinois has more than 47 percent of the high-sulfur coal, with an additional 42 percent being found in Ohio, Kentucky, West Virginia, Indiana, and Missouri. Overall, 80 percent of the high-sulfur coals are underground mineable and almost 99 percent are bituminous.

For those coals whose sulfur content is unknown, over three-fourths are recoverable by underground methods. Both the underground and surface mineable coals of unknown sulfur content are predominantly bituminous.

SUMMARY OF DEMONSTRATED RESERVE BASE   Much of the information presented earlier in tables and maps is summarized here in a series of bar graphs which provide a quick reference to relative amounts. In addition, these graphs emphasize the geographic dimension: first by distinguishing western from eastern states; and subsequently by treating each state separately.

Figure 1–25 shows the underground and surface mineable coals in the United States divided by rank and

Table 1–7   Amounts of Medium-Sulfur Coal by Rank and Mining Method, 1974 (in millions of short tons)

STATE	\multicolumn{4}{c}{SURFACE MINEABLE}				\multicolumn{4}{c}{UNDERGROUND MINEABLE}				TOTAL MINE-ABLE
	Lignite	Sub-bituminous	Bituminous	Anthracite	Lignite	Sub-bituminous	Bituminous	Anthracite	
Alabama			83				1,017		1,100
Alaska	21					163			184
Arizona		177							177
Arkansas			153				233	77	463
Colorado			146			100	540		786
Illinois			691				6,279		6,970
Indiana			549				2,060		2,609
Iowa							227		227
Kansas			309						309
Kentucky			1,108				2,778		3,886
Maryland			67				624		691
Michigan							85		85
Missouri			48				134		182
Montana	1,791	384				1,170	770		4,115
New Mexico		583	21			177	36	1	818
North Dakota	10,325								10,325
Ohio			991				5,450		6,441
Oklahoma			88				238		326
Pennsylvania			717	1			16,013	182	16,913
South Dakota	288								288
Tennessee			163				370		533
Texas	1,885								1,885
Utah			149				1,398		1,547
Virginia			218				945		1,163
Washington		308				908	50		1,266
West Virginia			1,423				12,583		14,006
Wyoming		10,122				2,306	2,229		14,657
UNITED STATES TOTALS	14,310	11,574	6,924	1	0	4,824	54,059	260	91,953

Sources: Bureau of Mines, *Circulars 8680* and *8693,* 1975.

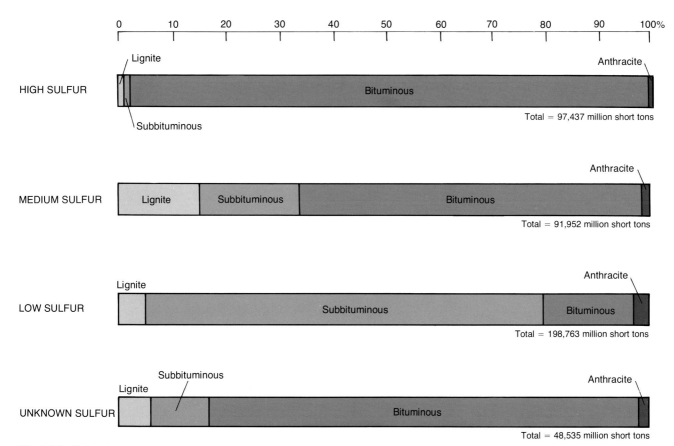

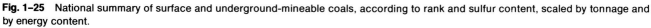

Fig. 1–24 National summary of mineable coals showing which ranks account for the low, medium, and high-sulfur coals.

Fig. 1–25 National summary of surface and underground-mineable coals, according to rank and sulfur content, scaled by tonnage and by energy content.

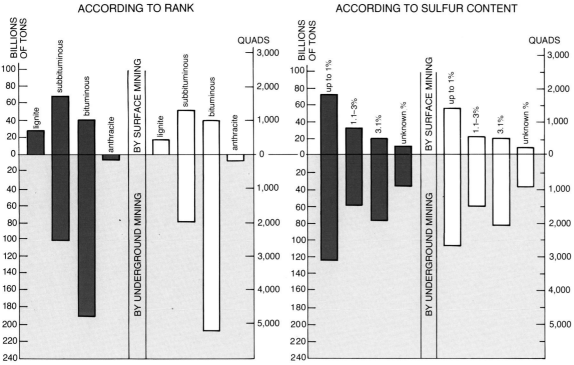

by sulfur content. For each of these two characteristics the relative amounts are measured in both tonnage and in Quads.

In Figs. 1–26 and 1–27 and Table 1–10, the prevalence of the four ranks and the four sulfur classes in surface mineable and underground mineable coals for eastern and western states is shown. Figure 1–27 compares western and eastern states in regard to the prevalence of the four ranks, the four sulfur classes, and the portions mineable by surface and underground methods.

Figures 1–28 and 1–29 reveal the amounts of surface and underground mineable coals for each state, and within each of those categories the proportions that are in the four ranks and the four sulfur classes.

Figure 1–30 adds a new perspective to the Demonstrated Reserve Base. It shows the amounts in each state that are Identified but not assigned to Reserve Base. As noted before, these coals are excluded from the Demonstrated Reserve Base because of depth, thickness of the bed, or conflicts with land use. Total length of the bar for each state represents all Identified coals in the state (Averitt, 1975). When arranged according to all Identified amounts, the states assume an order that is different from a ranking based on Reserve Base. North Dakota becomes the second leading western state on the basis of the tonnage of coal located in the state. This consists of more than 350 billion tons of lignite, only a small portion of which is considered mineable (Reserve Base). Similarly,

Table 1–8   Amounts of High-Sulfur Coals by Rank and Mining Method, 1974 (in millions of short tons)

STATE	SURFACE MINEABLE				UNDERGROUND MINEABLE				TOTAL MINE-ABLE
	Lignite	Sub-bituminous	Bituminous	Anthracite	Lignite	Sub-bituminous	Bituminous	Anthracite	
Alabama			2				15		17
Arkansas			17				18	11	46
Colorado							47		47
Illinois			10,206				36,166		46,372
Indiana			997				5,628		6,625
Iowa							2,106		2,106
Kansas			696						696
Kentucky			2,104				7,439		9,543
Maryland			16				171		187
Michigan							21		21
Missouri			1,636				3,590		5,226
Montana	46						456		502
North Dakota	269								269
Ohio			2,525				10,109		12,634
Oklahoma			39				203		242
Pennsylvania			232				3,568		3,800
South Dakota	36								36
Tennessee			55				102		157
Texas	284								284
Utah			43				7		50
Virginia			2				12		14
Washington		26				13			39
West Virginia			270				6,553		6,823
Wyoming		425				235	1,040		1,700
UNITED STATES TOTAL	635	451	18,840	0	0	248	77,251	11	97,436

Source: Bureau of Mines, *Circulars 8680 and 8693*, 1975.

Note: Estimated amounts of coal recoverable are probably high in those states with substantial amounts of subbituminous coal, because beds thicker than 10 feet common in that coal make underground recovery of 50 percent impossible by current United States technology. For the following six states the proportion of subbituminous coal in thick beds is indicated.

Alaska	97 percent	New Mexico	5 percent
Colorado	50 percent	Washington	30 percent
Montana	41 percent	Wyoming	58 percent

Colorado shows large amounts of Identified coal, much of it bituminous, which is not part of the Reserve Base.

The wide variation among states in their Identified amounts of coal necessitates the use of a *logarithmic scale* rather than a linear scale to show relative magnitudes. The relative amounts of Demonstrated Reserves and other Identified coals are therefore not immediately evident when examining Fig. 1–30. Most identified coals not in the Reserve Base are excluded because of their excessive depth. For this reason, the non-Reserve Base Identified coals are shown at the "bottom end" of each bar. Actually, some of this Identified coal may be in beds too thin for mining but at depths within the mineable realm, or some may even be in shallow beds that could be extracted by surface mining if the beds were thick enough.

RECOVERABLE COAL RESERVES

Coal Reserves are defined as the quantity of the Demonstrated Reserve Base that can actually be recovered, given present technology, and any economic or legal constraints (Bureau of Mines, *Information Circular 8680*, 1975). Specific factors influencing the amounts of coal that can be recovered are the mining method, environmental aspects, and the quality of the coal.

Initially, the quantity of the Demonstrated Reserve Base that can be recovered depends on whether or not the coal bed is suitable for underground or surface mining. As stated before, in surface mining the recovery of coal depends primarily on the ratio of the thickness of the overburden to that of the coal bed. Normally, this ratio should be around 15:1. In addition, the local topography determines whether the coal can be recovered through

Table 1–9 Amounts of Coal of Unknown Sulfur Content by Rank and Mining Method, 1974 (in millions of short tons)

STATE	SURFACE MINEABLE				UNDERGROUND MINEABLE				TOTAL MINE- ABLE
	Lignite	Sub-bituminous	Bituminous	Anthracite	Lignite	Sub-bituminous	Bituminous	Anthracite	
Alabama	1,027		37				176		1,240
Arkansas	32		23				19		74
Colorado						1,616	4,931		6,547
Illinois			1,323				10,878		12,201
Indiana			110				1,180		1,290
Iowa							549		549
Kansas			383						383
Kentucky			2,624				2,921		5,545
Maryland			35						35
Michigan							7		7
Missouri			1,730				2,350		4,080
Montana	1,464	702							2,166
New Mexico							28		28
North Dakota	15								15
Ohio			118				1,754		1,872
Oklahoma			186				264		450
Pennsylvania			84	6			2,215	649	2,954
South Dakota	1								1
Tennessee			34				54		88
Texas	444								444
Utah			18				460		478
Virginia			47				108	85	330
Washington						21	22		43
West Virginia			510				4,143		4,653
Wyoming		105				2,933	22		3,060
UNITED STATES TOTAL	2,983	807	7,262	6		4,570	32,171	734	48,533

Sources: Bureau of Mines *Circulars 8680* and *8693*, 1975.

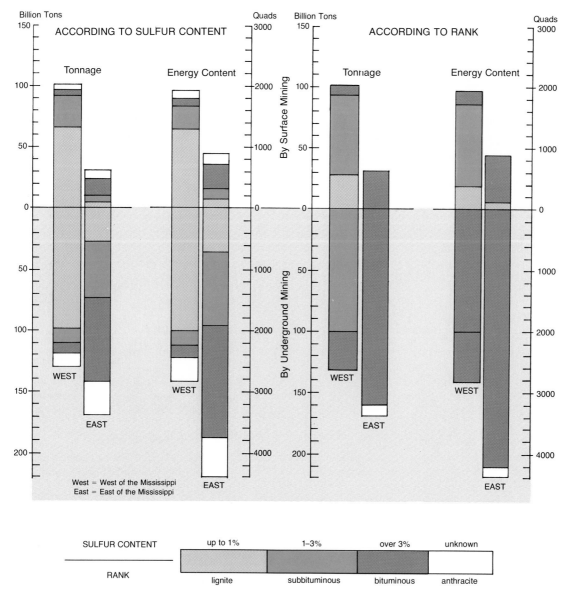

SULFUR CONTENT — up to 1% | 1–3% | over 3% | unknown

RANK — lignite | subbituminous | bituminous | anthracite

Fig. 1–26 Summary of western and eastern states, showing prevalence of various ranks and sulfur contents in coals mineable by surface and underground methods, scaled by tonnage and energy content.

Table 1–10 Summary of Western and Eastern States Showing Prevalence of Various Ranks and Sulfur Contents in Coals Mineable by Surface and Underground Methods, 1974.

	EASTERN UNITED STATES				WESTERN UNITED STATES			
	Surface Mineable		Underground Mineable		Surface Mineable		Underground Mineable	
BY SULFUR CONTENT	Billions of Tons	Quads	Billions of Tons	Quads	Billions of Tons	Quads	Billions of Tons	Quads
To 1 percent	5.2	136.3	26.2	683.5	67.9	1,310.1	99.5	2,042.7
1–3 percent	6.0	156.3	48.4	1,258.1	26.8	455.6	10.8	250.7
Over 3 percent	16.4	426.6	69.8	1,814.4	3.5	81.1	7.7	199.4
Percent unknown	6.0	142.5	24.3	631.1	5.1	104.4	13.2	316.2
TOTAL	33.6	861.7	168.7	4,387.1	103.3	1,951.2	131.2	2,809.0
BY RANK								
Lignite	1.0	14.4	0.0	0.0	27.1	379.7	0.0	0.0
Subituminous	0.0	0.0	0.0	0.0	68.1	1,362.1	100.2	2,004.6
Bituminous	32.5	844.9	161.5	4,197.9	8.1	209.3	30.8	801.1
Anthracite	0.1	2.4	7.2	189.2	0.0	0.0	0.1	3.3
TOTAL	34.4	861.7	168.7	4,387.1	103.3	1,951.1	131.1	2,809.0

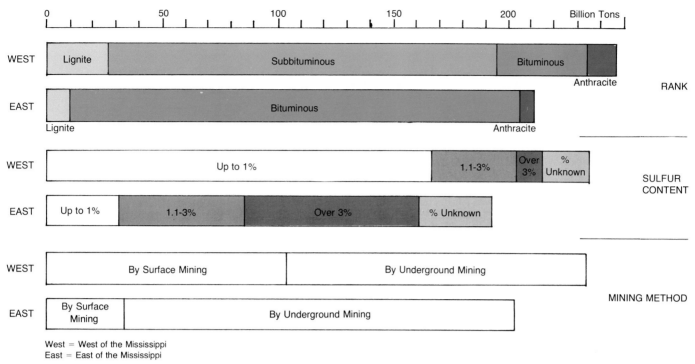

Fig. 1–27 Western and Eastern states: summary of mineable coals according to rank, sulfur content, and mining method.

contour stripping or area stripping. On the average, the recovery ratio for strip mining ranges from 80 to over 90 percent. For underground mining, Bureau of Mines studies indicate that average recovery is about 50 percent. The lower recovery ratio for underground mining is primarily due to the coal left unmined in order to support the roof.

Conflicting land use is another factor which may inhibit the mining of certain coal beds. When coal lies beneath urban areas, public facilities, such as parks and airports, or waterways, it is clearly impossible for the coal to be mined. In addition, coal beds overlying or underlying other worked-out beds can be hazardous and expensive to mine (Bureau of Mines, *Information Circular 8680*, 1975).

In the West, the underground mining of coals is often constrained by the bed thickness. Although approximately 35 percent of the total underground Reserve Base of the western states is found in coal beds greater than 10 feet thick (Bureau of Mines, *Information Circular 8678*, 1975), much of it is considered unavailable. Domestic underground mining technology and equipment are generally limited to face heights of 10 feet or less. In a 20-foot seam, for example, only a 10-foot high slice is mined, while the remaining 10 feet is not touched. Assuming that room and pillar mining can recover 57 percent of the 10-foot slice, less than 30 percent of the 20-foot seam is mined. If the longwall mining method is used (see earlier section), 85 percent of the 10-foot slice is recovered, and therefore 43 percent of the 20-foot bed.

Environmental factors can seriously inhibit the mining of certain coal beds. Land disturbance due to surface mining, subsidence or sinking of parts of the earth due to underground mining, and the effect of mine drainage on water quality can all prevent mining. Furthermore, the chemical and physical properties of coal such as excessive sulfur, ash, and volatile materials may restrict the use of certain coals and reduce the amounts actually recovered.

ESTIMATES OF RECOVERABLE COAL Due to conflicting land use, bed thickness, depth, and environmental factors, extreme caution should be exercised in interpreting coal Reserves. The estimates in Fig. 1–30 and Table 1–11 can therefore only be considered as rough approximations. In these estimates a 50 percent recovery factor was assumed for underground mining and a 90 percent factor for surface mining.[5] Based on these assumptions, recoverable reserves may be as high as 273 billion tons. Surface mining would account for 45 percent of the total and underground mining for 55 percent. About 58 percent of the reserves lie west of the Mississippi River.[6] Approximately 56 percent of the total underground mineable reserves are located in the eastern states while 75 percent of the surface mineable reserves are in western states.

In the east, Illinois, West Virginia, Pennsylvania, and Kentucky contain 33 percent of the total reserves, 45 percent of the underground reserves, and 19 percent of the surface reserves. West of the Mississippi River, Montana and Wyoming have 39 percent of the total reserves, 32 percent of the underground reserves and 48 percent of the surface reserves.

Furthermore, the Demonstrated Reserve Base, on which these estimates are based, is itself a conservative

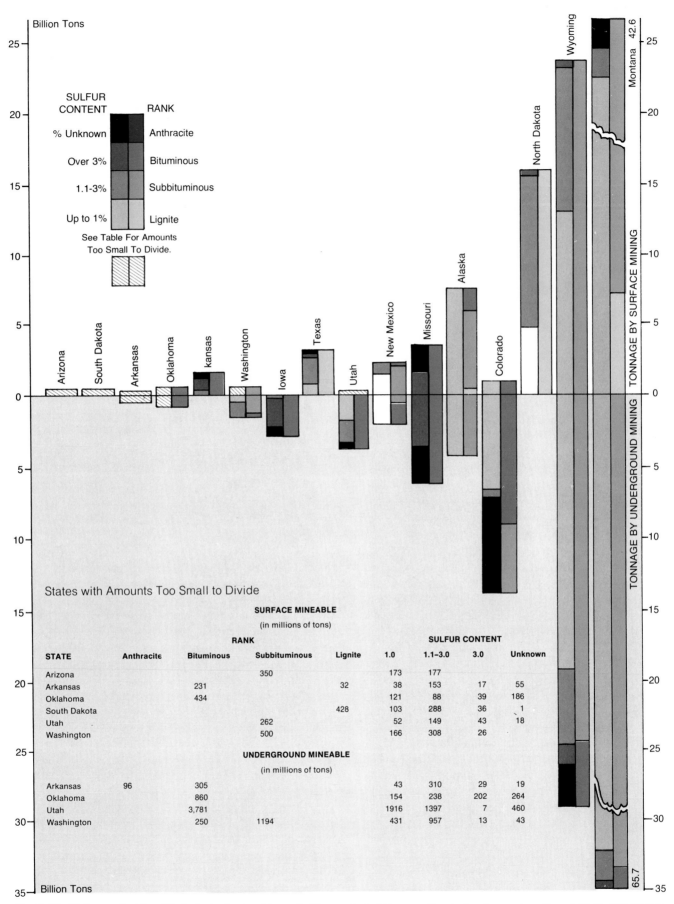

Billion Tons

SULFUR
CONTENT   RANK

% Unknown   Anthracite

Over 3%   Bituminous

1.1-3%   Subbituminous

Up to 1%   Lignite

See Table For Amounts
Too Small To Divide.

TONNAGE BY SURFACE MINING

TONNAGE BY UNDERGROUND MINING

States with Amounts Too Small to Divide

SURFACE MINEABLE

(in millions of tons)

	RANK				SULFUR CONTENT		
STATE	Anthracite	Bituminous	Subbituminous	Lignite	1.0	1.1-3.0	3.0	Unknown
Arizona			350		173	177		
Arkansas		231		32	38	153	17	55
Oklahoma		434			121	88	39	186
South Dakota				428	103	288	36	1
Utah			262		52	149	43	18
Washington			500		166	308	26	

UNDERGROUND MINEABLE

(in millions of tons)

STATE	Anthracite	Bituminous	Subbituminous	Lignite	1.0	1.1-3.0	3.0	Unknown
Arkansas	96	305			43	310	29	19
Oklahoma		860			154	238	202	264
Utah		3,781			1916	1397	7	460
Washington		250	1194		431	957	13	43

Billion Tons

Fig. 1–28 States west of the Mississippi: all mineable coals (Demonstrated Reserve Base) showing rank, sulfur content, and mining method.

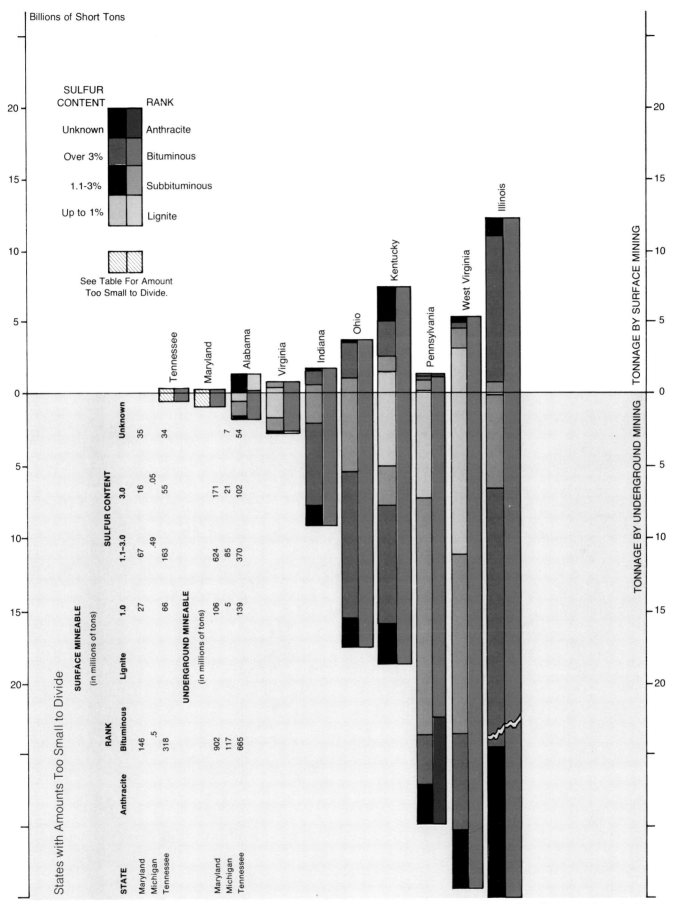

Fig. 1–29 States east of the Mississippi: all mineable coals (Demonstrated Reserve Base) showing rank, sulfur content, and mining method.

estimate of which beds are mineable according to geologic criteria of bed thickness and depth. Advances in mining technology may make possible the mining of Identified coals now *excluded* from the Reserve Base—especially those at depths greater than 1,000 feet. While it is impossible to make a careful estimate of additional amounts that may be recoverable, one approach is to simply estimate that perhaps 50 percent of all Identified coals in the nation may someday be recovered. The resulting amount, 865 billion tons, is shown on Fig. 1–31 for comparison with the 273 billion thought to be recoverable from Reserve Base alone.

Figure 1–32 shows that the 273 billion tons recoverable from Demonstrated Reserve Base, and the 865 possibly recoverable, together exhaust all Identified coals. There is, of course, the possibility that improved mining techniques will allow for higher recovery rates than 50 percent. Coals with overburden greater than 3,000 feet are classified as Hypothetical: with overburden 3,000 feet or less are 1,850 billion tons, roughly the same amount as all Identified; and with overburden ranging from 3,000 to 6,000 feet, only 388 billion tons. Recent revision of these 1975 estimates by the Department of Energy (see Table 1–2) has doubled the total resource amount. Although

Table 1–11 Estimates of Coal Recoverable from the Demonstrated Reserve Base, 1974 (in millions of short tons)

STATE	SURFACE MINEABLE DRB Amounts	SURFACE MINEABLE Amounts Recoverable Assuming 90% Recovery	UNDERGROUND MINEABLE DRB Amounts	UNDERGROUND MINEABLE Amounts Recoverable Assuming 50% Recovery	TOTAL RECOVERABLE
Alabama	1,183	1,065	1,797	899	1,964
Alaska	7,399	6,659	4,244	2,122	8,781
Arizona	350	315			315
Arkansas	263	237	402	201	438
Colorado	870	783	13,998	6,999	7,782
Illinois	12,223	11,001	53,442	26,721	37,722
Indiana	1,674	1,507	8,958	4,474	5,981
Iowa			2,883	1,442	1,442
Kansas	1,388	1,249			1,249
Kentucky	7,352	6,617	18,181	9,091	15,708
Maryland	146	131	902	451	582
Michigan[1]			117	59	59
Missouri	3,414	3,073	6,075	3,038	6,111
Montana	42,571	38,314	65,860	32,930	71,244
New Mexico	2,285	2,057	2,137	1,069	3,126
North Dakota	15,998	14,398			14,398
Ohio	3,653	3,288	17,428	8,714	12,002
Oklahoma	434	391	860	430	821
Pennsylvania	1,178	1,060	29,808	14,904	15,964
South Dakota	428	385			385
Tennessee	318	286	665	333	619
Texas	3,272	2,945			2,945
Utah	262	236	3,781	1,891	2,127
Virginia	678	610	2,969	1,485	2,095
Washington	500	450	1,445	723	1,173
West Virginia	5,208	4,687	34,366	17,183	21,870
Wyoming	23,846	21,470	29,485	14,743	36,213
UNITED STATES TOTAL[2]	136,894	123,215	299,794	149,903	273,118

[1]Michigan also has .54 million tons of surface mineable DRB not reported but included in totals.

[2]Oregon's 1.21 million tons of recoverable coal is included in the totals.

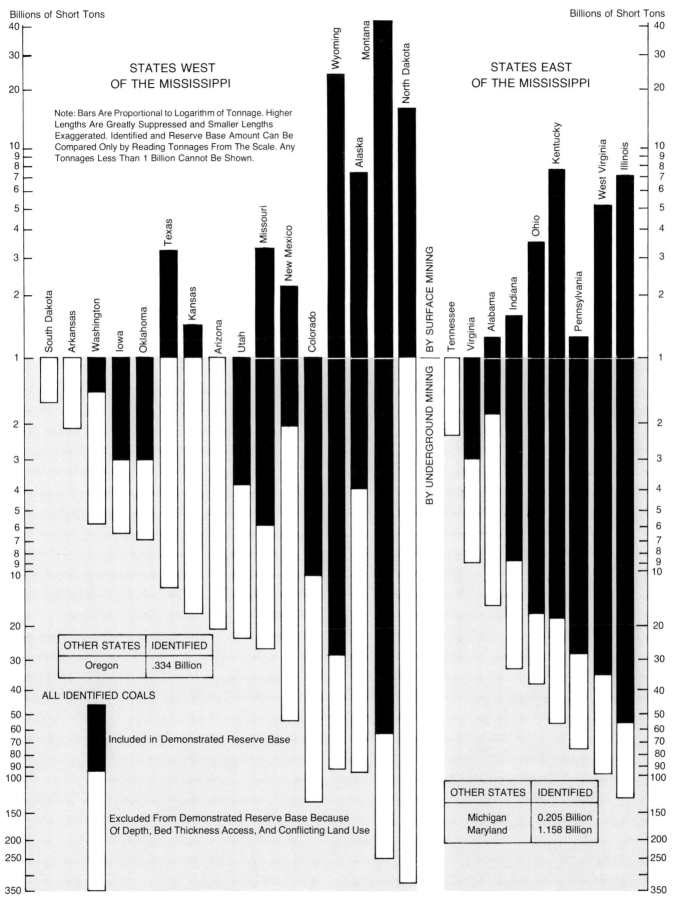

Fig. 1–30 Western and Eastern states: all identified coals, showing portions that are mineable (Demonstrated Reserve Base).

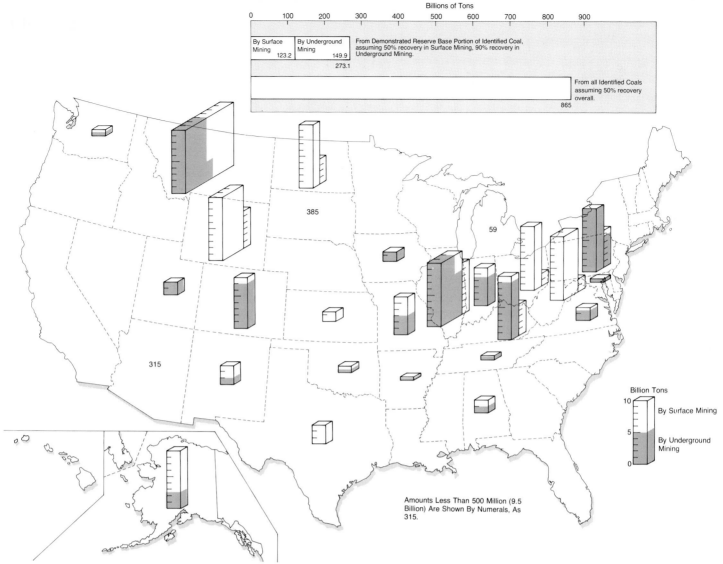

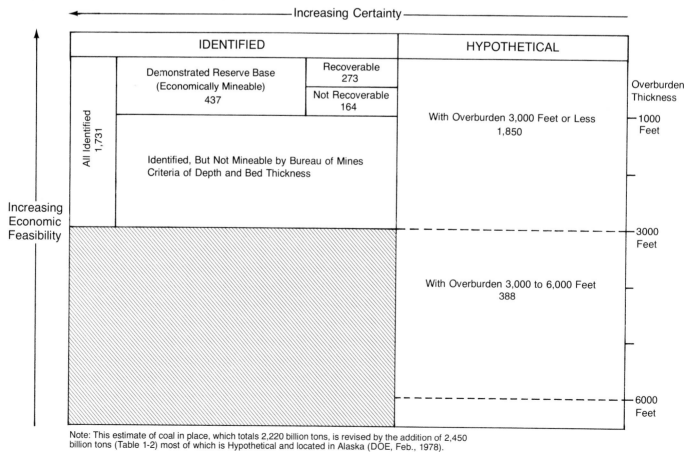

Fig. 1–31 Coal of all ranks recoverable from Demonstrated Reserve Base, assuming 50 percent recovery rate in underground mining and 90 percent in surface mining.

Fig. 1–32 Coal resource amounts, according to Averitt, 1975, in framework of varying geologic certainty and economic feasibility. Except where noted, the amounts are *coal in place*.

most of the additional coal is in Alaska, substantial amounts are in Montana and Wyoming and are being considered for in-place coal gasification (see section on coal gasification).

Coal Utilization

Although coal comprises an estimated 90 percent of the total United States' fossil fuel reserves, it meets less than 20 percent of the Nation's energy needs. Coal has not been used extensively because until recently oil and natural gas were both cheaper and cleaner sources of energy. Domestic shortages of oil, however, and the nearly ten-fold increase in the cost of a barrel of imported crude oil, have made coal a competitive source of energy. Faced with a certain shortage of domestic oil, President Carter called for a doubling of coal production by 1985 to meet the country's future energy needs. Most of the increased amounts of coal would be used in electric power plants and as liquid or gaseous fuels. The degree to which coal can meet future energy needs depends on the coal industry's ability to increase production, the cost of delivering the coal to the major markets of the East and Midwest, and technological developments that would allow coal to be burned without undue cost to the environment. The following discussion will focus on current and future production, and coal movement patterns.

CURRENT PRODUCTION

In 1978, total United States coal production was approximately 660 million tons (Table 1–12), a decrease, caused by the prolonged United Mine Workers' strike, of about 37 millions from 1977 production. The production by rank was 78 percent bituminous, 16 percent sub-bituminous, 6 percent lignite, and 1 percent anthracite (Table 1–13 and Fig. 1–33). Approximately 80 percent of the production took place in eastern states and 20 percent in the West. In the East, bituminous coals accounted for 99 percent of the production with the remainder consisting of Pennsylvania anthracite. The division by rank in the West was 76 percent subbituminous and lignite and 24 percent bituminous.

Figure 1–33 shows the extent to which surface and underground mining have played a part in recent production. In 1978, surface mining accounted for 63 percent of all production.

Three eastern states, Kentucky, West Virginia, and Pennsylvania, produced 45 percent of the Nation's coal and 62 percent of the total mined in eastern coal fields. Among the western states, Montana and Wyoming led in production, accounting for roughly 13 percent of the national total and 46 percent of the coal produced in western coal fields.

Since 1973, coal production in the states west of the Mississippi has been increasing at a substantial rate. In 1970, roughly 75 million tons of coal, or 11 percent of the national total, was produced in the West. In 1978, this in-

creased to 185 million tons, or 28 percent of the national total. Some analysts have predicted that, by 1985, western states will be supplying 50 percent of the Nation's coal. Several factors account for the rapid rise in production of western coals: they are low in sulfur and produce a minimum amount of sulfur dioxide pollution; much of the coal is strip mined, thereby lowering the cost and minimizing safety problems; and finally, most western mines are not unionized thereby minimizing labor disputes.

IMPACT OF PAST, PRESENT AND FUTURE PRODUCTION ON RECOVERABLE RESOURCES

With the rising interest in increased coal production to meet energy needs, it is important to examine the impact

Table 1–12 United States Coal Production for the Years 1976, 1977, and 1978 by State (in thousands of tons)

STATE	1976	1977	1978[1]
Alabama	21,537	21,545	20,110
Alaska	706	705	730
Arizona	10,420	11,059	9,290
Arkansas	534	563	630
Colorado	9,437	11,989	14,310
Georgia	186	226	335
Illinois	58,239	53,493	48,865
Indiana	25,369	27,797	23,940
Iowa	616	513	470
Kansas	590	897	615
Kentucky	143,972	146,262	131,215
Maryland	2,830	3,036	2,855
Missouri	6,075	6,366	5,845
Montana	26,231	27,226	26,680
New Mexico	9,760	11,083	12,365
North Dakota	11,102	12,028	14,625
Ohio	46,582	47,918	43,510
Oklahoma	3,635	5,978	5,425
Pennsylvania[2]	92,005	90,500	81,450
Tennessee	9,283	9,433	9,750
Texas	14,063	15,865	21,000
Utah	7,967	8,581	8,865
Virginia	39,996	37,624	29,750
Washington	4,109	5,057	4,695
West Virginia	108,834	95,433	84,700
Wyoming	30,836	46,028	58,175
UNITED STATES TOTAL	684,913	697,205	660,200

[1] These are preliminary production levels subject to final adjustment.
[2] Includes anthracite production.
Sources: *Coal Age*, February 1977, and Department of Energy, *Energy Data Reports*, March 30, 1979 and April 20, 1979.

Table 1-13   Coal Produced during 1978 Showing Tonnages by Mining Method and Rank[1] (in thousands of tons)

STATE	TOTAL PRODUCTION	PRODUCTION BY RANK				PRODUCTION BY MINING METHOD	
		Lignite	Sub-bituminous	Bituminous	Anthracite	Surface	Under-ground
Alabama	20,110			20,110		13,935	6,175
Alaska	730		730			730	
Arizona	9,290		9,290			9,290	
Arkansas	630			630		605	25
Colorado	14,310		8,443	5,867		9,110	5,200
Georgia	335			335		335	
Illinois	48,865			48,865		23,890	24,975
Indiana	23,940			23,940		23,355	585
Iowa	470			470		250	220
Kansas	616			615		615	
Kentucky	131,215			131,215		72,415	58,800
Maryland	2,855			2,855		2,475	380
Missouri	5,845			5,845		5,845	
Montana	26,680		26,680			26,680	
New Mexico	12,365		8,532	3,833		11,565	800
North Dakota	14,625	14,625				14,625	
Ohio	43,510			43,510		31,210	12,300
Oklahoma	5,425			5,425		5,425	
Pennsylvania	81,450			75,050	6,400	48,265	33,185
Tennessee	9,750			9,750		5,375	4,375
Texas	21,000	21,000				21,000	
Utah	8,865			8,865			8,865
Virginia	29,750			29,750		9,220	20,530
Washington	4,695		4,695			4,695	
West Virginia	84,700			84,700		19,340	65,360
Wyoming	58,175		48,285	9,890		57,475	700
UNITED STATES TOTAL	660,200	35,625	106,655	511,240	6,400	417,725	242,475
PERCENT	100	6	16	77	1	63	37

[1]These are preliminary production levels subject to final adjustment.   Source: Department of Energy, *Energy Data Reports*, March 30, 1979.

of past, present and future production on the nation's recoverable coal supply. Table 1–14 is a very important summary that offers: 1) comparison of cumulative production through 1978 with remaining reserves; 2) estimates of production capacity in 1985; and 3) life of remaining coal reserves. Through 1978, cumulative production of coal in this country amounted to 39.8 billion tons. Past production has consumed approximately 13 percent of the estimated original coal reserves of the United States.

Figure 1–34 shows the impact of past production on the estimated original reserves of each state. As would be expected, more coal in eastern regions has been produced. From 11 to 47 percent of original reserves have

been used in these states. In Tennessee, over 47 percent of the estimated original reserves have been mined and 39 to 41 percent in the states of Alabama, Maryland, Pennsylvania, and Virginia. By contrast, no state in the West has consumed as much as 22 percent of its estimated original reserves and most have consumed less than 11 percent. In Montana, less than one percent of the original reserves have been mined and in Wyoming only 1.8 percent.

COAL MINE DEVELOPMENT AND EXPANSION TO 1985 According to a survey of the coal industry recently completed by *Keystone Coal Industry Manual*, coal mines in development, in the process of expansion, or in the

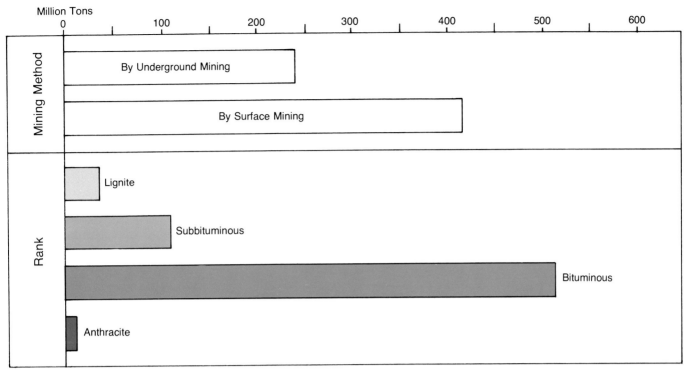

Fig. 1–33 U. S. coal production during 1978, showing tonnages according to mining method and rank.

Fig. 1–34 Cumulative production through 1978 as a proportion of estimated original reserves.

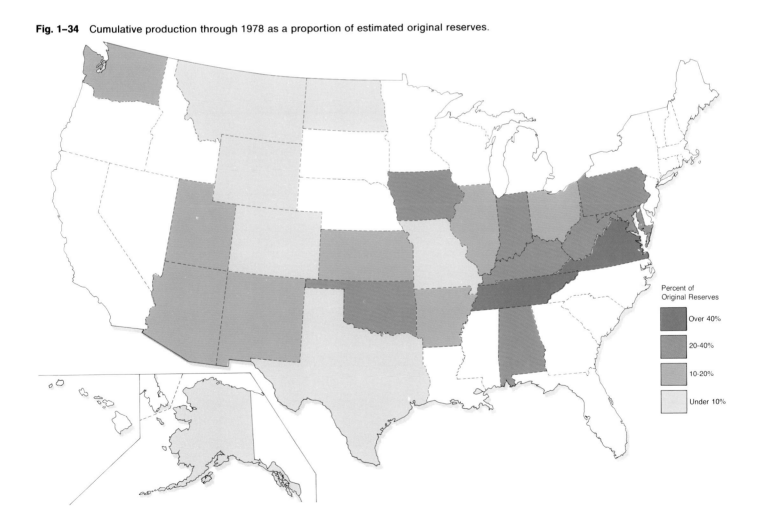

planning stages will increase coal production capacity from 0.685 billion in 1976 to over 1.4 billion tons by 1985 (*Coal Age*, February, 1977). The Keystone survey covered 315 mines in twenty-one states. Figure 1–35 shows the estimated 1985 mining capacity by state, and the proportion due to expansion between 1976 and 1985. For the country, new mining capacity will account for 53 percent of the projected 1985 capacity, of which 65 percent will be surface mining and 35 percent underground mining. States east of the Mississippi River account for 30 percent of the planned expansion and western states for 70 percent. For eastern mines, 25 percent of the new capacity will be strip mining and 75 percent underground mining. By contrast, 85 percent of the new capacity in the West will be surface mining.

The bulk of the planned expansion in the East is in West Virginia, Kentucky, Illinois, and Pennsylvania. Wyoming has the largest capacity, accounting for over 28 percent of the national total and 40 percent of the total for the West.

ESTIMATE OF LIFE OF REMAINING RESERVES   To provide some insight into the impact of future coal usage on remaining reserves, an estimated life was computed on the basis of estimated production capacity in 1985 (Table 1–14 and Fig. 1–36).[7] The estimated life of the remaining reserves for the country as a whole is 186 years. The estimated life among states ranged from 15 years in Arizona to over 12,000 years in Alaska. For the major producing states in the East, West Virginia, Kentucky, and Pennsylvania, the estimated life of the remaining reserves is 126, 80, and 134 years respectively. In the West, the estimated life in Montana is 703 years and 144 years in Wyoming.

The situation in Wyoming deserves special mention. At present only 1.8 percent of the estimated original reserves of Wyoming have been mined; yet, at the 1985 massive production rate the life of remaining reserves is only 144 years. By contrast, Pennsylvania has, in a century of mining, used up 38 percent of estimated original reserves, but the life of remaining reserves is only 9 years less than that of Wyoming.

COAL MOVEMENT AND CONSUMPTION PATTERNS

Although it is very important to have a careful estimate of the remaining coal reserves, and to determine how much can be mined, it is equally important to understand the cost of using coal to meet energy needs.

Since the energy crisis of 1973, and with the encouragement to use coal, the question has been raised of whether western coal can be transported to midwestern and eastern markets at a cost competitive with coals produced in these regions. The major reason for the keen interest in coals in the West, as noted before, is their low sulfur content. Nonetheless, the lower energy content of

Fig. 1–35   Estimated annual mining capacity in 1985, showing amounts of expansion during 1976-1985.

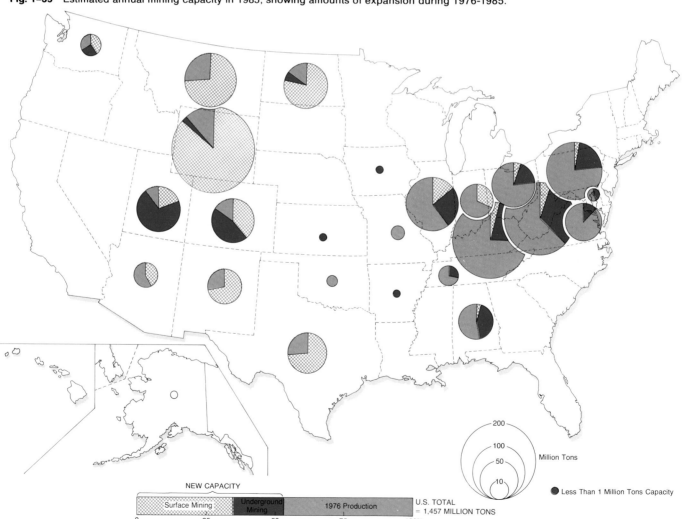

western coals and the expense of transportation to markets could offset the advantage of burning low-sulfur coals.

On average, the heating value of western coal is about 9000 BTUs per pound compared with about 12,000 BTUs per pound for eastern coal (Bureau of Mines, *Information Circular 8690*, 1975). In addition, they must often be transported 1,200 to 1,400 miles to the East before reaching major market areas. On first examination, the low energy content and distance handicaps of western coals should favor sales of coal from eastern mines; western coals, however, have other important features which help them compete. In many areas of the West, such as the Powder River Basin in Wyoming, seams 100 feet thick lie just below the surface and favor cheap strip mining methods. In 1973, it was estimated that coal could be

Table 1–14 Past, Present, and Future Coal Production

STATE	Cumulative Production through 1978	Remaining Reserves[1]	Cumulative Production as Proportion of Estimated Original Reserves[2]	Production during 1978[3]	Estimated Production Capacity in 1985[4]	New Capacity as Proportion of 1985 Capacity	Life of Remaining Reserves at 1985 Mining Rates
	(thousands of tons)			(thousands of tons)			(years)
Alabama	1,274,876	1,922,345	.399	20,110	40,387	.47	48
Alaska	6,883	8,779,565	.001	730	730	.00	12,027[6]
Arizona	51,682	294,651	.149	11,059	19,126	.42	15
Arkansas	104,918	436,807	.194	630	630	.00	693[6]
Colorado	625,801	7,755,701	.075	14,310	58,867	.84	132
Georgia				335	936	.80	
Illinois	4,695,134	37,619,642	.111	48,865	96,639	.40	389
Indiana	1,536,743	5,929,263	.206	23,940	37,569	.32	159
Iowa	367,787	1,441,017	.203	470	470	.00	3,066[6]
Kansas	293,277	1,247,872	.193	615	740	.20	1,686
Kentucky	4,688,538	15,430,523	.233	131,215	192,522	.25	80
Maryland	297,819	576,109	.341	2,855	4,630	.39	124
Missouri	365,132	6,098,089	.056	5,845	5,845	.00	1,043[6]
Montana	320,930	71,190,094	.004	26,680	101,331	.74	703
New Mexico	233,064	3,102,552	.070	12,365	35,160	.72	88
North Dakota	204,574	14,371,337	.014	14,625	69,002	.84	208
Ohio	2,894,052	11,910,572	.195	43,510	60,582	.23	197
Oklahoma	220,232	809,597	.214	5,425	5,425	.00	149[6]
Pennsylvania	9,700,470	15,792,050	.381	81,450	118,155	.22	134
Tennessee	535,049	599,187	.472	9,750	12,738	.27	47
Texas	80,603	2,908,135	.027	21,000	54,663	.74	53
Utah	362,072	2,109,554	.146	8,865	67,167	.88	31
Virginia	1,427,475	2,027,626	.413	29,750	45,546	.12	45
Washington	178,183	1,163,248	.133	4,695	12,106	.66	96
West Virginia	8,718,321	21,689,867	.287	84,700	172,484	.37	126
Wyoming	654,536	36,106,797	.018	58,175	251,336	.88	144
UNITED STATES TOTAL	39,844,266	271,760,595	.128	660,200	1,462,413[5]	.53	186

[1]Reserves are estimated tonnage recoverable from the Demonstrated Reserve Base, 1974 assuming 90 percent recovery in surface mining and 50 percent recovery in underground mining.

[2]Estimated original reserves are Cumulative Production plus remaining reserves.

[3]These are preliminary production levels subject to final adjustment. Overall production in 1978 appears to be down approximately 40 million tons due to the United Mine Workers' strike.

[4]Production in 1978 plus expansions in mining capacity according to Nielsen, 1977.

[5]Estimated annual loss of 15 million tons capacity due to mine closings in eastern United States will reduce national capacity by approximately 135 million tons.

[6]Using 1978 production rates.

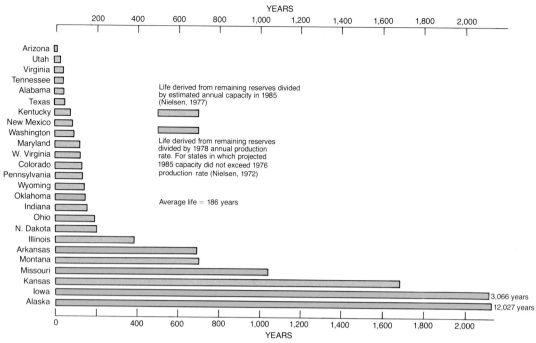

YEARS

Life derived from remaining reserves divided
by estimated annual capacity in 1985
(Nielsen, 1977)

Life derived from remaining reserves
divided by 1978 annual production
rate. For states in which projected
1985 capacity did not exceed 1976
production rate (Nielsen, 1972)

Average life = 186 years

Arizona
Utah
Virginia
Tennessee
Alabama
Texas
Kentucky
New Mexico
Washington
Maryland
W. Virginia
Colorado
Pennsylvania
Wyoming
Oklahoma
Indiana
Ohio
N. Dakota
Illinois
Arkansas
Montana
Missouri
Kansas
Iowa — 3,066 years
Alaska — 12,027 years

YEARS

Fig. 1–36 Estimated life of remaining coal reserves in producing states based for some states on an estimate of 1985 production capacity or for others on the 1978 production rate.

Loaded coal barges on the Monongahela River near Morgantown, West Virginia. Such barges move coal from the fields of West Virginia downstream on the Monongahela to Pittsburgh, and beyond Pittsburgh on the Ohio River. *Courtesy Consolidation Coal Company.*

strip mined in the West for 3 to 5 dollars a ton, compared with 9 to 14 dollars a ton in the East (Bureau of Mines, *Information Circular 8690*, 1975). It would appear that lower mining costs and a greater output per man-day are significant factors in allowing western coals to reach eastern markets.

Even though eastern coals are much closer to major market areas, environmental restrictions imposed upon electrical generating stations have made it more costly to burn the medium- and high-sulfur coals found in the Appalachian and Mid-Continent coal fields. These costs have, to a large degree, nullified the expense of long-distance transport of western coals. Table 1–15 and Fig. 1–37 show costs of burning various coals in compliance with state air quality standards. The cost added to the delivered cost for cleaning stack gas in installations burning central and eastern medium- and high-sulfur coals should

be noted. The comparison reveals that western coals burned in existing installations in Chicago are generally competitive with medium-sulfur eastern coals and high-sulfur central coal. If burned in new installations, western coal is only slightly more expensive in Chicago than central high-sulfur coal and considerably cheaper than eastern low- and medium-sulfur coals.

In western Kentucky, central high-sulfur coal is overall the cheapest to burn. In new installations, however, western coal is cheaper to burn than eastern coal. Columbus, Ohio, is the furthest western point where eastern coals are consistently cheaper to burn than those produced in the West. As the volume of coal moving from the West increases, a further decrease in the delivered cost to eastern markets may be expected.

MOVEMENT PATTERNS   Figure 1–38 shows total interstate coal movement in the United States for 1974. The

Table 1–15   Cost Comparison of Burning Various Coals to Comply with State Implementation Plan for Air Quality Standards (in cents/million BTUs)

	NORTHERN PLAINS	CENTRAL	EASTERN		
	Low-and Medium-Sulfur Coal	High-Sulfur	Low-Sulfur	Medium-Sulfur	High-Sulfur
Transportation					
To Chicago	44	9	23	23	n.a.
To W. Kentucky	47	nil	17	18	n.a.
To Columbus	55	n.a.	8	8	4
Delivered Cost					
To Chicago area	60	36	81	60	n.a.
To W. Kentucky	63	27	75	53	n.a.
To Columbus area	71	n.a.	66	43	30
Boiler modification to permit burning low-sulfur					
Dry bottom boilers	3	n.a.	3	3	3
Wet bottom boilers	10	n.a.	10	10	n.a.
Cost to clean stack gas					
Retrofit installation	n.a.	32	n.a.	n.a.	32
New installation	n.a.	23	n.a.	23	23
STATE IMPLEMENTATION PLAN ACCEPTABLE COMBUSTION COSTS					
Chicago Area					
Existing installations					
Dry bottom boilers	63	68[1]	84	63	n.a.
Wet bottom boilers	70	68[1]	91	70	n.a.
New installations	60	59[1]	81	83	n.a.
Western Kentucky					
Existing installations					
Dry bottom boilers	66	59[1]	78	56	n.a.
Wet bottom boilers	73	59[1]	85	63	n.a.
New installations	63	50	75	76	n.a.
Columbus Area					
Existing installations					
Dry bottom boilers	74	n.a.	69	46	62
Wet bottom boilers	81	n.a.	76	53	62
New installations	71	n.a.	66	66	53

n.a.—not applicable.
[1]Break-even cost 5 to 10 cents lower.

Source: United States Senate, Ninety-Fourth Congress.

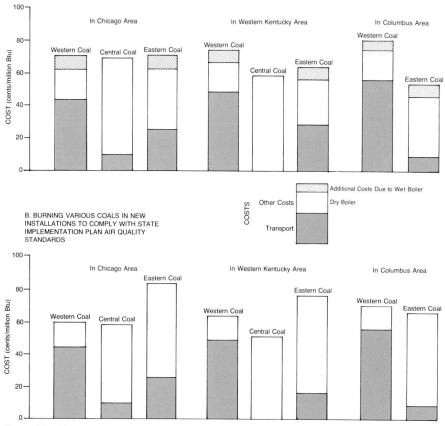

A. BURNING VARIOUS COALS IN EXISTING
INSTALLATIONS TO COMPLY WITH STATE
IMPLEMENTATION PLAN AIR QUALITY
STANDARDS

COSTS

Additional Costs Due to Wet Boiler
Other Costs   Dry Boiler
Transport

B. BURNING VARIOUS COALS IN NEW
INSTALLATIONS TO COMPLY WITH STATE
IMPLEMENTATION PLAN AIR QUALITY
STANDARDS

Fig. 1–37 Western coals competing with Central and Eastern: cost comparisons assuming coals burned in three different locations.

Fig. 1–38A. Interstate coal movement, 1974, by highway.

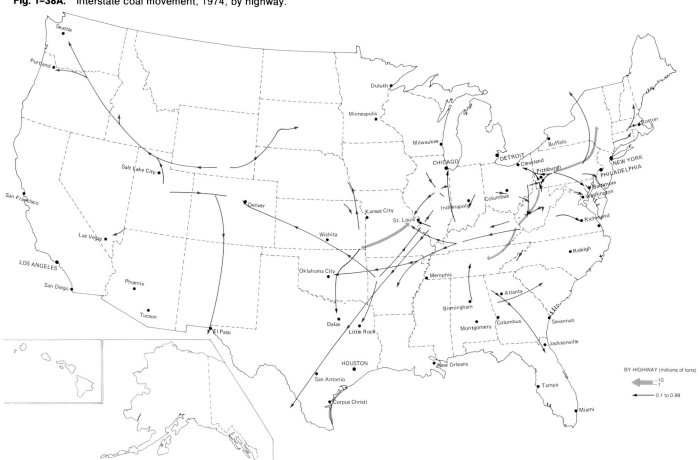

BY HIGHWAY (millions of tons)
10
1
0.1 to 0.99

Fig. 1–38B. Interstate coal movements, 1974, by water.

Fig. 1–38C. Interstate coal movements, 1974, by rail.

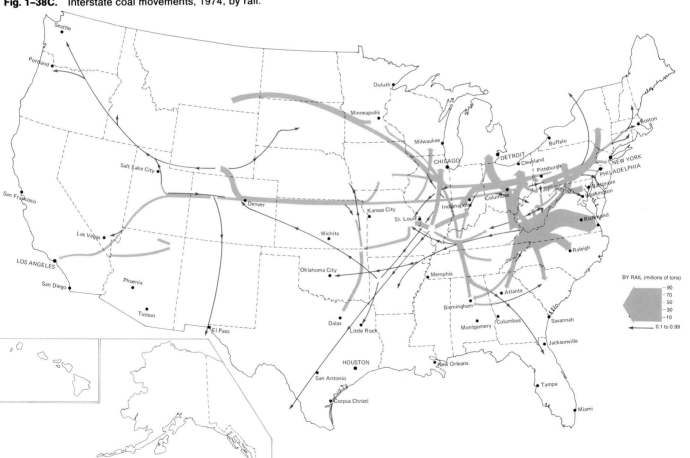

A coal-fired electrical generating plant at Morgantown, Maryland. Roughly half the nation's coal output is used now for power generation. *Courtesy Len Henig.*

greatest amount of coal traffic was generated in the East. Coal moves by rail, water, and highway from the major mining districts in the East to interstate and foreign markets. The largest volume of coal moves by rail from mining districts in West Virginia, Virginia, and Kentucky to markets throughout the East and South as well as to foreign markets in Europe, Asia, and South America from the port of Norfolk, Virginia. A substantial volume of eastern coal is moved by water through the Ohio and Mississippi River Systems and the Great Lakes. Coal movement along these waterways is destined for markets in the South, Midwest, and foreign countries.

In the West, significant amounts of coal are moved by rail from major mining districts in Wyoming and Montana to destinations in the upper and lower Midwest. Smaller amounts are transported by rail from Montana and Wyoming to destinations in the western states. In addition, approximately 5 million tons of coal are moved each year by way of the Black Mesa Slurry Pipeline (Bureau of Mines, *Information Circular 8690*, 1975). (Slurry is finely ground coal which is mixed with water so that it can be moved through a pipeline). This long-distance slurry line extends 273 miles from the Black Mesa Coal Mine in Arizona to the Mojave Electrical Power Generating Plant on the Colorado River near Bullhead City, Nevada. There are proposals for additional slurry pipelines exceeding 1000 miles in length connecting the Powder River Basin in Wyoming to Arkansas, and Craig, Colorado to Houston, Texas. Cost analysis studies have indicated that long-distance movement of coal by slurry pipeline is economically feasible (Bureau of Mines, *Information Circular 8690*, 1975).[8]

CONSUMPTION PATTERNS Figure 1–39 reveals the coal consumption patterns in the United States during 1976. It shows the amounts consumed in nine economic regions and the origin of those amounts in mining districts both within and beyond the consuming region. In general, most coals consumed within a given economic region are also produced in that region.

The East-North Central Region is by far the largest consumer of coal in the United States, accounting for one-third of the total. Approximately one-half (100 million tons) of the coal consumed in this region is also produced there. Coal entering the East-North Central Region from outside comes from more than ten states, including West Virginia and Kentucky in the East, and

Fig. 1–39 Self-sufficiency of economic regions: coal and lignite consumed during 1976, showing amounts from mining districts within and outside of the consuming regions.

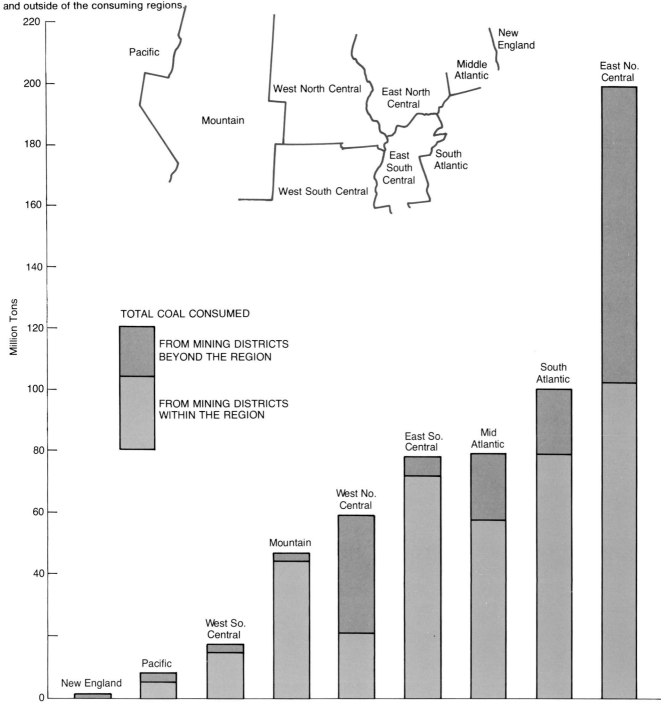

Montana and Wyoming in the West. Much of the coal consumed in this region is used in the generation of electricity; in fact, Ohio and Indiana depend on coal for over 95 percent of their electric generation capacity.

The South Atlantic Region consumes about 100 million tons of coal annually, with three-fourths being produced within the region. Most of the coal moving into this region from outside mining districts originates in western Kentucky and Pennsylvania. Generally, states in this region are less dependent on coal for electric generation than those in the East-North Central Region, because of the more extensive development of hydroelectricity, but, nevertheless, coal is the major fuel used in the production of electricity.

Together, the Middle Atlantic and East-South Central

regions consume over 150 million tons of coal annually. West Virginia and Virginia mines supply most of the coal entering the region from outside districts.

In the West-North Central states about two-thirds of the approximately 60 million tons of coal consumed annually has its origin in mining districts beyond the region. The major outside suppliers of coal to this region are mines located in Illinois, Wyoming, and Montana. In the remaining regions west of the Mississippi River, almost all of the coals consumed are produced in western mines. Only very small amounts entering the West-South Central Region are from Kentucky and West Virginia. Much of the coal consumed in the Mountain states is used for producing electricity, whereas coal moving to the Pacific Coast states is used mainly for non-electrical purposes.

Fig. 1–40 Coal gasification and liquefaction processes (at surface installations).

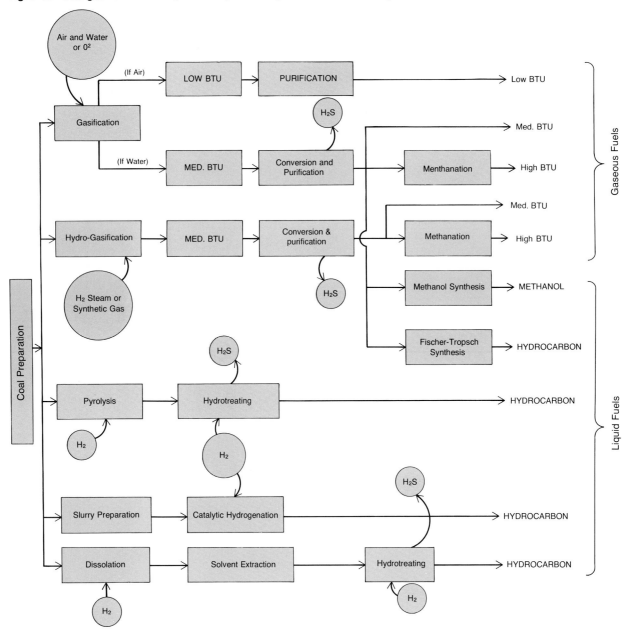

Coal Liquefaction and Gasification

In his July 1979 energy speech, President Carter called for the establishment of a government-run Energy Security Corporation that would supervise the spending of 88 billion dollars over a twelve-year period. The main purpose of the corporation would be to bring about, by 1990, the production of fuels from domestic resources in amounts sufficient to replace 2.5 million barrels per day of imported petroleum (*Chemical and Engineering News*, August 27, 1979). Much of the 2.5 million barrels per day of replacement fuels would be synthetics produced from coals, oil shales, peat, biomass (the organic matter in animal wastes and plants), and organic waste.[9]

Up to now, coal conversion projects have received much attention and most of the federal dollars. Not only is coal the United States' most abundant fossil fuel resource, coal conversion processes are, in addition, generally more advanced than other synthetic fuel technologies. The general objectives of the coal conversion proposal are to develop and demonstrate the technologies necessary to allow coal and coal-derived fuels to be substituted for oil and gas. Moreover, these objectives would be pursued in ways and at rates that are economically, environmentally, and socially acceptable (Department of Energy, March 1978).

The specific products that can be derived from coal conversion processes include: (1) crude oil, (2) fuel oil, (3) distillates, (4) chemical feedstock, and (5) pipeline quality gas (high-BTU) and fuel gas (low- and inter-mediate-BTU). Figure 1–40 depicts the various coal conversion processes. Basically, liquid or gaseous fuels are produced by decreasing the carbon to hydrogen ratio of solid coal. Coal has a carbon to hydrogen (C/H) weight ratio ranging from 12 for lignite to 20 for bituminous grades (Office of Fossil Energy, 1976). Either by addition of hydrogen or by rejection of carbon, the C/H ratio can be lowered to 10 to produce, by molecular weight, synthetic crude oil. If the C/H ratio is further decreased to 3, methane gas can be produced.

LIQUEFACTION

The history of coal liquefaction began in 1913 when work on the Bergius concept of direct hydrogenation of coal was undertaken in Germany (Office of Fossil Energy, 1976). During World War II, the Germans produced a major part of their aviation gasoline using liquefaction technology. Today there are several liquefaction plants in operation around the world. South Africa, for instance, produces 20,000 barrels a day of synthetic oil called *Sasol*.

Although there are no commercial coal liquefaction plants presently operating in the United States, the Department of Energy (D.O.E.) lists four liquefaction pilot plants that are planned or under construction (Fig. 1–41 and Table 1–16). According to the D.O.E., the first com-

Fig. 1–41 Proposed coal liquefaction and gasification pilot plants, and underground *(in situ)* coal conversion test sites (1979).

Table 1–16   Coal Gasification and Liquefaction Research and Development Facilities

COAL GASIFICATION PILOT PLANTS	UNDERGROUND COAL CONVERSION TEST SITES	COAL LIQUEFACTION PILOT PLANTS
Chicago, Illinois (2)	Hanna, Wyoming	Catlettsburg, Kentucky
Grand Forks, North Dakota	Hoe Creek, Wyoming	Cresap, West Virginia
Homer City, Pennsylvania	Pricetown, West Virginia	Fort Lewis, Washington
Pekin, Illinois		Baytown, Texas
Pittsburgh, Pennsylvania		
Windsor, Connecticut		

Source: Department of Energy, *Fossil Energy Research and Development Program*, DOE/ET-0013, March 1978.

mercial scale modules for two synthetic crude oil processes should be in operation by late 1983. A commercial five-module plant to produce 100,000 barrels a day of low-sulfur fuel oil could be on stream in 1987. Plant costs for coal liquefaction are estimated from 1 billion to 3 billion dollars each. The cost of the synfuel is estimated at 29 dollars per barrel from the first modules and 21 dollars per barrel from full-scale plants.

Coal liquefaction is an enormous undertaking. Each full-scale plant producing 50,000 barrels of boiler fuel or syncrude per day would require 13,200 to 17,500 tons per day of bituminous coal.[10] The assumed fuel efficiency of the plant is 60 to 75 percent and water requirements are estimated at 10,000 acre feet per year (Corey, 1976).

Up to now, coal liquefaction research indicates that high yields of syncrude can be achieved from bituminous, subbituminous, and lignite coals. Most of the research in progress appears to use high-sulfur eastern coals as feedstock.

Before commercialization of coal liquefaction, several environmental concerns must be dealt with. Beyond the heavy burden placed on the environment by the coal and water demand, liquefaction plants produce carcinogenic organic compounds in the coal residue and products, and trace metal pollutants.

GASIFICATION

Gas made from coal had its first commercial success in the early 1800s when it was used in the cities of London and Baltimore in street lights. With the rapid exploitation of natural gas since World War II, gas derived from coal has played a rather minor role in commerce and industry. Although coal gasification processes have been available for many years, present technology is expensive, thermally inefficient, and, in many ways limited in the kinds and sizes of processable coal (Department of Energy, March 1979).

Coal gasification research and development in the United States is focused on three main activities: (1) high-BTU gasification for clean pipeline quality gas, (2) low-BTU gasivation for clean industrial and utility fuel gas, and (3) underground *in situ* coal gasification technol-

ogy for on-site conversion of coal to gas in unmineable seams.

SURFACE GASIFICATION   As indicated in the upper half of Fig. 1–40, the initial step in the conversion of coal to gas may be either simple gasification or hydrogasification. The former primarily involves the direct reaction of steam with coal and the latter brings hydrogen produced elsewhere into contact with coal. The resultant products, low-BTU gas, medium-BTU gas, and high-BTU gas, are then purified and ready to be used as fuel. Low-BTU gas, with a heating value of 100 to 500 BTUs per cubic foot, is suitable for use as a fuel feedstock or for power generation in combined gas-steam turbine power cycles. It is believed that low-BTU gas can be produced at a competitive cost if the gasifier is built on the premises of a power generating station, thereby eliminating long-distance pumping costs. Medium-BTU gas, with heating value of 500 to 950 BTUs per cubic foot, is usually a feed gas for production of high-BTU gas. High-BTU gas, with a heating value of 750 to 1000 BTUs per cubic foot can be used in the same way as natural gas (Corey, 1976).

Presently, the only gasification process commercially available is the German developed Lurgi Gasifier. This process is already widely used in many countries to produce synthetic gas and is used in most commercial gasification projects throughout the world. Figure 1–41 shows the location of United States gasification pilot plants under construction or planned. The Department of Energy indicates that the Lurgi Gasifier planned at the Grand Forks, North Dakota location could be the first coal gasification facility in the United States using commercial scale equipment. The project is expected to cost about 11.5 million dollars and will be designed to produce 125 million standard cubic feet per day (SCFD) of pipeline quality gas. The Lurgi Process works well with non-caking coals (lignite and subbituminous) found in the western states. Eastern bituminous coals, however, which have a tendency to cake, clog the gasifier. Although caking tendencies of eastern coals can be lessened by pretreating the coal, this additional step reduces the thermal efficiency of the process (*Chemical and Engineering News*, August 27, 1979).

Because of the problems in using eastern coals in the Lurgi Gasifier, the Department of Energy is supporting work on more advanced gasification processes that might overcome some of the shortcomings of first generation gasifiers. Pilot plants for some of these advanced processes are planned for Chicago, Pittsburgh and Homer City, Pennsylvania (Figure 1–41).

Like coal liquefaction plants, coal gasification plants make heavy demands on coal resources and water. A supply of 250 million standard cubic feet per day of synthetic natural gas would require 14,700 to 17,900 tons per day of bituminous coal or 19,600 to 23,800 tons per day of lignite. Water requirement estimates range from 10,000 to 45,000 acre feet per year. The assumed fuel efficiency of such a plant is 56 to 58 percent, somewhat less than the fuel efficiency of a coal liquefaction plant.

UNDERGROUND COAL GASIFICATION As has already been indicated, coal is the Nation's most abundant energy resource, with an estimated 6.4 trillion tons within 6000 feet of the earth's surface, including Alaska's estimated coal resources (Department of Energy, February 1978). Because of the depth and thickness of the beds, however, more than 85 percent of these coal resources are not recoverable by conventional underground or strip mining methods. In order to recover the energy in the deep and thick coal seams of the United States, the Department of Energy, in partnership with private industry, is working to develop new recovery techniques. One of the more promising is underground coal gasification (UCG). This technique involves the injection of air at a rapid rate into holes drilled into the coal bed. The bed is then set afire and the fire reacts with unburned coal and water to produce methane, carbon dioxide, and hydrogen. These gases are removed through a second hole called a production well. This process is called the Linked Vertical Wells (Fig. 1–42). Using this process, a coal bed covering a four square mile area and averaging 30 feet in thickness could supply enough energy to meet the electrical needs of a city of a million people for twenty years (*UA Journal*, March 1979).

With such techniques, the Department of Energy believes that *as much as 1.8 trillion tons of unmineable coal* could be utilized by converting it in-place to clean-burning gases (Fig. 1–43 and Table 1–17). Comparing this figure with the 437 billion tons of Demonstrated Reserve Base that can be mined economically using current technology, underground coal gasification could potentially triple the recoverable energy in coal for the forty-eight contiguous states (Fig. 1–44). Using another comparison, the energy potentially recoverable by UCG is slightly less than that recoverable from all Identified coal to a depth of 3000 feet assuming a 50 percent recovery ratio.

The products from underground coal gasification process are low-BTU gas suitable for local (on-site) use in electrical power generation; medium-BTU gas suitable for chemical feedstock; and high-BTU gas that can be transported through pipelines directly into existing natural gas transmission systems.

The Department of Energy lists a number of advantages that underground coal gasification has over conventional coal mining techniques. In addition to tripling the recoverable energy from coal, it is expected to minimize health and safety problems associated with

Fig. 1–42 Underground coal gasification by linked vertical wells process.

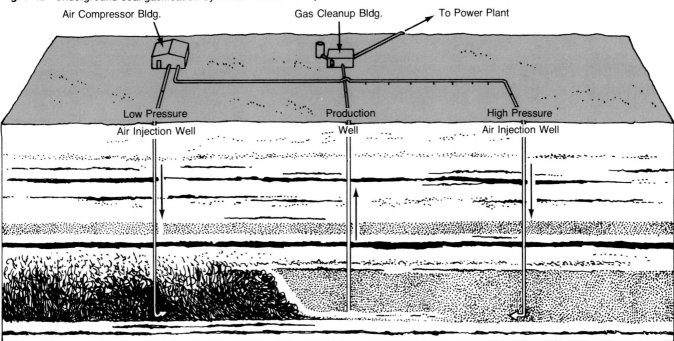

Forward Gasification   Reverse Combustion Linking

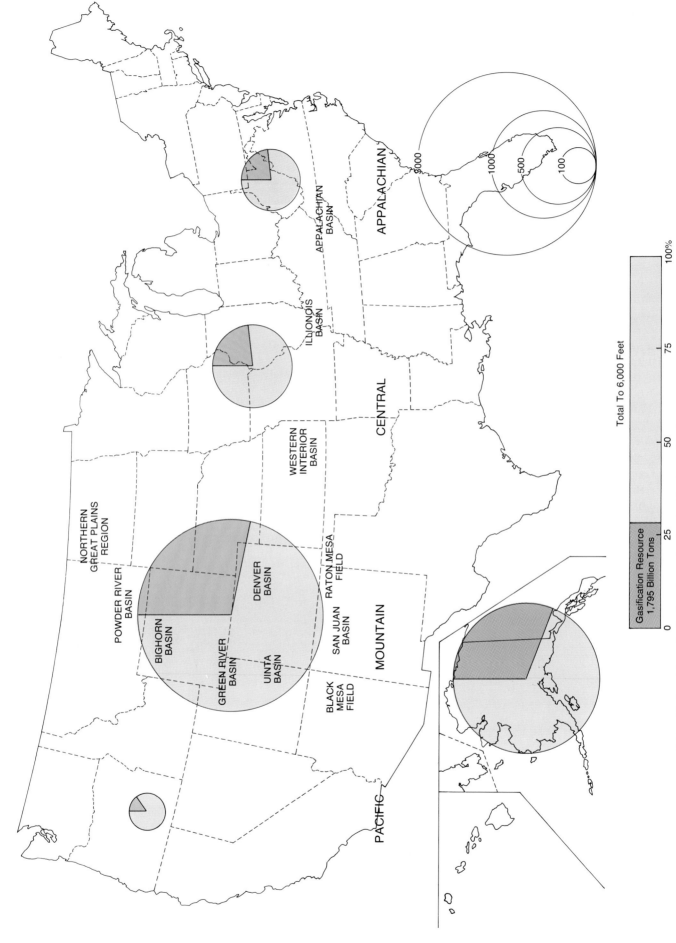

Fig. 1-43 Recent estimates of all identified and hypothetical coals to depth of 6,000 feet, showing the portion in beds suitable for underground coal gasification.

conventional coal mining; produce less surface disruption than strip mining; consume less water; generate less air pollution; and reduce both the capital investment and gas cost by at least 25 percent.

Suitable Coals. There is some overlap between coals suitable for underground gasification and those suitable for conventional mining. An important exception is the coal found in steeply dipping beds. These deposits are difficult to mine through conventional techniques but can be gasified successfully. In fact, they appear to have advantages over flat-lying seams. In any case, the total amount of recoverable coal appears to be so large that underground coal gasification and conventional mining should not have to compete for resources. For example, based on assumptions in Table 1–18, an underground coal gasification-fired 500 Megawatt electric power plant would require only 50 million tons of bituminous coal and 70 million tons of subbituminous coal over a 20 year period. The volume of coal required is inversely proportional to the overall UCG efficiency.

In general, western coals of low rank (lignite and subbituminous) are the easiest to gasify underground (Fig. 1–43). Bituminous coal is more difficult to gasify in-place because, when heated, it swells and produces viscous tars which plug natural or induced permeability escape channels (Department of Energy, February 1978). Because of the proximity of Appalachian and midwestern bituminous coals to market, a large effort is being made to overcome these difficulties.

The best prospects for underground coal gasification east of the Rocky Mountains are the bituminous coals in the Illinois Basin and the lignites of the Gulf Coast. The coal seams of the Illinois Basin are consistently thicker than 5 feet and swell less than Appalachian coals when

Table 1–17 Coal Resource for Underground Gasification in Five Regions of the Country, Showing Total of Identified and Hypothetical Coals to Depth of 6,000 Feet, Coal Suitable for Underground Gasification, and Estimated Recoverable Energy

REGION (see map)	Total of Identified and Hypothetical Coals to 6,000 Feet Depth	Tonnage Considered Resource for Underground Gasification[1]	Estimated Energy Recoverable[2]
	(in Billions of tons)		(in Quads)
Alaska	2,029	640	6,400
Pacific	103	15	250
Mountain	3,263	920	9,200
Central	620	140	1,820
Appalachian	352	80	1,040
TOTAL	6,367	1,795	18,610

[1]According to Department of Energy, February, 1978 (Vol. III: *Resources*).
[2]On the basis of one-half of the coal energy in-place. For each region half the in-place tonnage was multiplied by an energy value dictated by rank of coal in the UCG resource: for the three western regions a value of 20 million BTUs per ton; for Central and Appalachian regions, a value of 26 million BTUs per ton.

heated. Gulf Coast lignites are the thickest coals east of the Rockies, highly reactive, and shrink upon heating, making them easier to gasify than bituminous.

With their vast resources of thick bituminous, subbituminous, and lignite coal, the Mountain states and those of the Northern Great Plains contain over one-half

Fig. 1–44 Estimated energy recoverable from underground coal gasification resource base (to depth of 6,000 feet) compared with energy recoverable by two approaches to mining of identified coal to depth of 3,000 feet.

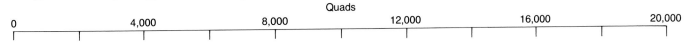

Quads

| 0 | 4,000 | 8,000 | 12,000 | 16,000 | 20,000 |

From Demonstrated Reserve Base[1]

From underground coal gasification, assuming recovery of half the energy content in suitable beds to depth of 6,000 feet

Possible recoverable coal from all identified coal to depth of 3,000 feet, assuming 50% recovery from underground and surface mining

[1]Assuming 50% in Underground Mining; 90% in surface mining from Demonstrated Reserve Base

Table 1–18 Assumptions Used in Calculating the Volume of Coal Required to Support a UCG Product Fired Power Plant

1. POWER PLANT

Capacity (Megawatts)	500
Operating Life (years)	20
Operating Factor (percent)	70
Power Plant Efficiency (Kilojoules/Kilowatt hour)	10,551
Power Plant Efficiency (BTU/Kilowatt hour)	10,000

2. UCG

Overall Conversion Efficiency (Percent)	50
(net Megajoules of gas produced/Megajoules available in coal)	

3. COAL

	Bituminous (45° Dip)	Subbituminous (Horizontal)
Heating Value In-Place (Kilojoules/kg)	30,211	20,125
Heating Value In-Place (BTU/lb.)	13,000	8,860
Density In-Place (kg./cu. meter)	1,411	1,313
Density In-Place (lb./cu. ft.)	87	81

Source: Department of Energy, *Underground Coal Conversion Program*, Volume III, *Resources*, (Washington: Government Printing Office, February, 1978).

GASES	FREE GAS OR GAS IN SOLUTION IN VARIOUS KINDS OF ROCK	Conventional natural gas in porous and permeable reservoir rocks
		In tight gas sands of western states
		In Devonian shales of Appalachians
		In geopressured reservoirs
		In un-mined coal beds
	SYNTHETIC GASES	From coal gasification at mine-head or in place underground
		From biomass

CRUDE OILS	CRUDE OIL 'AS SUCH' IN VARIOUS RESERVOIR ROCKS	Conventional crude recovered by primary, secondary, or tertiary (enhanced) recovery techniques
	SYNTHETIC CRUDE OILS	From bitumen in tar sands
		From kerogen in oil shales
		From coal liquefaction
		From biomass

Conventional and synthetic crude oils and gases, highlighting the synthetic gas and crude oil that can be obtained from coal.

the coal of potential interest for near-term underground gasification. Markets will, undoubtedly, determine which of these deposits will be exploited earliest (Fig. 1–43).

Washington contains lignite, subbituminous, and bituminous coal of interest for underground conversion to gas. Much of the coal located there is in steeply dipping beds. Although Alaska contains huge coal resources of all ranks which would be suitable for underground gasification, its remoteness from the major markets makes exploitation unlikely.

Department of Energy Test Sites. Technical feasibility of underground coal gasification has been proven by the British and, most notably, by the Russians, who have had commercial underground coal gasification plants in operation for twenty years (Department of Energy, March 1978). Proving economic feasibility is the key to commercialization of underground coal gasification in the United States.

At the present, several major underground coal gasification process options are being developed by the Department of Energy and its subcontractors. At Hanna, Wyoming, tests have already been conducted using the Linked Vertical Wells process to produce a low-BTU gas for utility power generation or utility use from a seam of subbituminous coal 30 feet thick (Fig. 1–41). A second test at Hanna conducted in 1976 produced an average of 8.5 million standard cubic feet (SCF) per day of 172 BTU/SCF gas, the amounts equivalent to electrical needs

of a town of 6,000 people. Present plans at Hanna call for a pilot plant to be built in the early to mid-1980s. The plant would have a capacity of 50 to 60 Megawatts, large enough to supply a city of 50 to 60 thousand people with its electrical needs (*UA Journal*, March 1979).

A second process has been tested by Lawrence Livermore Laboratory at Hoe Creek, Wyoming. The Hoe Creek project uses underground explosions to fracture the coal and create permeable zones. Up to now, these tests have produced medium-BTU gas suitable for upgrading to methanol, gasoline, ammonia, or pipeline quality gas. Following further tests, the Hoe Creek project may move to a pilot-testing stage.

A third underground coal gasification process is in the very early stages of development at Pricetown, West Virginia. At this site the objective is to develop a method of gasification for eastern coals that up to now have been difficult to gasify because of their thin seams, low permeability and high-swelling characteristics.

Based on the work conducted to date, it would appear that underground coal gasification has the potential of being an important energy option for the future. One future coal-based energy scenario puts the potential contribution of underground coal gasification in perspective by suggesting that electrical and synthetic natural gas mine-mouth UCG plants alone could comprise about 15 percent of the total coal usage by the year 2000 and 35 percent by 2050 (Department of Energy, March 1978).

NOTES

[1] The above assumes that all sulfur is converted to sulfur dioxide (SO_2); but, while 95 to 100 percent of the sulfur in bituminous coal is converted to that gas, only 72 percent of sulfur in subbituminous coal becomes SO_2, which suggests the requirement in Fig. 1–5 may be modified or suggests that subbituminous coal can be used with less expensive SO_2 removal devices than would be required if all its sulfur were converted to SO_2.

[2] Measured reserves are roughly equivalent to the Bureau of Mines' Demonstrated Reserve Base; Indicated and Inferred resources are roughly equivalent to Averitt's Hypothetical resources.

[3] Statement is based on Averitt's 1974 estimate of United States coal resources.

[4] Amounts are not mapped for California, Idaho, Nebraska, Nevada, and Louisiana because the data were not available. The total tonnage of Identified coal for these states is estimated at 688 million tons.

[5] The Bureau of Mines usually bases its estimates for surface mineable coal on an 80 percent recoverability factor.

[6] Estimate may be high because of the thick seams of subbituminous coal in the West.

[7] Four states noted in Table 1–14 reported no planned expansion of capacity, so 1976 production rates were used to estimate the life of remaining reserves.

[8] Proposed slurry pipelines may not materialize because of opposition from railroad interests and environmental groups. A bill that would have empowered the Secretary of the Interior to rule on applications for pipeline rights-of-way for slurry lines was defeated by the House of Representatives July 19 (*Philadelphia Inquirer*, July 20, 1978). Railroaders feared competition from the pipelines, and environmentalists logically questioned pumping water from arid areas for the sake of delivering coal (and water) to markets in Arkansas and Texas.

[9] In June, 1980, both the House and Senate approved a bill that created a quasi-governmental Synthetic Fuels Corporation with 20 billion dollars to spend in the first phase of a synthetic fuels program. Through loans, loan guarantees, purchase guarantees, price guarantees, and joint ventures the Corporation will strive for synthetic fuels output of one-half million barrels per day by the year 1987. The first phase will be funded through the years 1981–1985. The second phase will be funded by 68 billion dollars in order to expand production to 2 million barrels per day.

[10] In the July 16, 1979 issue, *Newsweek* reported that this same plant would require 30,000 tons of coal a day to run.

2

Crude Oil and Natural Gas

←

Horsehead pumps unit at Huntington Beach, California, with offshore platform drilling rig in background. *Courtesy American Petroleum Institute.*

Crude oil, or petroleum, is a complex mixture of hydrocarbons (compounds of hydrogen and carbon) which, under normal conditions, exists as a liquid in the earth's crust. In chemical composition, an "average" crude oil is roughly 83 percent carbon and 12 percent hydrogen, with sulfur and oxygen making up 0.1 to 5.0 percent, and nitrogen accounting for 0.1 to 1.5 percent. No two crude oils ever contain the same mix of molecules. Some are far richer in carbon than others; some contain more sulfur than others. In physical composition, an important factor is the oil's viscosity (or the liquid resistance to flow). Some crude oils are black, thick and tarry (asphaltic) while others are lighter in color, thinner, and more volatile. The asphaltic crudes are referred to as "heavy"; but, in fact, their specific gravity is lower than that of the less viscous "lighter" crude oils.

In refining, the first step is distillation, or the vaporization and condensation which separates the lighter components, such as gaseous hydrocarbons, gasoline, and kerosene, from the heavier components, such as lubricating oils and residual fuel oil. The crude oils that are considered light will yield more gasoline per barrel than the heavy crudes which will yield greater amounts of lubricants (Fig. 2–1). In subsequent processes, some of the heavier and less volatile products, such as fuel oil, may be subjected to *catalytic cracking,* a process which alters their molecular structure and yields some lighter products such as gasoline. Similarly, some of the very light products may be treated by *polymerization* in order to produce gasoline. Thus, the ultimate output of a refinery may be adjusted according to season or the demands of local markets in order to produce more or less of products such as gasoline and heating oil.

About 6 percent of the output from refineries is in the form of petroleum gases. When combined with natural gas, these products provide the petrochemical industry with raw materials, such as methane, ethane, propane, butane, xylene, benzene, and toluene which are used in the production of fertilizers, antifreeze, plastics, solvents, synthetic rubber, polyester fibers, nylon, and a host of other products. Considering the output from refineries and from the petrochemical industry, roughly 15 percent of all the crude oil consumed in the United States goes into products that are not used for their energy content.

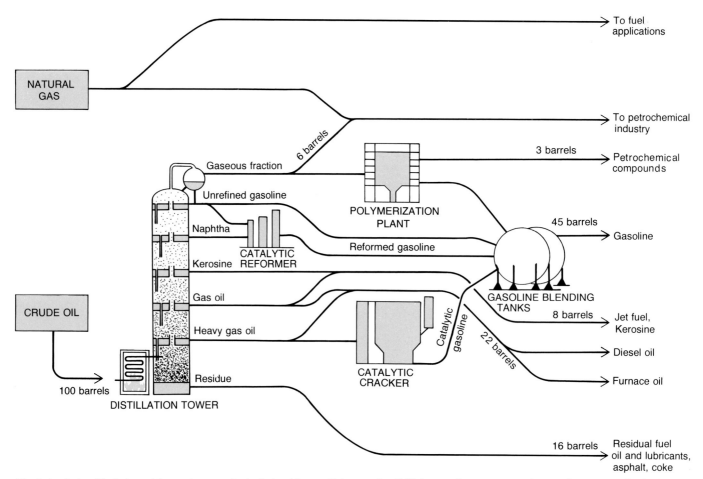

Fig. 2–1 A simplified view of the various products derived from refining crude oil. Volumes shown represent a rough average of output from U. S. refineries.

The Nature of Oil and Gas Accumulations

Complete agreement has not been reached as to how liquid petroleum is produced in nature. Microscopic plant and animal remains, usually of marine origin, are the source of organic materials found in certain dark shales (Fig. 2–2). These organic materials are converted to liquid or gaseous hydrocarbons when the rocks are subjected to the heat and pressure of deep burial. Considerable time (perhaps a half-million years), moderate temperatures (100°F), and moderate pressure (1,000 pounds per square inch or 10 kilograms per square centimeter) are required to transform the organic matter into the hydrocarbons commonly found in crude petroleum.

Squeezed out of the shales, that is, the source rocks, the liquid petroleum finds its way into adjacent porous and permeable rocks, such as sandstones or limestones, which allow the liquid to migrate upward as buoyancy and differential pressure dictate. It appears that migration was necessary to produce accumulations because nearby source beds are either absent or inadequate to account for the volumes of oil or gas found in a given reservoir.

The petroleum liquids accumulate in a *reservoir rock,* often the same rock unit through which migration has taken place. Sandstones and carbonates (limestone and dolomite) frequently serve as reservoir rocks because they have pore spaces among grains (porosity) and the ability to transmit fluids (permeability).

If the upward migration of crude oil or natural gas were not interrupted in some way, the fluids would escape at the surface. Natural gas, for instance, escapes in spectacular flares at the Eternal Fires of Iraq. Crude oil escaping makes obvious oil seeps such as those at Baku on the Caspian Sea or the La Brea tar pits in California. The existence of large accumulations of oil (or gas) at depth indicates that some migrations have been interrupted. Indeed, a study of every oil and gas field shows that the reservoir rock holding the oil or gas is deformed, interrupted, or discontinuous in such a way that migration of fluids through it appears impossible. The arrangement of rocks that frustrates migration is called a *trap,* and, whatever the type of trap, it always entails some impermeable *cap rock,* such as shale, salt, or dense limestones, overlying the reservoir rock and sealing in the petroleum (Fig. 2–3).

A

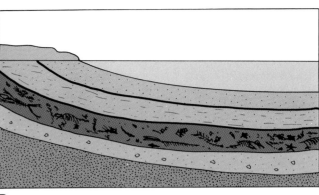

D

B

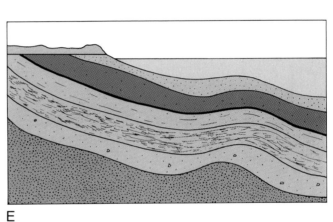

E

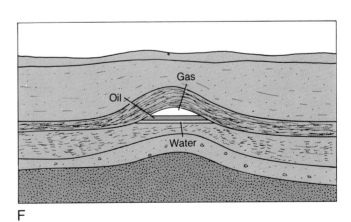

C

F

Fig. 2–2 A. Some scientists believe petroleum formation began millions of years ago, when tiny marine creatures abounded in the seas.
B. As marine life died, it settled to the sea bottom and became buried in layers of clay, silt, and sand.
C. The marine plants and animals, held in clay, sand, and silt, were changed to gas and oil, probably by gradual decay, heat, and pressure, and possibly bacterial and radioactive actions.
D. As millions of years passed, pressure compressed the deeply buried layers of clay, silt, and sand into layers of rock.
E. Earthquakes and other earth forces buckled the rock layers.
F. The petroleum migrated upward through porous rock until it became trapped under nonporous rock.

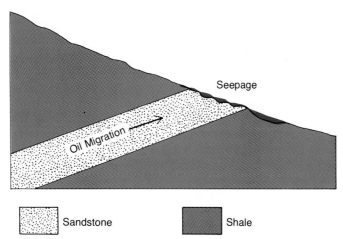

Sandstone   Shale

Fig. 2–3 Oil migrating through permeable rock to surface seep.

There are two forms of traps: structural and stratigraphic. Structural traps, which account for roughly 70 percent of the oil in United States reservoirs, involve some sort of deformation or breaking of the reservoir rock, due sometimes to large movements within the crust, and sometimes to nothing more than settling and compaction of sediments. The best-known structural traps occur when reservoir rock is warped into a dome or elongated domal form (anticline) as in Fig. 2–4C. Traps of this sort are responsible for most of the world's largest oil fields, such as Prudhoe Bay, in Alaska, and Burgan, in Kuwait. Another structural trap occurs with the faulting or displacement of a rock series as in Fig. 2–4B. This often occurs in combination with anticlinal uplifts, as in fields of the Los Angeles Basin. A large faulted block by itself is responsible for the Officina field in Venezuela.

Stratigraphic traps are the result of a change in permeability or a discontinuity of the reservoir rock, rather than structural deformation. This form of trapping accounts for 30 percent of United States crude oil accumulation. A body of sand, such as an ancient sandbar, that gives way to a less permeable rock (Fig. 2–4A) is exemplified in the shoestring sands of eastern Kansas, and in the Pembina field in the Canadian province of Alberta. A buried coral reef, can create a trap if enclosed by less permeable rock (Fig. 2–4D). Reefs are responsible for

Fig. 2–4 Four types of traps resulting in petroleum accumulations.

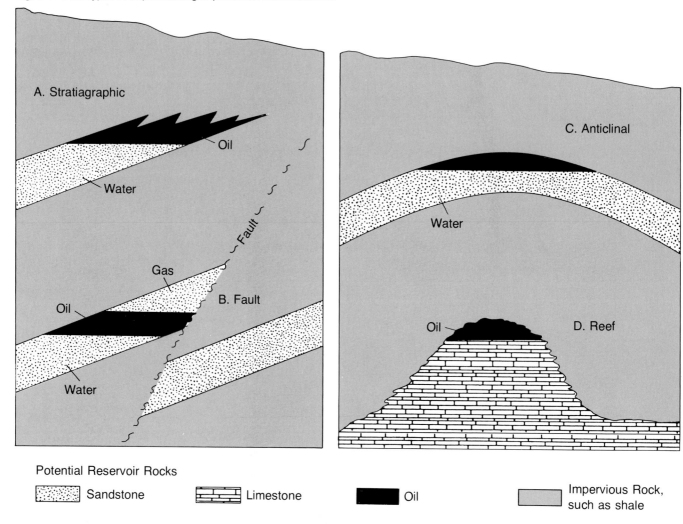

Potential Reservoir Rocks

Sandstone   Limestone   Oil   Impervious Rock, such as shale

some important reservoirs, such as the Leduc chain in Alberta, the Sirte Basin fields of Libya, and—in a more widespread form—some large fields in West Texas.

It appears that there are five factors necessary for the occurrence of crude oil in quantities sufficiently large to warrant commercial exploitation. The first requirement is the existence of a source rock which is sufficiently rich in organic remains—preferably remains of marine origin and not continental. Deep burial of these organisms must occur to provide the temperatures and pressures needed to change the organic material into liquid hydrocarbons. In addition, porous and permeable rocks must exist for the migration and accumulation of the liquids. A suitable trap must be formed in order to cause the accumulation. And finally, the timing must be right: if the trap is formed *after* the period of migration then it will be barren.

Exploration for oil or gas begins by identifying regions that will meet these five requirements. More specific identification of drilling sites is based largely on geologic or geophysical evidence that suggests a trap of some kind is likely in a certain area. Whether the trap exists, and whether it holds hydrocarbons or just salt water, can be determined only by drilling.

An oil or gas field is essentially a porous and permeable rock saturated with oil, gas, or both. The dimensions of the field and the amount of oil or gas in it can be estimated, as shown in Fig. 2–5. The portion of the reservoir rock that is saturated with oil is termed the *pay zone*. Its thickness, measured in drill holes, often will vary across the field because of the inclination of the reservoir rock. The bottom of the pay zone is defined by the top of the water-saturated zone. Areal extent of the pay zone will sometimes depend on changes in permeability. More often, the pay zone simply gives way to a reservoir rock that is too low in elevation to be oil-saturated. Drill holes that are too far from the crest of the structure (Fig. 2–5) will encounter just salt water and will indicate the limits of the field. From its estimated thickness and area, the volume of saturated rock can be calculated. From core samples, the percentage of pore space can be measured. From these two numbers, the volume of *oil in-place* or *gas in-place* can be calculated.

Around 85 percent of the natural gas in a reservoir can be easily recovered. Its extremely low viscosity allows it to flow through the reservoir to lower pressures in the borehole. Crude oil, and especially the more viscous "heavy" oil, adheres to the many surfaces in the reservoir rock and is not so easily recovered. In the United States an average of only 35 percent of the crude oil in-place has been recovered. More can be recovered by using special techniques. The amounts of oil that may be recovered from existing and future oil fields depends therefore upon whether prices justify the application of special recovery techniques. This factor is taken into account in the following review of resource terminology.

Resource Terms Applicable to Crude Oil and Natural Gas

A number of special terms are used by the oil and gas industry and by the federal government to refer to: amounts that have been produced; resource amounts re-

An oil derrick.

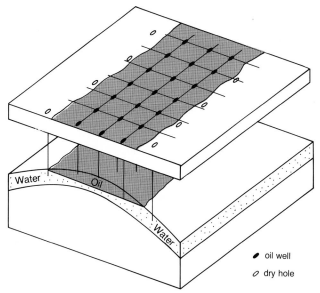

Fig. 2–5   Dimensions of an oil accumulation.

maining; and techniques for recovery of crude oil. The terms that deal with the resource amounts appear in Fig. 2–6, which is a variation on the general scheme for understanding amounts of mineral resources (see Introduction). The resource amounts that are more certain are placed toward the left of that diagram; and the amounts more easily attained are placed toward the top.

ORIGINAL OIL (OR GAS) IN-PLACE   This is the amount of oil in known reservoirs prior to production. The term may be applied to reservoirs that currently are productive, to those that are known but not yet exploited, or to those that have been depleted. The estimation of original oil in-place is made from calculations of the volume of saturated reservoir rock and the proportion of it that is pore space.

ALL REMAINING RECOVERABLE   In this atlas, this term is used for amounts of oil or gas that are recoverable, given present economic limitations, from reservoirs that are either Identified or Undiscovered. Varying degrees of certainty are expressed by the categories below.

PROVED, OR MEASURED RESERVES   These are amounts shown by geologic and engineering data to be *Recoverable* from Identified reservoirs under existing economic and operating conditions. Because both the reservoir's existence and the economic feasibility of recovery are certain, these amounts occupy the extreme upper-left corner of the resource diagram (Fig. 2–6).

INDICATED (OR INDICATED ADDITIONAL) RESERVES Applicable to crude oil, not natural gas, these are amounts (beyond the Proved reserves) that may be re-

Fig. 2–6   Categories of crude oil resources, recognizing variations in geologic certainty and economic feasibility.

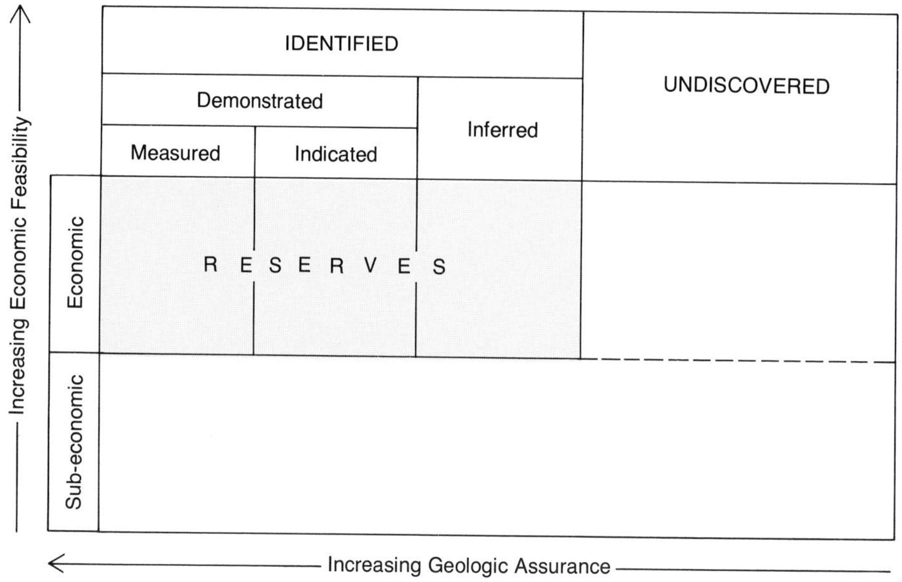

	IDENTIFIED			UNDISCOVERED
	Demonstrated		Inferred	
	Measured	Indicated		
Economic	R E S E R V E S			
Sub-economic				

Increasing Economic Feasibility →

← Increasing Geologic Assurance →

covered through application of modest improved recovery techniques such as injection of water or gas to maintain reservoir pressures. For some fields these secondary recovery techniques are likely to be successful and economical; but since they have not yet been applied, the anticipated additional crude is not considered Proved. These amounts tend to be *relatively* large in old productive regions or states whose Proved reserves have been severely depleted, leaving substantial amounts of crude in the reservoirs.

DEMONSTRATED RESERVES This term refers to the sum of Proved (Measured) and Indicated reserves. It is used largely by the U.S. Geological Survey.

INFERRED RESERVES Used by the U.S. Geological Survey, this category refers to amounts of oil or gas in identified reservoirs near the edges of fields that have not yet been fully defined by drill holes. It is known that young fields tend to be underestimated by Proved reserves which consider only the amounts in reservoir rock that has been drilled. Inferred amounts thus anticipate what is reported annually by the American Petroleum Institute as *extensions* and *revisions* (though some revisions are downward). Since the amounts are expected in identified reservoirs but are not yet measured, the Inferred category occupies a position on the resources chart (Fig. 2–6) that is intermediate between Demonstrated and Undiscovered.

UNDISCOVERED RECOVERABLE AMOUNTS Applicable to both oil and gas, these are amounts thought to be recoverable, usually under today's economic conditions, from reservoirs not yet discovered. The existence of additional reservoir rocks and traps is postulated on the knowledge that sedimentary rock sequences similar to those already productive do exist and are either unexplored or inadequately explored. Frontier areas untouched by the drill, such as the Atlantic continental shelf, are prime targets for such speculation; but also considered are remote or deep parts of sedimentary regions whose nearby and shallower beds have been explored. Amounts may be estimated on the basis of historic recovery rates, and are represented by position X in Fig. 2–6.

CUMULATIVE PRODUCTION As the name suggests, this refers to amounts of oil or gas (or natural gas liquids) that have been produced to date in a given field, state, or region for which statistics are gathered. It is an indication of the richness of an area—or rather of its past richness—and is, when combined with information on amounts remaining, a valuable guide to the status of an area.

ULTIMATE RECOVERY As used by the American Petroleum Institute, for either oil or gas, this is simply the sum of Cumulative Production to date and current Proved reserves. It expresses, therefore, "ultimate recovery" *only from identified reservoirs* and only as determined by current technology and economics. It should not be construed as the ultimate ever to be recovered for an area or region, since it does not consider either undiscovered reservoirs or additional amounts that may be obtained from identified reservoirs by enhanced recovery techniques.

World Resources of Crude Oil

Since the time of early civilizations crude oil from natural seepages has been used for fuels, mortar, and medicines. Some oil was distilled from shales in Scotland in the 17th century, but it was not until 1857 in Rumania that a successful hole was drilled expressly to seek petroleum underground. Two years later, one of the first American oil wells was completed at Titusville in Northwestern Pennsylvania. These discoveries of liquid petroleum, a versatile and transportable fuel and raw material, stored naturally at depth, unaltered by surface

---

The Commodities: Crude Oil, Natural Gas, and Natural Gas Liquids

Crude Oil A complex mixture of hydrocarbons, varying in composition from one field to another, existing as a liquid underground and remaining liquid at the surface.

Conventional Crude Oil This term distinguishes oil extracted as a liquid (through a borehole) from the crude that can be obtained by processing oil shales or tar sands. Both of these alternatives are considered here to be unconventional crude or synthetic crude from mineral sources. (They are dealt with in Chapter 3.)

Natural Gas A mixture of hydrocarbons, dominantly methane, which may exist in the reservoir in the gaseous phase or in solution with oil, but at surface pressure is in the gaseous state.

Natural Gas Liquids These consist of propane and similar heavier hydrocarbons which separate from natural gas through condensation in the reservoir or are separated from gas at the surface through condensation, absorption, or related methods in separators at or near the well-head. Since these liquids are produced only when the host gas is produced, their amounts produced and remaining are directly tied to the production and the reserves of natural gas. The amounts are expressed in barrels, and are often added to crude oil amounts to arrive at Total Petroleum Liquids.

Crowded derricks in Spindletop Field, Texas, around 1903. *Courtesy American Petroleum Institute.*

oxidation, and under pressures that eased its recovery through a drillhole, was a revelation that began a remarkable exploitation of world resources (Fig. 2–7A).

The growth in production for different regions (Fig. 2–7B) illuminates the present status of the United States in relation to other countries. For the first 100 years of the oil industry's existence, North America (dominated by the United States) was the world's leading producer. In fact, the United States accounted for 60 to 70 percent of world production through 1950, being surpassed only in the post-war years as the fields in the Middle East, Eastern Europe, and the Soviet Union expanded rapidly.

The differing rates for development among countries have led to extremely interesting differences in amounts of oil produced to date, that is, Cumulative Production for world regions, as shown on the *left side* of Fig. 2–8. Clearly the Western Hemisphere has produced more than any other region; and the United States alone has produced more crude oil than the entire Middle East.

All crude oil remaining in the world, divided by regions, is on the *right side* of Fig. 2–8 where most recent reserve estimates (Table 2–1) are combined with one estimate of Undiscovered Recoverable amounts. The substantial reserve amounts in the United Kingdom are due to North Sea discoveries. Mexico's reserves have doubled since the end of 1976 because of development on the

Chiapas-Tobasco area.[1] Although recoverable reserves in the U.S.S.R. are shown to be 71 billion, they may be only 60 billion, because Soviet reports include some anticipated reserves.

The Middle East clearly dominates in current reserves. In addition, the Middle East reserves are much greater than oil produced. Nonetheless the total crude oil remaining (reserves plus Undiscovered Recoverable amounts) does not show overwhelming concentration in the Middle East. If the estimate of Undiscovered amounts is valid, then the Middle East has been rather thoroughly explored already, whereas the U.S.S.R., China, and the Western Hemisphere can expect to find larger amounts in the future. For any region, the total length of bar on the illustration approximates all the crude oil that has been and will be recoverable. On that basis, the Middle East, the Soviet Union, China, and the Western Hemisphere appear roughly equivalent.

The amounts of crude oil remaining in the world may suggest the supply is quite comfortable: only 328 billion barrels have been produced, and perhaps 1,500 billion barrels remain, 642 billion of which are certain. Nonetheless there is reason for concern about world supplies meeting demand, especially in those regions where demand is great but assured resources are relatively small (Fig. 2–9).

A. WORLD

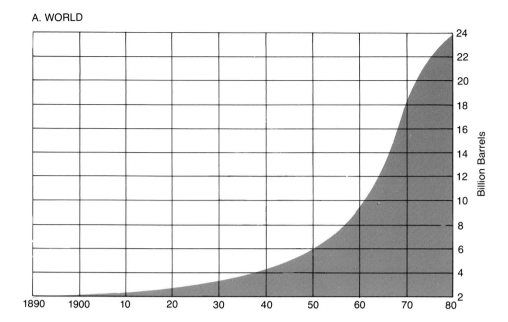

Fig. 2–7 A. (Above) and B. (Below) Annual crude oil production in United States, 1890 to 1978, showing roles of different regions.

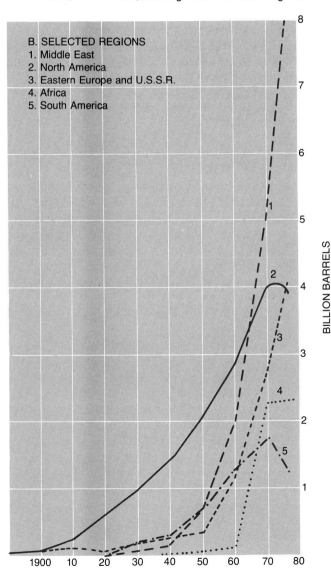

B. SELECTED REGIONS
1. Middle East
2. North America
3. Eastern Europe and U.S.S.R.
4. Africa
5. South America

One projection of demand and supply (Fig. 2–10) shows demand doubling in roughly twenty-three years, that is, rising at an average of 3 percent per year. This rate of increase is far more modest than that experienced during the 1960s when annual demand doubled in only *ten years* (see Fig. 2–7). The projection shows that production from current reserves will fall after 1985 and be supplanted by production from *future* reserves that are assumed to be 900 billion barrels in non-Communist areas alone (Granville, p. 59). The projection assumes more undiscovered amounts than those shown in Fig. 2–8. According to this projection, production from all sources will peak in 1990, and then fall short of the projected demand in the following years. This view is consistent with the much-publicized CIA report of early 1977 (*Oil and Gas Journal,* April 25, 1977) which foresaw a world shortage by 1985.

Despite the long-range certainty of petroleum shortage, there have actually been periods of world oversupply, such as the brief glut in 1978 that caused Saudi Arabia's production to fall to a two-year low. Early in 1979, however, the halt to Iran's exports deprived world markets of roughly 1.8 billion barrels per year, forcing various importing nations to scramble for other supplies (*Oil and Gas Journal,* Feb. 26, 1979).

Future abundance of crude oil for world markets will be affected by the following factors: 1) whether the U.S.S.R. becomes a net exporter of crude or an importer competing with other nations for Middle East supplies; 2) whether nations of the Middle East, especially Saudi Arabia, see expansion of production more desirable than letting the oil appreciate in the ground; 3) the reality and size of undiscovered amounts thought to exist in various regions of the world; 4) the actual size of the newly

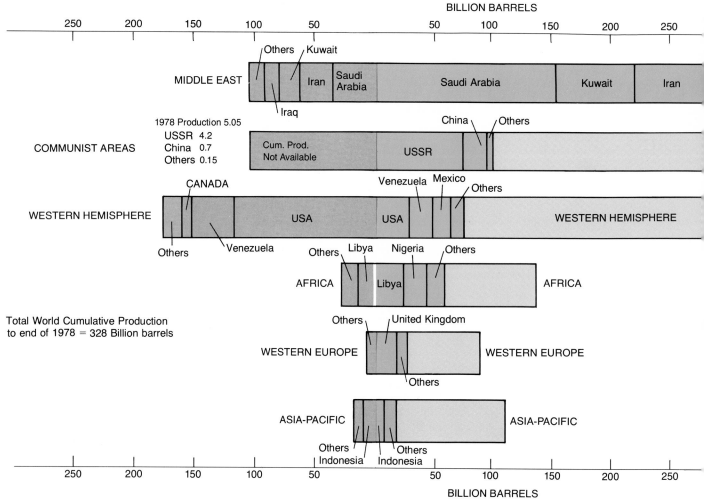

BILLION BARRELS

Fig. 2–8   World regions: Cumulative production of crude oil to end of 1978, versus recoverable amounts remaining.

discovered Mexican reserves, and the rate at which PEMEX, the Mexican national oil and gas corporation, develops production capacity.

United States Crude Oil Resources

Conventional crude oil resources in the United States are reviewed in two parts. First, a broad view takes into account both identified and undiscovered reservoirs and makes use of past production and resource amounts assigned to fifteen regions of the country. A later section focuses upon identified reservoirs only, and uses current production and reserve data for states.

RESOURCES IN IDENTIFIED AND UNDISCOVERED RESERVOIRS

This section is based almost entirely upon one report by the U.S. Geological Survey which provides a unique overview of oil and gas amounts of varying degrees of certainty for onshore and offshore regions (U.S.G.S. *Circular 725*). Estimates in that report are all as of the end of 1974. Although the amounts of Cumulative Production

and Proved reserves are now out of date, the estimates of Inferred and Undiscovered amounts are very useful, and have been cited by Survey spokesmen as late as 1979 (Richard F. Meyer, 1978, and Charles D. Masters, 1979). The fifteen regions that are referred to throughout this section are outlined in Fig. 2–11.

Figure 2–12 shows for each region the total of all crude oil produced and remaining to be recovered as of the end of 1974. In those total amounts, Alaska, Pacific Coastal, West Texas, the Gulf of Mexico, and the Mid-Continent regions are dominant. Circle segments show that many regions approximate the national average of 43 percent in Cumulative Production. One exception is Alaska, where production from large reserves has only recently begun. Other exceptions are the Pacific, Gulf, and Western Rockies regions, along with the Atlantic Coastal region where drilling began only in 1978.

Figure 2–13 shows amounts of crude oil produced in different areas through the industry's history. Four regions have together produced 80 percent of the total as of December 31, 1974: Western Gulf Basin, West Texas and

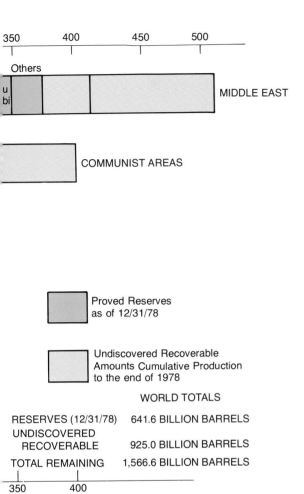

350　　　400　　　450　　　500

Others

u
bi

MIDDLE EAST

COMMUNIST AREAS

Proved Reserves
as of 12/31/78

Undiscovered Recoverable
Amounts Cumulative Production
to the end of 1978

WORLD TOTALS

RESERVES (12/31/78)	641.6 BILLION BARRELS
UNDISCOVERED RECOVERABLE	925.0 BILLION BARRELS
TOTAL REMAINING	1,566.6 BILLION BARRELS

350　　　400

The actual amounts of crude oil in Demonstrated and Inferred classes is shown for the fifteen regions in Fig. 2–15. Since United States domestic production is around 3 billion barrels annually, the Demonstrated reserves in Alaska, West Texas, and the Gulf regions are each equivalent to three-years' production. Inferred reserves are substantial only in the Gulf regions and do not exist in the Atlantic offshore region because there are no fields there.

UNDISCOVERED RECOVERABLE AMOUNTS, AND FRONTIER AREAS As suggested earlier, there are amounts thought to be recoverable, according to present-day economics, from reservoir rocks not yet discovered. The basis for all such speculation is analogy: where sedimentary sequences exist and are similar to known productive rocks, additional fields may be expected. In making estimates, basic information, such as the total *volume* of unexplored sedimentary rocks, is augmented by clues to the presence of source rocks and the existence of structures or discontinuities that might lead to traps.

For undiscovered amounts reported here, volumes of sedimentary rock were estimated from geophysical and drilling data that indicated depth to the crystalline basement (see Introduction). Excluded are any rocks thought to be severely deformed or metamorphosed. Also excluded as inaccessible are rocks deeper than 30,000 feet, and any offshore areas under ocean water whose depth exceeds 200 meters (656 feet).

For each of the geologic provinces making up a region, committees of industry and government geologists agreed on a *low* estimate, consistent with a 95 percent probability that a certain amount of oil or gas exists, a *high* estimate, consistent with only a 5 percent probability, and a third estimate considered the most probable amount. The three were added, and divided by three to yield the *statistical mean for the province*.

Table 2–3 lists the low, high, and statistical mean amounts for each region. Figure 2–16 provides the probability distribution for the national total. It shows a 75 percent probability for at least 65 billion barrels, a 50 percent chance (statistical mean) for 82 billion, but only a 25 percent chance that 100 billion barrels are recoverable from undiscovered reservoirs.

Earlier estimates made by the Geological Survey were much more optimistic than those made by other authorities: in fact the U.S.G.S. estimate in 1965 was over three times the figure cited here. But the present figure used by the U.S.G.S. is lower than the totals predicted by the American Association of Petroleum Geologists in 1971 and by the National Petroleum Council in 1973 (Fig. 2–17). The Survey's estimate is lower partly because it considered offshore areas only to a depth of 200 meters whereas the other two authorities considered depths up to 2,500 meters, including the continental *slope* areas as well as the *shelf*.

Undiscovered Recoverable amounts within the United States are portrayed in Fig. 2–18 and Table 2–3 in which all figures, except those for Atlantic offshore areas, are the statistical mean of the range of estimates

Eastern New Mexico, Mid-Continent, and the Pacific Coastal region in which California is the only producing state.

ALL REMAINING RECOVERABLE CRUDE The total of all remaining recoverable crude oil is divided according to degree of certainty in Fig. 2–14 and Table 2–2. These two illustrations use, in order of diminishing certainty, Demonstrated (Proved plus Indicated) reserves, Inferred reserves, and Undiscovered Recoverable amounts (see box on p. 75). For most regions, the amounts that *may* remain are greater than those that are certain, because Demonstrated reserves have been largely depleted. This is not the case in Alaska onshore, where the large reserves have barely been touched, or in the Pacific Coastal, West Texas, and Western Gulf regions. Generally, where the resources remaining are thought to be large, a substantial portion is in the Demonstrated class. The relatively larger share assigned to Inferred reserves in Alaska onshore and in the Gulf regions indicates the prevalence of young fields whose reserves are likely to be expanded by development drilling.

Table 2–1 World Conventional Crude Oil Reserves, December 31, 1978, Showing Individual Countries Whose Reserves Are 1 Billion Barrels or Greater (in millions of barrels)

REGION	Country	Estimated Proved Reserves	REGION	Country	Estimated Proved Reserves
ASIA-PACIFIC	Australia	2,100	AFRICA	Algeria	6,300
	Brunei	1,480		Angola-Cabinda	1,115
	India	2,900		Egypt	3,200
	Indonesia	10,200		Gabon	1,970
	Malaysia	2,800		Libya	24,300
	Others	527		Nigeria	18,200
	REGION TOTAL	20,007		Tunisia	2,300
WEST EUROPE	Norway	5,900		Others	507
	United Kingdom	16,000		REGION TOTAL	57,892
	Others	2,066	WESTERN HEMISPHERE	Argentina	2,400
	REGION TOTAL	23,966		Brazil	1,200
MIDDLE EAST	Abu Dhabi	30,000		Ecuador	1,170
	Dubai	1,300		Mexico	16,000[1]
	Iran	59,000		Venezuela	18,000
	Iraq	32,100		United States	28,500
	Kuwait	66,200		Canada	6,000
	Divided Zone	6,480		REGION TOTAL	75,746
	Oman	2,500		NON-COMMUNIST TOTAL	547,608
	Qatar	4,000			
	Saudi Arabia	165,700	COMMUNIST AREAS	U.S.S.R.	71,000[2]
	Syria	2,080		China	20,000
	Others	636		Others	3,000
	REGION TOTAL	369,996		COMMUNIST AREAS TOTAL	94,000
				WORLD TOTAL	641,608[3]

[1] More recently, reserves of 25,600 million barrels (crude oil plus NGL) have been reported for Mexico (*Oil & Gas Journal,* Aug. 20, 1979, p. 78).
[2] U.S.S.R. amounts include Proved reserves plus Probable, and some Possible. For all other countries amounts shown are Proved reserves only.
[3] *i.e.,* 641.6 billion barrels.
Source: *Oil and Gas Journal,* Dec. 25, 1978, p. 102–103.

for each region. Alaska offshore is thought to have the greatest untapped potential, Alaska onshore being next, while five onshore regions in the contiguous forty-eight states have 6 to 8 billion barrels anticipated, that is, two to three years national production at the country's present rate of 3 billion barrels per year.

The amount estimated in offshore areas in the contiguous states is not large. In the cases of the Pacific Coast and the Gulf of Mexico, the resources have already been partly exploited. Off the Atlantic coast, the area is untouched; yet the U.S.G.S. (*Circular 725*) predicts only *3 billion barrels* as the mean estimate for the entire shelf. In preparation for offshore lease sales, the Survey has provided the Bureau of Land Management with more recent estimates for the sale areas, made with the benefit of seismic and test hole information not available earlier. The arithmetic means of estimates for the three areas,

George's Bank, Baltimore Canyon, and the Southeast Georgia Embayment (as shown on Fig. 2–19) total 1.88 billion barrels—over half the 3 billion estimated earlier for the whole shelf.

The George's Bank lease sale scheduled for January 31, 1978, was blocked by a Federal court injunction granted in response to environmental concerns. After much controversy the sale was held December 18, 1979, and attracted bids on 73 of the 116 tracts. The Southeast Georgia Embayment sale took place as scheduled, March, 1978, but bidding was not enthusiastic. The Baltimore Canyon area has attracted the most attention because its crude oil potential was thought to be large. Although the sale of leases in that area was completed in 1976, drilling was delayed by a series of legal actions until March of 1978. As of early 1980, fifteen holes have been drilled in the area: thirteen have been dry, and two

World Oil Consumption

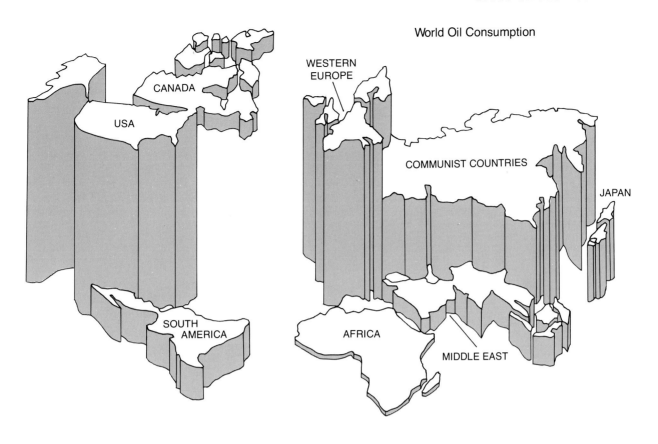

The World's Oil

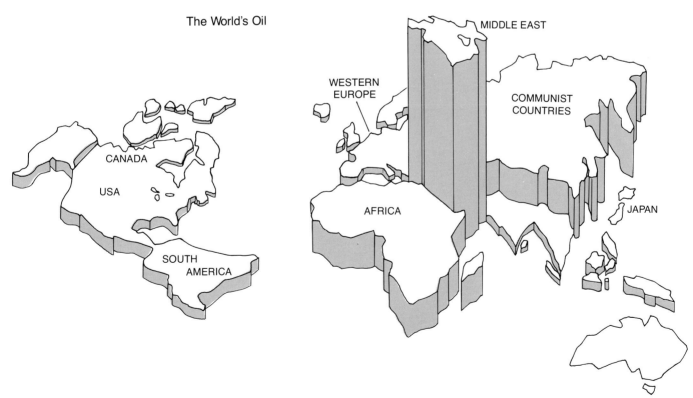

Fig. 2–9 Regional differences in demand for oil and consumption of oil.

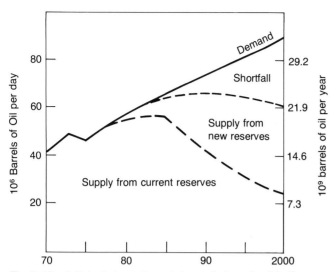

Fig. 2–10 Anticipated supply and demand of crude oil in the non-communist world.

Fig. 2–11 Fifteen regions used to summarize U. S. crude oil and natural gas resources as of end 1974.

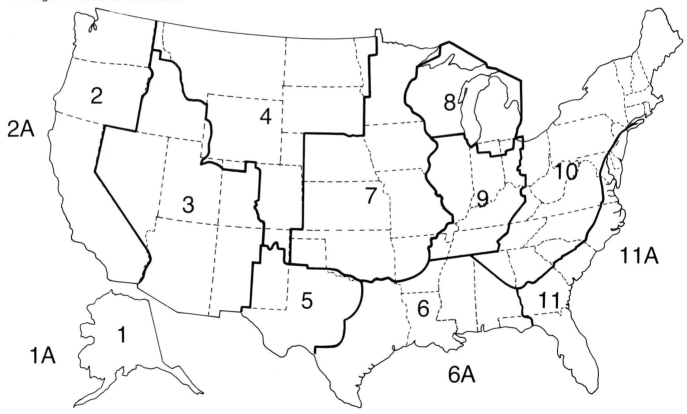

1. Alaska
1A. Alaska Offshore
2. Pacific Coastal States
2A. Pacific Coastal Offshore
3. Western Rocky Mountains
4. Northern Rocky Mountains
5. West Texas and East New Mexico
6. Western Gulf Basin

6A. Gulf of Mexico
7. Mid-Continent
8. Michigan Basin
9. Eastern Interior
10. Appalachians
11. East Gulf and Atlantic Coastal Plain
11A. Atlantic Coastal States Offshore

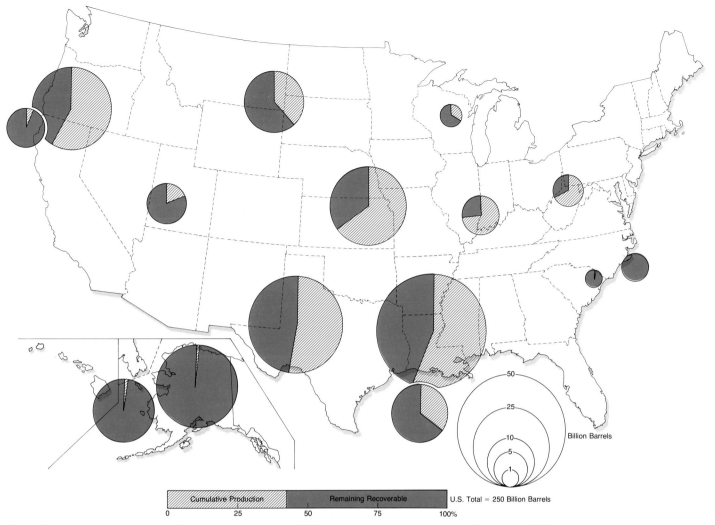

Cumulative Production   Remaining Recoverable   U.S. Total = 250 Billion Barrels
0   25   50   75   100%

50
25
10
5
1
Billion Barrels

Fig. 2–12 Cumulative production of U. S. crude oil through 1974 versus remaining recoverable oil as of end 1974.

Fig. 2–13 Cumulative crude oil production in the U. S. through 1974 in 15 regions.

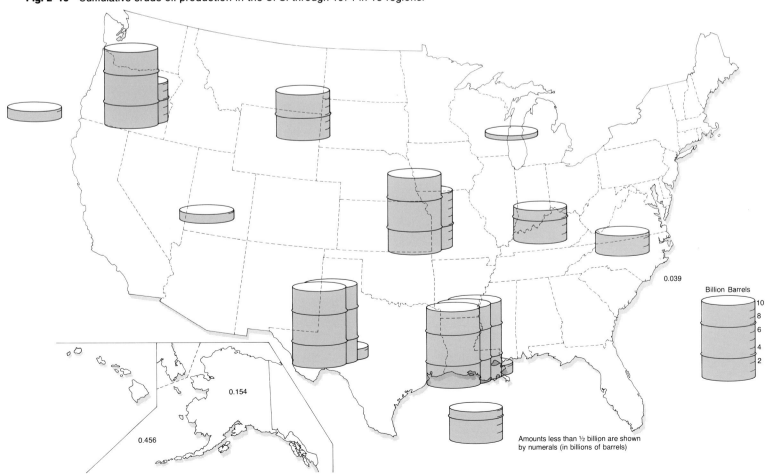

0.039

0.154

0.456

Billion Barrels
10
8
6
4
2

Amounts less than ½ billion are shown
by numerals (in billions of barrels)

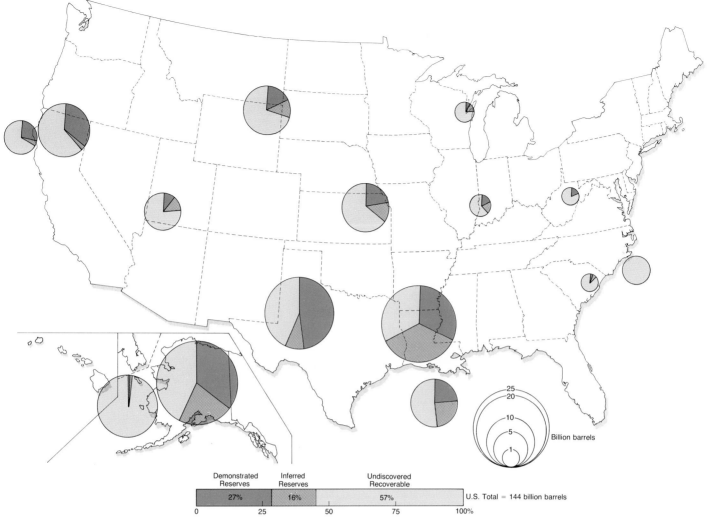

Fig. 2–14 All remaining crude oil estimate recoverable divided by degree of certainty, as of end 1974.

Table 2–2 Crude Oil in Identified and Undiscovered Reserves as of December 31, 1974 for Fifteen Regions of the United States Including Alaska and Offshore Regions (in billions of barrels)

REGION	Cumulative Production	As % of Grand Total	Demonstrated Reserves	ALL REMAINING RECOVERABLE As % of Total Remaining	Inferred Reserves	As % of Total Remaining	Undiscovered Recoverable[1]	As % of Total Remaining	Total Remaining	GRAND TOTAL Total Remaining
1	0.154	1	9.957	35	6.1	22	12	43	28.057	28.211
1a	0.456	3	0.150	1	0.1	1	15	98	15.250	15.706
2	15.254	58	3.790	34	0.3	3	7	63	11.090	26.344
2a	1.499	06	1.160	27	0.2	5	3	68	4.360	5.859
3	1.115	18	0.506	10	0.7	13	4	77	5.206	6.321
4	6.021	38	1.717	17	1.2	12	7	71	9.917	15.938
5	21.385	53	9.051	48	1.6	9	8	43	18.651	40.036
6	31.345	56	7.669	32	8.6	35	8	33	24.269	55.614
6a	4.135	38	2.262	23	2.4	25	5	52	9.662	13.797
7	17.203	65	2.016	22	1.3	14	6	64	9.316	26.519
8	0.645	33	0.090	7	0.2	16	1	77	1.290	1.935
9	4.346	73	0.292	18	0.3	19	1	63	1.592	5.938
10	2.539	67	0.222	18	Neg	0	1	82	1.222	3.761
11	0.039	03	0.048	4	0.1	9	1	87	1.148	1.187
11a	0.000	00	0.000	0	0.00	0	3	100	3.000	3.000
TOTAL	106.136	42	38.886	27	23.1	16	82	57	143.986	250.490

[1]Statistical mean of estimates.

Source: Derived from information in U.S. Geological Survey *Circular 725,* 1975.

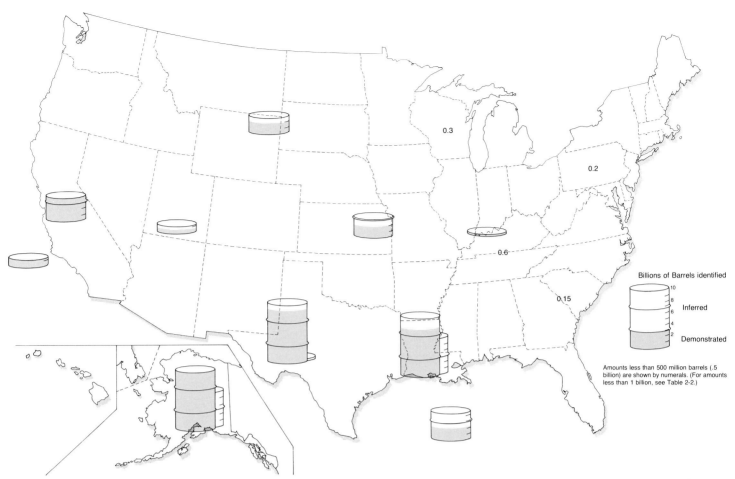

Fig. 2–15 Amounts of crude oil in identified reservoirs showing demonstrated and inferred reserves, as of end 1974.

Table 2–3 Undiscovered Recoverable Amounts of Crude Oil in Fifteen Regions of the United States as of December 31, 1974 (in billions of barrels)

REGION	Low Estimate (95% chance)	High Estimate (5% chance)	Statistical Mean	REGION	Low Estimate (95% chance)	High Estimate (5% chance)	Statistical Mean
1	6	19	12	TOTAL LOWER 48	36	81	55
1a	3	31	15				
2	4	11	7				
2a	2	5	3	TOTAL ALASKA	12	49	27
3	2	8	4				
4	5	11	7	TOTAL ONSHORE	37	81	56
5	4	14	8				
6	5	12	8				
6a	3	8	5	TOTAL OFFSHORE	10	49	26
7	3	12	6				
8	0.3	2	1				
9	0.6	2	1	TOTAL UNITED STATES	50	127	82
10	0.4	2	1				
11	0.2	2	1				
11a	2	4	3				

Source: U.S.G.S., *Circular 725.*

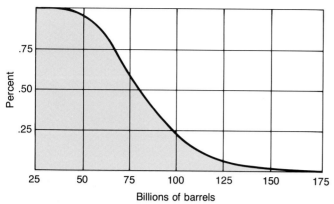

Fig. 2-16 Probabilities of occurrence of various amounts of undiscovered crude oil in the U. S.

Table 2-4 Estimates of Undiscovered Crude Oil at Three Sale Areas on the Atlantic Continental Shelf (in billions of barrels)

SALE AREA	Low Estimate (95% chance)	High Estimate (5% chance)	Arithmetic Mean
George's Bank	0.15	0.53	0.34
Baltimore Canyon	0.40	1.4	0.9
Southeast Georgia Embayment	0.28	1.0	0.64
TOTAL	0.83	2.93	1.88

Source: U.S. Department of Interior, Sales No. 40, 42, and 43, 1976 and 1977.

Fig. 2-17 Various estimates of undiscovered and inferred crude oil in the U. S.

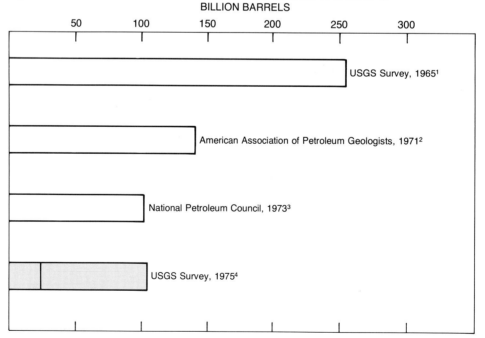

[1]See Hendricks, 1965
[2]See Cram, Ira, 1971
[3]See National Petroleum Council, 1973
[4]See USGS Circular 725, 1975

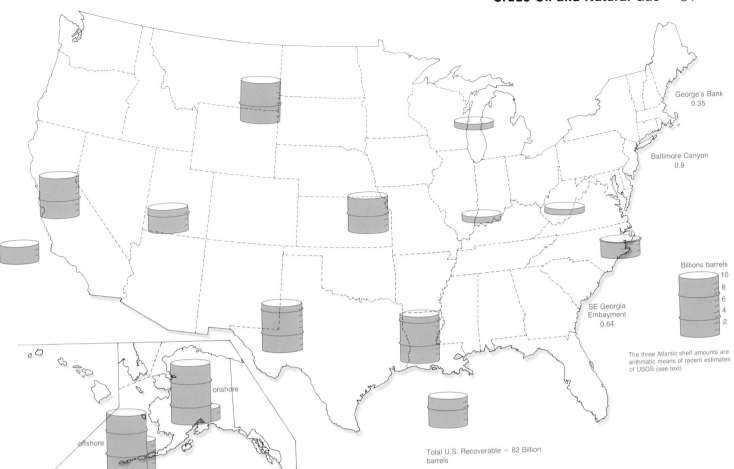

George's Bank
0.35

Baltimore Canyon
0.9

Billions barrels
10
8
6
4
2

SE Georgia
Embayment
0.64

onshore

offshore

The three Atlantic shelf amounts are
arithmatic means of recent estimates
of USGS (see text)

Total U.S. Recoverable = 82 Billion
barrels

Fig. 2–18   Amounts of undiscovered recoverable crude oil in 15 regions, as of end 1974.

(Texaco's) have found natural gas in small quantities. It is too early to write off the possibility of large amounts of crude oil in the continental shelf, especially as the more promising rocks farther offshore have yet to be explored. There is, however, some indication that the conditions required for generation and trapping of oil may have been lacking (Kerr, 1979).

SUMMARY OF AMOUNTS REMAINING   All conventional crude oil, known and thought to remain as of December 31, 1974, is shown in Figure 2–20, according to both its degree of certainty and location within broad regions of the country. The left-hand bar shows the mean estimate of Undiscovered Recoverable amounts (82 billion barrels) exceeds the more certain total of Demonstrated and Inferred reserves by about 20 billion barrels. The total of all three categories is 144 billion barrels, or roughly a forty-eight year supply at the national production rate of 3 billion barrels per year. Just as the locations are hidden in the left-hand bar, the degree of certainty is masked in the right-hand bar which summarizes totals by region. It is evident that the lower forty-eight states will have the lion's share; and in both Alaska and the lower forty-eight states, onshore amounts will dominate. Since *Undiscovered Recoverable* is the largest contributor for most regions (Table 2–2), great uncertainty accompanies this summary.

CRUDE OIL IN IDENTIFIED RESERVOIRS

Annual information published by the American Petroleum Institute and the American Gas Association makes possible a thorough dissection of the identified petroleum with the benefit of current data arranged by

Fig. 2–19   Leases awarded August, 1976, in Baltimore Canyon area of Atlantic continental shelf.

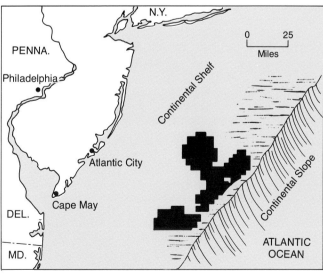

N.Y.

PENNA.

Philadelphia

0   25
Miles

Continental Shelf

Atlantic City

Cape May

DEL.

Continental Slope

MD.

ATLANTIC
OCEAN

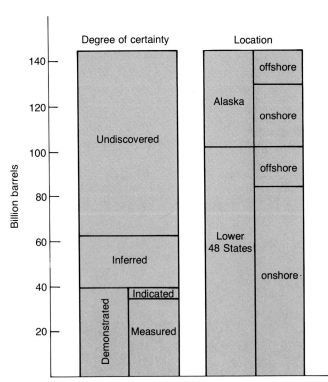

Fig. 2–20   (Above) Conventional crude oil summarized by degree of certainty and by location, as of end 1974.

state rather than region. Lacking, however, is information on Inferred and Undiscovered amounts. The following series of illustrations begins with the character of the accumulations, and proceeds to amounts produced, amounts remaining, and finally the effects of recovery techniques that may lead to expanded reserve estimates.

ESTIMATED ORIGINAL OIL IN-PLACE   The relative amounts portrayed by circle areas in Figs. 2–21 and 2–22, since they consider all fields, past and present, reveal the past importance of some states that would not be recognized if only current reserves were mapped. The amount assigned to California, for instance, is over twice that of Louisiana's—a contrast that is not consistent with the distribution of today's producing fields.

Circles are divided in the first two maps to show the nature of the reservoir rocks and traps in which the oil occurs. While the overall national proportion of sandstone as the reservoir rock (71 percent) is reflected in many states, there are some where either sandstones or carbonates are absolutely dominant. California's fields, through the San Joaquin and Los Angeles Basins, are virtually all in sandstones, as is the Prudhoe Bay field in

Fig. 2–21   (Below) Estimated original oil in place as of end 1978, showing portions in three classes of reservoir rock.

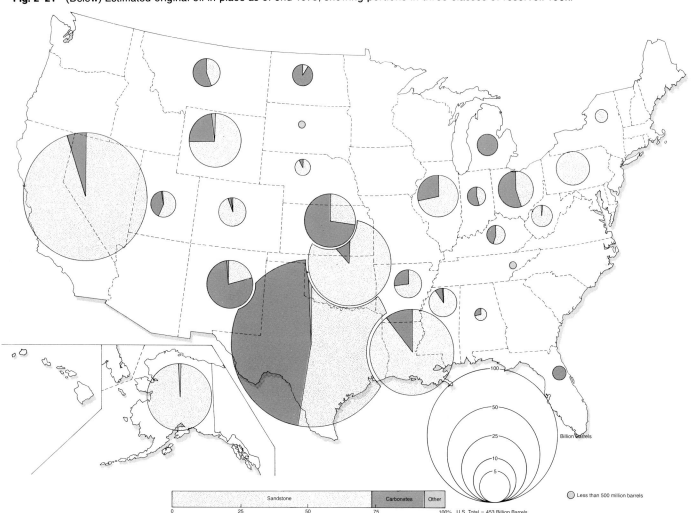

Alaska. The many fields of Texas, however, show a near-even split between sandstones and carbonates. Carbonates appear to dominate only in states whose total oil in-place is relatively small. The national average of 71 percent shows an importance of sandstones in United States reservoirs that is not found in all nations; in fields in Canada and in the Middle East for instance, carbonates play a much larger role.

The importance of structural rather than stratigraphic trapping mechanisms in past and present fields is apparent in most states (Fig. 2–22) and is striking in Alaska, where one field of the structural type accounts for most of the oil in-place. The states with smaller amounts of oil most strongly demonstrate the opposite character, for example, West Virginia, New York, Pennsylvania, and Indiana.

ESTIMATED ULTIMATE RECOVERY   Estimated Ultimate Recovery is the total of Cumulative Production and current Proved reserves, or the extent to which oil in-place in identified reservoirs has been, and will be, recovered. It reflects, therefore, the all-important factor of recovery rates. In 1978, the national average of 32 percent recovery was exceeded in states such as Louisiana,

which have either exceptionally permeable reservoir rocks, crude that is "light" (less viscous), or a combination of the two. Conversely, relatively "heavy" crude leads to lower recovery rates in Ohio, Pennsylvania, and New York (see Fig. 2–23 and Table 2–5). State averages, of course, do not reflect the much greater extremes of recovery which occur among actual reservoirs. The fact that 60 to 75 percent of the original oil in-place *remains in-place* in many reservoirs, means that large volumes might be recovered now from old fields, and also suggests that present and future fields might yield far more than 35 percent of their oil in-place if special techniques are employed (The possible amounts are discussed in a later section on higher recovery rates).

When ultimate Recovery for each state is apportioned to Cumulative Production and Proved reserves (Fig. 2–24) the relative size of these two components reveals the status of a state very clearly. In Alaska, for instance, Cumulative Production is relatively small in relation to the total, whereas in states such as Kansas and Oklahoma the reserves are trivial in comparison with past production. If reserves were added suddenly to a state through large discoveries, this picture could change

Fig. 2–22 Estimated original oil in place as of end 1978, showing portions in two types of trap.

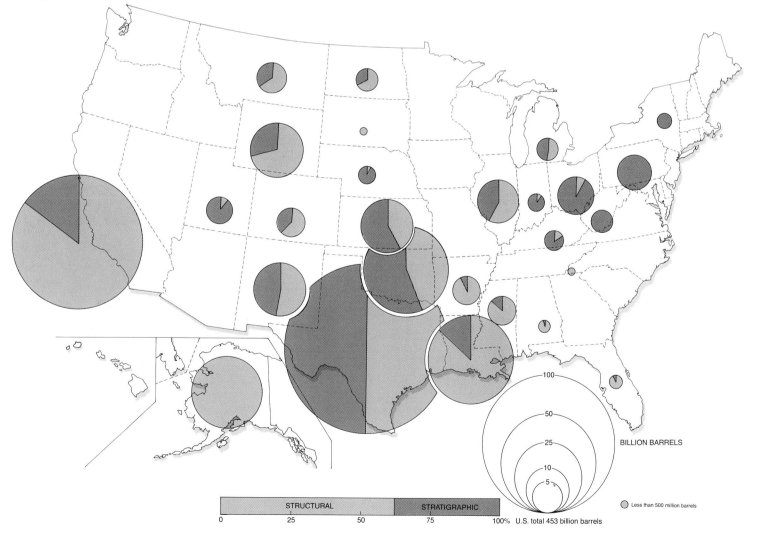

STRUCTURAL   STRATIGRAPHIC

0   25   50   75   100%   U.S. total 453 billion barrels

BILLION BARRELS

100
50
25
10
5

Less than 500 million barrels

drastically. (The United States average in this regard, in comparison with other countries and regions was shown earlier in Fig. 2–8.)

AMOUNTS OF CUMULATIVE PRODUCTION Figure 2–25 shows the contribution of various states to the total United States production through time. Texas, California, Louisiana, and Oklahoma have produced the most. Comparison with Fig. 2–6 puts these volumes in world perspective: Texas production to the end of 1978 exceeded Saudi Arabia's total production, and California's Cumulative Production is almost equal to that of Kuwait.

AMOUNTS OF PROVED RESERVES As shown in Fig. 2–26 and Table 2–6, the reserves (calculated at end of 1978) are concentrated in the same states as are the Cumulative Production amounts. There are two exceptions: 1) California and Oklahoma are less conspicuous than they were in past production, and 2) Alaska's reserves are roughly equal to those of Texas. Large as they are, Alaska's 9 billion barrels represent only three years of United States total production at the present rates. The

country's total of 29 billion barrels would, at current rates, last for ten years.

Although the reserves are located in thousands of fields throughout the country, a surprising proportion of the oil is concentrated in a small number of large fields. A listing of the 100 largest fields in the country, of which the smallest holds current reserves of 28 million barrels, accounts for 72 percent of all United States Proved reserves (Fig. 2–27). Twenty-nine fields have current reserves of over 100 million (0.1 billion) barrels, and account for 59 percent of the total. Moreover, three fields with over 1 billion barrels, Prudhoe Bay in Alaska, and the Yates and East Texas fields in Texas, hold 38 percent of the country's reserves.

United States reserves through the twentieth century increased steadily until an unprecedented plateau was reached in the 1960s, followed by a decline in 1968 that was arrested by the giant Prudhoe Bay discovery in 1970 (Fig. 2–28). Since then, reserves have declined to a level below that of 1969. Figure 2–29, which focuses on the

Fig. 2–23 Variation in recovery rates from known reservoirs: estimated ultimate recovery of crude oil (cumulative production plus proved reserves) as proportion of original oil in place, Dec. 31, 1978.

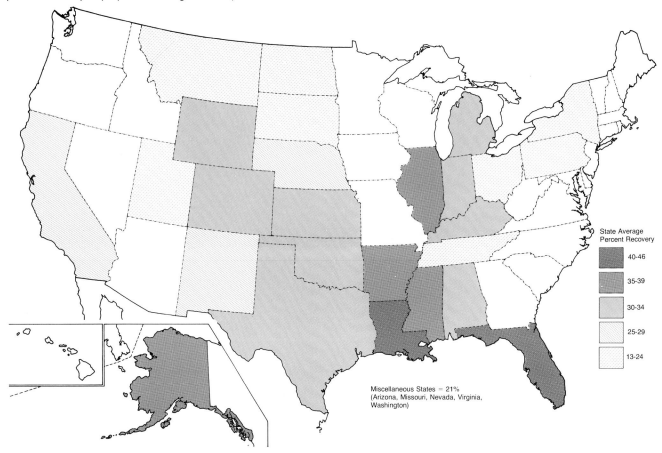

post-war period, shows changes in reserve amounts to be the result of *new oil added during the year* (through discoveries, extensions, and revisions) versus *oil produced during the year*. In any year with production exceeding new oil added, a drop in reserves occurs. While production in recent years has been about 3 billion barrels per year, new oil added each year has been closer to 2 billion and in the last few years has been between 1 and 2 billion barrels per year. Oil from Alaska will probably keep the annual production rate near 3 billion barrels for a few years. Unless substantial discoveries are made, reserves will continue to decline.

Demonstrated reserves are slightly different from Proved reserves. They include Indicated amounts, that is, amounts recoverable economically at present, with the application of secondary recovery methods. Because Indicated amounts are small and therefore difficult to map, Fig. 2–30 shows the relative contribution for each state. Indicated amounts tend to be relatively large in states where past years' production has reduced Proved reserves and left many old reservoirs holding oil that could be partially recovered. California, New Mexico, Pennsylvania, and New York are such states: over 30 percent of their Demonstrated reserves are in the Indi-

Table 2–5 Original Oil In-Place, Cumulative Production, Ultimate Recovery, and Proved Reserves as of December 31, 1978 (in thousands of barrels)

STATE	Estimated Original Oil In-Place	Cumulative Production	Estimated Ultimate Recovery	Proved Reserves	Ultimate Recovery as Percent of Original Oil In-Place
Alabama	775,461	200,747	233,854	33,107	30.2
Alaska	27,712,540	1,360,345	10,607,701	9,247,356	38.3
Arkansas	4,281,245	1,446,573	1,540,611	94,038	36.0
California	83,789,092	18,098,245	21,569,546	3,471,301	25.7
Colorado	4,294,244	1,230,506	1,428,518	198,012	33.3
Florida	1,054,065	285,606	454,967	169,361	43.2
Illinois	9,087,590	3,100,162	3,238,089	137,927	35.6
Indiana	1,638,055	464,690	491,205	26,515	30.0
Kansas	16,171,741	4,789,712	5,140,079	350,367	31.8
Kentucky	2,115,885	646,307	678,140	31,833	32.0
Louisiana	41,157,487	15,856,288	18,749,689	2,893,401	45.6
Michigan	2,904,343	769,008	959,172	190,164	33.0
Mississippi	4,931,960	1,636,855	1,824,442	187,587	37.0
Montana	4,666,879	987,950	1,128,416	140,466	24.2
Nebraska	1,408,047	370,871	400,162	29,291	28.4
New Mexico	14,859,136	3,395,296	3,880,936	485,640	26.1
New York	1,117,739	229,092	238,088	8,996	21.3
North Dakota	2,843,910	524,969	686,182	161,213	24.1
Ohio	7,192,243	844,348	975,542	131,194	13.6
Oklahoma	38,444,465	11,598,031	12,671,500	1,073,469	33.0
Pennsylvania	6,668,990	1,294,939	1,343,095	48,156	20.1
South Dakota	47,045	6,281	9,019	2,738	19.2
Tennessee	35,424	5,599	8,088	2,489	22.8
Texas	153,096,442	43,004,929	50,694,920	7,689,991	33.1
Utah	3,838,340	650,777	806,148	155,371	21.0
West Virginia	2,640,476	520,064	549,739	29,675	20.8
Wyoming	16,273,334	4,423,258	5,227,825	804,567	32.1
Miscellaneous[1]	193,434	24,766	34,301	9,535	17.7
TOTAL UNITED STATES	435,239,612	117,766,214	145,569,974	27,803,760	32.1

[1]Includes Arizona, Missouri, Nevada, Virginia, and Washington.
Source: American Petroleum Institute, *Vol. 33*, June, 1979.

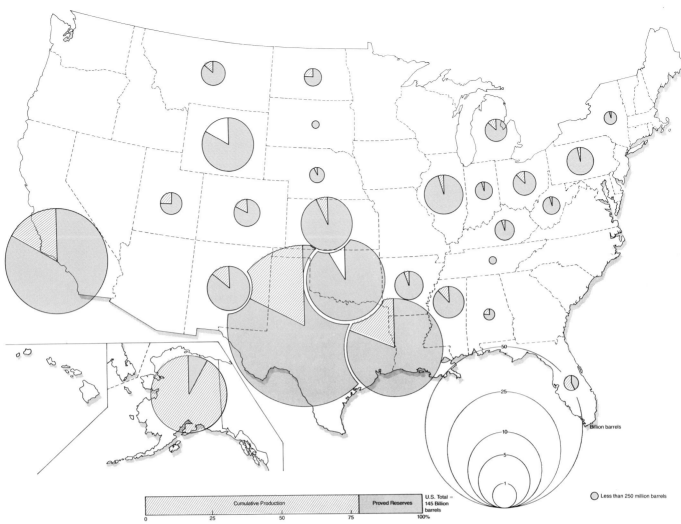

Fig. 2–24 Estimated ultimate recovery of crude oil from known reservoirs, showing proportions in cumulative production versus proved reserves, Dec. 31, 1978.

Fig. 2–25 Cumulative production of crude oil to Dec. 31, 1978.

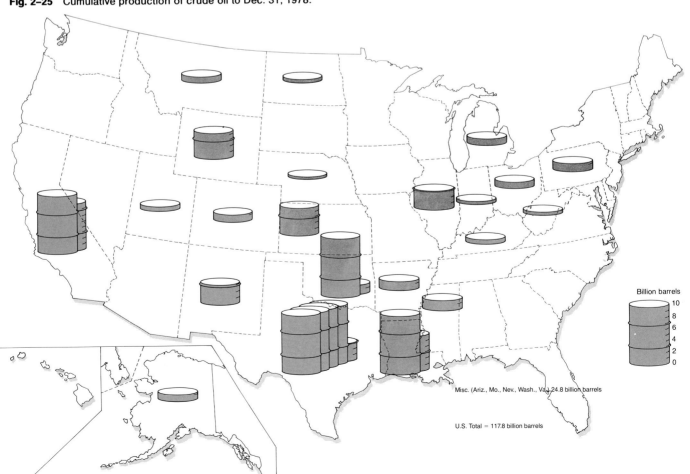

Misc. (Ariz., Mo., Nev., Wash., Va.) 24.8 billion barrels

U.S. Total = 117.8 billion barrels

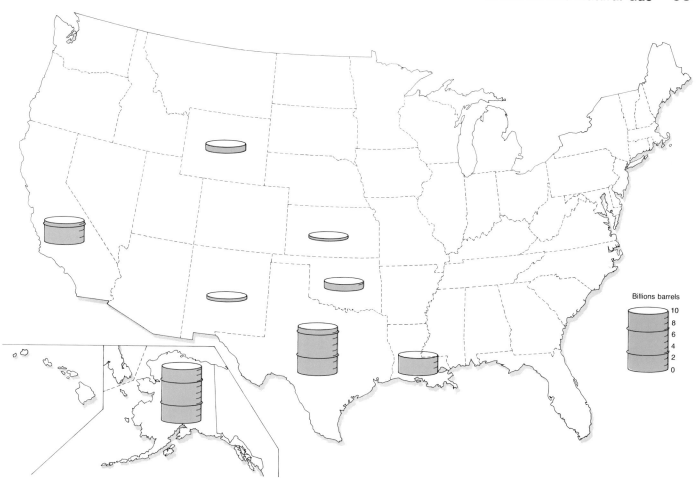

Fig. 2–26 Proved reserves of crude oil, Dec. 31, 1978.

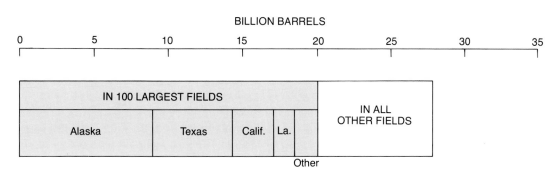

Fig. 2–27 The role of 100 largest fields in U. S. proved reserves of crude oil.

cated portion. Altogether, Indicated amounts for the country were 3.9 billion barrels at the end of 1978. The latest Proved and Indicated reserves are listed in Table 2–6, which shows a decline in reserves between 1977 and 1978 in most states.

NATURAL GAS LIQUIDS   Although derived from natural gas, not from crude oil, these hydrocarbons are liquid, measured in barrels, and therefore associated with crude oil in tabulations of total petroleum liquids. Figure 2–31 shows reserves concentrated in states with large reserves of natural gas. The contribution to the national total is 5.9 billion barrels in 1978.

IMPACT OF HIGHER RECOVERY RATES ON CRUDE OIL ESTIMATES   Improved recovery, often termed *tertiary* or *enhanced recovery,* could greatly affect estimates of oil recoverable from identified and undiscovered reservoirs. A variety of enhanced recovery techniques are being researched, some of which are essentially chemical in nature, and some of which entail steam injection or controlled combustion in the reservoir to lower the oil's viscosity and encourage its flow.

Additional amounts attainable by applying these techniques to *identified reservoirs* are the subject of a wide range of estimates, some of which are represented

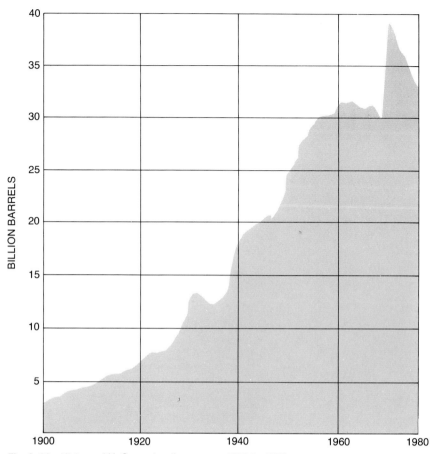

Fig. 2–28   History of U. S. crude oil reserves, 1900 to 1978.

This photo shows an example of enhanced oil recovery. A steam generator with a rated fuel input of 50 million BTU per hour produces the steam needed for the injection wells. The hot combustion gases, after flowing through the main cylindrical portion of the boiler feed water, are cooled and discharged to the atmosphere. *Courtesy Department of Energy.*

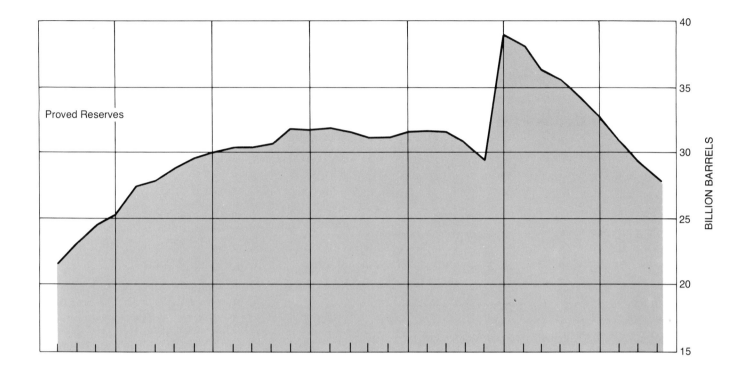

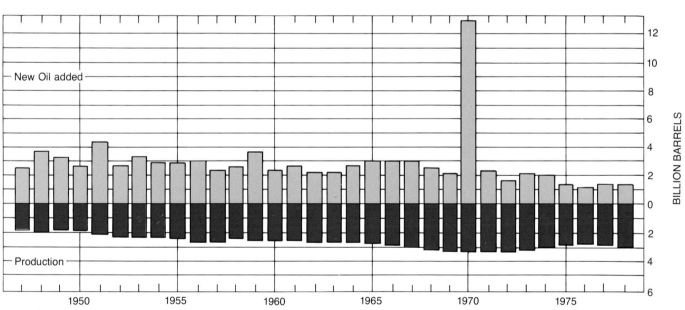

Fig. 2–29 History of U. S. crude oil reserves, 1945 to 1978, showing influence of annual production and annual additions to reserves.

Table 2-6 Proved and Indicated Reserves of Crude Oil as of December 31, 1978 (in thousands of barrels)

STATE	Proved Reserves as of 12/31/77	Proved Reserves as of 12/31/78	Net Changes During 1978[1]	Indicated Reserves as of 12/31/78
Alabama	44,174	33,107	11,067	4,000
Alaska	9,616,111	9,247,356	368,755	0
Arkansas	86,111	94,038	* 7,927	11,617
California	3,631,851	3,471,301	160,550	1,664,325
Colorado	241,910	198,012	43,898	37,300
Florida	208,749	169,361	39,388	1,050
Illinois	149,959	137,927	12,032	2,900
Indiana	25,783	26,515	* 732	1,600
Kansas	359,434	350,367	9,067	0
Kentucky	35,336	31,833	3,503	1,200
Louisiana	3,113,409	2,893,401	220,008	46,425
Michigan	133,228	190,164	* 56,936	20,000
Mississippi	202,632	187,587	15,045	27,255
Montana	151,601	140,466	11,135	25,800
Nebraska	31,010	29,291	1,719	0
New Mexico	490,508	485,640	4,868	354,685
New York	9,099	8,996	103	2,500
North Dakota	150,191	161,213	* 11,022	44,180
Ohio	129,148	131,194	* 2,046	0
Oklahoma	1,121,430	1,073,469	47,961	228,785
Pennsylvania	49,334	48,156	1,178	21,670
South Dakota	2,291	2,738	* 447	0
Tennessee	2,536	2,489	47	0
Texas	8,467,436	7,689,991	777,445	1,244,023
Utah	182,147	155,371	26,776	13,630
West Virginia	28,686	29,675	* 989	1,000
Wyoming	815,578	804,567	11,011	167,657
Miscellaneous	6,720	0	0	0
TOTAL UNITED STATES	29,496,402	27,803,760	1,682,642 (loss)	3,921,602

Source: American Petroleum Institute, *Vol. 33*, June, 1979
[1]Net losses in most states. Gains are asterisked.

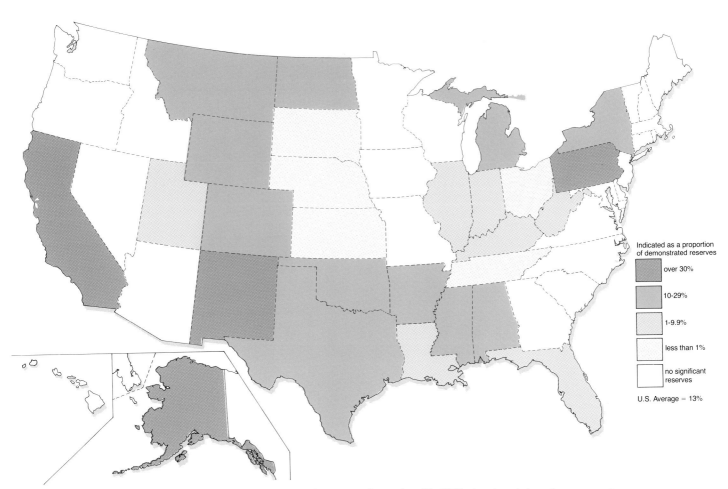

Indicated as a proportion
of demonstrated reserves

over 30%

10-29%

1-9.9%

less than 1%

no significant
reserves

U.S. Average = 13%

Fig. 2–30   Indicated reserves as proportion of demonstrated reserves, December 31, 1978, showing states where secondary recovery efforts will return relatively large amounts of crude oil.

Fig. 2–31   Proved reserves of total natural gas liquids, Dec. 31, 1978.

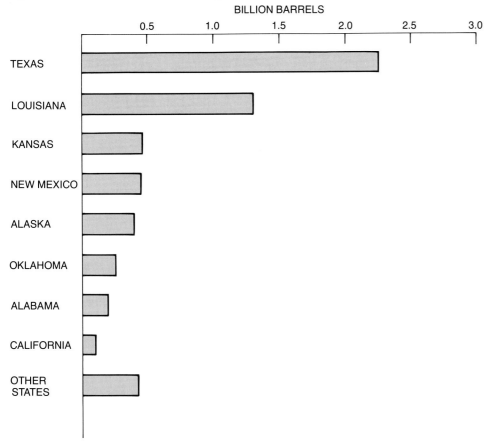

BILLION BARRELS

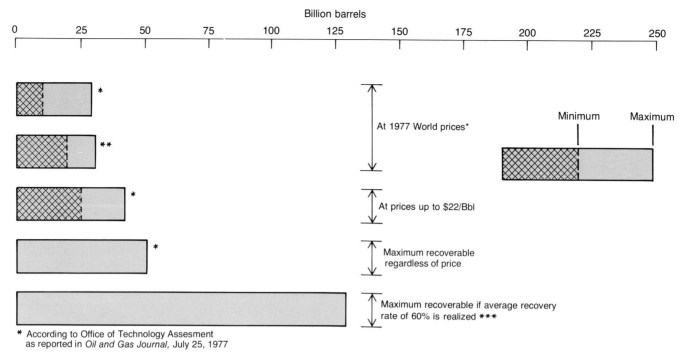

Billion barrels

* According to Office of Technology Assesment
  as reported in *Oil and Gas Journal,* July 25, 1977

** According to ERDA Study
  as reported in *Oil and Gas Journal,* July 25, 1977

*** According to U.S. Geological Survey, Circular 725
  as reported in *Oil and Gas Journal,* July 25, 1977

Fig. 2–32   Various estimates of presently uneconomic conventional crude oil, in identified reservoirs and recoverable through enhanced recovery techniques.

Fig. 2–33   Crude oil amounts through enhanced recovery, showing states with greatest potential.

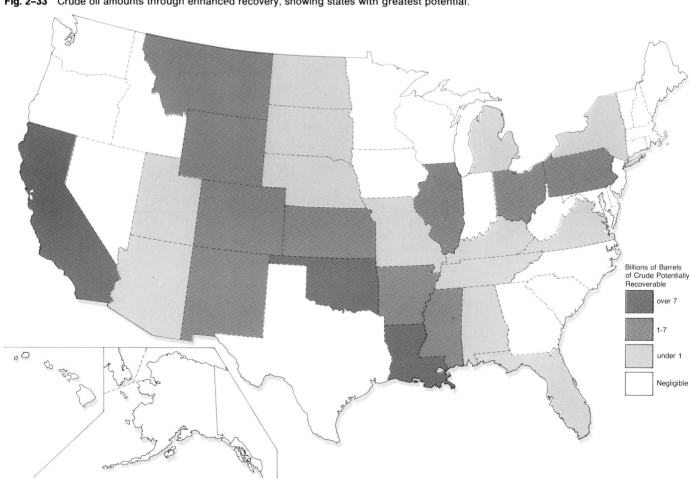

Billions of Barrels of Crude Potentially Recoverable

over 7

1-7

under 1

Negligible

	Identified			Undiscovered
	Demonstrated		Inferred	
	Measured	Indicated		
Economic	27.8[1]	4.2[1]	23.1[2]	50–127[2] (82)
Sub-economic	11–51 (see text) (31)			35 (assuming 50% recovery)
	120–140[2] (130)			44–111[2] (72)

[1] As of 1978
[2] As of December, 1974 (USGS Circular 725)
  Sub-economic amounts on grey assume very high recovery rate of 60%.

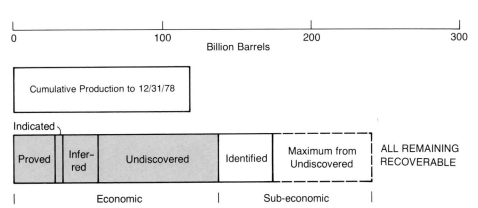

0   100   200   300
Billion Barrels

Cumulative Production to 12/31/78

Indicated

| Proved | Infer-red | Undiscovered | Identified | Maximum from Undiscovered | ALL REMAINING RECOVERABLE |

Economic | Sub-economic

Fig. 2–34 Summary of conventional crude oil resources of the U. S., identified and undiscovered, and economic and sub-economic.

in Fig. 2–32. Even the lowest estimate of 11 billion barrels is quite substantial. Progressively higher prices for oil will justify greater recovery efforts which may yield up to 42 billion barrels. According to a report from the Office of Technology, the maximum amount technically feasible is 51 billion barrels; but the graph includes, for completeness, an estimate by the U.S. Geological Survey of the extra amount attainable if an average of 60 percent recovery were realized for all the known reservoirs.

The target of enhanced (or tertiary) recovery techniques is the large volume of *oil in-place* that is left behind in most reservoirs after primary and secondary recovery methods have been applied. Because of their large fields and long production histories, Texas, Louisiana, and California account for roughly 65 percent of the national total of such oil (Fig. 2–33).

Figure 2–34 places all the United States conventional crude oil resources in the framework that shows degree of certainty and economic feasibility. The diagram shows a composite of resource estimates from different sources. Measured and Indicated reserves (as of the end of 1978) are from the American Petroleum Institute; Inferred and Undiscovered Recoverable amounts (as of the end of 1974) are from the U.S. Geological Survey. Subeconomic amounts in identified reservoirs are from the foregoing (Fig. 2–33). Shown in grey in the subeconomic categories are very high estimates made by the Geological Survey on the assumption that 60 percent recovery of oil in-place may be possible in the future. For identified reservoirs this assumption suggests that an additional 120 billion barrels may be recoverable. The Survey's estimate of 23 billion barrels in the Inferred category was based on the historic 32 percent recovery rate; assuming 60 percent recovery would add 20 billion barrels to that category. Similarly, the estimate of oil to be recovered from undiscovered reservoirs (50 to 127 billion barrels) would be augmented by 44 to 111 billion if 60 percent recovery were realized. While these amounts are not considered realistic (*Circular 725*, p. 17) they do show how enhanced recovery might alter the relative importance of identified and undiscovered reservoirs.

SUMMARY OF CRUDE OIL RESOURCES

Figure 2–34 compares Cumulative Production of crude oil (to the end of 1978) with all economically recoverable oil remaining, that is, Demonstrated reserves, as of December, 1978 plus estimates of Inferred and Undiscovered Recoverable amounts as of December, 1974. Since some of the amounts Inferred and Undiscovered in 1974 may have been realized as extensions, revisions, and discoveries in the ensuing four years, the summation of those anticipated reserves and the current reserves may inflate the resource. However, quite similar estimates of Inferred and Undiscovered amounts might well be made as of the end of 1978, so the distortion is not serious.

The amounts of oil produced and those remaining are roughly equal. This balance indicates that the United States has produced almost half its conventional crude

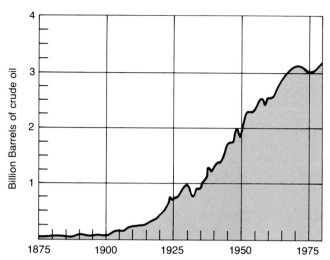

Fig. 2–35   Annual rates of U. S. crude oil production, 1875 to 1978.

Fig. 2–36   Crude oil production in leading states during 1978.

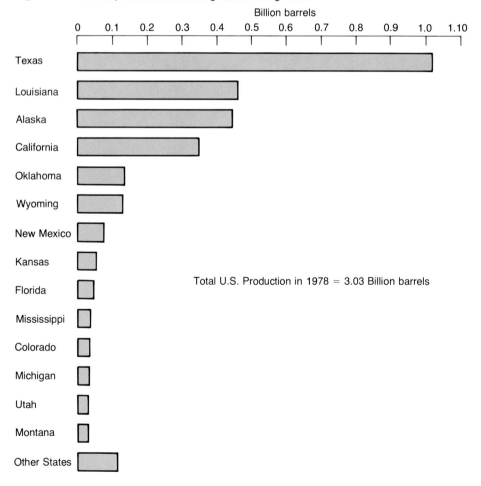

Total U.S. Production in 1978 = 3.03 Billion barrels

resources that ever existed, if amounts recoverable through enhanced recovery are ignored. By contrast, the world as a whole has produced only one-fifth of its crude resources (see Fig. 2–8).

A country with half its oil resources remaining can be regarded as either half empty or half full, depending on whether the view is optimistic or pessimistic. The realist, though, knows the first half was much easier to extract than will be the second. The reservoirs already discovered and exhausted were easily found, relatively shallow, and mostly in easily accessible areas. Much of the crude oil remaining to be discovered is likely to be in smaller accumulations, deeply buried, or in frontier areas, particularly offshore. The decreasing availability of United States oil is revealed by a continuing decline in the amount of crude discovered for every foot of exploratory hole drilled.[2]

Production, Transportation, and Imports of Crude Oil

Annual rates of crude oil production in the United States have grown throughout the twentieth century. Production peaked in 1972, and since then has dropped to about 3 billion barrels in 1978 (Fig. 2–35). Of all states, Texas produced the most crude oil during 1978 (Fig. 2–36). Louisiana, Alaska, and California produced roughly equivalent amounts, while Oklahoma and Wyoming were next in annual production. The pattern is generally consistent with the relative magnitude of oil reserves in these states, except that the Alaskan reserves exceed those of Texas. Production figures for 1977 and 1978 (Table 2–7) show that Alaska's production capacity is growing as the fields there are developed.

CRUDE OIL PIPELINES AND MOVEMENTS

Crude oil was transported from the country's earliest producing areas in Pennsylvania by means of barrels on wagons and flatboats. A railroad line to the fields was completed in 1862, and carried the crude at first in barrels, then, after 1865, in tankcars which consisted of two large upright barrels on a flatcar.

Pipeline technology was developing at the same time, and by the 1870s a 6-inch line was moving oil from the fields about 130 miles to Williamsport from which tankcars carried it to refineries. By 1880, pipelines were carrying the crude from Pennsylvania fields to Pittsburg, Cleveland, Buffalo, and New York City. As more oil fields were discovered in Ohio, Indiana, Illinois, Oklahoma, Kansas, and Texas a continental network of pipelines developed.

Recent movements of crude oil can be represented by information for 1974. In that year, a total of 621 million tons of crude oil was moved by domestic transportation, 75 percent by pipeline. The balance was moved by water carriers (13 percent) motor carriers (12 percent) and by rail (one-half percent).

Figure 2–37 plots oil fields, the pipelines and the vol-

Table 2–7 Crude Oil Production during 1977 and 1978 in thousands of barrels

STATE	1977	1978
Alabama	11,237	11,188
Alaska	169,198	447,541
Arkansas	19,385	19,419
California	352,622	347,102
Colorado	38,822	36,280
Florida	41,917	47,262
Illinois	24,664	22,620
Indiana	5,251	4,785
Kansas	57,151	55,163
Kentucky	6,511	5,685
Louisiana	492,543	458,755
Michigan	32,943	34,648
Mississippi	41,248	37,756
Montana	32,382	30,087
Nebraska	5,849	5,587
New Mexico	81,736	78,130
New York	819	853
North Dakota	23,769	24,965
Ohio	10,358	11,154
Oklahoma	144,338	135,326
Pennsylvania	2,659	2,820
South Dakota	558	815
Tennessee	820	593
Texas	1,097,044	1,044,936
Utah	32,589	31,481
West Virginia	2,518	2,382
Wyoming	129,466	130,878
Miscellaneous	1,147	1,687
TOTAL	2,859,544	3,029,898[1]

[1] i.e., 3.03 billion barrels.
Source: American Petroleum Institute, *Vol. 32*, June, 1978; and *Vol. 33*, June, 1979.

umes of flow through pipelines in 1974. The major movements are within Texas and from Texas to refineries in the Great Lakes area. Small amounts of Canadian crude can be seen flowing to the Great Lakes area; but these flows have diminished in subsequent years. Refineries in the Northeast receive their crude by domestic water transport not shown on the map, and by tanker directly from overseas. California refineries are supplied by domestic water transport (not shown) by tanker from overseas, and by pipeline from California fields.

REFINERY RECEIPTS Figure 2–38 shows, for various districts in the country, whether refineries rely more on domestic or foreign crude oils. On the East Coast, for instance, virtually all the crude oil fed to refineries is from overseas. This map is complementary to Fig. 2–37, since the bulk of crude moving by pipeline is domestic, and its destination is refineries.

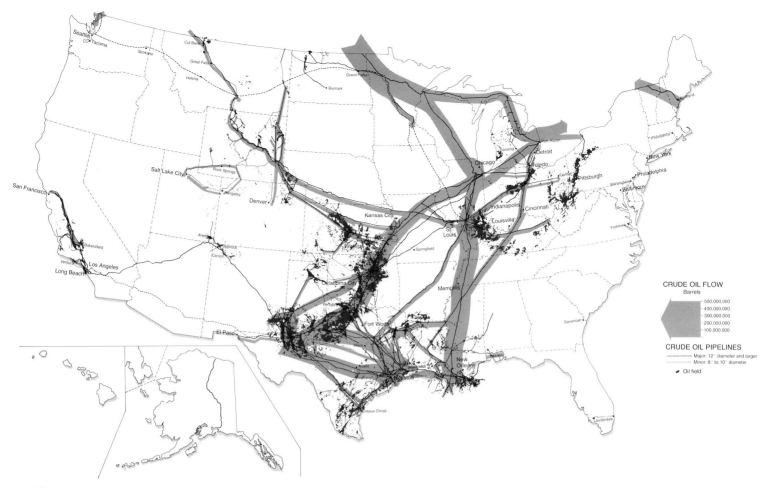

Fig. 2–37 Crude oil pipelines and movements, 1974.

Fig. 2–38 Refinery receipts during 1976, by PAD District, showing domestic and foreign contributions.

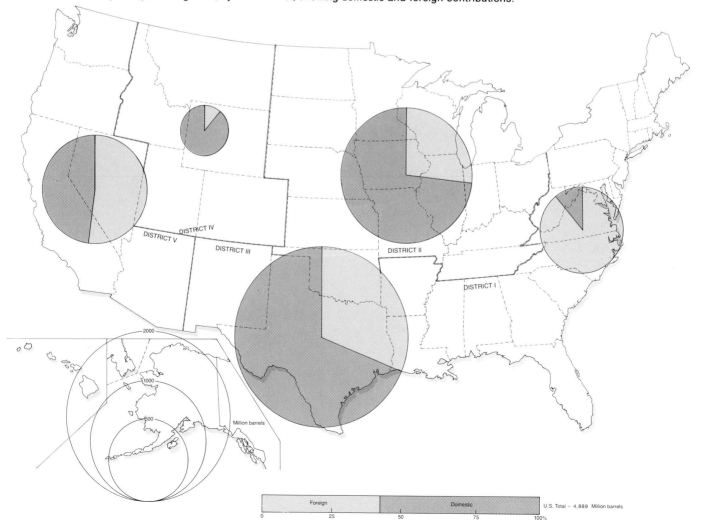

IMPORTS OF CRUDE OIL AND REFINED PRODUCTS

A growing dependence on imported rather than domestic crude oil has resulted from a continuing demand for crude in the face of declining productive capacity in the United States. Strategically, this dependence is unwise. It has led to a deterioration in both the United States' balance of payments and the value of the United States dollar in world money markets. Coincidentally, the ever-rising prices exacted by the Organization of Petroleum Exporting Countries (O.P.E.C.) have fueled inflation in this and other countries more than any other factor.

Demand for crude oil in the United States rose rapidly through the 1960s and 1970s until domestic production, which was peaking near 3 billion barrels per year, was unable to meet the demand. As a result, imports rose rapidly during the early 1970s, and accounted for 46 percent of demand in 1977. The major sources are no longer Canada and Venezuela but are in the Eastern Hemisphere (Fig. 2–39).

Figure 2–40 reveals how, since 1961, the source of imports has shifted to Africa and the Middle East (mainly Nigeria and Saudi Arabia) with Indonesia also providing a substantial portion. Canada has sharply reduced its exports of oil, leaving Venezuela the most important Western Hemisphere supplier of foreign oil. Table 2–8 shows amounts imported from foreign nations during 1976.

The industry projection of prospects in the United States to the year 2000 (Fig. 3–38) foresees demand rising steadily after 1980 at a reduced rate of roughly 1.5 percent per year. Domestic production is expected to hold steady at 3.8 billion barrels per year, and the increase in demand to be met by an increase in imports, which by the year 2000 will account for 56 percent of demand. If, however, imports must be reduced for economic and political reasons then a shortage of 2 billion barrels per year would be experienced. The projection used here, assumes *domestic production* will be maintained until the year 2000 at rates greater than those in the late 1970s. That expresses faith in presently undiscovered resources, in enhanced recovery, or in production of unconventional crude (see Chapter 3).

Natural Gas

Natural gas is a mixture of the lighter hydrocarbons which exists in a gaseous phase underground or may be in solution with crude oil at the pressures which occur in the reservoir. The principal hydrocarbons in the mixture are methane, ethane, propane, butanes and pentanes. Some non-hydrocarbon gases which may occur in the mixture are carbon dioxide, helium, hydrogen sulfide, and nitrogen.

Natural gas is often found trapped in regions (and in rock sequences) where crude oil is not abundant. In some cases the gas is in deeply buried rocks where the high temperatures at depth may have carried the conversion of organic materials beyond the crude oil stage until only gas remained. In other cases, it appears that the presence of *terrestrial* rather than *marine* organic matter in source rock led to formation of gas, not crude oil. Some occurrences of natural gas without oil may be explained by migration processes that separated the two fluids.

In the past a great deal of natural gas has been burned at the well-head (flared) when it was produced at the same time as crude oil. This was considered more desirable because of the problems of storage and transportation. Now, welded pipelines that can effectively move gas, storage systems, and even methods for liquefying and shipping gas in tankers have given this fuel new status.

World Resources of Natural Gas

Proved reserves of gas at the end of 1978, for various regions and leading nations, are summarized in Fig. 2–41 and Table 2–9. Although the Proved resources in the

---

Table 2–8 Refinery Receipts during 1976, Showing Domestic and Foreign Contributions

PETROLEUM ADMINISTRATION FOR DEFENSE DISTRICTS	DOMESTIC CRUDE[1]	FOREIGN CRUDE										TOTAL CRUDE[1]
		Proportions from Foreign Nations (percent)									Total[1]	
		Saudi Arabia	Nigeria	Libya	Algeria	Indonesia	Venezuela	Canada	Iran	Other Foreign	TOTAL FOREIGN	Total[1]	
I	61,474	27	18	0	12	0	10	0	0	22	89	517,840	579,314
II	955,268	5	4	6	0	0	0	6	0	7	28	362,739	1,318,007
III	1,450,084	8	9	3	2	0	0	0	0	9	31	651,281	2,101,465
IV	143,123	0	0	0	0	0	0	12	0	0	12	18,610	161,733
V	379,866	9	0	0	0	23	0	4	5	11	41	379,806	728,672
TOTAL	2,989,815											1,930,370	4,889,191

[1] In thousands of barrels.

Information deduced from Tables 9 and 14, U. S. Department of Energy, *Petroleum Statement Annual*, Jan. 31, 1978.

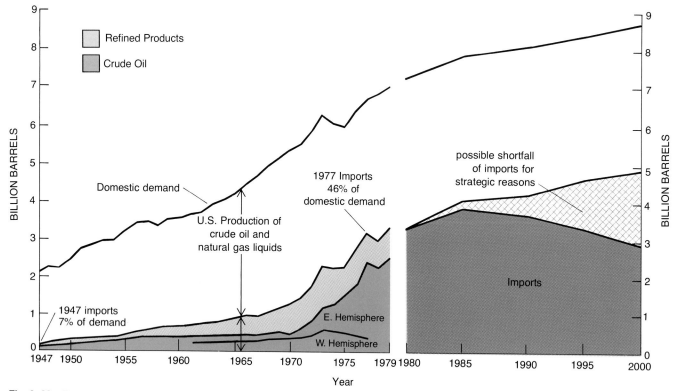

Refined Products

Crude Oil

Domestic demand

1977 Imports
46% of
domestic demand

possible shortfall
of imports for
strategic reasons

U.S. Production of
crude oil and
natural gas liquids

1947 imports
7% of demand

E. Hemisphere

W. Hemisphere

Imports

Year

Fig. 2–39 Thirty-year history of imported petroleum liquids as a portion of U. S. domestic demand.

An offshore drilling rig. *Courtesy Department of Energy.*

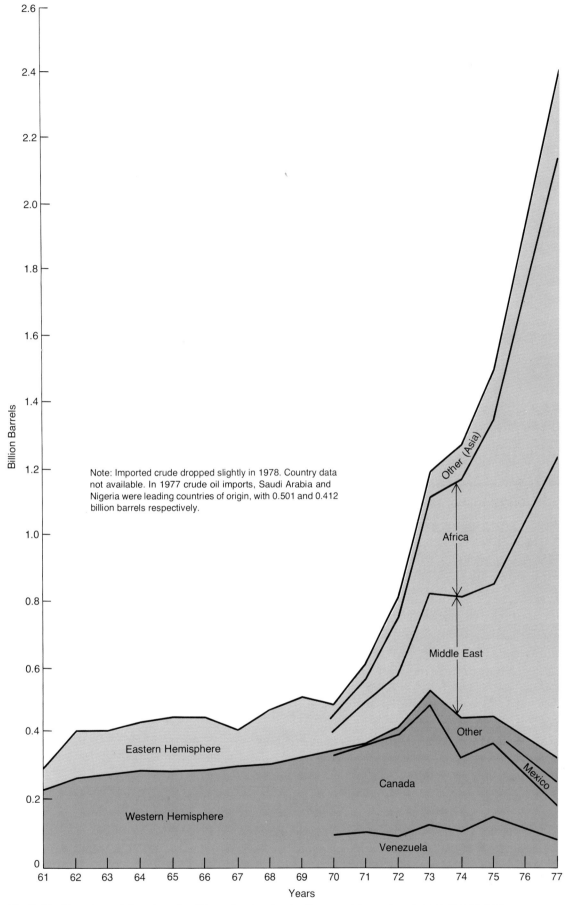

Note: Imported crude dropped slightly in 1978. Country data not available. In 1977 crude oil imports, Saudi Arabia and Nigeria were leading countries of origin, with 0.501 and 0.412 billion barrels respectively.

Fig. 2–40   Changing origins of crude oil imports, showing leading countries and regions in recent years.

Table 2-9 World Reserves of Conventional Natural Gas, as of December 31, 1978, Showing Individual Countries Whose Reserves Are 1 Trillion Cubic Feet or Greater (in billions of cu. ft.)

REGION	Country	Estimated Proved Reserves	REGION	Country	Estimated Proved Reserves
ASIA-PACIFIC	Australia	31,000	AFRICA	Algeria	105,000
	Bangladesh	8,000		Angola-Cabinda	1,200
	Brunei	8,000		Congo	2,260
	India	3,500		Egypt	3,000
	Indonesia	24,000		Gabon	2,400
	Malaysia	17,000		Libya	24,200
	New Zealand	6,000		Nigeria	42,000
	Pakistan	16,000		Tunisia	6,000
	Thailand	5,000		Others	230
	Others	1,350		REGION TOTAL	186,290
	REGION TOTAL	119,850			
			WESTERN HEMISPHERE	Argentina	12,000
WEST EUROPE	Denmark	2,500		Bolivia	6,000
	France	6,500		Brazil	1,500
	West Germany	6,300		Chile	2,500
	Greece	4,000		Colombia	4,800
	Ireland	1,000		Ecuador	4,000
	Italy-Sicily	8,000		Mexico	32,000[1]
	Netherlands	62,000		Peru	1,150
	Norway	24,000		Trinidad and Tobago	8,000
	United Kingdom	27,000		Venezuela	41,000
	Yugoslavia	1,340		United States	205,000
	Others	620		Canada	59,000
	REGION TOTAL	143,260		REGION TOTAL	376,950
				NON-COMMUNIST TOTAL	1,557,010
MIDDLE EAST	Abu Dhabi	20,000			
	Bahrein	7,000	COMMUNIST AREAS	U.S.S.R.	910,000
	Dubai	1,600		China	25,000
	Iran	500,000		Others	10,000
	Iraq	27,800		COMMUNIST AREAS TOTAL	945,000
	Kuwait	31,300		WORLD TOTAL	2,502,010[2]
	Divided Zone	5,000			
	Oman	2,000			
	Qatar	40,000			
	Saudi Arabia	93,900			
	Syria	1,500			
	Others	560			
	REGION TOTAL	730,660			

[1] More recently, reserves of 58,935 billion cubic feet have been reported for Mexico (*Oil & Gas Journal*, Aug. 20, 1979, p. 74).
[2] *i.e.*, 2,502 trillion cu. ft.
Source: *Oil and Gas Journal*, Dec. 25, 1978, p. 101–103.

Middle East are impressive because of Iran's recently expanded reserves, it is the U.S.S.R. that holds the overwhelming share of the world's proven gas. This is true even when its reserves are adjusted from 920 to 736 trillion cubic feet to compensate for the fact that Soviet estimates include some anticipated reserves (*Oil and Gas Journal,* December 26, 1977, p. 102).

In the Western Hemisphere, the United States maintains its leading position, though Mexico's reserves have more than doubled in the past year due to development in the Chiapas-Tobasco region. Algeria dominates African states in gas reserves, and is preparing to export large quantities in liquefied form. In Europe, the North Sea discoveries have enhanced the status of the United Kingdom and Norway; while in the Asia-Pacific area, Australia leads the region.

As with crude oil supplies, reserves of natural gas are conservative estimates based on amounts recoverable from well-defined reservoirs. It is possible to approximate the *entire* supply by adding amounts thought to be

recoverable from reservoirs as yet undiscovered. One recent estimate of such amounts is added to reserves in Fig. 2–42, which also shows total Cumulative Production of gas as of the end of 1978. The *total* length of the bar for each region represents all the gas that has been and will ever be produced—subject to great uncertainty in the undiscovered portions. From this estimate, it appears that most regions have produced both small amounts and small proportions of the total gas anticipated, with the Western Hemisphere and Europe being obvious exceptions. While figures for the amounts produced in the Western Hemisphere are probably realistic, the small amounts shown for the Middle East and Africa are due partly to under-recording of gas produced in association with crude oil production, then lost through flaring.

Not reflected on the bar graph are the recent shifts in gas production capability. Production in the Western Hemisphere has essentially leveled off, while that in the U.S.S.R. has been climbing steadily. In Africa, production has increased more than six times since 1974 due to

Fig. 2–41   World Regions: reserves of natural gas, Dec. 31, 1978.

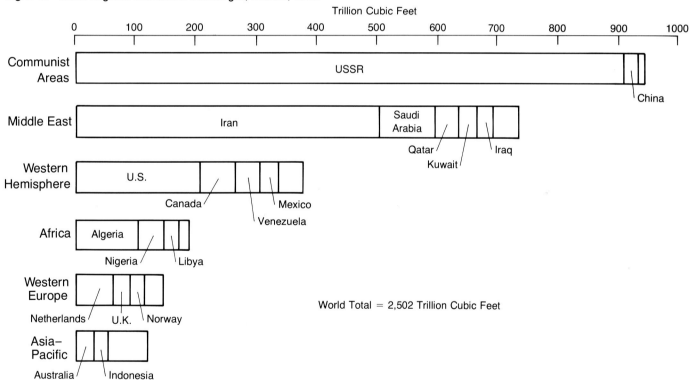

World Total = 2,502 Trillion Cubic Feet

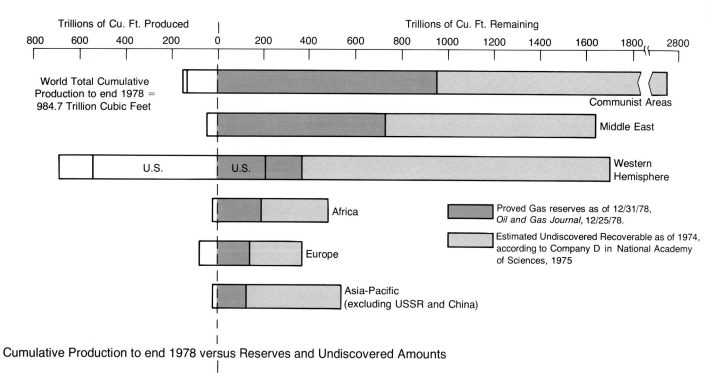

Trillions of Cu. Ft. Produced | Trillions of Cu. Ft. Remaining

800   600   400   200   0   200   400   600   800   1000   1200   1400   1600   1800   2800

World Total Cumulative
Production to end 1978 =
984.7 Trillion Cubic Feet

Communist Areas

Middle East

U.S.   U.S.   Western Hemisphere

Africa

Proved Gas reserves as of 12/31/78,
Oil and Gas Journal, 12/25/78.

Europe

Estimated Undiscovered Recoverable as of 1974,
according to Company D in National Academy
of Sciences, 1975

Asia-Pacific
(excluding USSR and China)

Cumulative Production to end 1978 versus Reserves and Undiscovered Amounts

Cumulative production through 1973 from National Academy of Sciences, 1975. Revision through 1978 from annual production figures in *Oil and Gas Journal*, e.g. 1978 cumulative reproduction in Dec. 25, 1978 issue.

In the revision, Communist Areas of the *Oil and Gas* Journal are equated with the continental Asia region used by National Academy of Sciences.

Fig. 2–42   World regions: cumulative production of natural gas to Dec. 31, 1978 versus recoverable amounts remaining.

Fig. 2–43   Fifteen regions used to summarize U. S. natural gas resources as of Dec. 31, 1974.

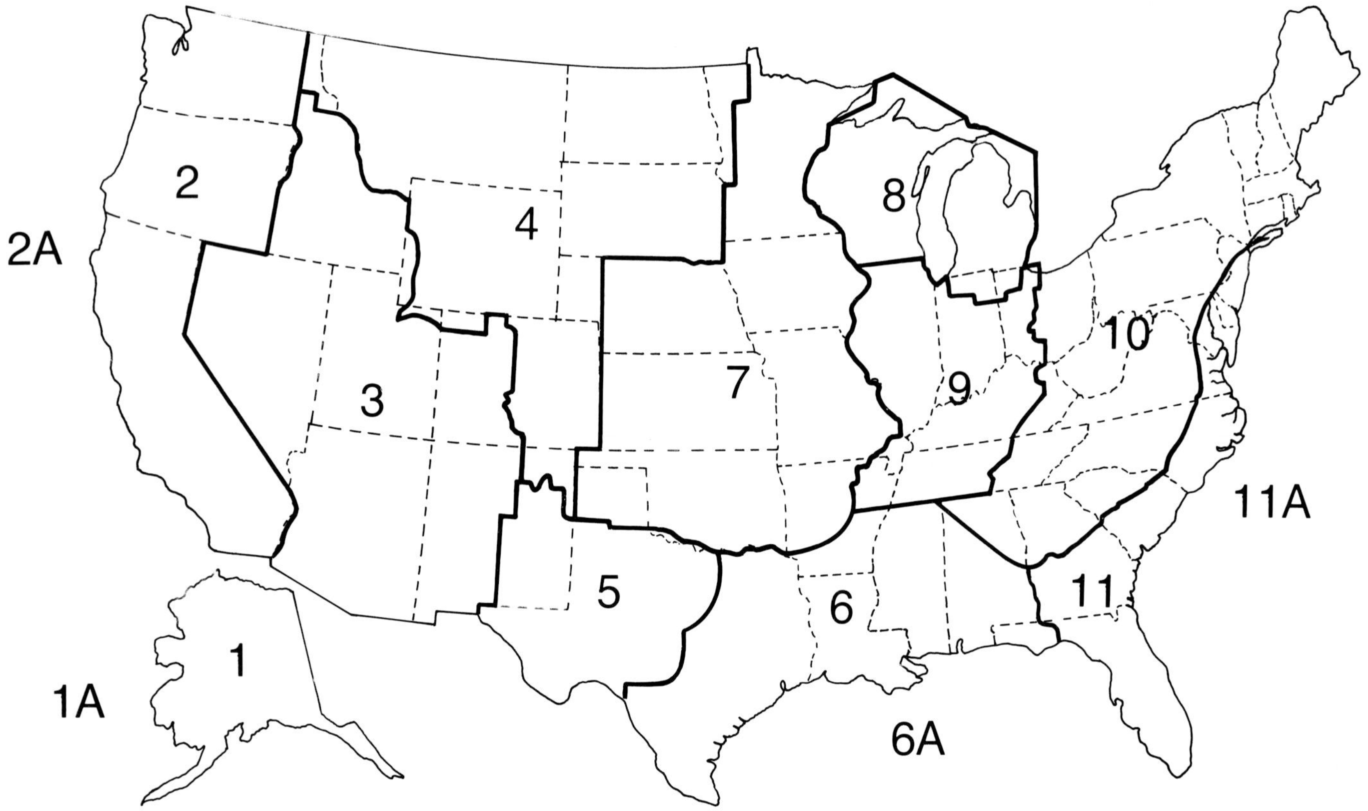

1. Alaska
1A. Alaska Offshore
2. Pacific Coastal States
2A. Pacific Coastal Offshore
3. Western Rocky Mountains
4. Northern Rocky Mountains
5. West Texas and East New Mexico
6. Western Gulf Basin

6A. Gulf of Mexico
7. Mid-Continent
8. Michigan Basin
9. Eastern Interior
10. Appalachians
11. East Gulf and Atlantic Coastal Plain
11A. Atlantic Coastal States Offshore

great expansion in Nigeria, Libya, and Algeria. Now that liquefaction of gas makes possible its shipment by tanker, the production and export of gas from African and Middle Eastern nations is expected to grow rapidly.

The world gas resource, if assessed on the basis of current use rates, will have a longer life than that of crude oil. Assuming the 5,200 trillion undiscovered amount to be valid, the total world resource is near 7,719 trillion cubic feet. At current world annual production of 54 trillion cubic feet per year, this supply would last 143 years. A corresponding assessment of world crude resource suggests a life of only 68 years. Current use rates, of course, are not the most relevant: world use rates since 1973 have *increased* at an average rate of 3 percent per year.

United States Resources of Natural Gas

Natural gas is, by a small margin, the most significant *domestic* fuel in the country. In 1976 it contributed 28 percent of the total raw energy need. Domestic crude oil supplied 26 percent, imported oil 19 percent, and coal 20 percent.

As in the foregoing section on United States crude oil, this review of natural gas in the country is in two parts. First, a broad survey considers both identified and undiscovered reservoirs and makes use of information assigned to fifteen regions of the country. A second part considers only the identified reservoirs, and uses current reserve and production data for states, not regions.

GAS IN IDENTIFIED AND UNDISCOVERED RESERVOIRS

One report from the U.S. Geological Survey provides a unique summary of gas amounts of varying degrees of certainty for both onshore and offshore portions of fifteen regions (U.S.G.S. *Circular 725*). It is used here, despite the fact that its estimates are dated December 31, 1974, because its version of Inferred and Undiscovered amounts by region is the best available. The fifteen regions referred to are defined on Fig. 2–43.

TOTAL OF ALL NATURAL GAS PRODUCED AND REMAINING   When mapped for the fifteen regions (Fig. 2–43 and 2–44) the total of natural gas produced and remaining reveals that the Gulf Coast regions (onshore and offshore), West Texas, the Mid-Continent region, and

Fig. 2–44 Cumulative production of natural gas versus all remaining recoverable as of Dec. 31, 1974.

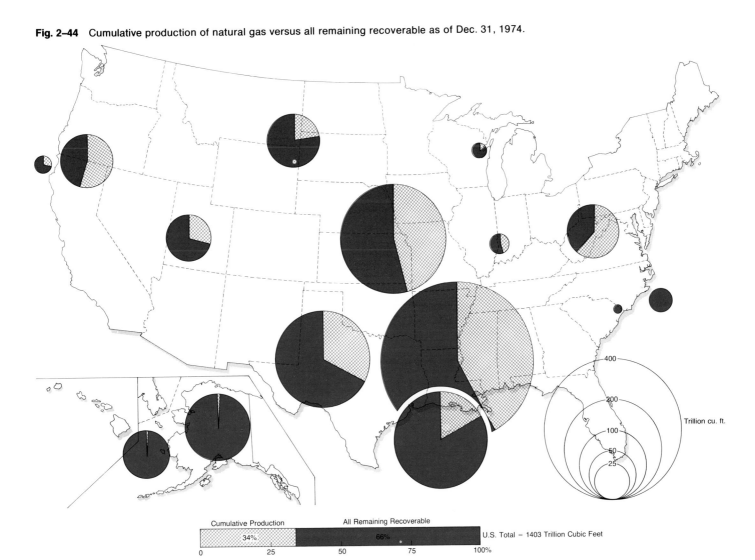

Cumulative Production   All Remaining Recoverable

| 34% | 66% | U.S. Total = 1403 Trillion Cubic Feet |

0   25   50   75   100%

Trillion cu. ft.

Alaska are the major gas regions. In proportions of gas produced versus gas remaining, the frontier areas—Alaska and the Atlantic offshore, along with the Gulf Coast offshore—show the most favorable ratios. Amounts of gas production through 1974 (Fig. 2–45, Table 2–10) are clearly concentrated in the Gulf Coast area.

ALL NATURAL GAS REMAINING Amounts of natural gas thought to remain are divided in Fig. 2–46 according to *degree of certainty*. Demonstrated reserves, Inferred reserves, and Undiscovered Recoverable amounts are as defined earlier in this chapter. One difference is the lack of an Indicated reserves category for natural gas. For oil, that category reflects the possibility of near-economic oil that can be recovered by use of secondary recovery techniques. For natural gas this is not relevant because normal gas reservoirs yield 85 percent of their gas without need for special recovery techniques. As for crude oil, in most regions the undiscovered amounts are estimated to be greater than the more certain amounts.

The actual amounts of Demonstrated and Inferred natural gas (Fig. 2–47, Table 2–10) are made more meaningful by comparison with United States annual production of about 20 trillion cubic feet. The Western Gulf and Gulf of Mexico regions hold most of the Demon-

strated reserves, but the Mid-Continent region assumes a more important role than for crude oil reserves—holding more Demonstrated reserves of gas than the West Texas and Eastern New Mexico region. Inferred amounts are relatively large in the Gulf of Mexico, indicating young fields whose reserves will grow with further drilling.

Undiscovered Recoverable Amounts. These are estimated using the logic described earlier for Undiscovered Recoverable amounts of crude oil. With the exception of three areas on the Atlantic shelf, the amounts are represented in Fig. 2–45 by the statistical mean of estimates whose range is shown on Table 2–11. Expected amounts are greatest in the same regions that lead in Demonstrated and Inferred reserves, namely the Western Gulf Basin, Gulf of Mexico, Mid-Continent, and West Texas-Eastern New Mexico.

In the Atlantic offshore region, the U.S. Geological Survey has made recent estimates of undiscovered gas for the Bureau of Land Management in connection with lease sales (Table 2–12). Amounts shown for the three sale areas are simply the means of the range of estimates obtained from the Bureau of Land Management (Department of Interior, Sale 40). Their total, 12.6 trillion cubic feet, exceeds the mean estimate for the whole Atlantic coastal region made by the Survey in 1975

Table 2–10 Natural Gas in Identified and Undiscovered Reserves as of December 31, 1974, for Fifteen Regions of the United States Including Alaska and Offshore Regions (in trillions of cu. ft.)

REGION	Cumulative Production	As % of Grand Total	Demonstrated Reserves	As % of Total Remaining	Inferred Reserves	As % of Total Remaining	Undiscovered Recoverable[1]	As % of Total Remaining	Total Remaining	GRAND TOTAL
1	.482	1	31.722	40	14.7	19	32	41	78.422	78.904
1a	.423	1	0.145	00	0.1	01	44	99	44.245	44.668
2	25.455	54	4.732	22	4.0	18	13	60	21.732	47.187
2a	1.415	27	0.463	12	0.4	10	3	78	3.863	5.278
3	10.728	29	9.081	35	2.9	11	14	54	25.981	36.709
4	11.485	22	6.754	16	5.3	13	29	71	41.054	52.539
5	58.686	33	24.624	21	23.3	19	70	60	117.924	176.610
6	197.899	42	81.903	30	58.7	21	133	49	273.603	471.502
6a	32.138	17	35.348	23	67.0	44	50	33	152.348	184.486
7	107.700	46	34.150	27	20.6	16	72	57	126.750	234.450
8	0.558	15	1.458	45	0.8	24	1	31	3.258	3.816
9	2.797	46	0.766	23	0.5	16	2	61	3.266	6.063
10	31.057	62	5.985	31	3.3	17	10	52	19.285	50.342
11	0.001	0	0.001	00	Neg	00	1	100	1.001	1.002
11a	0.000	0	0.000	0	0.0	0	10	100	10.000	10.000
TOTAL	480.824	34	237.132	26	201.6	22	484	52	922.732	1,403.556

[1]Statistical mean of estimates.

Source: Derived from information in U.S. Geological Survey *Circular 725*, 1975.

Fig. 2–45 Amounts of cumulative production of natural gas in fifteen regions as of Dec. 31, 1974.

Fig. 2–46 All remaining natural gas thought to be recoverable as of Dec. 31, 1974, divided by degree of certainty.

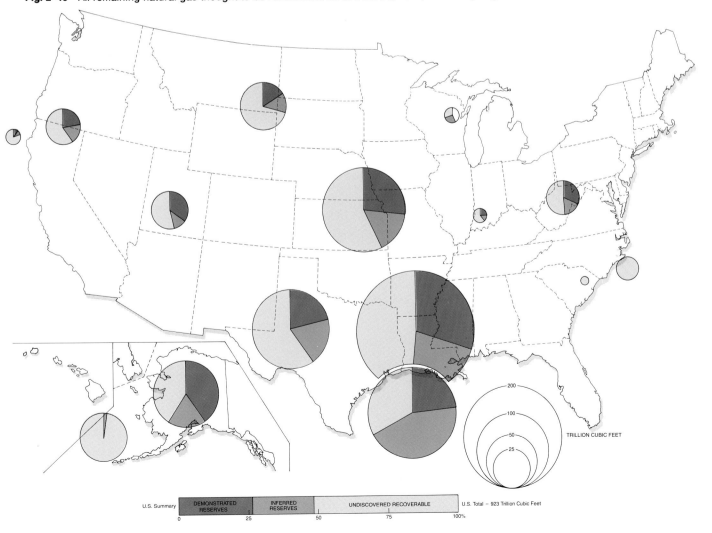

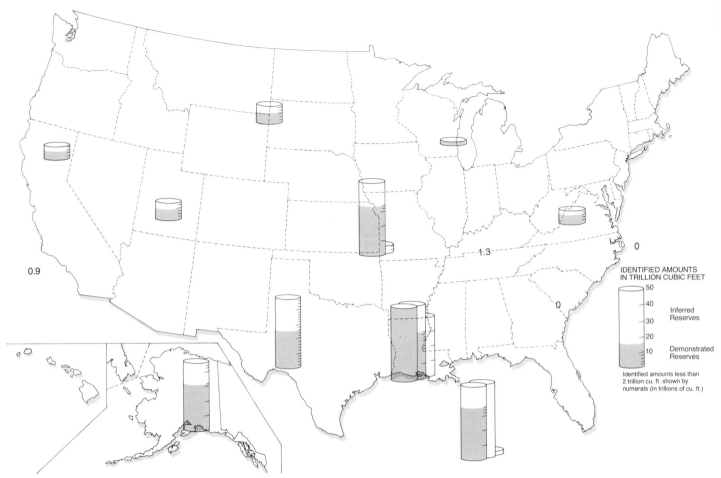

Fig. 2–47 Natural gas in identified reservoirs, showing demonstrated and inferred reserves as of Dec. 31, 1974.

Fig. 2–48 Amounts of undiscovered natural gas as of Dec. 31, 1974.

(U.S.G.S. *Circular 725*). Because of the evident revision, the earlier mean figure for the region as a whole, 10 trillion cubic feet, is not shown on Fig. 2–48 but appears in Table 2–11, while the revision appears in Table 2–12. The total mean amount for the three sale areas, 12.6 trillion cubic feet, if discovered and developed, would be a little over one-half of one year's production for the whole country.[3]

For the country as a whole, the mean estimate of Undiscovered Recoverable gas is 484 trillion cubic feet. While the probability of that amount occurring is 50 percent, amounts at other levels of certainty can be derived from the probability distribution in Fig. 2–49. For instance, there is roughly a 75 percent chance that 375 trillion cubic feet remain to be recovered, but only a 25 percent chance for amounts of 550 trillion cubic feet.

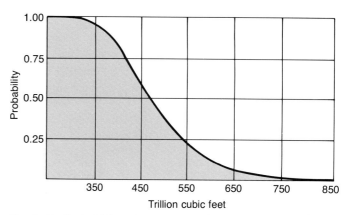

Fig. 2–49 Probabilities of occurrence of various amounts of undiscovered natural gas in the U. S.

Table 2–11 Undiscovered Recoverable Amounts of Natural Gas in Fifteen Regions of the United States as of December 31, 1974 (in trillions of cu. ft.)

REGION	Low Estimate	High Estimate	Statistical Mean
	(95% chance)	(5% chance)	
1	16	57	32
1a	8	85	44
2	8	20	13
2a	2	6	3
3	6	25	14
4	18	47	29
5	35	101	70
6	85	196	133
6a	18	91	50
7	50	101	72
8	0.8	2	1
9	0.7	4	2
10	5	17	10
11	0.4	2	1
11a	5[1]	14[1]	10[1]
TOTAL LOWER 48	286	529	408
TOTAL ALASKA	29	132	76
TOTAL ONSHORE	264	506	377
TOTAL OFFSHORE	42	181	107
TOTAL UNITED STATES	322	655	484

[1]Now revised: see following table.
Source: U.S. Geological Survey *Circular 725*, 1975.

Estimates for the whole country (Fig. 2–50) have become more conservative through the years, though the trend was reversed by the Potential Gas Committee in 1977. Its resource terms are different from those of the Survey; but it is reasonably clear that the Committee's "Probable" is the Survey's "Inferred" and that "Possible" and "Speculative" are different degrees of "Undiscovered." An estimate that supports the U.S.G.S. figures mapped here was presented by John Moody, a Mobil Oil geologist and 1977 President of the American Association of Petroleum Geologists. His undiscovered gas estimate for the country is 500 to 600 trillion cubic feet (*Oil and Gas Journal*, June 20, 1977). The estimates in Fig. 2–50 consider only conventional gas that will flow freely from a reservoir rock. Additional amounts possible from unusual reservoirs are dealt with in a later section headed *Unconventional Natural Gas*.

SUMMARY OF AMOUNTS REMAINING All conventional natural gas known and thought to remain as of December, 1974 is summarized in Fig. 2–51 according to degree of certainty and location within regions. The rel-

Table 2–12 Estimates of Undiscovered Natural Gas at Three Sale Areas on the Atlantic Continental Shelf (in trillions of cu. ft.)

SALE AREA	Low Estimate	High Estimate	Arithmetic Mean
George's Bank	1.0	3.5	2.25
Baltimore Canyon	2.6	9.4	6.0
Southeast Georgia Embayment	1.9	6.8	4.35
TOTAL	5.5	16.2	12.60

Source: U.S. Department of Interior, Sale No. 40, 1976.

ative magnitudes of the divisions resemble those on the corresponding summary for crude oil: on the left side, undiscovered amounts overshadow the more certain; and on the right side, the lower forty-eight states (especially onshore) dominate the resources. In the right-hand bar of the graph a large part of each regional total is very uncertain because it consists of undiscovered amounts (represented here for every region simply by the statistical mean of the high and low estimates) (see Table 2–10).

NATURAL GAS IN IDENTIFIED RESERVOIRS

Because most of the gas in a reservoir is easily recovered, it is not usual to refer to Original Gas In-Place. Unlike the corresponding section on crude oil, therefore, this review of gas resources in identified reservoirs begins with Estimated Ultimate Recovery, that is, the total of all gas produced and known to be producible from identified reservoirs.[4]

NON-ASSOCIATED VERSUS ASSOCIATED-DISSOLVED NATURAL GAS Non-associated and associated-dissolved are terms which recognize that some gas occurs with crude oil, whereas some gas occurs alone, and may be produced without regard to oil production. *Non-associated* is the latter type: it occurs simply as free gas trapped in a reservoir over water. *Associated-dissolved* refers to gas that occurs either (1) as a ''gas cap'' overlying oil (associated) or (2) as gas dissolved in crude oil at the pressures that obtain in the reservoir. Production of gas from a gas cap lowers formation pressure and can affect oil recovery. Production of gas-rich oil will inevitably release dissolved gas at the surface.

The two modes of gas occurrence are not equally prevalent. *Non-associated* accounts for 71 percent of Ultimate Recovery in the country, according to data as of December, 1976 (Fig. 2–52). Non-associated gas is the dominant form in states with smaller Ultimate Recovery amounts, while states with larger amounts often display a mixture of these two forms. Alaska's gas is 80 percent associated-dissolved, whereas the large amounts in Texas and Louisiana are mostly non-associated.

CUMULATIVE PRODUCTION AND PROVED RESERVES The general pattern (Fig. 2–53) is that past production from identified reservoirs exceeds current Proved re-

Fig. 2–50 Various estimates of undiscovered and inferred natural gas in the U. S.

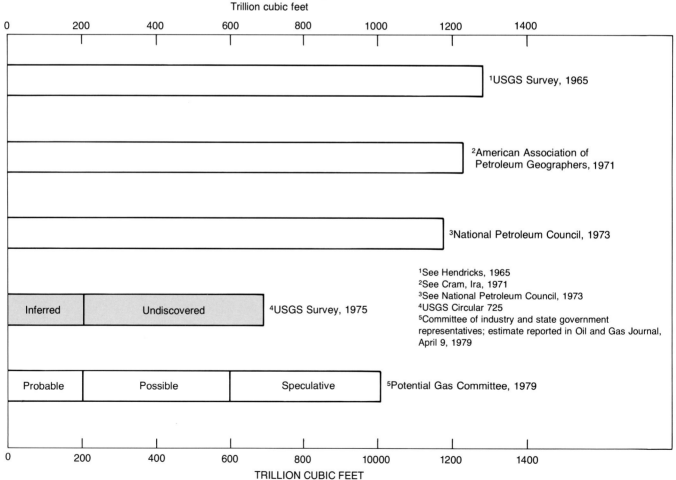

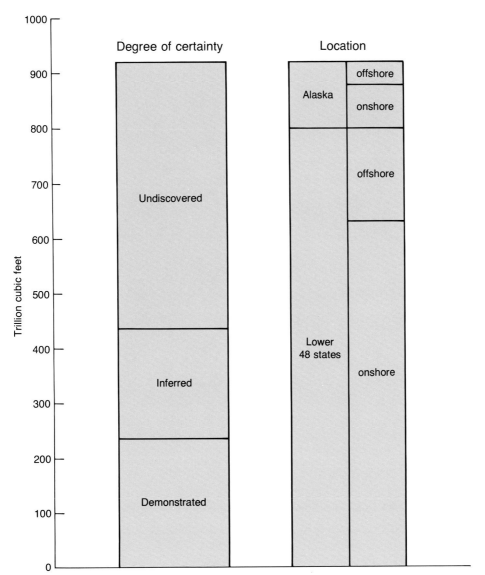

Fig. 2–51   Recoverable natural gas remaining in the U. S. as of Dec. 31, 1974, showing degree of certainty and location.

serves. Exceptions to this are Alaska, Alabama, and Michigan, whose gas discoveries have been recent. On this map, the two modes of gas occurrence, non-associated and associated-dissolved, are combined as *total gas*. Table 2–13 shows that, on a national basis Cumulative Production exceeds reserves to the same extent for both non-associated and associated-dissolved types of gas.

Proved reserves of total gas (Fig. 2–54 and Table 2–14) are largest in Texas and Louisiana, which embrace very thick sedimentary rocks of the Gulf Coast and their offshore extensions into the Gulf of Mexico. Alaska's reserves are also very large.

Proved reserves of associated-dissolved and non-associated gas shown on Table 2–14 are too small to be mapped separately. In keeping with the proportions observed earlier, the reserve amounts of non-associated ex-

ceed the associated-dissolved amounts in most states. In recent years, the annual production of natural gas in the United States has been approximately 20 trillion cubic feet. In non-associated reserves, Louisiana and Texas each hold just over a two-year national supply, whereas in total gas, their amounts represent a three-year supply.

The corresponding map for crude oil reserves (Fig. 2–26) shows these two Gulf states with unequal amounts of crude: Texas with a three-year supply at current national production rates, and Louisiana with a one-year supply. Another variation is seen in California, which in crude oil has reserves equivalent to more than a year's supply for the country, but in total gas has only one-quarter of one year's production. Despite the related origins for oil and natural gas and the similarity of their traps, there are certain geologic "provinces" more favorable for crude oil and others that favor natural gas.

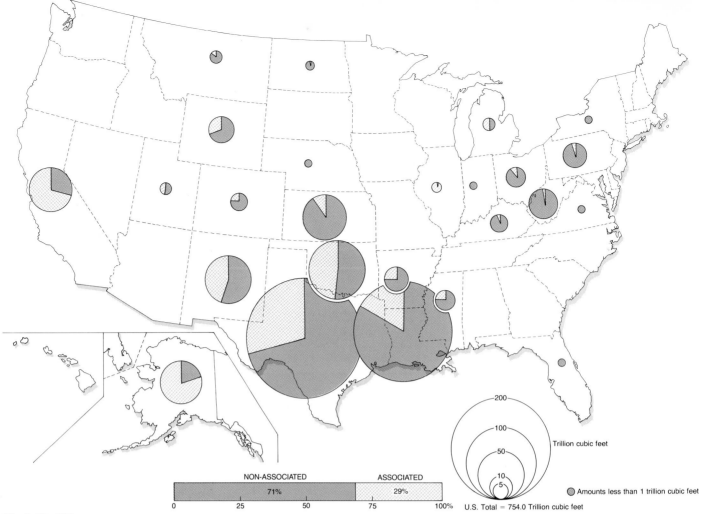

NON-ASSOCIATED 71% | ASSOCIATED 29%

0   25   50   75   100%

Trillion cubic feet

Amounts less than 1 trillion cubic feet

U.S. Total = 754.0 Trillion cubic feet

Fig. 2–52   Ultimate recovery of natural gas in identified reservoirs as of Dec. 31, 1978, showing portions that are associated-dissolved gas versus non-associated gas that can be produced independently of crude oil.

Fig. 2–53   Ultimate recovery of natural gas in identified reservoirs showing cumulative production versus proved reserves as of Dec. 31, 1978.

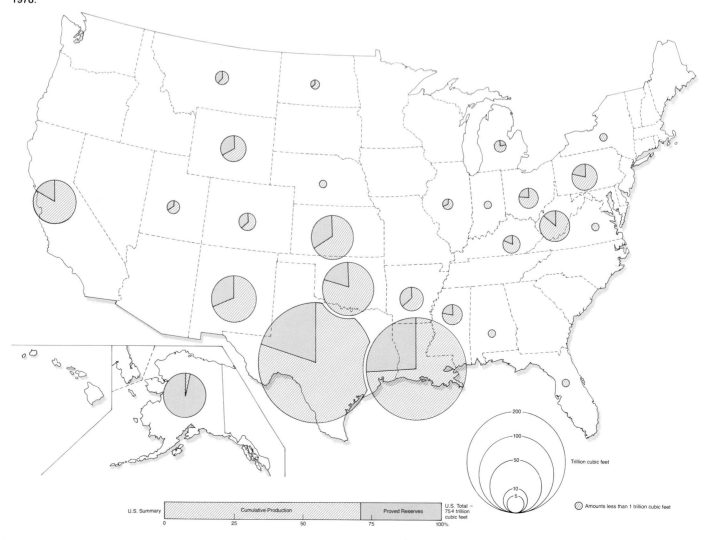

Trillion cubic feet

Amounts less than 1 trillion cubic feet

U.S. Summary | Cumulative Production | Proved Reserves | U.S. Total = 754 trillion cubic feet

0   25   50   75   100%

Table 2-13 Cumulative Production to December 31, 1978 versus Proved Reserves for Natural Gas of *Associated-Dissolved* and *Non-Associated* Types (in millions of cu. ft.)

	Associated-Dissolved	Non-Associated
Proved Reserves	56,661,626	138,991,322
Cumulative Production	161,646,661	376,292,524
Ultimate Recovery	221,021,020	583,003,432
Cumulative Production as Proportion of Ultimate Recovery	0.7314	0.7059

Source: Deduced from information from American Gas Association, June 1979.

NATURAL GAS LIQUIDS (NGL) As defined earlier, Natural Gas Liquids, such as propane, are recovered at the well-head when natural gases rich in these heavier components are produced. The mode of occurrence, non-associated or associated-dissolved, is not important. Reserves of NGL from both modes of gas occurrence, summed up in Table 2–15, occur in a pattern that coincides roughly with gas reserves, since estimates are based on reserves of natural gas and on the amounts of NGL recovered per cubic foot from gases in various reservoirs. While the national average is around 33 barrels per million cubic feet of gas, the ratio is around 50 barrels per cubic foot in the West Texas-Eastern New Mexico region and as low as 8 barrels per cubic foot in the Appalachians (U.S.G.S. *Circular 725*, p. 45). As liquids expressed in barrels, these hydrocarbons are usually added to crude oil resources to represent total petroleum liquids.

Table 2-14 Natural Gas Reserves at End of 1977 and 1978, Showing Total Gas and Non-Associated and Associated-Dissolved Components (in millions of cu. ft.)

STATE	TOTAL GAS RESERVES AS OF 12/31/77	RESERVES AS OF 12/31/78		
		Total Gas	Non-Assoc.	Assoc.-Dissolved
Alabama	745,538	751,219	725,596	25,623
Alaska	31,832,616	31,612,295	5,324,733	26,287,562
Arkansas	1,680,336	1,627,064	1,446,011	155,533
California	5,198,348	5,095,082	1,942,374	2,778,912
Colorado	2,025,987	1,965,765	1,712,594	217,681
Florida	215,323	160,296	0	160,296
Illinois	417,783	420,437	2,447	1,401
Indiana	53,249	56,710	1,068	13
Kansas	11,926,191	12,287,341	12,032,661	142,934
Kentucky	745,046	718,929	541,697	41,502
Louisiana	52,685,970	49,674,148	42,213,444	7,140,089
Michigan	1,791,200	1,768,581	607,556	562,190
Mississippi	1,307,133	1,410,514	1,251,184	68,060
Montana	1,043,982	991,668	768,683	65,779
Nebraska	68,122	72,839	4,907	13,299
New Mexico	11,931,112	13,261,489	10,978,198	2,255,303
New York	247,303	262,711	150,099	114
North Dakota	392,066	411,485	4,075	407,410
Ohio	1,459,430	1,560,478	1,037,333	140,178
Oklahoma	11,712,342	11,463,291	9,994,502	2,210,919
Pennsylvania	1,884,932	2,093,516	1,499,965	11,291
Texas	62,157,836	54,600,235	41,340,511	13,007,175
Utah	748,907	698,655	392,197	304,360
Virginia	69,041	79,064	79,064	0
West Virginia	2,375,855	2,683,136	2,253,645	47,626
Wyoming	3,962,850	4,315,775	3,649,637	613,204
Miscellaneous	199,380	258,984	37,141	3,172
TOTAL UNITED STATES	208,877,878	200,301,707	138,991,322	56,661,626

Source: American Gas Association, *Vol. 32*, 1978, and *Vol. 33*, 1979.

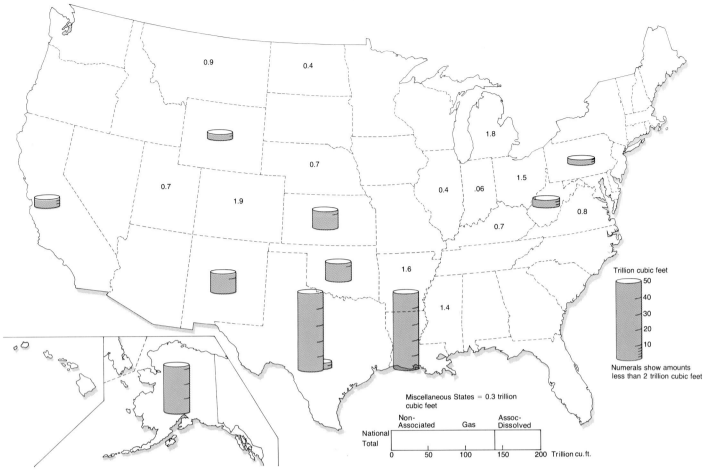

Fig. 2–54 Proved reserves of natural gas as of Dec. 31, 1978.

Table 2–15 Proved Reserves of Natural Gas Liquids, December 31, 1978 (in millions of cu. ft.)

STATE	TOTAL NATURAL GAS LIQUIDS	SOURCE GAS Non-Associated	SOURCE GAS Associated-Dissolved	STATE	TOTAL NATURAL GAS LIQUIDS	SOURCE GAS Non-Associated	SOURCE GAS Associated-Dissolved
Alabama	204,065	202,042	2,023	Nebraska	457	199	258
Alaska	407,791	0	407,791	New Mexico	451,672	326,450	125,222
Arkansas	5,361	738	4,623	North Dakota	54,126	0	54,126
California	102,321	1,537	100,784	Ohio	0	0	0
Colorado	61,199	42,849	18,350	Oklahoma	268,670	173,410	95,260
Florida	21,490	0	21,490	Pennsylvania	323	323	0
Illinois	0	0	0	Texas	2,268,284	1,137,639	1,130,645
Indiana	0	0	0	Utah	31,714	617	31,097
Kansas	463,761	455,051	8,710	West Virginia	96,279	96,279	0
Kentucky	39,253	39,253	0	Wyoming	51,842	28,920	22,922
Louisiana	1,315,545	1,186,555	128,990				
Michigan	65,693	20,857	44,836	TOTAL UNITED STATES	5,925,852	3,726,022	2,199,830
Mississippi	12,707	10,757	1,950				
Montana	3,299	2,546	753				

Source: American Gas Association, *Vol. 33*, June, 1979.

THE HISTORY OF NATURAL GAS RESERVES Figure 2–55, like the corresponding figure for crude oil, shows how annual reserves change in response to annual production and annual additions to reserves through discoveries, extensions, and revisions. Reserves grew every year during the 1940s, 50s and early 60s as additions exceeded production. With greatly increased demand and production in the late 1960s and a sudden decline in discoveries, reserves fell abruptly, with only a minor respite due to Alaskan discoveries in 1970. Total reserves, being about 200 trillion cubic feet at the end of 1978, represent only ten year's production at recent rates, and on that basis have never been lower in the period illustrated. For the world as a whole, however, the life of gas reserves at the current annual production rate is nearly forty years.

SUMMARY OF CONVENTIONAL NATURAL GAS RESOURCES

Figure 2–56 shows the amounts of gas produced from all United States reservoirs (Cumulative Production to end of 1978) to be 60 percent of all amounts thought to remain in identified and undiscovered reservoirs. Dem-

Fig. 2–55 History of natural gas reserves, 1947 to 1978, showing influence of annual production and additions to reserves.

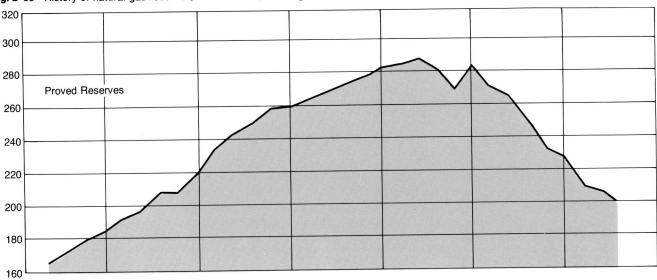
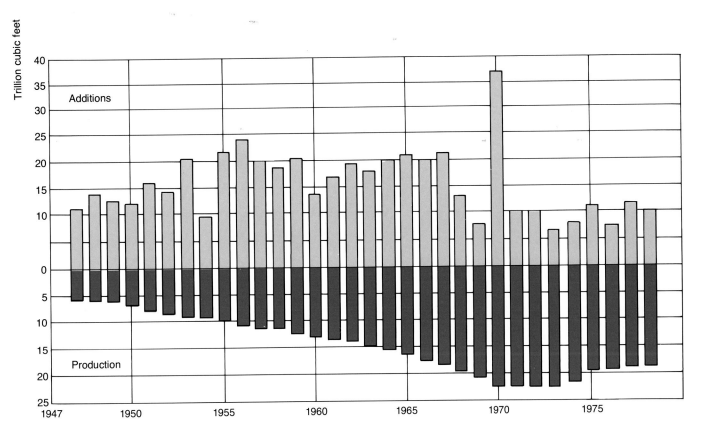

onstrated reserves as of end of 1978 are identical to Proved reserves, there being no Indicated component for natural gas; and reserves were augmented by 4 trillion cubic feet of gas in underground storage. Inferred and Undiscovered Recoverable amounts are those explained earlier in this chapter. Together, the Demonstrated, In-

ferred, and Undiscovered amounts approximate all that remains economically recoverable.

About 80 percent of conventional natural gas in place is recovered normally, and it is unlikely that additional amounts will be recovered through special recovery techniques. Nonetheless the U.S.G.S. estimates addi-

Fig. 2–56 Summary of conventional natural gas resources in the U. S., in trillions of cubic feet, showing cumulative production through 1978 versus all remaining recoverable gas.

	Identified		Undiscovered
	Demonstrated	Inferred	
economic	200 [1]	202	322-655 (mean 484)
Sub-economic	90-115 (mean 102)		40-82 (mean 61)

[1] In trillions of cubic feet as of end 1978 according to
American Gas Association, June, 1979.
All other figures as of end 1974, according to USGS Circular 725.

Trillion cubic feet

100   200   300   400   500   600   700   800   900   1000

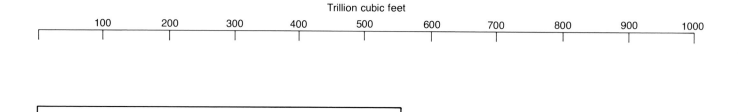

Cumulative production to December 31, 1978[1]

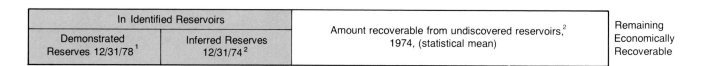

In Identified Reservoirs		Amount recoverable from undiscovered reservoirs,[2] 1974, (statistical mean)	Remaining Economically Recoverable
Demonstrated Reserves 12/31/78[1]	Inferred Reserves 12/31/74[2]		

Undiscovered

Identified		Non-economic, possible recoverable[3]

[1] Current reserves and recent production from AGA, 1977, 1978, 1979. Cumulative production from USGS Circular 725.
[2] USGS Circular 725, Recovery rate of 80% is assumed.
[3] USGS Circular 725. Recovery rate of 90% is assumed.

tional amounts that would be recovered if recovery rates of 90 percent were realized from all identified and undiscovered reservoirs. The additional 163 trillion cubic feet of gas would be almost as large as current Demonstrated Reserves.

Figure 2–56 places the natural gas resource amounts in the context of their geologic certainty and the economic feasibility of their recovery. Figure 2–57 shows that past production and remaining amounts of Natural Gas Liquids are in the same proportions as for natural gas.

UNCONVENTIONAL NATURAL GAS

Just as oil shales and oil sands offer potential for extraction of "unconventional" crude that will not flow into a borehole (see following chapter), there are possibilities for additional natural gas beyond the amounts considered Proved, Inferred, or Undiscovered in reservoirs of the usual kind. (Possible amounts are compared with estimates of conventional gas in Fig. 2–58.)

The occurrences of unconventional natural gas are quite varied. Two occurrences entail natural gas of normal origin which does not flow easily into a drill hole because reservoir rocks are not sufficiently permeable. Another is methane associated with coal beds. A fourth is gas of normal origin that is involved in a geothermal occurrence.

TIGHT GAS SANDS In the Rocky Mountain areas of Wyoming, Utah, Colorado, and New Mexico (Fig. 2–59) these sands occur in sequences of clay, chalk, and sandstone interbedded with shale. Commercial flows of gas cannot be obtained unless the sandstone is fractured to open communication routes to the borehole. Experimental *nuclear* fracturing was tried in the Gasbuggy,

Rulison, and Rio Blanco projects in 1967, 1969, and 1973, with disappointing results. Massive hydraulic fracturing of the sort often used to stimulate conventional oil or gas flows is being attempted in projects initiated in 1979 by the Department of Energy. These are a seven-well project in the Uintah basin of Utah, and a one-well project in the Piceance Creek area of Colorado (DOE, March, 1978, p. 404–405). Gas in-place in such reservoirs in the Rocky Mountain region may amount to 600 trillion cubic feet, while amounts recoverable may be 300 trillion cubic feet (Fig. 2–58). A recent study, however, suggests gas in-place is only 400 trillion cubic feet, while recoverable amounts are 100 to 180 trillion cubic feet at gas prices of 3 dollars per thousand cubic feet (Kuuskraa, V. A. *et al.*).

DEVONIAN SHALES OF THE APPALACHIANS These shales contain natural gas in an occurrence similar to that in the Rockies, that is, in a rock which yields adequate flows only when fractured. Gas has already been produced from these formations with the aid of natural fractures, as in the Big Sandy Field of eastern Kentucky. Other areas of potential extraction are in the states of New York, Pennsylvania, Ohio, West Virginia, Kentucky, Michigan, Indiana, Illinois, and Tennessee. The federal Department of Energy in 1979 began a demonstration project for hydraulic fracturing of Devonian age shales and associated sandstones in each of these states (DOE, March, 1978, p. 398–403). Gas in-place in such occurrences through the Appalachians may amount to 490 trillion cubic feet, with 8 to 16 trillion cubic feet recoverable (Kuuskraa, V. A. *et al.*).

METHANE FROM COAL BEDS Methane occurs in coal deposits. While methane is known as the dominant constituent of conventional natural gas, it is derived in this case from peat materials as they matured and altered

Fig. 2–57 Natural gas liquids: cumulative production and recoverable amounts remaining, Dec. 31, 1978.

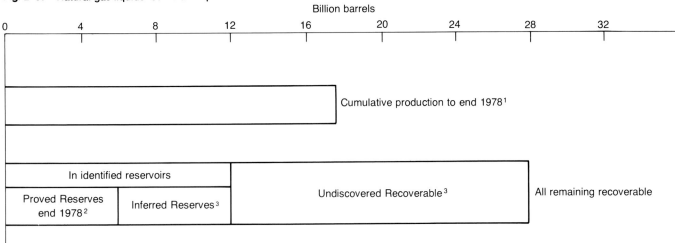

[1]USGS Circular 725 for recent years production
[2]AGA, June 1979
[3]As of end of 1974, according to USGS Circular 725. Based on inferred and undiscovered amounts of natural gas, and assuming national average of 33 barrels of NGL for every million cubic feet of natural gas.

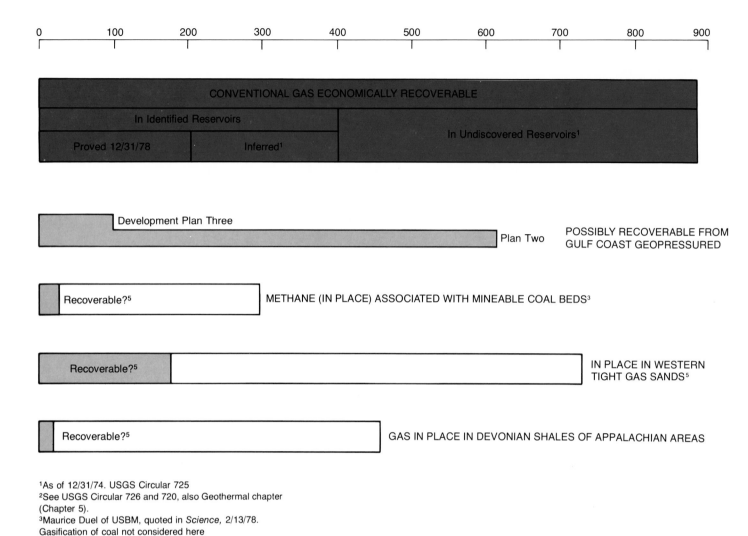

[1]As of 12/31/74. USGS Circular 725
[2]See USGS Circular 726 and 720, also Geothermal chapter
(Chapter 5).
[3]Maurice Duel of USBM, quoted in *Science,* 2/13/78.
Gasification of coal not considered here
[4]Department of Energy, March, 1978, pp. 400-404
[5]Kruskraa, V.A., Et. al., 1978.

Fig. 2–58   Unconventional natural gas presently uneconomic or undeveloped compared with conventional gas remaining.

chemically to become coal. Because methane is a highly noxious gas, it must be vented from mines at some expense, nearly 0.1 trillion cubic feet being wasted each year. In untouched Identified coal deposits, some of which are too deep for mining, there is perhaps 300 trillion cubic feet *in-place* (Maurice Deul, quoted by *Science,* Feb. 13, 1976). Recoverable amounts are quite uncertain, but may be as high as 20 to 25 trillion cubic feet (Kuuskraa, V. A. *et al.*).

GAS IN GEOPRESSURED ZONES OF THE GULF COAST The nature of geopressured zones is elaborated in the chapter on geothermal resources; but here it can be noted that natural gas dissolved in high temperature brines under great pressure at depths from 6,000 to 22,500 feet could be available at the surface if the heat and the hydrostatic pressure of those brines were ever exploited. The amounts of gas liberated would depend upon the specific plan for development of the geothermal re-

sources. Two plans described by the U.S.G.S. (*Circular 726,* p. 136–138) would be consistent with recovery of 97 or 617 trillion cubic feet.

SUMMARY The amounts of supplemental or unconventional gas, summarized in Fig. 2–58, are quite uncertain. In addition, it should be realized that some of the amounts illustrated are *gas in-place.* Because the tight gas sands and the Devonian shales are rock of extremely low permeability, the difference between gas in-place and gas recoverable is very great.

The unconventional natural gas discussed above occurs naturally in the form of gas. There are also a number of processes by which gas can be manufactured. Among these are gasification of coal, and derivation of *biogas* from various organic sources such as municipal and agricultural wastes. Any such gas produced from some other material or fuel is not considered here in a review of natural gas resources.

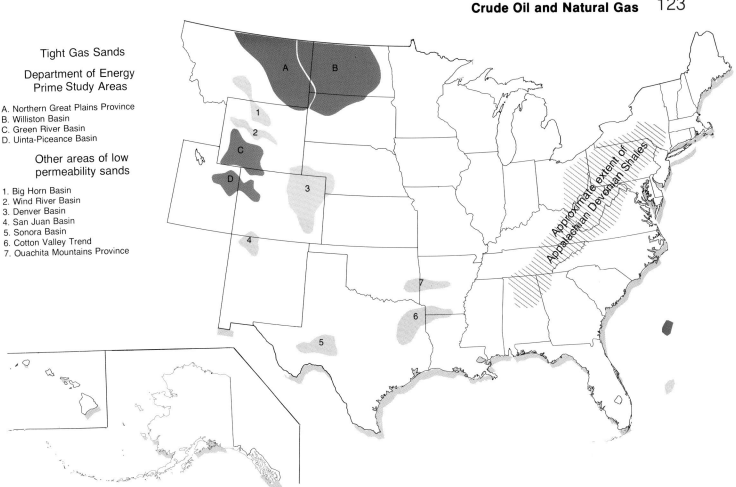

Tight Gas Sands

Department of Energy
Prime Study Areas

A. Northern Great Plains Province
B. Williston Basin
C. Green River Basin
D. Uinta-Piceance Basin

Other areas of low
permeability sands

1. Big Horn Basin
2. Wind River Basin
3. Denver Basin
4. San Juan Basin
5. Sonora Basin
6. Cotton Valley Trend
7. Ouachita Mountains Province

Fig. 2–59 Areas of potential for unconventional natural gas from western tight gas sands and Appalachian shales.

Natural Gas Applications

PRODUCTION OF ALL GAS DURING THE YEAR 1976

The production of all gas during 1976 is shown on Fig. 2–60 and Table 2–16 which distinguishes proportions of non-associated or associated-dissolved gas. For the country as a whole, the production is 81 percent non-associated, although this type of gas accounts for only 69 percent of Proved reserves. Production, as expected on the basis of reserves, is concentrated very heavily in the Gulf states, Texas and Louisiana, and the neighboring states of Oklahoma and New Mexico.

CONSUMPTION OF NATURAL GAS

Nationwide, gas consumption is roughly equivalent to production, with imports constituting less than 5 percent of demand (see following section). Because consumption amounts are small and widely distributed throughout forty-nine states Fig. 2–61 classifies states according to their shares of the national consumption in 1976. It is striking that the states producing the most gas also consume a very large proportion of it. Texas, for instance, consumes 21 percent of the national total, and Louisiana 11 percent, whereas some states in the lowest of the four categories consume less than half of 1 percent. Higher

prices allowed until recently by the Federal Power Commission for gas sold *within* a producing state have discouraged producers from contracting for out-of-state deliveries.

GAS MOVEMENTS

Much of the gas produced in states such as Texas and Louisiana is sold under contract to pipeline or transmission companies who supply industries or utilities in various parts of the nation. One aspect of these gas movements can be seen in records of gas piped into a state, gas piped out, and the net results of the two. Figure 2–62 notes, state by state, the amounts that are either *net receipts* or *net deliveries*. If the outflow exceeds the inflow, then the state is one of net deliveries. This is typical of the leading producing states. In the states which consume more than they produce—which is by far the majority of states—the net receipts are mapped.

UNDERGROUND STORAGE Twenty-six states have storage capacity mostly in the reservoir rock of exhausted gas fields (see Fig. 2–63 and Table 2–17). Gas delivered during the season of low demand is pumped into reservoirs by the pipeline or gas companies, and withdrawn during periods of high demand. This storage, as demonstrated during the severe winter of 1976/77, is a

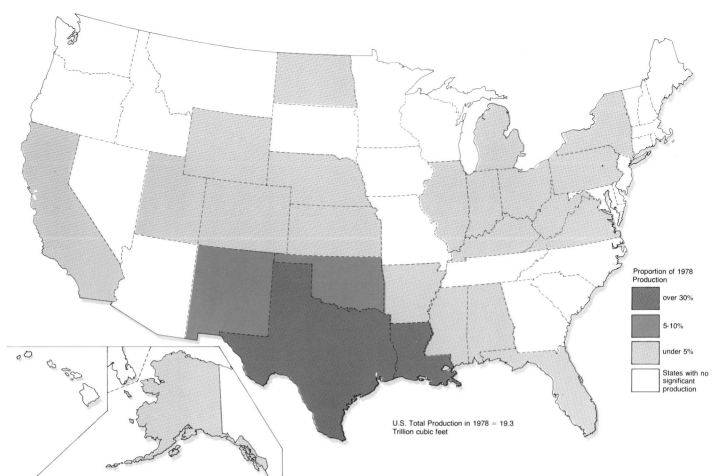

Proportion of 1978
Production

over 30%

5-10%

under 5%

States with no
significant
production

U.S. Total Production in 1978 = 19.3
Trillion cubic feet

Fig. 2–60   State shares of natural gas production during 1976.

Fig. 2–61   State shares of natural gas consumption during 1976.

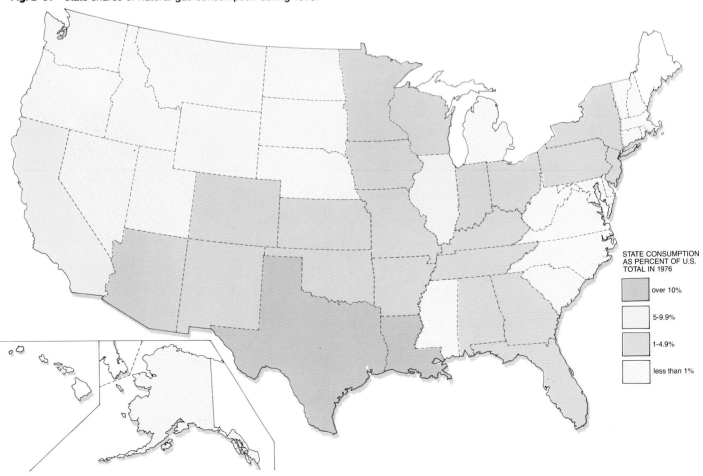

STATE CONSUMPTION
AS PERCENT OF U.S.
TOTAL IN 1976

over 10%

5-9.9%

1-4.9%

less than 1%

Table 2-16   Natural Gas Production during 1977 and 1978 (in millions of cu. feet)

STATE	PRODUCTION OF TOTAL GAS DURING 1977	PRODUCTION DURING 1978		
		Non-Associated	Associated-Dissolved	Total Gas
Alabama	37,581	48,874	7,052	55,926
Alaska	187,375	166,033	47,977	214,010
Arkansas	107,083	95,780	9,967	105,747
California	331,850	144,311	160,679	304,990
Colorado	186,795	133,325	52,287	185,612
Florida	43,571	0	50,987	50,987
Illinois	1,559	183	606	789
Indiana	200	171	2	173
Kansas	771,143	815,889	28,038	843,927
Kentucky	56,710	56,819	517	57,336
Louisiana	7,029,367	6,346,023	715,538	7,061,561
Michigan	130,951	95,722	63,502	159,224
Mississippi	89,439	119,144	12,473	131,617
Montana	53,831	47,845	9,542	57,387
Nebraska	3,241	671	2,233	2,904
New Mexico	1,141,798	806,771	262,162	1,068,933
New York	10,000	13,305	42	13,347
North Dakota	25,493	387	29,440	29,287
Ohio	99,656	103,715	11,524	115,239
Oklahoma	1,662,576	1,227,898	419,002	1,646,900
Pennsylvania	92,293	97,623	140	97,763
Texas	6,827,303	4,966,537	1,562,180	6,528,717
Utah	62,880	23,228	36,590	59,818
Virginia	8,202	8,377	0	8,377
West Virginia	147,337	145,434	1,107	146,541
Wyoming	338,062	269,685	92,794	362,479
Miscellaneous	754	638	279	917
TOTAL UNITED STATES	19,447,050	15,734,388	3,576,660	19,311,048

Source: American Gas Association, *Vol. 32,* 1978, and *Vol. 33,* 1979.

critical factor in a winter's gas supply. The capacity of pipelines and the productive capacity of gas fields are often not adequate to meet sudden large demand. In the winter of 1976/77, it became apparent that larger storage capacity was needed to avert shortages. The total capacity at present is 6.579 trillion cubic feet (D.O.E., Feb. 7, 1978) while gas in storage at the end of 1976 was 4.053 trillion cubic feet and at the end of 1978 was 4.6 trillion cubic feet (Table 2–17), an amount added to Proved reserves if a comprehensive total of recoverable gas is desired.

GAS PIPELINES AND MOVEMENTS[5]   The center of gas production and transport has not always been the southwestern states. The earliest development and transportation of natural gas was in Pennsylvania, Ohio, and West Virginia, and made use, incidentally, of gas lines that existed for coal gas used in lighting. Additional small gas production was developed later in Indiana, New York, Illinois, and Kansas. Gas discoveries that dwarfed any previous finds were made in Louisiana, Texas, Kansas, and Oklahoma during the first decade of this century— just as production capacity in the Northeast was beginning to dwindle. The need for a connection between the large fields of the Southwest and the markets in northern and northeastern states became apparent, but the southwestern and northeastern networks were not linked until 1930. By the year 1950, gas from the Southwest was being piped to thirty-three states, Canada, and Mexico.

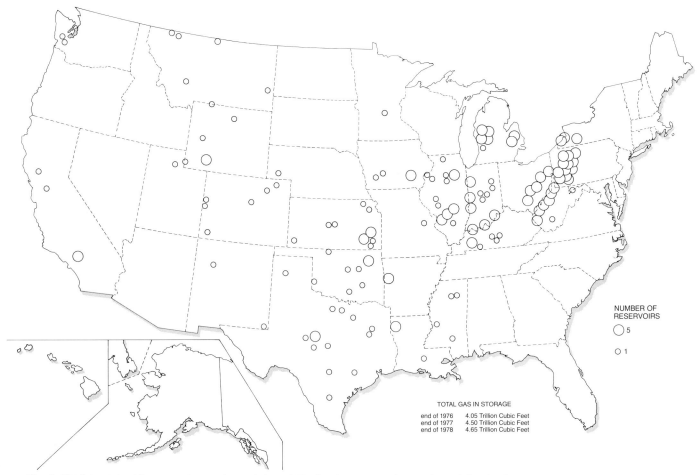

Fig. 2–62 Interstate shipments of natural gas during 1976, showing net receipts and net deliveries by state.

Fig. 2–63 Natural gas storage reservoirs.

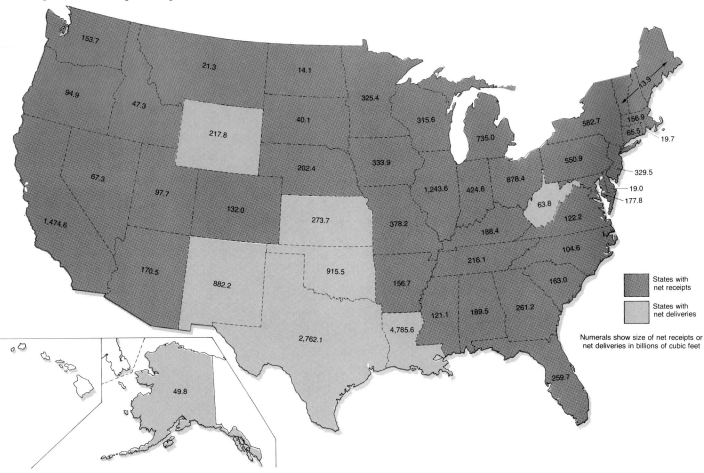

NUMBER OF RESERVOIRS

◯ 5

◯ 1

TOTAL GAS IN STORAGE

end of 1976	4.05 Trillion Cubic Feet
end of 1977	4.50 Trillion Cubic Feet
end of 1978	4.65 Trillion Cubic Feet

States with net receipts

States with net deliveries

Numerals show size of net receipts or net deliveries in billions of cubic feet

Table 2-17 Natural Gas in Underground Storage at End of 1978 (in millions of cu. ft.)

STATE	
Alabama	0
Alaska	0
Arkansas	25,520
California	373,796
Colorado	35,490
Florida	0
Illinois	416,589
Indiana	55,629
Kansas	111,746
Kentucky	135,730
Louisiana	320,615
Michigan	598,835
Mississippi	91,270
Montana	157,206
Nebraska	54,633
New Mexico	27,988
New York	112,498
North Dakota	0
Ohio	382,967
Oklahoma	257,870
Pennsylvania	582,260
Texas	252,549
Utah	2,098
Virginia	0
West Virginia	381,865
Wyoming	52,934
Miscellaneous	218,671
TOTAL UNITED STATES	**4,648,759**

Source: American Gas Association, *Vol. 33,* 1979.

A number of these specially designed natural gas tankers are already in service transporting natural gas in liquefied form at minus 260 degrees Fahrenheit. *Courtesy Brooklyn Union Gas Company.*

Figure 2–64 maps gas fields, pipelines, and the magnitudes of flow accomplished by pipeline. The largest flows originate in the Texas-Louisiana area and travel northeastward to the Mid-Atlantic region. A small component continues to New England. The next largest flow begins in the Oklahoma and Texas panhandle region and serves Chicago and the Great Lakes regions. Smaller flows move westward to California by two different routes and eastward to Florida. The southernmost of the flows to California is now obsolete. In fact that pipeline has been abandoned and has been considered as one leg of a crude oil route from the West Coast to Texas (see later section on Alaska's crude oil).

IMPORTS OF GAS Imports have never formed a large proportion of the gas used in the United States. In 1977, imported crude oil and refined products accounted for 46 percent of demand in that year, while gas imports were roughly 5 percent of total demand. As in the past decade, virtually all imports are Canadian, arriving via pipelines through Idaho, Montana, and Minnesota (Fig. 2–64). Gas from Mexico was first imported in 1977, and contributed 0.012 trillion cubic feet, while Liquefied Natural Gas entering Massachusetts from overseas jumped from 0.005 in 1976 to 0.015 trillion cubic feet in 1977 (*Oil and Gas Journal*, Jan. 31, 1978, p. 126).

Figure 2–65 offers a comparison between domestic production and imports during the past three years. In addition, it provides estimates of future production and imports. The American Gas Association sees domestic production, bolstered by Alaskan gas, continuing near the 20 to 21 trillion cubic foot level (Lawrence, *Oil and Gas Journal*, Aug., 1977, p. 71). A projection by Exxon, however, estimates that domestic production will fall to about 14 trillion cubic feet per year by 1980 (*USA's Energy Outlook 1978–1990*, p. 15). The two sources agree that annual imports may expand to 3 or 4 trillion cubic feet by 1990. According to Lawrence's version in Fig. 2–65 Liquefied Natural Gas (LNG) from overseas will be the largest part of imports.

Liquefied Natural Gas from overseas is a controversial element in the natural gas future. Importation implies a further dependence on foreign supplies, and raises the question of safety at the terminals where tankers deliver the product. Small amounts (16 billion cubic feet per year) of LNG from Algeria have regularly entered at Everett, Massachusetts, near Boston, since 1971. Recently, approval was granted by the Federal Power Commission for two additional terminals, owned by the El Paso Company, to be established at Cove Point, Maryland, and Elba Island, near Savannah, Georgia (Fig. 2–66). The first shipments arrived from Algeria at Cove Point in March of 1978, and at Elba Island in mid-year. Whereas the total LNG imports were only 11.9 billion cubic feet in 1977, they totalled 84.4 billion cubic feet in the year 1978—all arriving from Algeria at Everett, Cove Point, and Elba Island (DOE, June 8, 1979).

If LNG imports are expanded, tankers will dock at terminals proposed in California, Texas, and Louisiana (Fig. 2–66). Approval for these terminals depends upon price rulings to be made by the new Economic Regulatory Administration of the Federal Department of Energy. If companies purchasing the costly imported gas are allowed to sell it at a price that is an average of expensive imported and cheaper domestic gases, then it will find a market. If, as some public interest groups suggest, the imported gas should be sold at the price that reflects its own cost, then it is likely not to be competitive, and the proposed expansion will not materialize. These groups fear that customers will install gas-burning equipment on the strength of the artificially low initial prices, and then will be faced by drastic price hikes if the company loses access to cheap domestic gas.

Transfer of Alaska's Crude Oil and Natural Gas

CRUDE OIL SHIPMENTS AND DISPOSITION

Oil from the large Prudhoe Bay fields on the North Slope began to flow in the summer of 1977 through the Trans-Alaska Pipeline. When operating at its initial capacity the line delivers 1.2 million barrels of oil per day (bpd), that is, 0.438 billion barrels per year, to Valdez, Alaska, where it is loaded onto tankers of participating companies. From the point of view of the West Coast, and particularly of the Petroleum Administration for Defense District V, most of this crude oil is surplus to demand in the region (Fig. 2–68).

Fig. 2–64 Natural gas pipelines and movements, 1974.

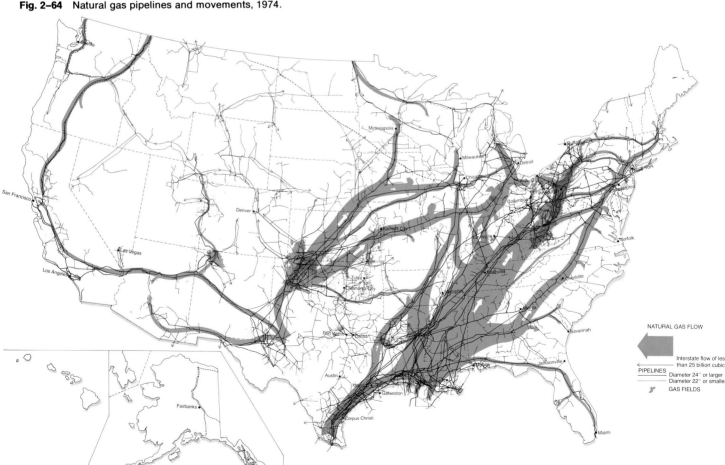

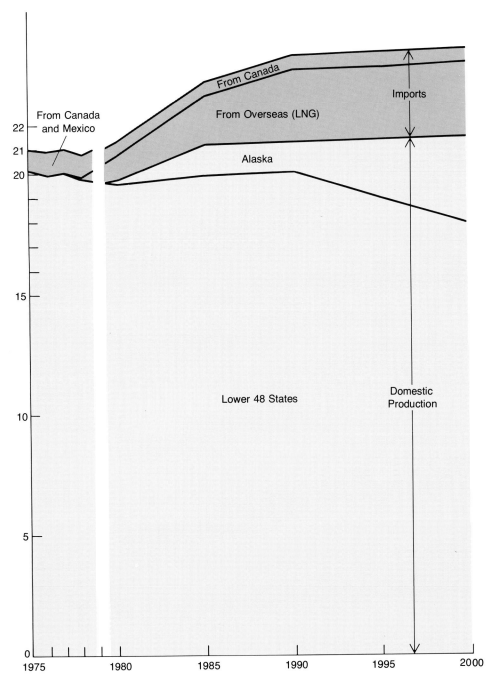

Fig. 2–65 Domestic production potential and anticipated imports of natural gas.

Major producers are Arco, Exxon, and Sohio. Arco's share of the crude oil from Alaska is refined at its Cherry Point location near Seattle, and at its Watson refinery near Los Angeles, each of the two absorbing about 125 thousand barrels per day (bpd). Of Exxon's share, about 100 thousand bpd enters the Benecia refinery near San Francisco, and the remaining 150 thousand bpd are shipped through the Panama Canal to refineries on the Gulf Coast (*Oil and Gas Journal*, July 18, 1977). Sohio's 53.7 percent share of the 1.2 million bpd flow to Valdez amounts to 644 thousand bpd, but on some days can amount to as much as 700 thousand bpd (Department of

Interior, *Environmental Impact Statement,* Exec. Summary, p. I–1). Sohio proposed that most of this be transferred by pipeline from the West Coast to Midland, Texas, a major distributing center, from which it could be piped to areas of the country deficient in crude oil. Alternatives considered include two shown in Fig. 2–67: a pipeline from the British Columbia port of Kitimat through Edmonton to Clearbrook, Minnesota,[6] and a line from Puget Sound to that same distribution center in Minnesota.

The Sohio project was approved by the Secretary of the Interior late in 1977 (*Department of Interior Press*

Fig. 2–66 Existing and proposed terminals for liquified natural gas (LNG) imports.

Fig. 2–67 Disposition of crude oil from Alaska.

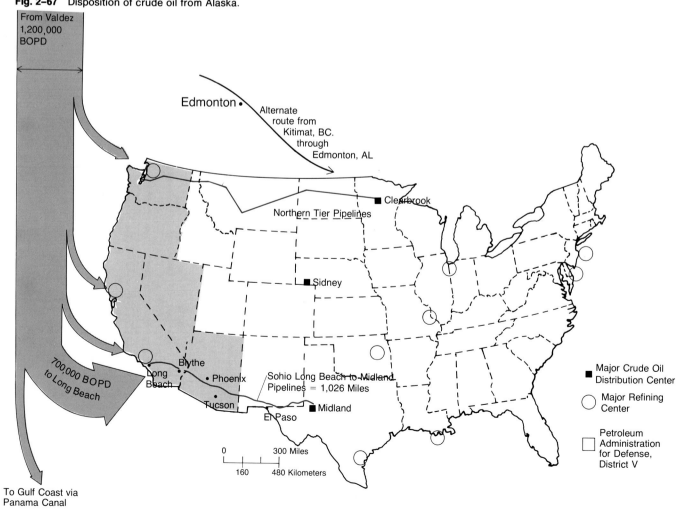

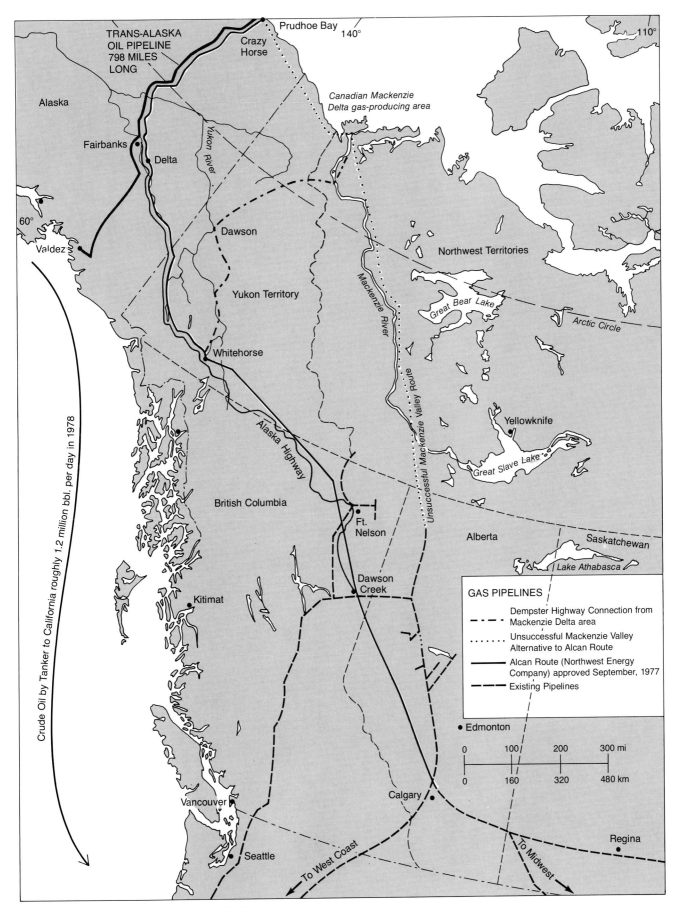

Fig. 2–68 Routes for Alaskan gas and crude oil.

Release, Dec. 2, 1977). According to the plan, surplus crude was to be piped from Long Beach to Midland in a line whose initial capacity would be 500,000 bpd. The project would have required 237 miles of new pipeline construction, and utilized 789 miles of abandoned natural gas pipelines (see Fig. 2–63). The plan for this pipeline was scrapped in the summer of 1979 because the anticipated surplus became less certain. The West Coast demand for Alaskan crude has grown since the pipeline was conceived in 1974. Sohio believes that West Coast refiners may increase their capacity and be able to accommodate even more than at present. In addition, the rate of production from Alaskan North Slope fields is expected to decline during the 1980s. For these reasons, Sohio has decided against constructing its own pipeline, and may, instead, participate in one of the alternative lines being planned (*Oil and Gas Journal,* June 4, 1979, p. 56).

One alternative line in advanced stages of planning is the Northern Tier Pipeline (Fig. 2–67) which would serve the northern tier and upper midwest states where a shortage exists because of reduced Canadian exports (*Oil and Gas Journal,* Nov. 20, 1978).

NATURAL GAS FROM ALASKA

Demonstrated gas reserves in northern Alaska and also in the Mackenzie Delta area of the Northwest Territories (Fig. 2–68) have led the United States and Canada to consider a pipeline route that would serve both areas. Since the late 1960s a number of competing schemes have been proposed, including (1) a line to Valdez where gas would be liquefied for tanker shipment to the West Coast of the United States, and (2) two overland lines whose composite route from Prudhoe Bay to the Mackenzie Delta and south up the Mackenzie Valley is shown in Fig. 2–65 as an unsuccessful alternative.

The Alcan route proposed by the Northwest Energy Company was approved in September, 1977. It will follow the Trans-Alaska oil pipeline as far as Delta, near Fairbanks, where it will leave the pipeline and parallel the Alaska Highway to Dawson Creek, British Columbia. From there it will follow the foothills of the Rockies to Calgary, Alberta, where connections will be made to existing pipelines that now carry Canada's exports to the West Coast and the Midwest. The new route, originally to be completed by mid-1983, will have a capacity of 2.4 billion cubic feet per day, that is, 0.876 trillion cubic feet per year—roughly equal to the annual amount of gas imported from Canada in recent years. Start of construction has been delayed until November, 1984, partly because Dome Petroleum and Esso are interested in reviving the McKenzie Valley alternative (*Oil and Gas Journal,* March 5, 1979).

GASES	FREE GAS OR GAS IN SOLUTION IN VARIOUS KINDS OF ROCK	Conventional natural gas in porous and permeable reservoir rocks
		In tight gas sands of western states
		In Devonian shales of Appalachians
		In geopressured reservoirs
		In un-mined coal beds
	SYNTHETIC GASES	From coal gasification at mine-head or in place underground
		From biomass

Conventional, unconventional, and synthetic gases, showing the unconventional gases reviewed in this chapter.

The Alaskan pipeline and adjacent haul road. *Courtesy American Petroleum Institute.*

NOTES

[1] While Proved Reserves are estimated at 16 billion at end of 1978, Probable reserves are near 60 billion (*World Oil*, Aug. 15, 1978). Speculation that Undiscovered Recoverable amounts may be 200–300 billion barrels suggests that Mexico may be a most significant non-OPEC supplier of crude oil in the future.

[2] Statistics released by the American Association of Petroleum Geologists show that the success rate for exploratory holes has *risen* from about 17 percent in 1971 to 27 percent in the year 1977. This means that in 1977 roughly one of every five exploratory holes was completed as a producer, a very encouraging ratio until it is realized that far more holes were completed as *marginal* wells than in the past. Another positive-sounding indication is the fact that 1,004 new oil or gas fields were discovered in the year 1977—the largest number in a single year since 1956. Few of those fields, however, hold significant amounts of oil or gas. Whereas 26.4 percent of fields discovered in 1971 were of significant size (greater than 1 million barrels of oil or 6 billion cubic feet of gas) this ratio dropped to 19.4 percent in 1972, and has stayed at 12 or 13 percent in the years since. As suggested by an earlier graph, the actual amount of reserves in fields discovered in 1977 was just over half a billion barrels. Production in that year was 2.9 billion. (All data here reported by *Oil and Gas Journal*, July 3, 1978, pp. 32–33.)

[3] During 1978 and 1979 two Texaco holes encountered some possibly commercial flows of natural gas in the Baltimore Canyon area.

[4] Recent data by state which is presented here is virtually all from the American Gas Association, an industry group which publishes annual summaries which appear together with similar information for crude oil (American Petroleum Institute, 1977). Some revisions are made from tabulations in the weekly *Oil and Gas Journal*.

[5] This material was derived almost entirely from a most thorough report on energy transportation by John Jimison (Jimison, 1977).

[6] This alternative was vetoed by the Canadian government early in 1978 (*Oil and Gas Journal*, Mar. 6, 1978).

3

Oil Shales and Tar Sands

Oil Shales

Oil shale represents one of the largest untapped sources of hydrocarbon energy known to man. Deposits occur throughout the world, on every continent, and constitute a greater potential energy source than any other natural material with the exception of coal. The Rocky Mountain area of the United States alone contains oil shale representing billions of barrels of oil. This resource, however, has remained virtually undeveloped because supplies of conventional crude have been available at lower development costs.

Oil shales and tar sands potentially represent two of the major sources of synthetic crude oil in the United States. Next to coal, oil shales comprise the second largest fossil fuel resource in the country. The major oil shale deposits are found in the western states: Colorado, Wyoming, and Utah. For more than fifty years, oil companies have been experimenting with extracting oil from shale, but up to now, economic realities have prevented commercialization.

In Canada, the bitumen content of the well-known Athabasca Tar Sands is already being converted to syncrude on a commercial scale. Although the tar sands of the United States are not nearly as extensive or rich as those in Canada, scattered rich deposits do occur in the West, principally in Utah. However, none of these deposits are anywhere near commercial exploitation.

Technically, oil shale contains no oil; instead, it contains an organic-rich material known as *kerogen*. Because the organic material in these shales may be converted to liquid oil by heating, the shales may be thought of as *source rocks* for petroleum (see Chapter 2). Apparently these shales never were subjected to enough natural heat to accomplish the conversion process. In a mining and retorting operation, the shale is subjected to destructive distillation or 'retorting' in a large pot (retort). High temperatures in that process convert the waxy organic *kerogen* to a liquid hydrocarbon, thereby finishing the evolution of oil that was begun by natural processes. During the distillation process, oil vapors begin to appear at temperatures between 480 and 660 degrees Fahrenheit. Practically all of the *kerogen* is converted to oil when the temperature approximates 900 degrees Fahrenheit (Colorado Conservation Board, 1957).

In comparison with conventional crude, shale oil exhibits a number of important physical differences. It usually has a high viscosity due to wax content and does not flow freely at room temperature. In order to facilitate movement through a pipeline it must be heated to a temperature over 90 degrees Fahrenheit (Welles, 1970). In

Colony Development Project. *Courtesy American Petroleum Institute*

135

addition, shale oil commonly contains excessive impurities, such as sulfurs, nitrogen, and ammonia. When these are removed, however, shale oil can be refined into a full range of petroleum products.

Oil shale deposits of various sizes and grades are found throughout the world in unmetamorphosed sedimentary rocks dating geologically from the Cambrian to Tertiary ages. The organic-rich material present in these rocks was derived from aquatic plants and animals deposited in water. The principal environments ranged from small bodies of water such as lakes and lagoons near coal-forming swamps, to large marine and lake basins (Duncan and Swanson, 1965). Most of the world's oil shale resources are found in only a few locations—the Tertiary deposits of the western United States, the Permian deposits of southern Brazil, the Cambrian and Ordovician deposits in northern Europe and Asia, in the Jurassic rocks of western Europe and Russia, and the Triassic oil shale of central Africa.

RESOURCE TERMINOLOGY

As with crude oil and natural gas, some shale oil resources are geologically better known and others are less certain. Those shale oil amounts based on deposits whose magnitude and richness can be established within reason-

able limits are considered *Identified*. In contrast, *Hypothetical* resources of shale oil, which may be loosely compared with undiscovered amounts of conventional oil and gas, are in deposits whose approximate dimensions can be predicted by geologic inference, but whose richness or yield is based on very scanty data. In assessing Identified resources, estimates can be made of oil (kerogen) in place in the deposits. Recoverable amounts of oil can then be derived by estimating what proportion of oil in-place can be extracted. This estimate must take into consideration the thickness and continuity or persistence of beds and how accessible the beds are to mining or other recovery methods.

World Shale Oil Resources

The total Identified shale oil resources in the world are estimated to be over 3,000 billion barrels, 22 percent (678 billion barrels) being in high-grade deposits averaging 25 to 100 gallons per ton, and the balance of 2,421 billion barrels being in lower-grade deposits averaging 10 to 25 gallons per ton (Fig. 3–1 and Table 3–1). North America houses 62 percent of the identified high-grade deposits and 66 percent of the lower-grade deposits. Almost all are located in the United States. Identified shale

A truckload of oil shale rock. *Courtesy of API.*

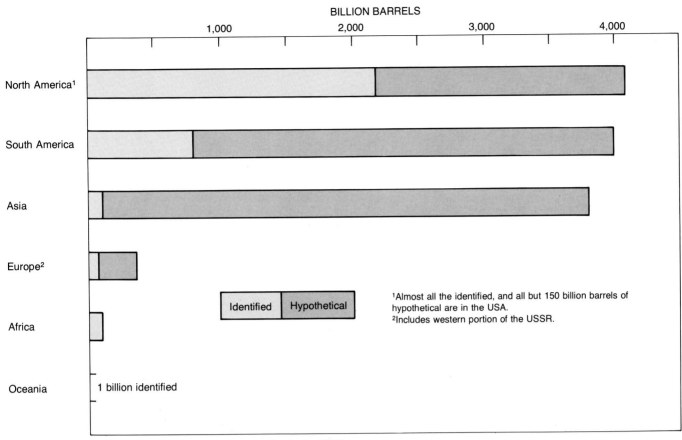

BILLION BARRELS

Fig. 3–1 World regions: shale oil amounts in place, estimated in 1973.

Table 3-1 World Regions: Shale Oil Amounts In-Place, Identified and Hypothetical, Showing Richness Categories, as Estimated in 1973 (in billions of barrels of oil in-place)

CONTINENT	IDENTIFIED[1]		HYPOTHETICAL[2]		TOTAL
	25–100 Gal/ Ton	10–25 Gal/ Ton	25–100 Gal/ Ton	10–25 Gal/ Ton	
Africa	100	small	0	0	100
Asia	90	14	2	3,700	3,806
Europe[3]	70	6	100	200	376
North America[4]	418	1,600	350	1,700	4,068
Oceania	small	1	0	0	1
South America	small	800	0	3,200	4,000
TOTAL	678	2,421	452	8,800	12,351

[1]Identified Resources: Specific, identified mineral deposits that may or may not be evaluated as to the extent and grade, and whose contained minerals may or may not be profitably recoverable with existing technology and economic conditions.

[2]Hypothetical Resources: Undiscovered mineral deposits, whether of recoverable or sub-economic grade, that are geologically predictable as existing in known districts.

[3]Includes the western portion of the U.S.S.R.

[4]Practically all of the Identified and all except 150 billion barrels of the Hypothetical resources are located in the United States.

Source: Culbertson and Pitman, *U.S. Geological Professional Paper,* 820.

oil resources outside of the United States are estimated at slightly over 1,000 billion barrels, of which approximately 26 percent are in high-grade deposits. Most of the high-grade deposits outside of the United States are located in Africa, Asia, and Europe (which includes the western portion of the Soviet Union). Large amounts (800 billion barrels) of low-grade deposits are reported to exist in South America, principally in Brazil. (It should be noted that the Duncan and Swanson estimates of Identified resources are fifteen years old and may not accurately reflect foreign shale oil resources).

The amount of Hypothetical shale oil resources of the world is enormous, over 9,000 billion barrels of oil in-place. Approximately 22 percent of the Hypothetical resources are found in North America, where all but 150 billion barrels are in the United States. Almost three-fourths of the world's Hypothetical resources are located in Asia and South America.

It is virtually impossible to make accurate comparisons of *recoverable* shale oil resources because of the lack of equivalent data for most nations of the world. Accurate estimates require extensive geological information and the evaluation of numerous samples. Because this

Table 3–2 World Regions: Shale Oil Amounts Recoverable, According to Two Estimates (in billions of barrels)

| | RECOVERABLE SHALE OIL | |
CONTINENT	Duncan and Swanson[1]	World Energy Conference[2]
Africa	10	12
Asia	20	160
Europe	30	43[3]
North America	80	703[4]
Oceania	small	2
South America	50	4
TOTAL	190	924

[1]Duncan, Donald C. and Vernon E. Swanson. "Organic-Rich Shale of the United States and World Land Areas," *Geological Survey Circular, 523* (Washington: Government Printing Office, 1965).
[2]World Energy Conference. *Survey of Energy Resources,* (New York: United States National Committee–World Energy Conference, 1974).
[3]U.S.S.R. included, and accounts for approximately 25 billion barrels of the total.
[4]The United States' portion of the North American total is 610 billion barrels. This value was derived by assuming a 33 percent recovery of total known in-place oil shale in Colorado, Wyoming, and Utah as suggested by John Donnell of the U.S. Geological Survey (Donnell, 1976).

Fig. 3–2 World regions: shale oil amounts recoverable according to two estimates.

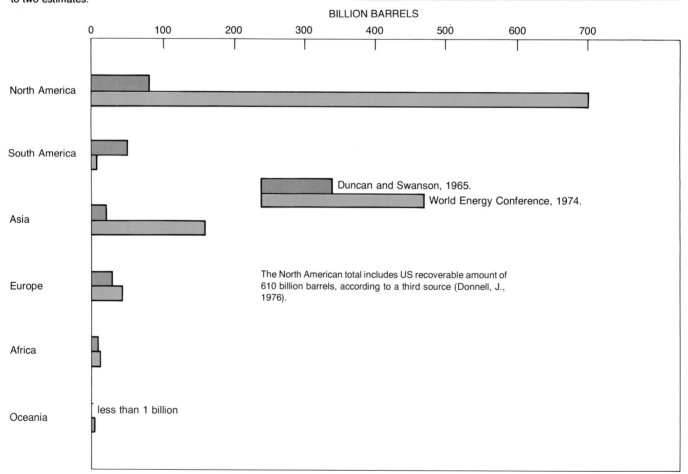

BILLION BARRELS

Duncan and Swanson, 1965.
World Energy Conference, 1974.

The North American total includes US recoverable amount of 610 billion barrels, according to a third source (Donnell, J., 1976).

less than 1 billion

information does not exist for most countries outside of the United States, assessments concerning recoverable shale oil resources in foreign deposits must be tentative.

Nevertheless, Fig. 3–2 and Table 3–2 indicate recoverable shale oil resources for world regions according to two sources: Duncan and Swanson, 1965, and the National Committee, World Energy Conference, 1974. According to estimates made by Duncan and Swanson, the total recoverable shale oil resources for world regions is estimated at 190 billion barrels of oil, 42 percent of which, or 80 billion barrels, are located in the United States.[1] South America, Europe (including the Soviet Union), and Asia were reported to have 50, 30, and 20 billion barrels of shale oil respectively. In contrast, the World Energy Conference reported world recoverable shale oil resources of 314 billion barrels excluding the United States. If an estimate of 610 billion barrels of recoverable shale oil for the United States is included (Donnell, 1976) the world total amounts to more than 923 billion barrels.

In areas outside of the United States, the major discrepancies between these two estimates are in South America, Asia, and North America. The World Energy Conference reported only 4.2 billion barrels of recoverable shale oil in South America compared to the 50 billion barrels reported by Duncan and Swanson. On the other

hand, Duncan and Swanson's estimate for Asia was only 20 billion barrels, while 160 billion barrels were estimated by the more recent World Energy Conference report. In addition, the World Energy Conference estimated Canadian shale oil resources at 93 billion barrels whereas Duncan and Swanson estimated no recoverable resources for Canada.

United States Shale Oil Resources

Oil shale occurs in marine and lacustrine (sediments produced by lakes) deposits in three principal areas in the United States: (1) in the Green River Formation in Colorado, Utah, and Wyoming, (2) in the Chattanooga Shales and equivalent deposits in central and eastern states, and (3) in marine shales in Alaska.[2]

For these areas the approximate amounts of shale oil *in-place* in beds of richness over 15 gallons per ton are shown in Fig. 3–3 according to their status as Identified or Hypothetical amounts. Table 3–3 roughly divides the in-place amounts according to richness of the deposits. From these two summaries it is apparent that the western states have the major portion, two-thirds of which is in the Identified category. The total deposits in central and eastern states are smaller; most are Hypothetical in

Fig. 3–3   Shale oil amounts in place in U. S. deposits of richness 15 gallons per ton and over.

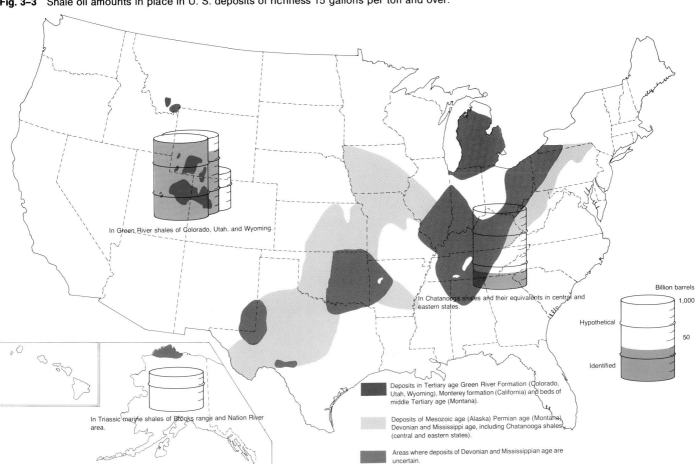

In Green River shales of Colorado, Utah, and Wyoming.

In Chatanooga shales and their equivalents in central and eastern states.

In Triassic marine shales of Brooks range and Nation River area.

Billion barrels

1,000

Hypothetical

50

Identified

Deposits in Tertiary age Green River Formation (Colorado, Utah, Wyoming), Monterey formation (California) and beds of middle Tertiary age (Montana).

Deposits of Mesozoic age (Alaska) Permian age (Montana) Devonian and Mississippi age, including Chatanooga shales (central and eastern states).

Areas where deposits of Devonian and Mississippian age are uncertain.

Table 3-3   United States' Identified and Hypothetical Shale Oil Resources In-Place, Showing Richness Categories, as Estimated in 1973 (in billions of barrels in-place)

DEPOSIT	IDENTIFIED[1]		HYPOTHETICAL[2]		TOTAL
	25–100 Gal/ Ton	10–25 Gal/ Ton	25–100 Gal/ Ton	10–25 Gal/ Ton	
Green River Formation, Colorado, Utah, and Wyoming	418[3]	1,400	50	600	2,468
Chattanooga Shale and equivalent formations, Central and Eastern United States	0	200	0	800	1,000
Marine Shale, Alaska	small	small	250	200	450
TOTAL	418	1,600	300	1,600	3,918

[1]Identified Resources: Specific, identified mineral deposits that may or may not be evaluated as to the extent and grade, and whose contained minerals may or may not be profitably recoverable with existing technology and economic conditions.

[2]Hypothetical Resources: Undiscovered mineral deposits, whether of recoverable or subeconomic grade, that are geologically predictable as existing in known districts.

[3]The 25–100 gallons per ton category is considered virtually equivalent to the category average of 30 or more gallons per ton.
Source: Culbertson and Pitman, *U.S. Geological Professional Paper, 820.*

character, and are spread over a very large area. Alaskan deposits, according to present knowledge, are relatively small and almost entirely Hypothetical.

CHATTANOOGA SHALES AND EQUIVALENT FORMATIONS   Black shale deposits of Mississippian and Devonian age underlie vast areas in the east-central United States between the Appalachian and Rocky Mountains (Fig. 3–3). They yield so little oil by conventional recovery methods, 1 to 15 gallons per ton, however, that it is unlikely that recovery of oil from these shales will be considered economically feasible anytime in the forseeable future (Duncan and Swanson, 1965, and Culbertson and Pitman, 1973). Using Duncan and Swanson's minimum thickness and grade criteria of *5 feet of shale yielding 10 gallons of oil per ton*, the Identified resources of these deposits are estimated to be 200 billion barrels of oil and the Hypothetical resources 800 billion barrels. Two of the more interesting formations containing marine black shales are the Chattanooga Black Shales in the central part of the eastern Highland Rim of Tennessee, and the New Albany Shale in Indiana and Kentucky. The Chattanooga Shales are on the average 15 feet thick, and yield an average of 10 gallons of oil per ton. Some areas of the Albany Shales contain zones 20 to 100 feet thick and yield an average of 10 to 12 gallons of oil per ton of shale.

ALASKAN MARINE SHALES   Oil shales of marine origin underlie substantial areas in both northern and eastern Alaska, in rock sequences ranging from Mississippian to Cretaceous in age. As of yet, none of the Alaskan deposits have been appraised carefully enough to allow estimates of Identified amounts. In eastern Alaska near the

Nation River, however, local outcrops of Triassic oil shales reportedly are 200 feet thick and yield 30 gallons of oil per ton (Duncan and Swanson, 1965). Triassic and Cretaceous oil shales on the north slope of the Brooks Range are reported to contain thin zones, less than 5 feet thick, that yield as much as 160 gallons of oil per ton of rock. Other even thinner zones are known to yield 15 gallons of oil per ton. Duncan and Swanson estimated that Hypothetical resources of Alaska total 450 billion barrels of oil of which 200 billion barrels would be in high-grade deposits.

THE GREEN RIVER FORMATION   By far the richest and most extensive oil shale deposit in the United States, and the world, is in the Green River Formation, which covers roughly 17,000 square miles in northwestern Colorado and adjacent areas in Utah and Wyoming (Fig. 3–4). This formation was deposited about 50 million years ago in two large lakes of the Eocene age: Lake Uinta, which covered northwestern Colorado and part of Utah, and Lake Gosiute, which covered southern Wyoming.[3] The formation ranges in thickness from a few feet to several thousand feet, and consists of marlstone interbedded with various amounts of siltstone, sandstone, halite, nahcolite, and dawsonite.[4] If the estimated 1800 billion barrels of Identified oil in-place is to be properly assessed, its varying richnesses and the amounts recoverable must be understood.

Table 3–4, and Figures 3–4 and 3–5 show the locations of various components of the total according to both the oil yield per ton of rock, and the amounts recoverable, which in every case are simply *one-third* of the estimated amounts in-place. This one-third ratio, similar to the average primary recovery rate for conventional crude oil, is based on very limited experience in recovering shale oil, but is a reasonable approximation to apply to large areas with deposits of varying character (Donnell, J., 1976).

As with many mineral deposits, there are large amounts of lower-grade shale, and much smaller amounts of the higher-grade. If shales yielding only 15 gallons of oil per ton are considered, there is a total of slightly over 1,840 billion barrels of oil in-place and 610 billion barrels recoverable, two-thirds of which is in Colorado (see Figs. 3–4 and 3–5). If a more rigorous criterion demanding over 25 gallons per ton is applied, there are 731 billion barrels in-place and 243 billion recoverable, about 90 percent of which is in Colorado. In the richest category, 30 gallons per ton or over, there are only 418 billion barrels of oil in-place, and 139 billion barrels recoverable. Of these high-grade resources, 85 percent are in Colorado, 12 percent in Utah, and 3 percent in Wyoming.

Colorado leads all other states in volume of shale oil because of the thick and rich deposits in the Piceance Basin identified in Fig. 3–4. This geologic basin lies at the center of the ancient Lake Uinta, contains beds of high kerogen content, and varies from a few feet at its edges to 2,300 feet in the basin's center. The overburden is quite thick, about 1000 feet (Welles, 1970).

Two principal layers of rich shale occur in the Piceance Basin: 1) the Mahogany Zone, which is 50 to 200 feet thick and averages 30 gallons per ton, and 2) a deeper zone, much thicker than the Mahogany, but less rich and restricted to the center of the basin. Considering both of the rich zones together, a single square mile at the center of the basin may contain as much as 2.5 billion barrels of shale oil in-place (Welles, 1970). Because the rich Mahogany Zone is exposed in the accessible cliffs of the Colorado River and its tributaries, such as the one near Rifle, Colorado, oil shale research operations have been concentrated in such locations.

Table 3–4 and Figure 3–4 single out particular mineable beds which average over 30 gallons of oil per ton, and are at least 30 feet thick everywhere. Therefore they are susceptible to about 60 percent recovery of the shale through underground room and pillar mining (National Petroleum Council, 1973).[5] (See Chapter 2 for description of these mining methods). Oil recoverable from these most attractive beds amounts to 48 billion barrels in Colorado and 6 billion barrels in Utah. So far, none have been identified in Wyoming because exploration has not been so extensive.

ASSESSMENT OF RESOURCES

Given the current status of shale oil development it is risky to assign recoverable amounts to *reserves*. Closest to the line of economic feasibility are the amounts just discussed. According to the National Petroleum Council, the only beds which are likely to be exploited in the near future are located in the thick and rich Mahogany Zone beds (National Petroleum Council, 1973). Progressively more remote in economic feasibility are the leaner beds

which do not yield as many gallons per ton. If *all* beds averaging over 15 gallons per ton were processed, 610 billion barrels would be recoverable. It is important to realize that the four recoverable estimates shown in Table 3–4 cannot be added. In each case, they indicate the total recoverable oil based on four different assumptions about which beds (rich or lean) will contribute to the total.

The recoverable amounts are, of course, only estimates derived from applying a "blanket" recovery ratio of one-third (with the exception of the higher recovery for 30-foot beds averaging over 30 gallons per ton). In specific locations this one-third ratio may not apply, as factors of bed thickness and continuity, overburden, or water in the deposit vary from place to place. If, for instance, the rich deposits in the Piceance Basin, which contribute heavily to the recoverable estimates, contain unmanageable amounts of water, then recovery will be less than estimated because certain beds will not be mined. If, on the other hand, *in situ* recovery processes are utilized, then oil may be recovered from deeply buried low-grade deposits not represented in the present estimates of amounts recoverable. Despite the uncertainty of the estimates, it is nevertheless interesting to note the following comparison: If high grade beds averaging 30 gallons per ton are exploited, the total recoverable oil in them (139 billion barrels) is roughly the same as the total of *conventional* crude oil recoverable from identified and undiscovered reservoirs on the basis of traditional recovery rates.

RECOVERY PROCESSES

Two major options are being considered for oil shale recovery: (1) mining followed by surface processing, and (2) *in situ* processing (Fig. 3–6). Mining followed by surface processing is better understood, but economic and environmental factors associated with it have raised serious doubts as to its use in the early years of commercial-scale production. The *in situ* process is presently being developed; and the modified version successfully tested by Occidental appears to be the favored recovery process at this time.

MINING AND SURFACE PROCESSING   The best-developed mining technique is the underground roof and pillar method developed by the Bureau of Mines between 1944 and 1956, and later improved by Union Oil, Colorado School of Mines, and the Colony Development Operation (Department of Interior, *Vol. III*, 1973). Simply put, the technique involves cutting a series of room openings and roof support pillars, both 60 feet square, in a mineable seam of shale. The Bureau of Mines has achieved a shale extraction ratio of 75 percent using this technique, the supporting pillars being the remaining 25 percent.

Surface or open-pit mining has not been tested on oil shale to any large degree. Since the technique is highly developed for mining other ores, it may be practical in areas where overburden is thin enough.

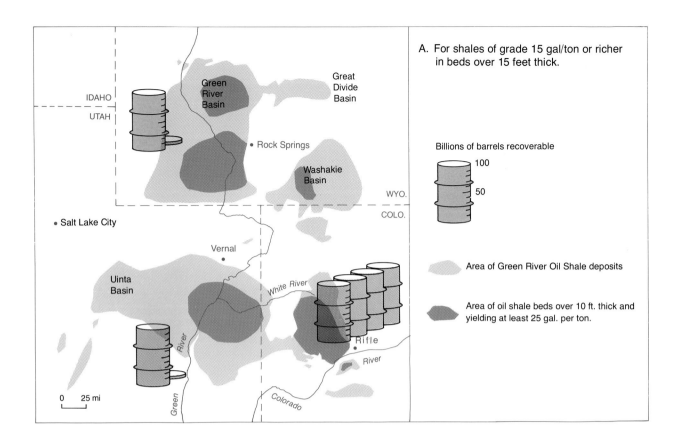

A. For shales of grade 15 gal/ton or richer in beds over 15 feet thick.

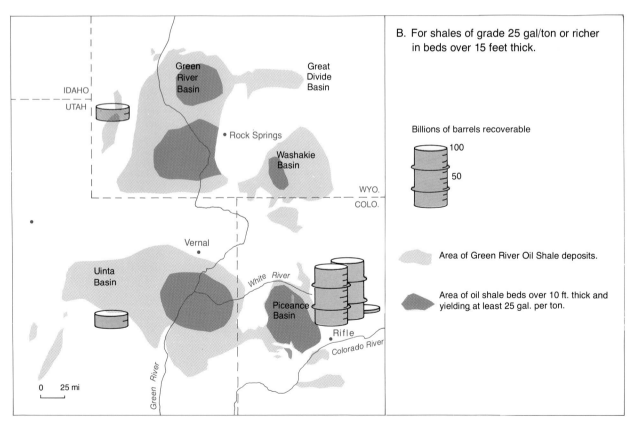

B. For shales of grade 25 gal/ton or richer in beds over 15 feet thick.

Fig. 3–4 Shale oil amounts recoverable from various beds in the Green River formation, showing areas of deposit and of richer beds.
A. From rock yielding 15 or more gallons per ton, in beds over 15 feet thick.
B. From rock yielding 25 or more gallons per ton, in beds over 15 feet thick.
C. From rock yielding 30 or more gallons per ton, in beds over 100 ft. thick.
D. From rock yielding 30 or more gallons per ton, in a selected 30-foot mineable seam.

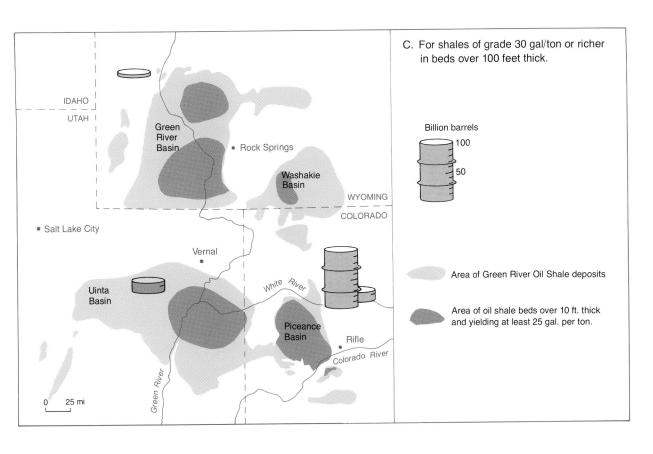

C. For shales of grade 30 gal/ton or richer in beds over 100 feet thick.

Billion barrels

100

50

Area of Green River Oil Shale deposits

Area of oil shale beds over 10 ft. thick and yielding at least 25 gal. per ton.

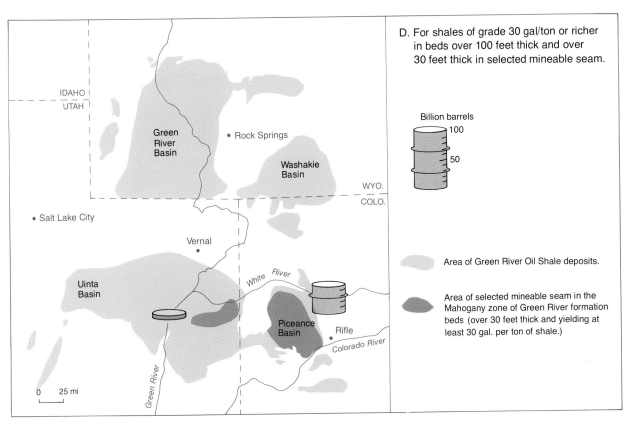

D. For shales of grade 30 gal/ton or richer in beds over 100 feet thick and over 30 feet thick in selected mineable seam.

Billion barrels

100

50

Area of Green River Oil Shale deposits.

Area of selected mineable seam in the Mahogany zone of Green River formation beds (over 30 feet thick and yielding at least 30 gal. per ton of shale.)

Table 3-4 In-Place and Recoverable Shale Oil Amounts in Identified Deposits of the Green River Formation (in billions of barrels)

LOCATION	In-Place	Recoverable
OVER 15 GALLONS/TON		
Colorado	1,200	400
Utah	321	105
Wyoming	321	105
TOTAL	1,842	610
OVER 25 GALLONS/TON		
Colorado	607	202
Utah	64	21
Wyoming	60	20
TOTAL	731	243
OVER 30 GALLONS/TON		
Colorado	365	118
Utah	50	17
Wyoming	13	4
TOTAL	418	139
OVER 30 GALLONS/TON IN A SELECTED MINEABLE SEAM		
Colorado		48
Utah		6
TOTAL		54

Note: Oil shale resources in the 15, 25, and 30 gallons per ton categories, estimated by John Donnell of the U.S.G.S for *Project Independence*, appeared in the February 1976 issue of *Shale Country*. The resources with over 30 gallons per ton in a selected mineable seam of the Mahogany Zone were reported by the National Petroleum Council in *U.S. Energy Outlook–Oil Shale Availability, 1973*.

After the shale is mined, it is crushed and then transported to surface retorts. These are of three types classified according to the method used to provide heat (Whitecombe, 1975). The crushed shale may be heated by (1) gas combustion, (2) recycled gas, or (3) recycled heat-carrying solids. The TOSCO II process developed by The Oil Shale Corporation supplies heat to the shale by direct contact with recycled, heat-carrying ceramic balls. This is one of the processes that would likely be used should surface techniques be used in recovering oil from shale (Fig. 3–7).

MODIFIED IN SITU PROCESSING   One of the major goals of oil shale research has been to recover shale oil without mining. *In situ* processing, or processing the shale while it is still underground, is the method by which the shale does not have to be mined. The major problem with the use of *in situ* processing, however, is the fact that shale is not a permeable rock. As a result, it is difficult to circulate the fluids necessary to heat the rock in place.

The modified *in situ* process developed by Occidental Oil relieves some of the difficulties. In this process, about 20 percent of the volume of shale that is to be tapped is first mined. This mining serves to form chambers in which the rock is fragmented by drilling and blasting in a procedure that ultimately leaves a series of chambers separated by undisturbed rock (Fig. 3–8). The removed shale can then be retorted at the surface. Occidental has produced about 30,000 barrels of oil using their modified *in situ* process. Five thousand barrels have been marketed through a Michigan utility where it was successfully burned in boilers without further refining.

The use of *in situ* recovery processes substantially reduces the environmental impact of a shale oil plant. It requires only one-third of the work force needed for mining and surface retorting techniques, only 25 to 33 percent of the water, and it reduces spent shale disposal by at least

A U.S. Department of Energy field site near Rocky Springs, Wyoming, where experiments are being conducted on in situ recovery of crude oil from the Green River oil shales. *Courtesy of Department of Energy.*

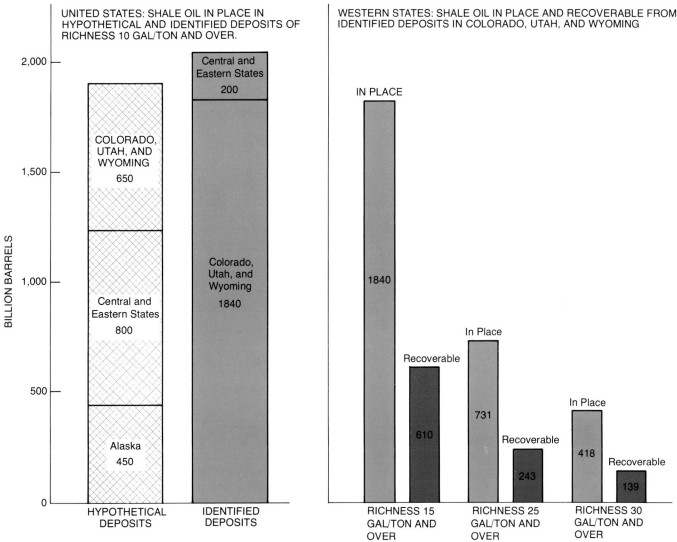

UNITED STATES: SHALE OIL IN PLACE IN HYPOTHETICAL AND IDENTIFIED DEPOSITS OF RICHNESS 10 GAL/TON AND OVER.

WESTERN STATES: SHALE OIL IN PLACE AND RECOVERABLE FROM IDENTIFIED DEPOSITS IN COLORADO, UTAH, AND WYOMING

Fig. 3–5 Summary of U. S. shale oil in place and recoverable.

Fig. 3–6 Various operations required for *in situ* processing and for conventional mining followed by surface retorting.

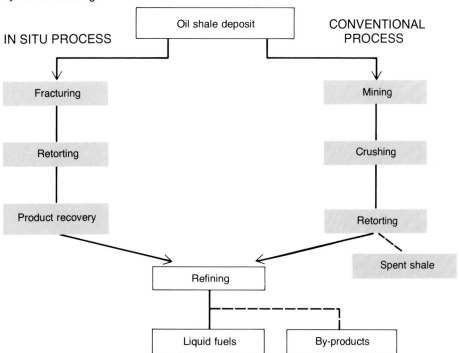

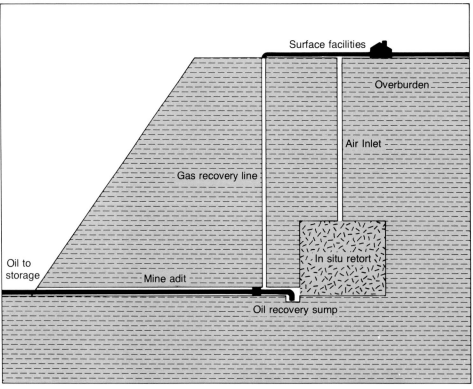

Fig. 3-7 In situ shale oil recovery process.

Davis Gulch, Colorado, property that is part of the Colony Development Project. *Courtesy of American Petroleum Institute.*

80 percent (*Science*, December 9, 1977). The modified *in situ* process recovers 60 to 70 percent of the organic material, about the same conversion efficiency as is achieved by surface retorting. This modified process may be practical only in thick beds close to the surface. In the long term, therefore, *in situ* conversion with no mining will be necessary.

HISTORY OF OIL SHALE DEVELOPMENT

Extraction of oil from shale is not a recent event. In the United States, lamp oil, lubricants, and medicines were produced from lean oil shales in the eastern Appalachians during the early 1800s. The 1859 discovery of oil at Titusville, Pennsylvania, however, destroyed hopes for a commercial oil industry because it offered a cheaper source of crude oil. Since that time, several attempts have been made to revitalize interest in oil shale.

Shortly after World War I, the U.S. Bureau of Mines constructed an oil shale experimental plant near Rifle, Colorado which operated as a research facility until 1929 (East and Gardner, 1962). The plant was closed with the discovery of large oil fields in California, Oklahoma, and Texas. In the years after World War II new interest was kindled by the fear of near-term energy shortages. Many individuals began to predict that the oil supply might not fulfill the demand for it. As a result of these concerns, interest in oil shale development continued at a slightly

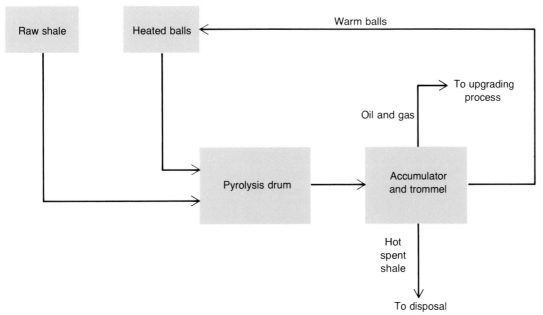

Fig. 3–8 TOSCO II retort.

accelerated pace throughout the 1950s and 1960s. During this period, the most publicized oil shale operation was one launched by a group of private companies in 1964 called the Colony Project.[6] This consortium began the construction of a pilot plant in western Colorado designed to establish the technical and economic feasibility of a commercial oil shale industry. After several years of successful work, a poor economic climate and the unwillingness of Congress to pass supportive legislation forced the suspension of the plan and operations by the Colony Group in 1974 (*Oil and Gas Journal*, October, 1974).

Although commercial oil shale technology was substantially developed during the 1960s, shale oil still could not be produced profitably at a price that would compete with domestic conventional crude oil and the still cheaper oil from the Middle East. But the 1973 Middle Eastern oil embargo brought on an instant collision of the supply and demand curves. Although the embargo lasted only a short time, it appears to have had a major impact on the United States energy outlook: it clearly illustrated the United States' growing dependence on foreign oil and ended decades of cheap oil and cheap energy. With foreign oil rising to more than 11 dollars per barrel, it appeared that the economic barrier to commercial oil shale development had been removed, and shale oil's entry into the commercial market was inevitable.

THE PROTOTYPE OIL SHALE LEASING PROGRAM
Approximately 72 percent of the land and 60 to 65 percent of the shale in the western United States is owned by the federal government. The government's renewed interest in shale development in the 1970s appeared to be the needed stimulus for establishment of a commercial oil shale industry. In 1971, President Nixon, in response to growing concerns regarding the adequacy of the Nation's energy resources, requested that the Department of Interior initiate a leasing program to develop oil shale resources on public lands, provided that environmental considerations could be satisfactorily resolved (Department of Interior, *Vol. III*, 1973). Initially, the prototype program was formulated to make six tracts of land of not more than 5,120 acres each available for private development. After gathering resource information on various federal sites, the Department of Interior announced the selection of six tracts of about 5,000 acres each, two in the Piceance Creek Basin, two in the Uinta Basin, and two in the Washakie Basin (Table 3–5 and Fig. 3–9). In early 1974, the six tracts were offered to private companies in lease sales. The four in Colorado and Utah were bid upon and the highest bids were accepted by the Department of the Interior, but no bids were made on the two Wyoming tracts. The total value of the accepted bids paid to the federal government amounted to almost 450 million dollars for the four tracts.

Under the prototype leasing program, development has not progressed as smoothly as might have been expected. Following a year or more of gathering environmental baseline data, the companies submitted detailed development plans to the Department of Interior. But in the face of the rising cost of construction, lack of congressional support, environmental objections, and prohibitive environmental regulations, the companies elected not to proceed with development and requested a one-year suspension of their leases. Subsequently, their request was granted by the Department of Interior.

Since the suspension of the leases, a number of events have brightened the prospects for commercial oil shale development. Occidental Petroleum, which has had a great deal of success in developing underground conversion techniques, joined Ashland Oil as an equal partner in

Table 3–5 Summary Data–Oil Shale Prototype Tracts

	ACRES	ESTIMATED RECOVERABLE OIL SHALE (in billions of tons)		EXPECTED RECOVERY METHOD	COST OF LEASE (millions of dollars)	LEASING COMPANIES
TRACT		30 or More Gal/Ton	20 or More Gal/Ton			
Colorado C-a	5,089	1.857		underground	210.3	Standard Oil of Indiana and Gulf
Colorado C-b	5,093	1.012		underground	117.8	Atlantic Richfield, The Oil Shale Corp., Shell Oil and Ashland Oil[1]
Utah U-a	5,120	.342		underground	75.6	Phillips Petroleum and Sun Oil
Utah U-b	5,120	.372		underground	45.0	Standard of Ohio, Phillips Petroleum, and Sun Oil
Wyoming W-a	5,111		.354	in situ	no bid	
Wyoming W-b	5,083		.352	in situ	no bid	

Sources: U.S. Department of Interior. *Final Environmental Statement for the Prototype Oil Shale Leasing Program,* Volume III of VI, *Specific Impacts of Prototype Oil Shale Development,* Washington: Government Printing Office, 1973 and *Department of Interior News Releases, 1974.*

[1] In the first year of the lease, Atlantic Richfield and the Oil Shale Corporation dropped out of tract C-b development. In the fall of 1976, Shell Oil dropped out of the lease and Occidental oil has become an equal partner with Ashland oil.

the Colorado Tract C-b venture. Shell Oil and the Oil Shale Corporation have dropped out of the lease. In 1978, Ashland Oil also dropped out of the C-b lease. Standard Oil of Indiana and Gulf Oil, the operators of Colorado tract C-a, have also filed new development plans calling for *in situ* recovery of the shale oil.

In 1978, government action further brightened the oil shale picture. The Department of Energy agreed to finance 71 percent of a 60.5 million dollar two-phase project to develop Occidental Petroleum's modified *in situ* shale oil production technique. In addition, the Colorado Air Pollution Commission has eased the standards for sulfur emissions, and the Environmental Protection Agency has issued permits allowing development on the two tracts to proceed (*Science*, December 9, 1977). Furthermore a 3 dollars a barrel tax credit for oil from shale was, in 1979, before the Congress with President Carter's blessings. Some members of the oil industry think this incentive could launch the development of a full-scale oil shale industry.

PROSPECTS FOR COMMERCIALIZATION

The basic problem that has long bothered the shale oil industry is that of economics. Can shale oil be produced at a cost competitive with oil imports? Because of the high rate of inflation since the 1973 Arab embargo, private investors have been reluctant to invest capital in a pioneer oil shale industry. Table 3–6 illustrates the effect of inflation on the prospects of commercial oil shale development. The data were derived for the Colony Project, a private venture of Atlantic Richfield and the Oil Shale Corporation (Whitecombe, 1975). The Colony data

Table 3–6 Cost Estimates–Colony Shale Oil Plant of 55,000 Barrels per Day Capacity

ESTIMATE DATE	INITIAL[1] CAPITAL ($mm)	DIRECT OPERATING COSTS ($/bbl)	OIL PRICE FOR 10% DCF RATE OF RETURN ($/bbl)
1972	255	1.82	4.40
March 1974	425	3.13	8.00
August 1974	653	4.30	11.50
September 1975	750	4.55	12.70

[1] The investments shown are for production of upgraded oil and include all on-site facilities and equipment required for the mining and processing operations. Also included are the costs for the plant access road; flood control and run-off dams; pipeline and pumping facilities to transport water from the catalysts, chemicals, spare parts and supplies; operator recruitment and training; environmental monitoring; plant owner's supervision of the design and construction; and working capital.

The capital investment estimates do not provide for the cost of the oil shale reserve, the cost of pipeline for the oil product or for recovery of past research, development and demonstration costs. Also, the investments shown are on an equity basis and do not provide for interest during construction.

Source: Whitecombe, John A. *Oil Shale Development: Status and Prospects,* prepared for the Society of Petroleum Engineers of Aime, 50th Annual Meeting. September 28–October 1, 1975, Dallas, Texas.

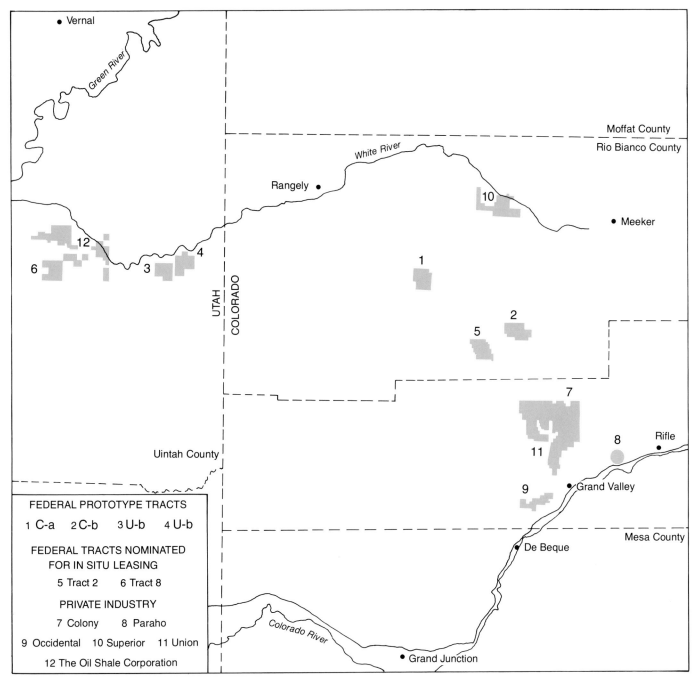

Fig. 3–9 Federal prototype and private oil shale tracts in Colorado and Utah, as of 1979.

were selected for analysis because they afford the best-developed base for which detailed cost estimates are available.

As the data indicate, the necessary initial capital investment for a Colony-type plant increased nearly three-fold between 1972 and 1974 and the required price for upgraded oil rose by the same ratio. More recent cost estimates derived by the Energy Research and Development Administration indicate that the present cost of oil produced by surface retorting ranges from 20 to 30 dollars per barrel (Table 3–7). The same data show that the cost of a barrel of oil using the modified *in situ* process is con-

siderably lower, ranging from 11 to 16 dollars per barrel. Although, cost estimates are extremely fluid, it would appear that the modified *in situ* process is emerging as a promising near-term approach to shale oil recovery.

CURRENT DEVELOPMENT ACTIVITIES   Taking into account recent events that have increased the likelihood of commercial oil shale development, the most optimistic 1979 estimates put shale oil production at 75,000 to 125,000 barrels per day by the mid-1980s, 300,000 to 500,000 barrels per day by 1990, and perhaps a million barrels per day by the end of the century (*Oil and Gas Journal*, June 18, 1979).

Oil shale facility in Rifle, Colorado. *Courtesy of American Petroleum Institute.*

The future of oil shale development would seem to rest with four major oil shale projects. These are operated by the following companies: Occidental Petroleum, Union Oil of California, Atlantic Richfield and TOSCO Corporation, and Rio Blanco Oil Shale Company owned by Gulf Oil and Standard Oil of Indiana (Fig. 3–9).

The Colony Project is seriously contemplating spending 1 billion (1979) dollars on a 45,000 barrels per day project at their site immediately east of Union Oil. They believe the cost of producing shale oil at their site using TOSCO II retorts will be 20 to 25 dollars per barrel. The Colony Project calls for six TOSCO II retorts capable of handling 66,000 tons per day of shale producing 45,000 barrels of oil per day. The oil will be upgraded on-site and moved to market by pipeline.

Rio Blanco has started what it calls a demonstration phase of its project. The plans call for both underground and surface retorting using the mined shale from the modified *in situ* operation. About 57,000 barrels per day of oil would be processed by underground retorts and 19,000 barrels per day from the surface retort. The earliest start-up for this commercial operation would be 1987 and the costs are estimated at over a billion dollars when completed.

Union Oil believes that their above ground retort developed twenty years ago at their Parachute Creek, Colorado site, can now become a commercial operation if the 3 dollars per barrel tax credit is approved by Congress. They feel they can produce 9000 barrels per day from a single retort at a cost of 15 dollars per barrel including 15 percent profit. Transport costs of the unrefined boiler fuel to Denver would be 3 dollars per barrel raising the total cost to 18 dollars per barrel. Union's retort is scheduled for start-up in early 1983.

The Occidental Project using the modified *in situ* recovery process is the furthest along in its development and appears certain of becoming a commercial venture. This plan on federal lease C-b calls for commercial production of 2,500 barrels per day by the mid-1980s and increasing to 57,000 barrels per day by 1990. Although Occidental does not consider the 3 dollars per barrel tax credit necessary for commercial development, it does consider it important.

Tar Sands

Another important source of crude is the *oil sands* of the world. On the one hand, they are similar to oil shales in the sense that their oil cannot be recovered through a drillhole by conventional petroleum production methods; on the other hand, they are quite different because hydrocarbons in oil sands are much more akin to conventional crude than is the kerogen in oil shales. In most cases, oil sands contain highly viscous asphaltic crude oil, often referred to as *bitumen*, in a sandstone that in some deposits is a consolidated rock, and in others is a relatively soft sand body held together by bitumen. In addition, asphaltic oil occurs in limestones, such as in the deposits in a number of European countries and in Oklahoma and Texas. The largest and most attractive occurrences in North America, however, are in sands.

Crude oil in oil sands is called *unconventional* crude oil because of the recovery techniques required to extract it. Any recoverable amounts are therefore considered additional to conventional crude. Nevertheless, in some deeply buried oil sands occurrences, there is no clear distinction between oil sands and reservoirs holding conventional crude of very high viscosity, that is, "heavy crude". As with shale oil extraction, recovery may be accomplished by mining, if deposits are accessible, or by *in situ* processes, if deposits are deeply buried. Some of the *in situ* processes for extraction of bitumen from oil sands make use of steam or combustion to heat the oil in-place, or solvents to flush it out, and are therefore quite similar to methods of enhanced recovery used to

Table 3–7 Shale Oil Costs

RECOVERY PROCESS	INVESTMENT[1] ($/B/D)	OPERATING ($/BBL)	PRODUCTION ($/BBL)
Modified *in situ*	5,000–7,000	3.50–5.00	8.00–11.00
Surface retorting	14,000–23,000	4.00–5.00	16.00–25.00

[1]1976 dollars for a 50,000 B/D plant. 100 percent equity and 15 percent discounted cash flow.
Source: *U.S. Energy Research and Development Administration, 1977.* (Dr. Philip White).

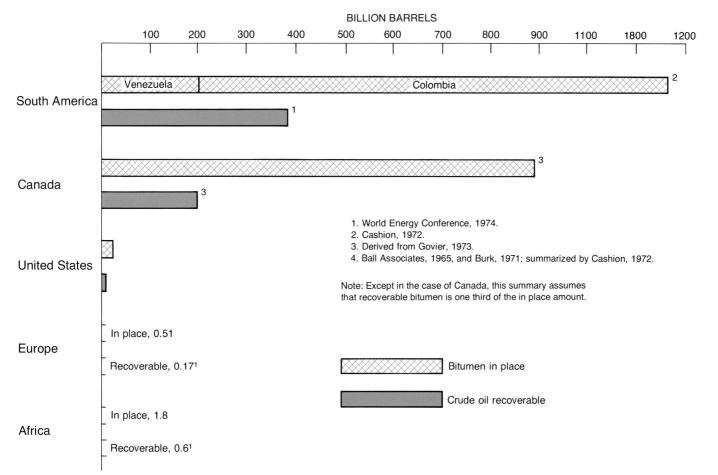

BILLION BARRELS

Fig. 3–10 Bitumen in place and crude oil recoverable: leading countries and regions.

1. World Energy Conference, 1974.
2. Cashion, 1972.
3. Derived from Govier, 1973.
4. Ball Associates, 1965, and Burk, 1971; summarized by Cashion, 1972.

Note: Except in the case of Canada, this summary assumes that recoverable bitumen is one third of the in place amount.

Bitumen in place

Crude oil recoverable

obtain oil left behind in conventional reservoirs (see Chapter 2).

The Western Hemisphere contains the largest share of estimated bitumen in-place in oil sands (see Fig. 3–10). Known Asian deposits are negligible, beyond those in the U.S.S.R., which is considered here to be part of Europe's total. Despite some deposits in European and Asiatic parts of the Soviet Union and some in Albania and Rumania, that total is barely over half a billion barrels. The known African total of 1.8 billion barrels is entirely in the Bemolanga deposit on the northwestern coast of Malagasy.

In South America, large deposits are in Colombia and eastern Venezuela. In North America, about 900 of the 923 billion barrels of bitumen in-place are in the Canadian province of Alberta.

In the United States many small deposits exist (Ball Associates) but only the largest of these are listed in Table 3–8. Three states hold most of these, but Utah accounts for virtually all the bitumen in-place by virtue of deposits in the vicinity of Vernal, just south of the Uinta Mountains (Fig. 3–11).

If the one-third recovery ratio assumed for large deposits in Utah (U.S. Department of Interior, Vol. 2, *En-*

Tar sands outcropping in Vernal. *Courtesy of American Petroleum Institute.*

Table 3–8   Bitumen-Bearing Deposits (Mostly Tar Sands) with In-Place Resources of at Least One Million Barrels, Ranked According to Size of Deposit

STATE	DEPOSIT	ROCK TYPE	BITUMEN IN-PLACE (millions of bbl.)
Utah	Tar Sand Triangle	sandstone	10,000–18,100
	Peor Springs	sandstone	3,700– 4,000
	Sunnyside	sandstone	2,000– 3,000
	Circle Cliffs	sandstone	1,000– 1,300
	Asphalt Ridge	sandstone	1,000– 1,200
	Whiterocks	sandstone	65– 125
	Hill Creek	sandstone	300– 400
	Lake Fork	sandstone	15– 20
	Raven Ridge	sandstone	100– 125
	Rim Rock	sandstone	30– 35
California	Edna	sandstone	141– 166
	Casmalia	mudstone	86
	Sisquoc	sandstone	26– 50
	Santa Cruz	sandstone	10
	McKittrick	sandstone	5– 9
	Point Arena	sandstone	1
Kentucky	Kyrock	sandstone	18
	Davis-Dismal	sandstone	7– 11
	Bee Spring	sandstone	8
New Mexico	Santa Rosa	sandstone	57
Texas	Uvalde	limestone	124– 141
TOTAL			18,693–28,862

Source: Cashion, W.B., p. 101.

Fig. 3–11   Some of the major tar sand deposits of Uinta Basin, Utah.

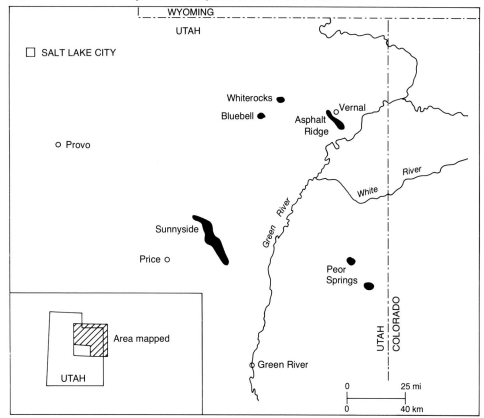

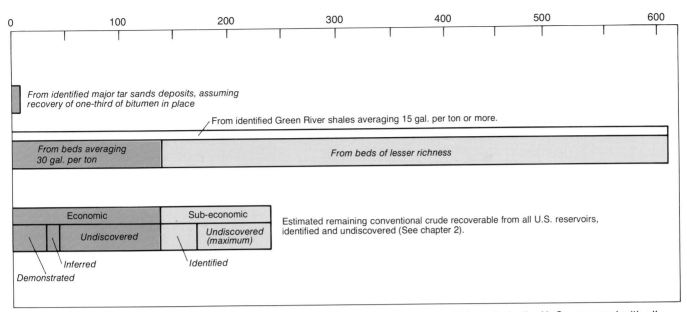

Fig. 3–12 Estimated synthetic crude oil recoverable from identified oil shale and tar sands deposits in the U. S. compared with all conventional crude oil thought to remain in identified and undiscovered reservoirs.

ergy Alternatives) is applied to all bitumen in Table 3–8, then about 8 billion barrels may be recovered if all deposits were mined. Comparison with potential crude from Green River oil shales and estimated crude, recoverable from identified and undiscovered conventional reservoirs (Fig. 3–12), shows the tar sands potential to be relatively small.

CRUDE OILS	CRUDE OIL AS SUCH IN VARIOUS RESERVOIR ROCKS	Conventional crude recovered by primary, secondary, or tertiary (enhanced) recovery techniques
	SYNTHETIC CRUDE OILS	From kerogen in oil shales
		From bitumen in tar sands
		From coal liquefaction
		From biomass

Conventional and unconventional (synthetic) oils, highlighting the synthetic crudes reviewed in this chapter.

NOTES

[1]This is in contrast to the 610 billion barrels estimated by John Donnell of the U.S.G.S. to be recoverable from all beds averaging over 15 gallons per ton.

[2]The age of the Green River Formation is Tertiary. The Chattanooga Shales are Devonian and Mississippian and the marine shales of Alaska are Triassic.

[3]Green River shales are preserved in seven geologic basins: the Green River, Great Divide, Washakie, and Fossil Basins in Wyoming; the Uinta Basin in Utah; and the Piceance and Sand Wash Basins in Colorado.

[4]Nahcolite is a sodium-bearing mineral, and dawsonite is an aluminum-bearing mineral that may be recovered in the future along with the shale oil.

[5]If 90 percent of the oil in the mined shale is recovered, this implies a recovery of 54 percent of the shale oil in-place.

[6]Standard Oil of Ohio, Cleveland Cliffs Iron Company, and the Oil Shale Corporation were the original partners. Atlantic Richfield took over as operator in 1970, and in 1974 Shell Oil and Ashland Oil joined the venture.

4

Nuclear Fuels

The use of nuclear materials for the purpose of generating electricity began in the 1950s as a by-product of the development of the atomic bomb. In 1954, the Atomic Energy Commission was given the right by Congress to encourage the development of nuclear power for electrical generation. By the 1960s, scores of plants were built, dozens planned, and, in 1968, the AEC predicted that one thousand plants would be operating by the year 2000.

Since these declarations were made, nuclear power has been surrounded by controversy. Well before the Three Mile Island accident in March, 1979, electric power companies had ceased requesting permits for nuclear power facilities. Between 1975 and 1979, less than half a dozen reactors were purchased.

Critics of nuclear power have attacked the industry for not providing safe, economic means of disposing of the hazardous waste products. In addition, they have maintained that the availability of nuclear materials throughout the world could result in the proliferation of atomic bombs. Proponents see nuclear power as a cleaner energy source than coal and one that is essential if the Nation is to reduce its dependence on foreign oil. They claim that the problems of toxic waste materials, leakages of contaminated gas, and technical and economic difficulties, can be solved. The outcome of this controversy will have a significant effect on the proportion of energy provided by nuclear power in the future.

The Nature of the Resource

A nuclear reaction is produced when the nuclei of atoms are changed, and produce an atom of a different element or new isotopes of the same element. (Isotopes are atoms of the same element differing only in the number of neutrons present in the atomic nucleus). Energy may be derived from nuclear materials by two different processes: *fission* and *fusion*. Energy is released through the fission process by splitting certain atomically heavy radioactive elements into two or more lighter radioactive elements. In the fusion process, energy is released by melting together two or more very light elements to produce heavier elements. Fission, the atomic bomb reaction, is the one presently being used in nuclear power reactors for the generation of electricity. Fusion, or the hydrogen bomb reaction, has yet to be advanced technologically to the stage where it can be used as a source of energy. The production of energy from fusion may be possible in the early years of the twenty-first century.

Cooling Tower at generation station, near Benedict, Maryland. *Courtesy American Petroleum Institute.*

The following discussion outlines the basic principles associated with the production of energy through the use of both fission and fusion, and presents a geographic evaluation of the United States' present nuclear fuel supplies as well as future requirements.

Fission

At present, most of our electricity is produced in fossil-fueled plants in which coal, oil, or natural gas is used as the energy source to produce steam in a boiler. The steam is then used to spin a turbine which drives a generator and produces electricity (Fig. 4–1). The steam is then cooled, condensed, and pumped back to the boiler for re-use.

In a nuclear power plant, a fission reactor is substituted for the fuel-fired boiler. The fuels for the nuclear reactor are the nuclei of uranium, plutonium, and thorium, all of which occur in ores of various types. The only naturally occurring fissionable material used in nuclear reactors is Uranium-235. Two artificial isotopes, Uranium-233 and Plutonium-239, can be produced from naturally occurring Thorium-232 and Uranium-238. In a nuclear reactor, the energy used for the production of

steam is released as shown in Fig. 4–2. The Uranium-235 reaction, shown in Figure 4–2A, the one commonly used in nuclear power plants, releases energy when a heavy atom of Uranium-235 is split apart by a slow-moving neutron, producing lighter fission fragments plus two or three additional neutrons. This reaction also allows the release of a large amount of energy.

The newly created neutrons, in turn, split additional Uranium-235 atoms and produce more energy and more neutrons. When the splitting of Uranium-235 atoms is repeated again and again under controlled conditions, the result is a self-sustaining chain reaction releasing very large amounts of energy. When it is produced in an uncontrolled manner, the result is an atom bomb.

Reactions shown in Fig. 4–2B and Fig. 4–2C illustrate how Uranium-233 and Plutonium-239 are "bred" from Thorium-232 and Uranium-238. Up to now, thorium has not been used extensively as a nuclear fuel; and Uranium-238, by far the largest component of natural uranium, has been considered a waste product. (An assessment of the potential role of Thorium-232 and Uranium-238 and how they may affect the United States' supply of fissionable fuels will be discussed in a later section).

Fig. 4–1 Electrical power generation by steam turbine.

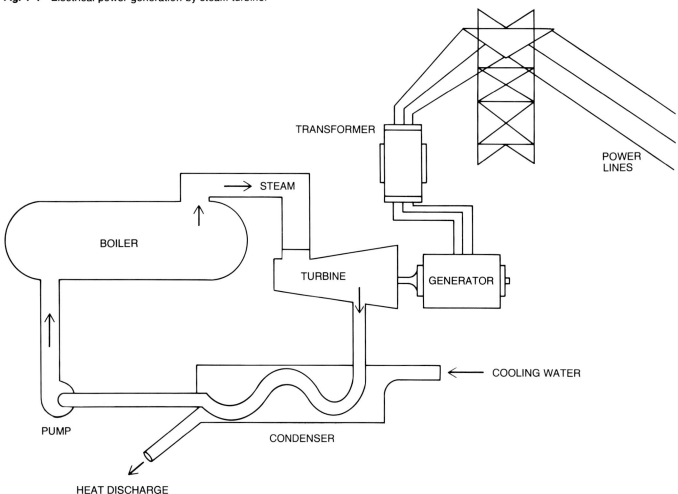

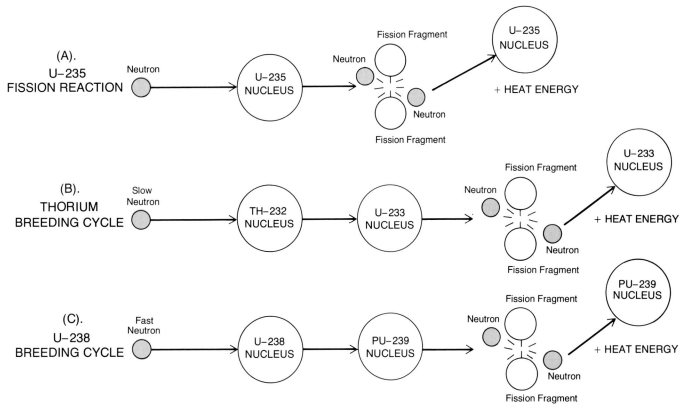

Fig. 4–2   Three nuclear reactions: one using naturally fissile U_{238} and two which breed fissile materials from U_{238} and Th_{232}.

The Nature and Occurrence of Uranium

Uranium, the basic raw material for nuclear energy, is a hard, nickel-white metal containing three radioactive isotopes: Uranium-238, Uranium-235, and Uranium-234.[1] Many nuclei, other than uranium, are radioactive, meaning that they are unstable and undergo changes, even when they are not being bombarded. Radioactive nuclei emit alpha particles (helium nuclei), beta particles (high energy electrons), or gamma rays (electromagnetic waves having greater energies than x-rays). When alpha and beta particles are emitted, a radioactive nucleus decays, becoming a different nucleus. Natural uranium contains only about 0.7 percent Uranium-235. Most of the remaining 99.3 percent is Uranium-238, which, under neutron bombardment yields Plutonium-239, a fissionable material. Plutonium-239 would become the major nuclear fuel if the United States pursued the development of breeder reactors which produce more nuclear fuel than they consume by converting nonfissionable Uranium-238 into fissionable Plutonium-239.

Natural uranium is not an abundant metal. It makes up only about two parts per million of the earth's crust, occurring in combination with other elements in ores of two broad types—those in igneous and metamorphic rocks, which account for 20 percent of the world resources, and those in sedimentary rock, which account for roughly 80 percent (Woodmanse).

Ores of the first type are "original" occurrences of a uranium compound as they were deposited in veins or in a pegmatite (very coarsely crystalline igneous rock) by hot solutions associated with an intrusion of magma. Such deposits, which tend to be rich but small, are found at Yellowknife on Great Bear Lake in the Canadian Shield, in the shield area of Western Australia, and in the Congo at Shinkolobwe.

Ores of sedimentary type occur in various rocks, such as bituminous black shales, sandstones, conglomerates, and phosphate rocks usually associated with vanadium. The age of the host rock ranges from Pre-Cambrian, as in the conglomerates of Eliot Lake, Ontario, through Devonian, as in the Chattanooga shales of the eastern United States, to Tertiary, as in some sandstones in the western United States.

Uranium oxide is the major compound in which uranium occurs. Sandstones in western states such as New Mexico, Colorado, and Wyoming, contain the primary ore minerals, uraninite, and coffinite. In some areas, weathering processes have produced carnotite and other secondary minerals which are very rich in uranium oxide. The ore minerals occupy pore spaces of the sandstones, or in some occurrences replace the sand grains or replace plant remains. Many hypotheses have been proposed for the genesis of these sandstone deposits. Generally, it is thought that the uranium compounds were formed by the leaching of original igneous sources

Uranium

A hard grey radioactive metal which occurs in a number of isotopic forms. As it is found in nature, uranium consists of three isotopes of mass numbers 238, 235, and 234. The important isotopes of uranium for use in the production of nuclear energy are Uranium-238, Uranium-235, and Uranium-233. The complete fission of all the nuclei in one pound of uranium releases the same energy as burning 1360 tons of bituminous coal.

Symbol	U
Atomic number	92
Atomic weight	238.0
Density	19.05 g/cm³
Melting point	1135°C
Boiling point	4000°C

in veins, pegmatites, or volcanic ash. The compounds were transported in ground water and precipitated when they encountered organic materials and the associated low-oxygen (reducing) environments.

Less rich sedimentary ores occur in deposits that lack the enrichment afforded by weathering. The ancient conglomerates of Eliot Lake and Blind River in Canada contain over 0.10 percent U_3O_8. Black bituminous shales range from 0.03 percent in an exceptionally rich Swedish example to 0.007 percent in the Chattanooga Formation of the eastern United States; and phosphate rocks in the Phosphoria Formation of Idaho and neighboring states contain around 0.012 percent (Finch, *et al.*).

Another type of low-grade deposit of uranium is found in the Conway granite in the White Mountains of New England which averages 0.001 to 0.003 percent. Whether these low-grade sedimentary or igneous deposits can be used as resources depends both on the future demand for uranium and on whether the more abundant isotope, U-238, is used as a fuel supply.

RESOURCE TERMS APPLICABLE TO URANIUM

The U.S. Department of Energy scheme for showing the relationship between uranium reserves and potential resources may be placed in the broad framework used by the U.S. Geological Survey (Fig. 4–3). *Reserves* are the most certain resources, constituting deposits that have been delineated by drilling or other sampling techniques. *Potential* resources are divided into three categories, *probable*, *possible*, and *speculative*, in declining order of certainty. These terms describe the uranium thought to be present in incompletely drilled or undiscovered deposits (National Uranium Resource Evaluation, 1976).

In addition to this classification, the Department of Energy also groups the resources into cost categories: under 15 dollars, 15 to 30 dollars, and 30 to 50 dollars per pound. The three cost categories are the cost required *to produce* one pound of U_3O_8 concentrate from these resources. These costs are operating and capital costs (or *forward* costs) and do not reflect the market price. These include costs of labor, materials, electricity, royalties, payroll, insurance, and applicable general and administrative costs (U.S. Department of Energy, 1979). It is im-

Fig. 4–3   Conceptual framework for organizing amounts of reserves and other resources of uranium oxide.

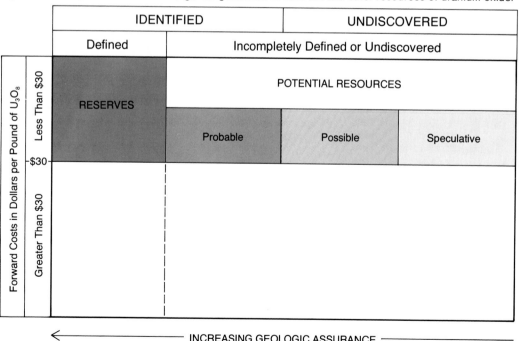

portant to note that resources in each cost category embrace the resources of all lower-cost categories. For example, resources in the 50 dollars per pound category also include resources in the 15 and 30 dollars per pound categories.

WORLD RESOURCES

Figure 4–4 and Table 4–1 provide a summary of world uranium resources up to 50 dollars per pound of U_3O_8. (The Soviet Union and China are excluded because data were not available).

World resources of uranium up to 50 dollars per pound of U_3O_8 amount to 5.7 million short tons, almost equally divided between Reserves (reasonable assured) and the estimated additional amounts. Most of the uranium Reserves, approximately 90 percent, are found in six countries: the United States, Canada, the Republic of South Africa, Niger, Australia, and Sweden. The United States alone houses almost 30 percent of the uranium Reserves. The same six countries also account for over 90 percent of the estimated Potential resources.

Most uranium resources outside the United States, found largely in Pre-Cambrian conglomerates and vein deposits, average more than 0.10 percent uranium. In Sweden, however, uranium is recovered from black shales containing only 0.03 percent U_3O_8. Although uranium resource data were unavailable for the Soviet Union and China, it is believed that both countries have ample resources to meet their needs (Nuclear Energy Policy Group Study 1977).

UNITED STATES RESOURCES

The cost of future development of nuclear energy in the United States will, to a degree, depend on the availability of fuel supplies. It is important therefore to assess the Nation's uranium resources as thoroughly as possible. A U.S. Department of Energy 1979 report estimates the United States' total uranium resources at more than 4.2 million tons of U_3O_8, including 120,000 tons of Reserves recoverable as a by-product of phosphate and copper mining (Table 4–2). In Fig. 4–5, this total of 4.2 million tons of uranium is broken down according to certainty and cost. This breakdown reveals the following points:

1). In descending order of certainty, Reserves make up 24 percent, (1.04 million tons of U_3O_8) of the total United States' resources, and Potential resources 76 percent. Probable resources, the most certain of the Potential category, account for 35 percent of the total. Possible and Speculative resources account for 27 and 13 percent of the uranium supply respectively.

2). A breakdown of the uranium resources by cost category reveals that 23 percent of the total resources are available at a forward cost of 15 dollars per pound of U_3O_8, 39 percent at 15 to 30 dollars per pound, and 35 percent at 30 to 50 dollars per pound. The remaining three percent of the resources are those recoverable as by-product of copper and phosphate mining.[2]

3). Of the one million tons of uranium resources in the 15 dollar cost category, 71 percent are classified as Potential, and the bulk of these are assigned to the Proba-

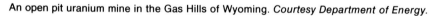

An open pit uranium mine in the Gas Hills of Wyoming. *Courtesy Department of Energy.*

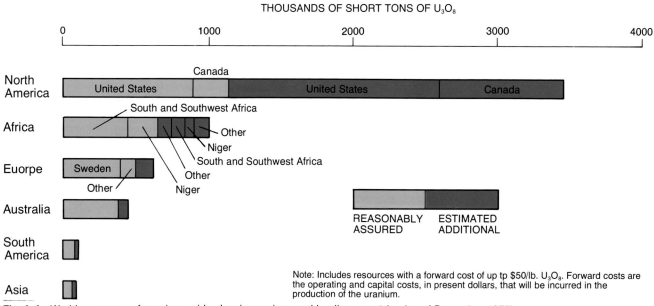

Fig. 4–4   World resources of uranium oxide showing regions and leading countries (as of December 1977).

Table 4–1   World Uranium Resources as of December 1977, Excluding Communist Areas (in thousands of short tons U_3O_8)

COUNTRY	Reasonably Assured $30/lb	Reasonably Assured $50/lb	Estimated Additional $30/lb	Estimated Additional $50/lb	COUNTRY	Reasonably Assured $30/lb	Reasonably Assured $50/lb	Estimated Additional $30/lb	Estimated Additional $50/lb
North America	*911*	*1,144*	*1,633*	*2,314*	**Europe**	*79*	*496*	*59*	*116*
United States	690	890	1,120	1,450	France	48	67	31	57
Canada	215	240	510	850	Spain	9	9	11	11
Mexico	6	6	3	3	Portugal	9	11	1	1
Greenland	0	8	0	11	Yugoslavia	6	8	7	27
					United Kingdom	0	0	0	10
Africa	*684*	*742*	*197*	*260*	Germany, F.R.	2	3	4	5
South and South West Africa	400	450	44	94	Italy	2	2	1	1
Niger	210	210	69	69	Australia	2	2	0	0
Algeria	36	36	65	65	Sweden	1	390	4	4
Gabon	26	26	7	13	Finland	0	4	0	0
C.A.E.	10	10	10	10					
Zaire	2	2	2	2	**Asia**	*54*	*58*	*30*	*30*
Somalia	0	8	0	4	India	39	39	30	30
Madagascar	0	0	0	3	Japan	10	10	0	0
					Turkey	5	5	0	0
Australia	*380*	*380*	*60*	*60*	Korea	0	4	0	0
					South America	*47*	*78*	*11*	*18*
					Brazil	24	24	11	11
					Argentina	23	54	0	0
					Chile	0	0	0	7
					TOTAL	2,155	2,898	1,990	2,798

Source: U.S. Department of Energy. *Statistical Data of the Uranium Industry, January 1, 1979.* Grand Junction, Colorado, 1979.

ble and Possible categories. For uranium for which forward costs are 15 to 30 dollars per pound and 30 to 50 dollars per pound, 76 and 84 percent respectively are Potential resources. Together, these amount to over 3 million tons of uranium, and account for three-fourths of the total uranium resources.

Considering that a large part of the total uranium resources of the United States is classified as either uncertain or low-grade (less than 0.10 percent U_3O_8), one might question the wisdom of recovering these ores in light of the high cost of energy extracted from these materials. The fuel for nuclear power plants, however, is a relatively small element in the total cost of nuclear-generated electricity. One calculation of fuel costs for several methods of power generation indicates the cost of uranium for fueling a light water reactor was .09 cents per

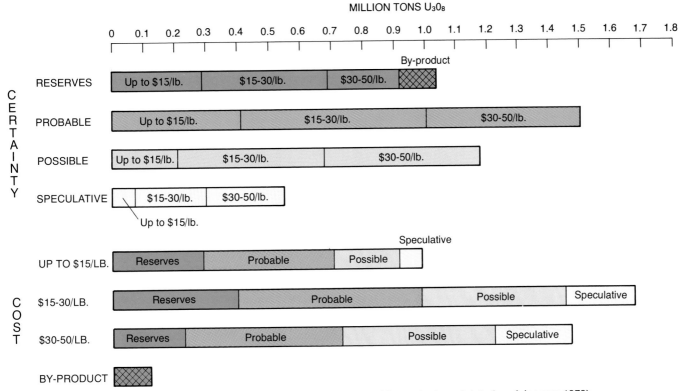

Fig. 4–5 National summary of uranium oxide resources arranged by cost and by geologic certainty (as of January 1979).

Table 4–2 United States' Uranium Resources as of January 1, 1979 (in thousands of tons U_3O_8)

U_3O_8 COST CATEGORIES ($/lb)	RESERVES	POTENTIAL RESOURCES			TOTAL
		Probable	Possible	Speculative	
$15 Average	290	415	210	75	990
$30 Average	690	1,005	675	300	2,670
$50 Average	920	1,505	1,170	550	4,145
By-product from Copper and Phosphate Production 1978–2000	120				4,265

Source: U.S. Department of Energy. *Statistical Data of the Uranium Industry, January 1, 1979*, Grand Junction, Colorado, 1979.

Note: Columns should not be added because resources with costs averaging $50/lb *include* the two lower cost categories. The total, in thousands of tons, is 4,145 plus 120 = 4,265.

Kilowatt hour at 1974 prices. For comparison, the cost of the fuel for a power plant using oil, in 1974, was 0.6 cents per Kilowatt hour (Sorenson, 1976). The report of the Nuclear Energy Policy Group concluded that assuming U_3O_8 costs at 30 dollars a pound (1976 dollars), a reactor coming into operation in 1985 would use uranium fuel costing only 0.25 cents per Kilowatt hour. These calculations indicate that the cost of U_3O_8 should not be a critical factor in determining the cost of nuclear energy in this century.

An additional consideration is the environmental impact of mining uranium. At the present time, land disturbance resulting from uranium mining is low compared to coal mining. As lower-grade uranium ore begins to be mined in greater volume, the impact will increase. A 1000-Megawatt coal-fired electrical generation plant consumes 3 million tons of coal per year, equivalent to 3.75 million tons of coal in-place. A nuclear power plant of the same capacity would not require this tonnage of ore to be mined until the U_3O_8 content declined from its present average level of 0.20 percent to a level of 0.006 to 0.007 percent, just about equal to that of Chattanooga Shales.

In fact, presently defined resources do not include such low-grade ores (Nuclear Energy Policy Group). It should be noted, however, that the volume of solid wastes from coal is small when compared with uranium whose ore averages around 0.2 percent, so that 99.8 percent of the material mined is discarded as tailings (waste).

REGIONAL DISTRIBUTION OF UNITED STATES' RESOURCES *According to Certainty*. The Department of Energy now recognizes the regional distribution of uranium resources within the framework of twenty uranium resource regions. Three regions, the Colorado Plateau, Wyoming Basins, and Western Gulf Coastal Plain, contain approximately 75 percent, or 3.1 million tons, of the Nation's 50 dollar per pound resources (Fig. 4–6 and Table 4–3). The Colorado Plateau contains 48 percent of the total resources followed by the Wyoming Basins with 18 percent. Most of these resources, found principally in sandstones, are located in areas of past and present production. In addition, these regions also contain proportionally more uranium Reserves, having more than 85 percent of the total 50 dollar per pound uranium (54 percent being on the Colorado Plateau and over 23

Fig. 4–6   Geologic certainty of uranium oxide resources at costs up to $50 per pound (1979) by DOE resource region.

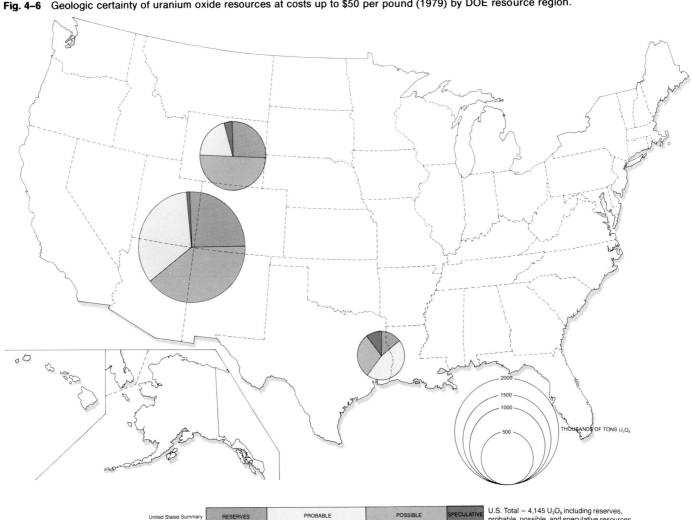

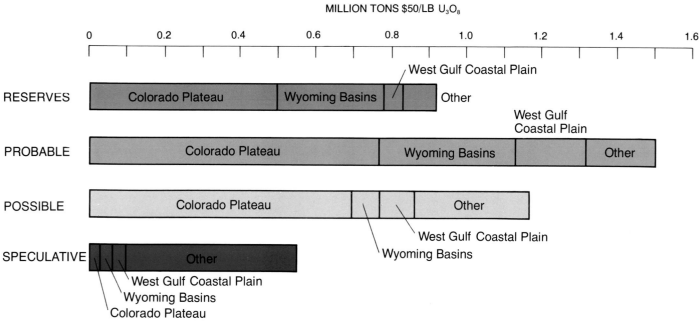

Fig. 4–7 National summary of uranium oxide resources by degree of certainty and by DOE resource region (1979).

percent in the Wyoming Basins (Fig. 4–7). The division of the resources into varying degrees of certainty indicates that regional shares are similar to those for the country as a whole. The Wyoming Basins, which have not been mined as extensively as the Colorado Plateau, have a larger share of total resources in the Reserve and Probable categories. The Western Gulf Coastal Plain province has relatively small Reserve amounts but a substantial amount in the Probable category.

Although Reserves are the most certain of the four resource categories, they are subject to some uncertainty, as shown in Fig. 4–8. The amount used here to represent national Reserves, 920,000 tons, is the amount whose probability is 50 percent. The curve and accompanying table point out that there is a 95 percent probability that

as little as 770,000 tons exist as Reserves, but only a 5 percent chance that as much as 1,080,000 tons of Reserves actually exist. Figures 4–9 to 4–12 indicate the actual amounts of uranium oxide for each region according to the categories of Reserves or Probable, Possible, or Speculative resources.

According to Cost. The regional shares of the three cost categories reflect the national pattern of 23, 39, and 35 percent (Fig. 4–13 and Tables 4–4 through 4–6). Over 56 percent of lowest-cost (15 dollars per pound) uranium resources are located in the Colorado Plateau, with 18 percent in the Wyoming Basins, and 10 percent on the West Gulf Coastal Plain (see bar graph, Fig. 4–14).

According to Certainty and Cost. Figures 4–15 and 4–16 bring together the information from individual maps

Table 4–3 Regional Summary of 50 Dollars per Pound Uranium Reserves and Potential Resources, as of January 1, 1979 (in thousands of tons U_3O_8)

REGION	RESERVES[2]	POTENTIAL RESOURCES			TOTAL
		Probable	Possible	Speculative	
Colorado Plateau	499.1	767	696	30	1,992.1
Wyoming Basins	208.4	364	73	32	749.4
Western Gulf Coastal Plain	49.6	190	93	35	367.6
Northern Plains	7.4	0	0	0	7.4
Others[1]	83.5	184	308	453	1,028.5
TOTAL	920.0	1,505	1,170	550	4,145.0

[1]Includes Colorado and Southern Rockies, Northern Rockies, Northern and Central Basin and Range, Northern Plains, Southern Plains, Sierra Nevada, Pacific Coast, Southern Basin and Range, Southeastern Basin and Range, and Columbia Plateau.

[2]Does not include Reserves recoverable as by-product of copper and phosphate production.
Source: U.S. Department of Energy. *Statistical Data of the Uranium Industry, January 1, 1979*, Grand Junction, Colorado, 1979.

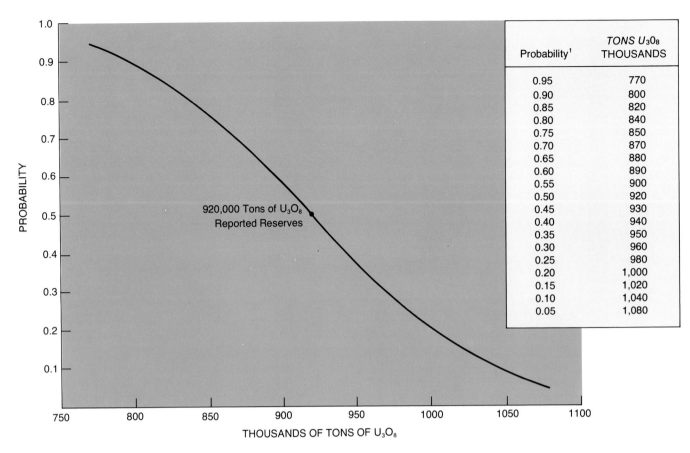

Probability[1]	TONS U_3O_8 THOUSANDS
0.95	770
0.90	800
0.85	820
0.80	840
0.75	850
0.70	870
0.65	880
0.60	890
0.55	900
0.50	920
0.45	930
0.40	940
0.35	950
0.30	960
0.25	980
0.20	1,000
0.15	1,020
0.10	1,040
0.05	1,080

[1]Probability of reserves exceeding given tonnage of U_3O_8.

Fig. 4–8   Probability of uranium oxide reserve tonnages at costs up to $50 per pound (1979).

and summarize all United States' resources by region and also by certainty and cost categories. The bar graph provides a comprehensive view of where the resources are located, their degree of certainty, and how much it is estimated to cost to produce one pound of U_3O_8 concentrate from these resources.

Application of Uranium Resources. Before considering the question of uranium, it is important to understand how uranium ores are used. There are two vital aspects of the use of uranium ores: the fuel cycle and the types of fission reactors.

THE NUCLEAR FUEL CYCLE   Before the mined uranium can be used as a nuclear fuel it must undergo several stages of processing and conversion. The process begins at mills located near areas where uranium is mined (Fig. 4–17). At these mills, the ore is concentrated into U_3O_8, a uranium-uranyl oxide called "yellowcake." Most concentrates prepared contain an average of 80 to 85 percent U_3O_8, but must contain a minimum of 75 percent (Woodmanse).

After being transported to refineries, the concentrate

Fuel rods containing uranium dioxide pellets are inspected at Westinghouse Electric Corporation's fuel fabrication plant, Columbia, South Carolina. *Westinghouse Photo by Jack Merhaut.*

Table 4–4 Fifteen Dollars per Pound U_3O_8 Reserves and Potential Resources by Physiographic Region and Geologic Certainty, January 1, 1979 (in thousands of tons U_3O_8)

REGION	RESERVES[2]	POTENTIAL RESOURCES			TOTAL
		Probable	Possible	Speculative	
Colorado Plateau	201.9	217.0	138.0	1.0	557.9
Wyoming Basins	57.0	93.0	10.0	14.0	174.0
Western Gulf Coastal Plain	10.2	62.0	22.0	1.0	95.2
Northern Plains	2.6	0	0	0	2.6
Others[1]	18.3	43.0	40.0	59.0	160.3
TOTAL	290.0	415.0	210.0	75.0	990.0

[1]Includes Rocky Mountains, Northern and Central Basin and Range, Great Plains, Pacific Coast, Sierra Nevada, Columbia Plateaus, Central Lowlands, Appalachians, and Alaska.

[2]Does not include Reserves recoverable as by-product of copper and phosphate production.
Source: U.S. Department of Energy. *Statistical Data of the Uranium Industry, January 1, 1979,* Grand Junction, Colorado, 1979.

Table 4–5 Thirty Dollars per Pound U_3O_8 Reserves and Potential Resources by Physiographic Region and Geologic Certainty, January 1, 1979 (in thousands of tons U_3O_8)

REGION	RESERVES[2]	POTENTIAL RESOURCES			TOTAL
		Probable	Possible	Speculative	
Colorado Plateau	395.2	557.0	428.0	16.0	1,396.2
Wyoming Basins	191.8	243.0	22.0	22.0	478.8
Western Gulf Coastal Plain	40.7	119.0	60.0	22.0	241.7
Northern Plains	6.0	0	0	0	6.0
Others[1]	56.3	86.0	165.0	240.0	547.3
TOTAL	690.0	1,005.0	675.0	300.0	2,670.0

[1]Includes Rocky Mountains, Northern and Central Basin and Range, Great Plains, Pacific Coast, Sierra Nevada, Columbia Plateaus, Central Lowlands, Appalachians, and Alaska.

[2]Does not include Reserves recoverable as by-product of copper and phosphate production.
Source: U.S. Department of Energy. *Statistical Data of the Uranium Industry, January 1, 1979,* Grand Junction, Colorado, 1979.

Table 4–6 Fifty Dollar per Pound U_3O_8 Reserves and Potential Resources by Physiographic Region and Geologic Certainty, January 1, 1979 (in thousands of tons U_3O_8)

REGION	RESERVES[2]	POTENTIAL RESOURCES			TOTAL
		Probable	Possible	Speculative	
Colorado Plateau	499.1	767.0	696.0	30.0	1,992.1
Wyoming Basins	280.4	364.0	73.0	32.0	749.4
Western Gulf Coastal Plain	49.6	190.0	93.0	35.0	367.6
Northern Plains	7.4	0	0	0	7.4
Others[1]	83.5	184.0	308.0	453.0	1,028.5
TOTAL	920.0	1,505.0	1,170.0	550.0	4,145.0

[1]Includes Rocky Mountains, Northern and Central Basin and Range, Great Plains, Pacific Coast, Sierra Nevada, Columbia Plateaus, Central Lowlands, Appalachians, and Alaska.

[2]Does not include Reserves recoverable as by-product of copper and phosphate production.
Source: U.S. Department of Energy. *Statistical Data of the Uranium Industry, January 1, 1979,* Grand Junction, Colorado, 1979.

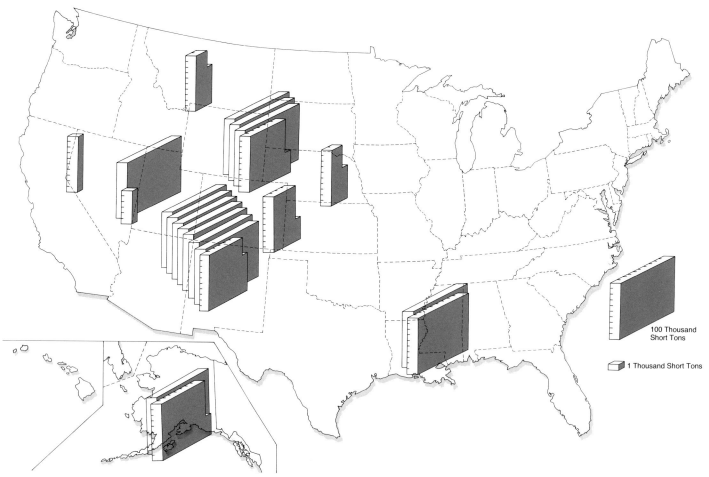

Fig. 4–9 Uranium oxide *reserve* tonnage at costs up to $50 per pound (1979).

Fig. 4–10 Uranium oxide *probable* resource tonnage at costs up to $50 per pound (1979).

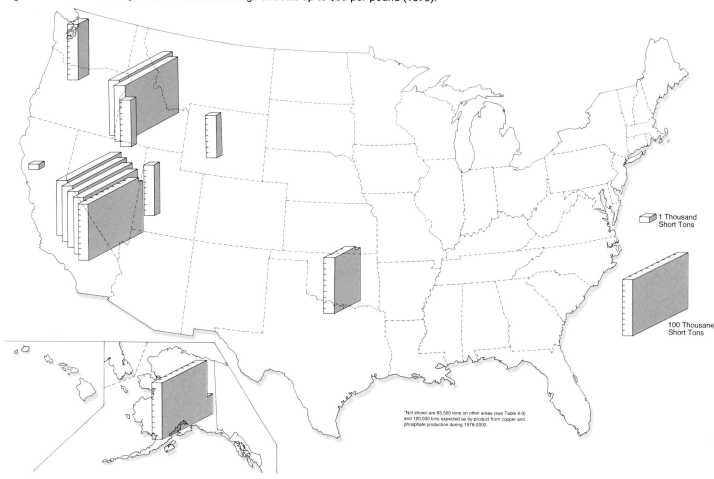

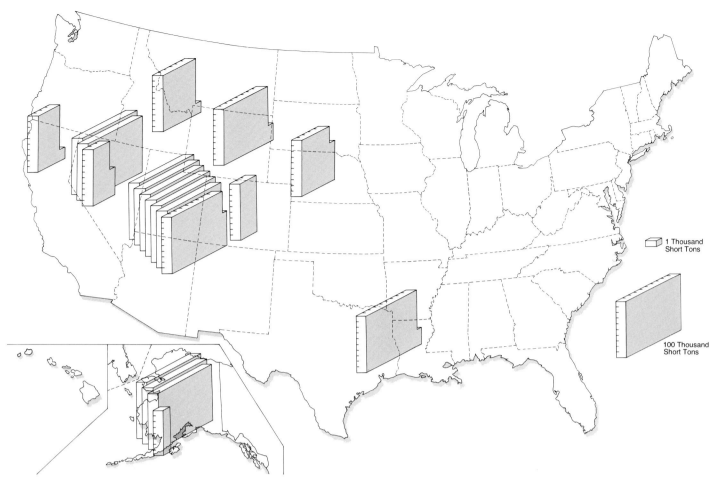

Fig. 4–11 Uranium oxide *possible* resource tonnage at costs up to $50 per pound (1979).

Fig. 4–12 Uranium oxide *speculative* resource tonnage at costs up to $50 per pound (1979).

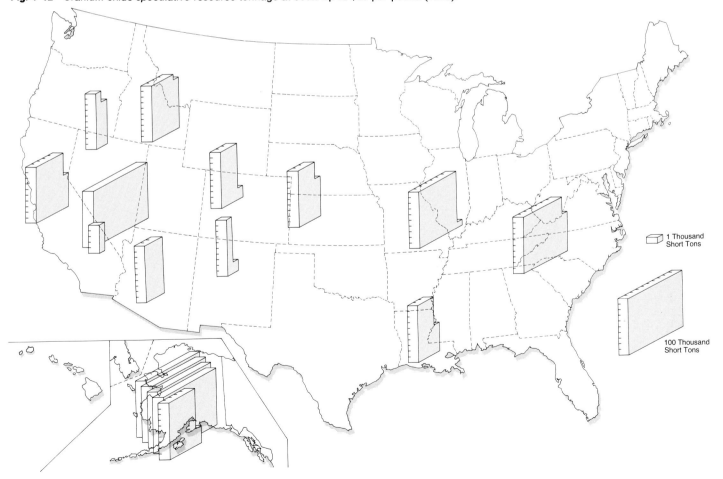

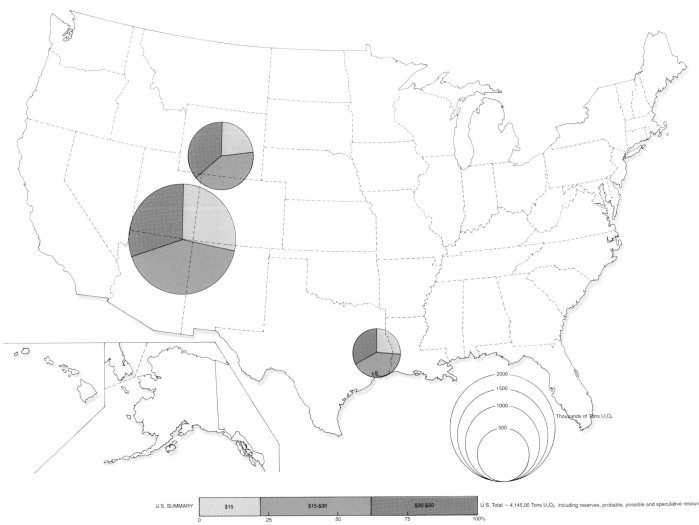

| U.S. SUMMARY | $15 | $15-$30 | $30-$50 | U.S. Total = 4,145,00 Tons U₃O₈ including reserves, probable, possible and speculative resources |

Fig. 4–13   Costs of total uranium oxide resources up to $50 per pound (1979).

Two of the thirty storage tanks for liquid high-level radioactive wastes at the Department of Energy's Savannah River facility at Barnwell, South Carolina. When completed, the double-walled steel tanks will be encased in two to three feet of concrete and covered with soil. Liquid wastes stored at such facilities are produced by uranium enrichment plants, reactors, weapons work, and reprocessing of spent fuel. *E. I. Dupont Denemours & Co., from Department of Energy.*

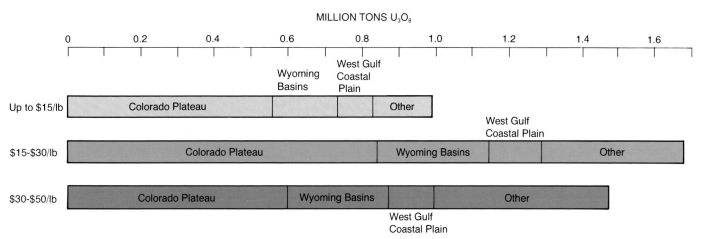

Fig. 4–14 National summary of uranium oxide resources by cost and by physiographic region (1979).

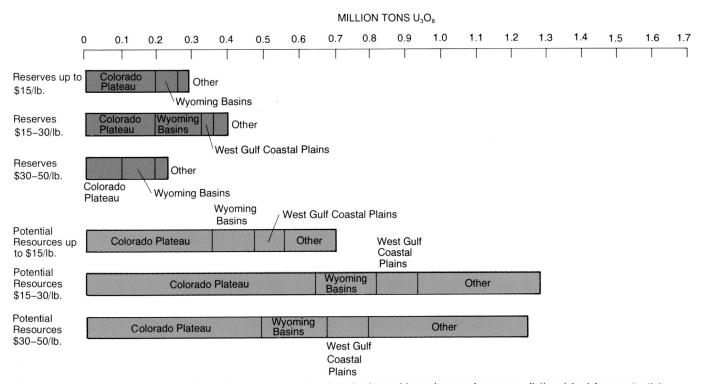

Fig. 4–15 National summary of uranium oxide resources, showing physiographic regions and reserves distinguished from potential resources (1979).

mill product is processed to remove impurities. After purification, the U_3O_8 is converted into a hexafluoride product, UF_6, which is gasified for enrichment at federally operated gaseous diffusion plants at Oak Ridge, Tennessee, Paducah, Kentucky, or Portsmouth, Ohio (Fig. 4–17). The enrichment process increases the content of the Uranium-235 isotope from 0.7 percent to specific levels depending on the end-use of the enriched product. For the boiling water reactor, enrichment is to 2.03 percent Uranium-235 and for the pressurized water reactor 2.26 percent. For other reactors, such as the high-temperature gas-cooled reactor and for use in weapons, enrichment may be to over 90 percent Uranium-235. At

this stage, about 98 percent of the natural uranium is left behind in the form of Uranium-238 and is stored at the gaseous diffusion facilities in stainless steel drums (Woodmanse).

After enrichment, the UF_6 is shipped to plants where it is converted into fuel form for use in the manufacture of fuel elements. The different forms include: uranium oxide (UO_2), metals, alloys, carbides, nitrides, and salt solutions. The most common form used in light water reactors is pelletized ceramic UO_2. The pellets are loaded into steel or zirconium cladding (covered) tubes and fastened together into fuel bundles.

During the operation of a nuclear reactor, fission

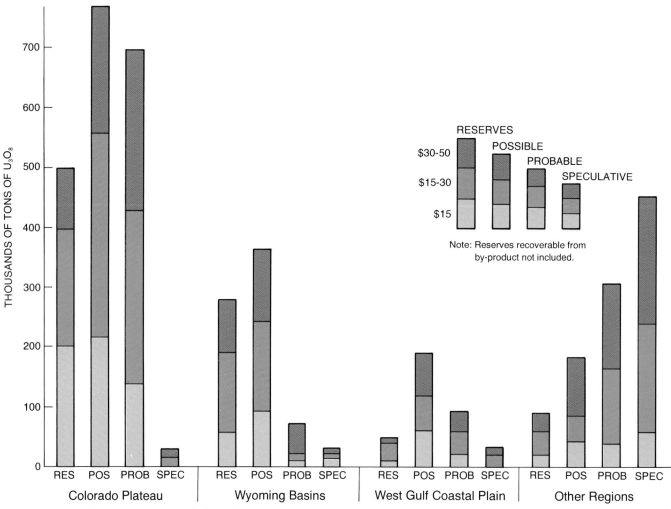

Fig. 4–16   Region by region summary of all uranium resources showing costs and degree of certainty (1979).

products accumulate in the fuel elements and reduce reactivity by absorbing neutrons. Because of this, over 25 percent of the reactor fuel core is removed and replaced on an annual basis. These highly radioactive spent fuel rods are submerged in water to allow for cooling and radioactive decay and are generally stored on the reactor site for several months. According to representatives of the Philadelphia Electric Company, 2 to 3 cubic feet of storage space is required for every metric ton of uranium loaded as fuel.

After the cool-down period, the spent fuel may be shipped in specially designed containers to reprocessing plants where uranium and plutonium are separated and recovered for re-use. There are presently no privately owned reprocessing plants operating in the United States, but some reprocessing is carried out at federal facilities. At one time, a privately owned plant operated for several years at West Valley, New York, but because of numerous containment and handling problems has been shut down. There are plans for additional reprocessing plants to be built in Washington and South Carolina but it is uncertain if and when these plants will be operable (Fig. 4–17).

Nuclear reactors accumulate radioactive waste and products in liquid, solid, and gaseous forms. These are differentiated as low-level and high-level wastes. The wastes are extracted from the spent fuels, transported, and disposed of by permanent burial or by storage. The low-level wastes include alpha materials contaminated with Plutonium-239, an isotope that does little damage on the outside of the body but is intensively radioactive when deposited on sensitive lung tissue. High-level wastes are accumulated during fuel reprocessing. Presently, there are liquid weapons' waste storage areas in Hanford, Washington, Aiken, South Carolina, and Idaho Falls, Idaho as well as solid waste burial grounds in such places as Oak Ridge, Tennessee, and Maxie Flats, Kentucky (Fig. 4–17).

As of yet, permanent storage methods for high-level radioactive waste have not been developed. The lack of safe methods of storing high-level waste products is a major reason why many people object to nuclear power development, and especially to the development of new-generation breeder reactors which produce highly toxic plutonium waste. The U.S. Department of Energy is considering permanent disposal techniques, such as burial in deep geological formations (salt beds, granites, and basalts), in polar areas under icecaps, or ejection to the sun or outer space (Woodmanse).

TYPES OF FISSION REACTORS Most of the nuclear reactors in commercial use in the United States are slow neutron, light water moderated, and use enriched Uranium-235 as fuels.[3] In the nuclear vernacular, these reactors are referred to as "converters." There are two basic types of light water reactors, the boiling water reactors (BWR) and the pressurized water reactors (PWR), whose components are illustrated in Figs. 4–18A and 4–18B.

Boiling water reactors were first used commercially in the United States at Commonwealth Edison's Dresden, Illinois plant in 1960. Inside a BWR the water is in contact with the hot fuel rods at a pressure of about 1000 pounds per square inch. The steam generated rises to the top of the reactor core, and after passing through steam separators and dryers, is piped directly to the steam turbine. Steam bubbles that form around the fuel rods aid in controlling the chain reaction by reducing the moderating effectiveness of water, thus slowing the growth of the chain reaction.[4] As the steam comes into direct contact with the fuel rods in the BWR, there is a possibility that radioactive materials may invade the turbine and leak to the environment. For this reason it is extremely important to seal the turbine.

In the pressurized water reactor (PWR), there are two separate water systems. Water circulates through the reactor core, under high pressure of 2000 pounds per square inch, and remains in the liquid form up to a temperature of 600 degrees Fahrenheit. Because the steam generator is completely separated from other parts of the system, it is easier to contain the radioactive materials. In order to compensate for the slowing down of the nuclear reaction in the core as the fuel is expended, boron is added to the cooling water of a PWR at the time the reactor starts up. As the power level decreases, the boron, which has been absorbing neutrons, is gradually removed from the coolant, releasing more neutrons to take part in the reaction (Fowler).

Figure 4–18C illustrates a relatively new type of reactor, the high-temperature gas-cooled reactor (HTGR). Although the HTGR is classified as a converter, it does produce new fissionable material. In this system, helium at 1400 degrees Fahrenheit is circulated through the reactor core. After passing through a heat exchanger the helium produces steam at approximately 1000 degrees Fahrenheit. In the HTGR, Thorium-232 is added to the reactor in order to create more fuel. Proponents of the

Fig. 4–17 Locations and flows in the uranium nuclear fuel cycle, showing material movement in truckloads (1975).

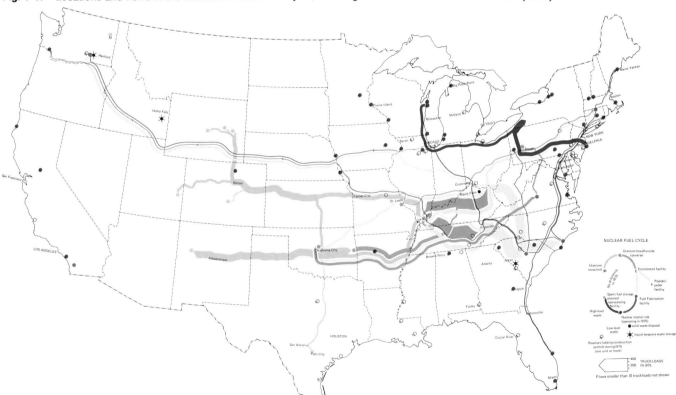

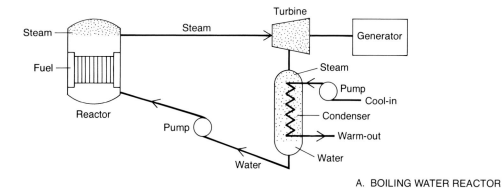

A. BOILING WATER REACTOR

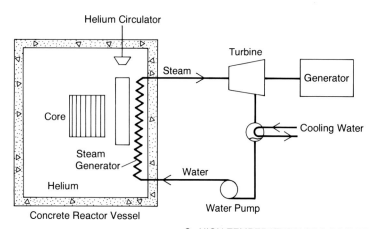

B. PRESSURED WATER REACTOR

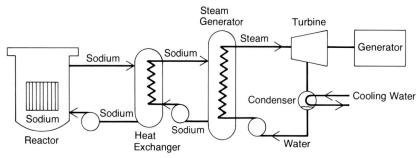

C. HIGH-TEMPERATURE GAS-COOLED REACTOR

D. LIQUID METAL FAST BREEDER REACTOR

Fig. 4-18 Four types of nuclear reactors for power generation by nuclear fission.

HTGR claim this reactor offers a thermal efficiency of 40 percent which would be equal to the best fossil-fuel plants in operation. In addition, representatives of Philadelphia Electric, which operated a small capacity HTGR for several years, indicate that fuel savings in this type of reactor could amount to about 30 percent because of its ability to produce new fuel from Thorium-232.

THE BREEDER Since the beginning of nuclear reactor development, a great deal of time has been spent considering ways of obtaining energy from the more abundant but non-fissionable Uranium-238. With the scarcity of uranium, attention has begun to focus on the breeder reactor, since it has the capability of converting non-fissionable Uranium-238 into the fissionable isotope, Plutonium-239 (Fig. 4–2C). There are two factors of major importance in a technical evaluation of breeder reactors: the "breeding ratio," or the ratio of Plutonium-239 produced to the nuclear fuel consumed; and the "doubling time," the number of years it takes to double the initial fuel load (Fowler).

There are two basic types of breeders: thermal breeders that operate with low-energy neutrons and fast breeders that operate with high-energy neutrons. At present, the liquid metal fast breeder reactor (LMFBR) illustrated in Fig. 4–18D has generated the most interest among both the proponents and opponents of breeder development. The LMFBR is attractive because its breeding ratio is estimated to be in the range of 1.4 to 1.5, which means that for every two fissionable nuclei burned as fuel three new ones would be created. In addition, doubling times in the order of eight to ten years are expected.

Since the LMFBR can be fueled with the enriched Uranium-238 now stored at enrichment plants, commercial operation could begin without further mining of uranium ore. Lapp reports that breeder power would allow the United States to tap the 200,000 tons of Uranium-238 presently stored in steel vessels at Oak Ridge, Tennessee. Furthermore, he argues that if the present supply of uranium resources were converted to plutonium, it would yield the energy equivalent of 400 billion tons of coal—the approximate amount the United States has underground in a mineable form (*Fortune*, October 1975).

A very strong case can be made for the fast breeder on the basis of its efficient use of uranium. Light water reactors (LWR) can use only two percent of uranium mined, whereas the fast breeder could use 50 to 70 percent (Fowler). In addition, the requirements for initial fuel are much lower for the fast breeder. Consider, for example, the fuel needs for a 1000-Megawatt generating plant operating at 80 percent capacity for coal-fired plants, nuclear light water reactors, and fast breeders. The initial fuel requirements for the coal-fired plant would be 200,000 tons; a light water converter would require approximately 200 tons, whereas the LMFBR would need only 1.4 tons (U.S. Atomic Energy Commission).

It seems clear that if our present assumptions regarding fast breeder reactors were shown to be correct and if the breeder could be developed safely, the United States would have a virtually unlimited supply of energy for several thousands of years.[5]

There are, however, a number of major problems associated with breeders that raise serious and legitimate doubts about a breeder reactor program. One of the prime concerns of opponents of fast breeders is the proliferation of highly toxic PU_{239} which has a half-life (the time period required for one half the atoms to decay into another isotope) of about 24,000 years. Developing ways to safely handle and store plutonium waste for such a long period is a problem yet to be solved. Furthermore, it has been argued that if the United States and other countries move ahead with fast breeder development, the large stockpiles of plutonium, capable of producing bombs would encourage production of nuclear weapons. Preventing the spread of plutonium throughout the world and protecting it from theft are problems of immense proportion.

CUMULATIVE AND CURRENT PRODUCTION After a period of rapid production spurred by weapons development and stockpiling from the mid 1950s to the early 1960s, the United States' output of uranium dropped dramatically until 1966. For nearly a decade following, uranium production remained fairly uniform. Recently, it has appeared to be entering a new growth period (Fig. 4–19). Through 1978, cumulative production amounted to approximately 333,000 tons of uranium taken from 157 million tons of uranium ore. The regional distribution of cumulative and 1978 production are shown in Figs. 4–20 and 4–21. The Colorado Plateau and Wyoming Basins have dominated past production and, in 1978, accounted for 87 percent of current production. The leading uranium-producing states, New Mexico and Wyoming, accounted for 47 and 27 percent respectively of the uranium mined, while Colorado, Florida, Utah, Washington and Texas produced the remaining 26 percent. Through 1978 major areas of production were in Ambrosia and Laguna in New Mexico, and Shirley Basin and Gas Hills in Wyoming (Fig. 4–22).

The Department of Energy reported that 55 percent more uranium ore was processed in 1978 than a decade before. This figure is deceptive, however, because the average grade of ore and the percentage of recovery of uranium contained in the ore have decreased, making the increase of uranium concentrate only 33 percent. Between 1965 and 1978 the average grade of processed ore decreased from 0.24 percent to 0.12 percent and the average recovery of the contained U_3O_8 declined from 95 to 93 percent.

STATUS OF U.S. NUCLEAR POWER GENERATION CAPACITY In the period between 1965 and 1979, the number of operable nuclear power plants in the United States grew almost six-fold, from twelve to seventy plants (U.S. Nuclear Regulatory Commission, 1979). During the same period the Nation's nuclear power

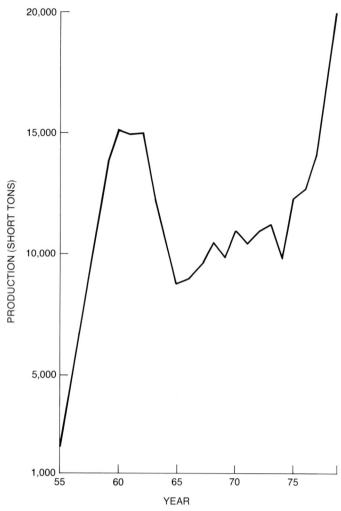

Fig. 4–19   U.S. annual production of uranium oxide 1955–1978.

generation capacity increased from a little more than 1000 Megawatts to over 51,000 Megawatts (MW). On the average, these plants have operated at about 60 percent of their designed capacity because of frequent breakdowns, lengthy maintenance operations and federal safety precautions (Fig. 4–23).

In addition to the seventy nuclear power plants that were operable as of January 1979, ninety-two plants with a generating capacity of 100,000 MW have been granted construction permits, and twenty-eight plants with a capacity of 32,000 MW are under construction permit review (see list). Four additional reactors with capacity of 2000 MW have been ordered and two reactors with a capacity of 2000 MW have been publicly announced (Nuclear Regulatory Commission). If all the plants under construction and those planned are completed, the nuclear power generation capacity of the United States will amount to 196,000 MW at the turn of the century, an almost three-fold increase over the 1979 capacity. In 1979, nuclear power produced 3 percent of the United States' energy; coal 19 percent; oil 49 percent; natural gas 26 percent; and hydropower and geothermal 3 percent.

Almost 75 percent of nuclear power capacity is located east of the Mississippi River, and one-half of that capacity is in the populous states of the northeast quadrant (Fig. 4–24). If and when those plants under construction and planned become operable, the states east of the Mississippi will still house approximately three-fourths of the nation's nuclear power capacity. Because of a large capacity planned for the southeastern states, however, the northeast quadrant's share will drop to about 40 percent.

PROSPECTS FOR DEVELOPMENT AFTER THREE MILE ISLAND   During the last week in March and the first week of April, 1979, the worst commercial nuclear accident in United States' history occurred at the Three Mile Island power plant near Harrisburg, Pennsylvania. A chain of mechanical and human failures triggered by a coolant water valve malfunction led to a series of radioactive steam leaks that spread over an area of up to twenty miles from the plant. There was widespread disagreement among the experts as to the hazard posed by the radiation leak. Many claimed that there was no threat to health, while others warned of the threat of cancer and possible genetic damage. Only time can determine whose claims were accurate.

In the aftermath of the Three Mile Island crisis, nuclear power faces a highly uncertain future. Because of growing anti-nuclear movements given impetus by the Three Mile Island incident, many of the ninety-two plants now under construction and the more than thirty that are planned could be delayed or even cancelled. At the very least, there will be a renewed national debate arguing the risks of nuclear power against the Nation's energy needs.

PROJECTED SUPPLY AND DEMAND FOR URANIUM Because of a number of uncertainties associated with the future development of nuclear power, it is difficult to estimate future demand for uranium resources. However, to test the range of the supplies of uranium discussed earlier, Fig. 4–25 offers a comparison of uranium supplies and requirements for two scenarios (Nuclear Energy Policy Study Group, 1977).

Scenario #1 is based on uranium requirements for the currently planned capacity of approximately 200,000 MW according to the assumptions outlined in Fig. 4–25. As indicated, the requirements for currently planned capacity would amount to approximately 1.3 million tons of U_3O_8 from now through the years 2015 to 2020.[6] As the bar graph shows, requirements for this scenario could be met by using all the presently defined Reserves, including those 120,000 tons of U_3O_8 recoverable as a by-product, and 17 percent of the Probable resources up to the 50 dollars per pound cut-off. Alternately, all Reserves, Probable, and Possible resources with a forward cost of 30 dollars per pound of U_3O_8 would be more than sufficient to meet the requirements.

Scenario #2 is the so-called ERDA "MID" scenario which calls for a 510,000-MW capacity in the year 2000 and cumulative uranium requirements of 1.2 million tons between the years 2000 and 2030. Total requirements for the 510,000-MW capacity up to the year 2030 would be almost 4 million tons of uranium, slightly less than the

Status of Licenses for New Reactors, 1979

	STATE	Name and Number of Reactors	Operating Utility	Nearest Settlement
■	Alabama	Bellefonte 1, 2	T.V.A.	Scottsboro
■	Alabama	Farley 2	Alabama Power	Dothan
●	Arizona	Palo Verde 1, 2, 3	Arizona Public Service	Wintersburg
●	California	Diablo Canyon 1, 2	Pacific Gas & Electric	Diablo Canyon
■	California	San Onofre 2, 3	Southern California Edison	San Clemente
●	Florida	St. Lucie	Florida Power & Light	Fort Pierce
●	Georgia	Vogile 1, 2	Georgia Power	Waynesboro
■	Illinois	Braidwood 1, 2	Commonwealth Edison	Braidwood
■	Illinois	Byron 1, 2	Commonwealth Edison	Byron
●	Illinois	Clinton 1, 2	Illinois Power	Clinton
■	Illinois	LaSalle 1, 2	Commonwealth Edison	Seneca
●	Indiana	Bailly	Northern Indiana Public Service	Westchester
●■	Indiana	Marble Hill 1, 2	Public Service of Indiana	Madison
●	Kansas	Wolf Creek	Kansas Gas & Electric	Burlington
●	Louisiana	River Bend 1, 2	Gulf States Utilities	St. Francisville
■	Louisiana	Waterford 3	Louisiana Power & Light	Taft
●	Massachusetts	Millstone 3	Northeast Nuclear Energy	Waterford
★	Massachusetts	Pilgrim 2	Boston Edison	Plymouth
■	Michigan	Fermi 2	Detroit Edison	Newport
■	Michigan	Midland 1, 2	Consumers Power of Michigan	Midland
■	Mississippi	Grand Gulf 1, 2	Mississippi Power & Light	Port Gibson
●	Mississippi	Yellow Creek 1, 2	T.V.A.	Corinth
●	Missouri	Callaway 1, 2	Union Electric	Fulton
●	New Hampshire	Seabrook 1, 2	Public Service of New Hampshire	Seabrook
●	New Jersey	Forked River	Jersey Central Power & Light	Forked River
●	New Jersey	Hope Creek 1, 2	Public Service Electric & Gas	Salem
■	New Jersey	Salem 2	Public Service Electric & Gas	Salem
●	New York	Jamesport 1, 2	Long Island Lighting	Jamesport
●	New York	Nine Mile Point 2	Niagara Mohawk Power	Scriba
■	New York	Shoreham	Long Island Lighting	Brookhaven
■	New York	Sterling 1	Rochester Gas & Electric	Oswego
●	North Carolina	Harris 1, 2, 3, 4	Carolina Power & Light	Bonsal
■	North Carolina	McGuire 1, 2	Duke Power	Cowans Ford Dam
★	North Carolina	Perkins 1, 2, 3	Duke Power	Davie County
★	Ohio	Davis-Besse 2, 3	Toledo Edison	Oak Harbor
★	Onio	Erie 1, 2	Ohio Edison	Berlin Heights
●	Ohio	Perry 1, 2	Cleveland Electric Illuminating	Perry
■	Ohio	Zimmer 1	Cincinnati Gas & Electric	Moscow
★	Oklahoma	Black Fox 1, 2	Public Service of Oklahoma	Inola
★	Oregon	Pebble Springs 1, 2	Portland General Electric	Arlington
●	Pennsylvania	Beaver Valley 2	Duquesne Light	Shippingport
●	Pennsylvania	Limerick 1, 2	Philadelphia Electric	Pottstown
■	Pennsylvania	Susquehanna 1, 2	Pennsylvania Power & Light	Berwick
●■	South Carolina	Catawba 1, 2	Duke Power	Lake Wylie
●	South Carolina	Cherokee 1, 2, 3	Duke Power	Cherokee County
■	South Carolina	Summer 1	South Carolina Electric & Gas	Broad River
●	Tennessee	Hartsville 1, 2, 3, 4	T.V.A.	Hartsville
●	Tennessee	Phipps Bend 1, 2	T.V.A.	Kingsport
■	Tennessee	Sequoyah 1, 2	T.V.A.	Daisy
■	Tennessee	Watts Bar 1, 2	T.V.A.	Spring City
★	Texas	Allens Creek 1	Houston Lighting & Power	Wallis
■	Texas	Comanche Peak 1, 2	Texas Utilities Generating	Glenrose
■	Texas	South Texas 1, 2	Houston Lighting & Power	Bay City
■	Virginia	North Anna 2	Virginia Electric & Power	Mineral
●	Virginia	North Anna 3, 4	Virginia Electric & Power	Mineral
■	Washington	W.P.P.S.S. 1, 4	Wash. Public Power Supply System	Richland
■	Washington	W.P.P.S.S. 2	Wash. Public Power Supply System	Richland
●	Washington	W.P.P.S.S. 3, 5	Wash. Public Power Supply System	Satsop
★	Wisconsin	Haven 1	Wisconsin Electric Power	Sheboygan County
●	Wisconsin	Tyrone 1	Northern States Power	Durand

■ Operating permit pending.   ● Construction permit granted.   Source: *New York Times*, October 23, 1979.
★ Construction permit pending.

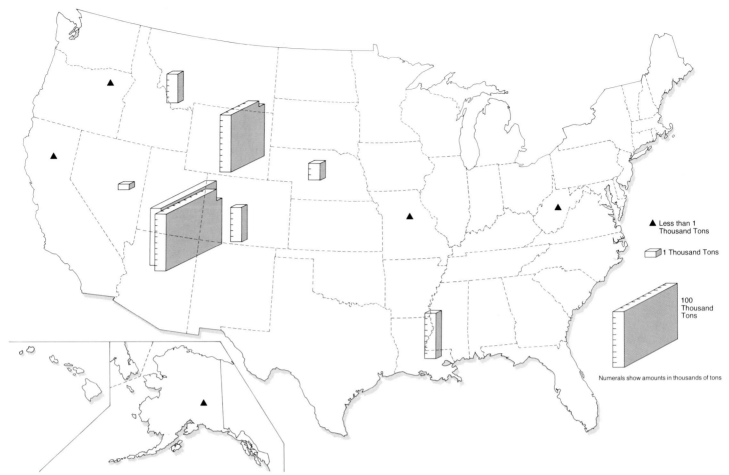

Fig. 4–20 Cumulative production of uranium oxide through 1978 showing dominance of the Colorado Plateau and Wyoming Basins.

currently estimated resource base at a forward cost of 50 dollars per pound.

Several factors could change this supply and demand situation. If many more fast breeder reactors were built after the mid 1980s, the demand for uranium in the year 2000 would be substantially reduced (Fig. 4–26). In addition, according to the Nuclear Energy Policy Study Group, recycling of uranium and plutonium could reduce the demand for U_3O_8 by 300,000 to 900,000 tons.[7]

As indicated earlier, it is possible that nuclear reactors could be fueled with 100 dollars per pound uranium and still produce electricity at a competitive price. Figure 4–27 indicates there are very large amounts of potential uranium resources, such as the Chattanooga shales, which would have forward costs of greater than 50 dollars per pound. Because of its low uranium oxide content, this shale may be most appropriate for supplying breeder reactors which employ the more abundant isotope, U_{238}. It may be possible, however, to derive fuel for nonbreeders from such a rock. In one area in eastern Tennessee, the Chattanooga shale contains 60 grams of uranium for every metric ton. Assuming a density of 2.5 metric tons per cubic meter, a vertical column of rock 5 meters tall and one square meter in cross-section would contain 12.5 tons of rock, and consequently 750 grams or 1.65

pounds of uranium (Hubbert, 1971). Since each pound of uranium is only 0.7 percent U_{235}, this 1.65 pounds contains 0. 0116 pounds of the fissionable isotope. Using the equivalence, 1 pound U_{235} to 1,360 tons of bituminous coal (Butler, 1967), this column of relatively rich shale 5

Fig. 4–21 Production of uranium oxide in 1978 showing contributions of leading regions.

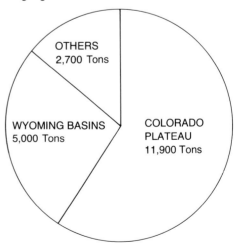

U.S. Total 20,200 Tons U_3O_8

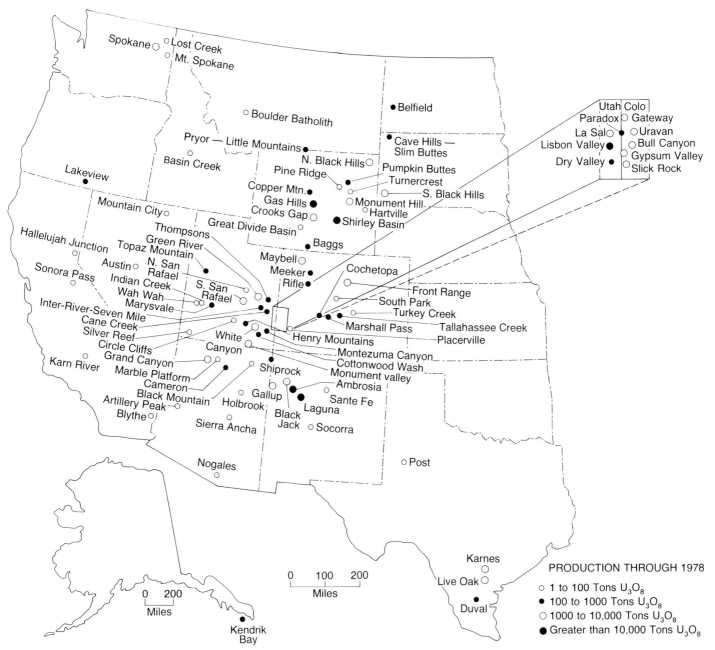

Fig. 4–22 Sites of uranium oxide production indicating tonnage produced through 1978.

feet tall and one square meter in area weighing 12.5 tons holds the energy of 15.77 tons of bituminous coal. Although a ton of the shale, used only for its U_{235} content, to fuel non-breeder reactors, contains slightly more energy than a ton of bituminous coal, there would be a great deal more waste rock from the shales, since only 0.006 percent of it is uranium.

The practicality of such low-grade uranium ore will depend not only upon volumes of rock to be mined, but upon the net energy gain, that is, whether the energy realized in a reactor exceeds energy expended in mining, concentrating, refining, and enriching the uranium compounds. It is interesting that the same Chattanooga shales, being rich in organic matter, are a low-grade source of *kerogen* or shale oil (averaging 10 gallons of oil per ton of shale) which presumably would be recovered if the rock were to be processed for its uranium content.

The Nature and Occurrence of Thorium

At present, thorium is an element of limited economic importance. It could, however, become an important nuclear fuel if the Nation's uranium resources become more scarce and more costly. As illustrated earlier, Thorium-232 can be transmuted under neutron bombardment into Uranium-233, a fissionable isotope in a chain reaction (see Fig. 4–2B). If thorium-fueled reactors, such as the high-temperature gas-cooled reactor (HTGR), the light water breeder reactor (LWBR), and the molten salt breeder reactor (MSBR), become technologically and economically feasible, there will be a greatly increased demand for thorium.

Thorium is a heavy, silver-grey metal widely distributed in nature, and usually found in association with uranium or other rare earth elements. Its geochemical abundance is estimated to range from 6 to 13 parts per million in the earth's crust, which is perhaps three times that of uranium (Staatz, *et al.*). Thorium concentrations of potential economic importance occur in beach sands (placers), in vein deposits in sedimentary rocks, such as thorium-bearing dolomite, and in conglomerates or quartzites enriched in thorium and uranium (Sondermayer). The chief ore minerals of thorium are monazite, thorite, uranothorite, and bannerite, of which monazite in beach sands is the major source of thorium.

The principal world thorium-bearing deposits are located in the United States (Atlantic Coastal Plain),

Thorium

Thorium is a grey radioactive metal which occurs in nature as Thorium-232. Other isotopes include Thorium-233 and Thorium-234.

Symbol	Th
Atomic number	90
Atomic weight	232.04
Melting point	1700°C
Boiling point	4500°C
Density	11.725g/cm³

Fig. 4–23   U.S. nuclear power plants 1965–1976 showing growth in capacity and overall capacity factor through time.

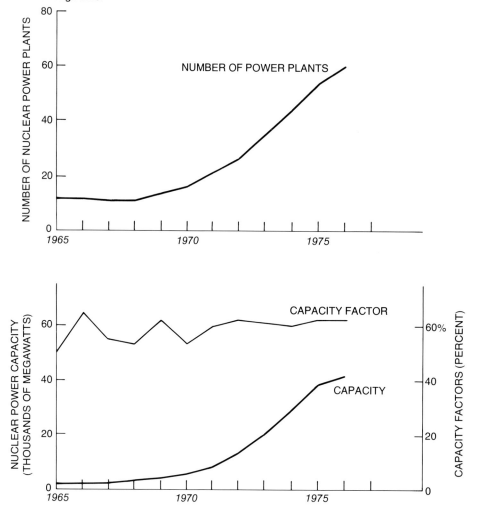

Capacity Factors: Calculated by dividing total electricity generated by that which would have been produced if all plants had operated continuously at full design capacity.

Three Mile Island nuclear generating station on the Susquehanna River near Middletown, Pennsylvania, operated by Metropolitan Edison Company. This photo of the station, viewed from the west, was taken before the accident in Spring of 1979 that put Unit Two (right half of photo) out of service. Unit One (located on the left half) was undamaged by the accident. The plant's generating capacity is 800 Megawatts. *Courtesy Metropolitan Edison Company, Reading, Pennsylvania.*

The Brown's Ferry Nuclear generating plant on the Tennessee River near Athens, Alabama. Its three generating units have a total capacity of 3.3 Megawatts, making it the largest nuclear generating plant in the country. *Courtesy Tennessee Valley Authority.*

Canada (Eliot Lake, Ontario), India (Bihar and West Bengal), and Brazil (Atlantic Coast). Excepting the conglomerate deposits of Canada, these deposits are *fluviatile*, or those produced by river waters, and *beach placers*, or those produced by ocean waters. Additional thorium deposits are found throughout the world in a variety of geological settings in Australia, Africa, Greenland, and Asia.

RESERVES AND RESOURCES

Total world resources of thorium amount to 2 million tons, with 780,000 tons of that in Reserves (Fig. 4–28). North America and Asia house 75 percent of the total Reserves and 72 percent of the total resources. Canada and India contain 29 and 26 percent of the Reserves and 33 and 22 percent of the total resources, respectively; the United States accounts for 18 percent of the thorium Reserves and 14 percent of the total resources.

Due to the limited demand, thorium resources of the United States are not well known. Presently identified thorium resource areas of the United States are shown in Fig. 4–29 and resource amounts are summarized in Table 4–7. At present, the only mineable reserves are localized beach placers along the Atlantic Coast, where monazite containing about 4 percent thorium oxide (ThO_2) is produced as a by-product of titanium mining (Staatz, *et al.*). These monazite sands in Green Cove Spring, Florida, Hilton Head Island, South Carolina, and Folkston, Georgia are estimated to contain 16,000 tons of ThO_2. In addition, there are approximately 28,000 tons ThO_2 that are recoverable as a by-product of rare earth mining in carbonatite deposits at Mountain Pass, California. Larger amounts of ThO_2, over 100,000 tons, are located in vein deposits at several locations in the western United States. The principal vein deposits, in the Lemhi District of Idaho and Montana, contain ore of more than 0.1 percent ThO_2 which is considered recoverable for its thorium content alone. Lower grade ThO_2 ores (less than 0.1 percent ThO_2) amounting to over 140,000 tons are considered recoverable from fluviatile placers in North Carolina, South Carolina, Idaho, and Montana.

Fig. 4–24 Nuclear reactor sites, and state nuclear generating capacity either (1) licensed and operating, or (2) under construction or planned as of January 31, 1979.

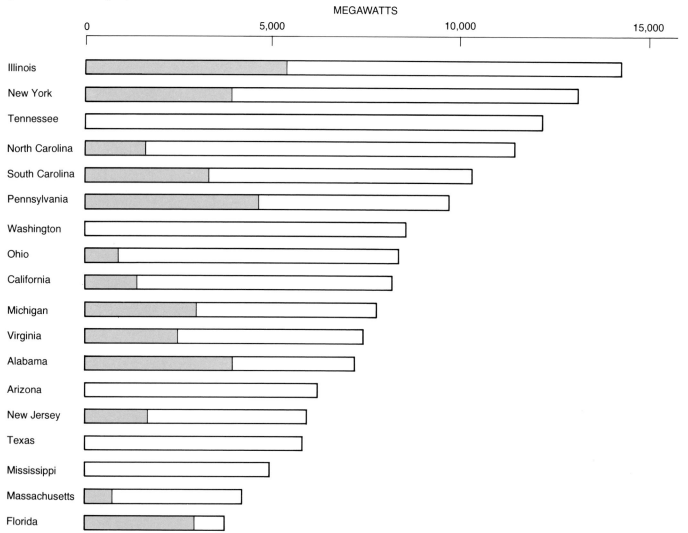

Considering the present low demand, it would appear that Atlantic Coast placer deposits will remain the major source of thorium in the United States for some time. It is also likely that new placer deposits, located below sea level and off present Atlantic Coast beaches, will be discovered in the future. One estimate of monazite content in undiscovered offshore deposits on the United States continental shelf is about 4 million tons (McKelvey, Ocean Industry, 1968). If these monazite sands contain 4 percent ThO_2 as do other known deposits of this type, undiscovered continental shelf deposits could amount to 160,000 tons of ThO_2.

UNITED STATES SUPPLY AND DEMAND

Total United States thorium consumption in all forms in 1974 was estimated at 80 short tons (Sondermayer). Most of this was used in the manufacture of incandescent gas mantles, magnesium-based alloys, dispersion-hardened metals, and for the creation of nuclear fuel. Future thorium demand will depend on the demand for thorium-fueled reactors in the electrical generating industry. Presently, there is only one thorium-fueled commercial nuclear reactor in the United States.[8]

Two other types of thorium-fueled reactors being considered for commercial application are the light water breeder (LWBR) and the molten salt breeder (MSBR). In the case of both the HTGR and LWBR, however, the breeding ratio is less than one. The use of the thorium cycle would only stretch the fuel life but not double it. The thorium-fueled MSBR is not, at present, being looked upon favorably by the nuclear power industry because when compared to the liquid metal fast breeder reactor (LMFBR), it has a relatively low breeding ratio (1.05) and a long doubling life (twenty years).

The future role of thorium-fueled reactors in the production of electrical energy is unclear. A recent study, however, attempted to project thorium demand based on the principal assumption that there would be a total nuclear generating capacity of 200,000 MW in 1985 and 700,000 MW in the year 2000 (Sondermayer). As indicated in Fig. 4–30, the estimated demand for thorium

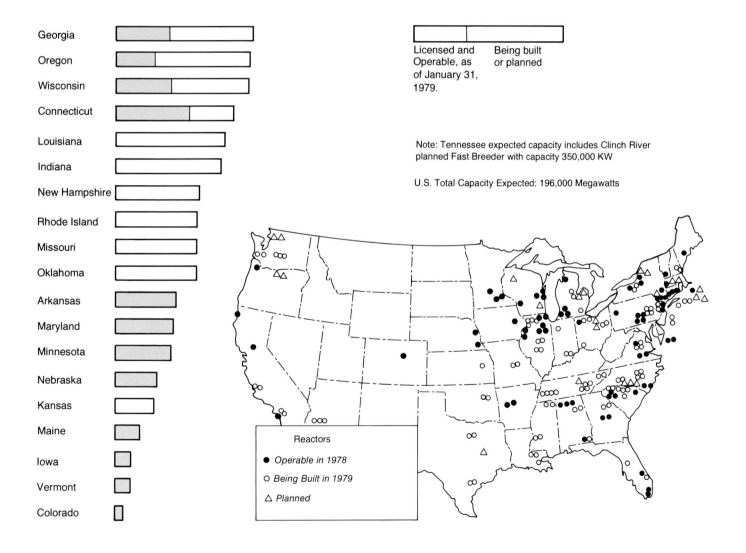

Licensed and Operable, as of January 31, 1979.　Being built or planned

Note: Tennessee expected capacity includes Clinch River planned Fast Breeder with capacity 350,000 KW

U.S. Total Capacity Expected: 196,000 Megawatts

Reactors
● Operable in 1978
○ Being Built in 1979
△ Planned

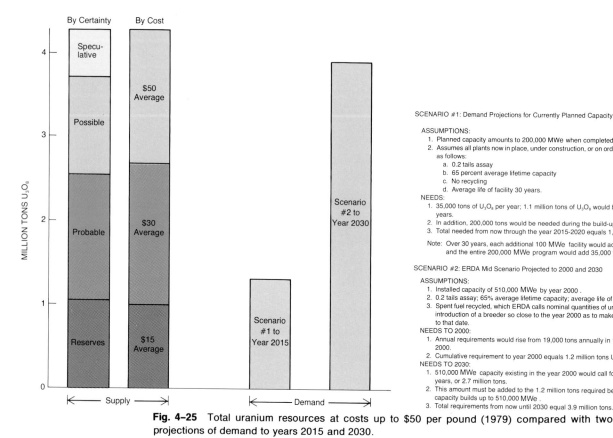

SCENARIO #1: Demand Projections for Currently Planned Capacity:

ASSUMPTIONS:
1. Planned capacity amounts to 200,000 MWe when completed.
2. Assumes all plants now in place, under construction, or on order (including deferred plants) operate as follows:
   a. 0.2 tails assay
   b. 65 percent average lifetime capacity
   c. No recycling
   d. Average life of facility 30 years.
NEEDS:
1. 35,000 tons of U_3O_8 per year; 1.1 million tons of U_3O_8 would be needed to support capacity for 30 years.
2. In addition, 200,000 tons would be needed during the build-up to 1985-1990.
3. Total needed from now through the year 2015-2020 equals 1,300,000 tons of U_3O_8.

Note: Over 30 years, each additional 100 MWe facility would add 5,300 tons to the total requirements; and the entire 200,000 MWe program would add 35,000 tons beyond 2015-2020.

SCENARIO #2: ERDA Mid Scenario Projected to 2000 and 2030

ASSUMPTIONS:
1. Installed capacity of 510,000 MWe by year 2000 .
2. 0.2 tails assay; 65% average lifetime capacity; average life of facility 30 years.
3. Spent fuel recycled, which ERDA calls nominal quantities of uranium, to be used in LWR's, and the introduction of a breeder so close to the year 2000 as to make little difference for requirements up to that date.
NEEDS TO 2000:
1. Annual requirements would rise from 19,000 tons annually in 1980 to 82,000 tons annually in year 2000.
2. Cumulative requirement to year 2000 equals 1.2 million tons U_3O_8.
NEEDS TO 2030:
1. 510,000 MWe capacity existing in the year 2000 would call for 90,000 tons of U_3O_8 annually for 30 years, or 2.7 million tons.
2. This amount must be added to the 1.2 million tons required between 1975-2000, the period when capacity builds up to 510,000 MWe .
3. Total requirements from now until 2030 equal 3.9 million tons.

Fig. 4–25 Total uranium resources at costs up to $50 per pound (1979) compared with two projections of demand to years 2015 and 2030.

Fig. 4–26 Projected annual domestic requirements for uranium oxide, 1972–2010.

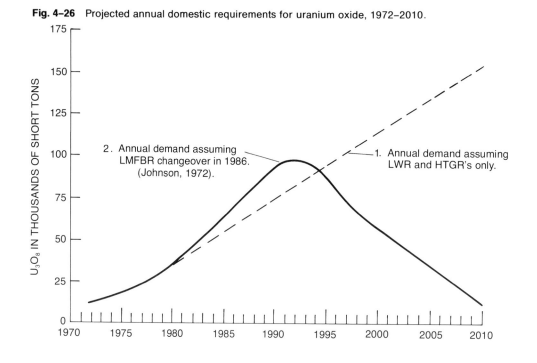

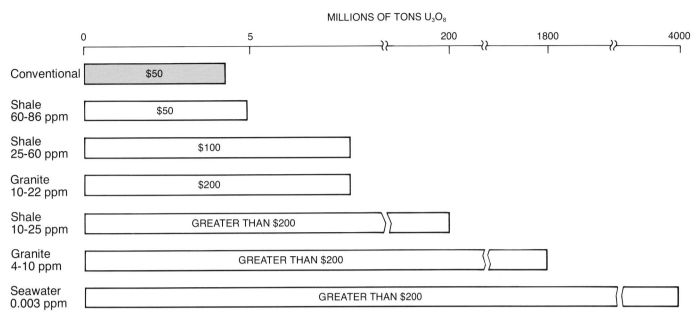

MILLIONS OF TONS U₃O₈

Note: Costs of non-conventional resources have risen since 1973, which is the date of these estimates.

Fig. 4–27 Potential uranium resources at costs greater than $50 per pound compared with resources at costs up to $50 per pound.

Fig. 4–28 World thorium resources (1975).

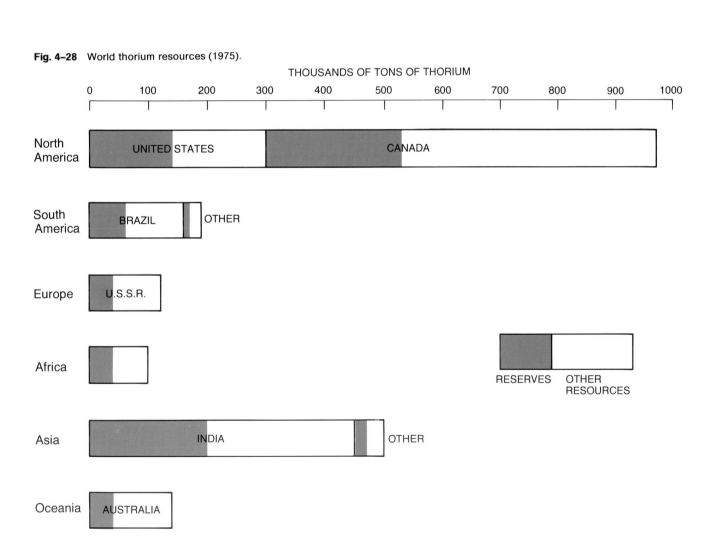

THOUSANDS OF TONS OF THORIUM

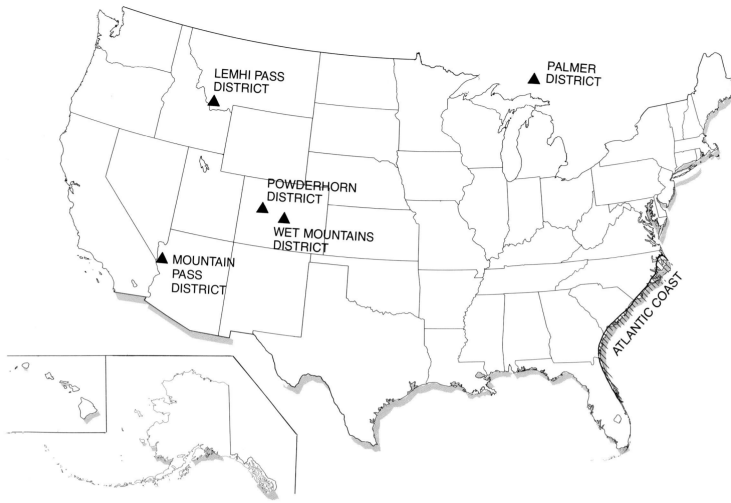

Fig. 4–29 Thorium resource areas in the United States (1975).

Table 4–7 Identified Thorium Resources of the United States as of 1975 (in thousands of short tons ThO$_2$ estimated recoverable)

AREA[1]	TYPE OF DEPOSIT	AS BY-PRODUCT	PRIMARILY FOR ThO$_2$ Ore Over 0.1%	Grade Under 0.1%
Atlantic Coast	Beach Placer	16	0	0
North and South Carolina	Fluviatile Placer	0	0	56
Idaho and Montana	Fluviatile Placer	2	0	38
Lemhi Pass District, Idaho and Montana	Veins	0	100	0
Wet Mountains	Veins	0	4.5	0
Powderhorn	Veins	0	1.5	0
Mountain Pass, California	Veins	0	0.5	0
Mountain Pass, California	Carbonatite	28	0	0
Palmer, Michigan	Conglomerate	0	0	46
Bald Mt., Wyoming[2]	Conglomerate	0	0	2
TOTAL UNITED STATES		46	106.5	142

[1]See accompanying map.

[2]Not shown on accompanying map.
Source: Sondermayer, Roman V., 1975.

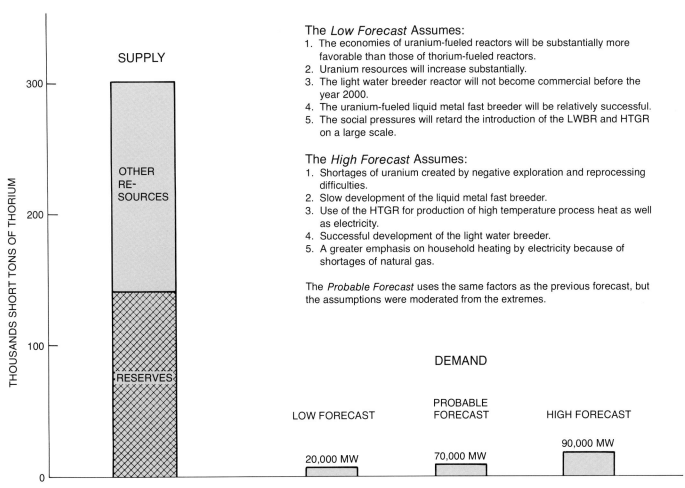

The *Low Forecast* Assumes:
1. The economies of uranium-fueled reactors will be substantially more favorable than those of thorium-fueled reactors.
2. Uranium resources will increase substantially.
3. The light water breeder reactor will not become commercial before the year 2000.
4. The uranium-fueled liquid metal fast breeder will be relatively successful.
5. The social pressures will retard the introduction of the LWBR and HTGR on a large scale.

The *High Forecast* Assumes:
1. Shortages of uranium created by negative exploration and reprocessing difficulties.
2. Slow development of the liquid metal fast breeder.
3. Use of the HTGR for production of high temperature process heat as well as electricity.
4. Successful development of the light water breeder.
5. A greater emphasis on household heating by electricity because of shortages of natural gas.

The *Probable Forecast* uses the same factors as the previous forecast, but the assumptions were moderated from the extremes.

Note: The demand projections are for all uses, however, nuclear uses will account for 90 percent of the cumulative needs.

Fig. 4–30 U.S. thorium supply and cumulative demand to the year 2000.

varies greatly due to the uncertainties of nuclear applications.

Based on this analysis of future thorium demand, it seems clear that the present supply of thorium is more than adequate to meet the needs of the United States well beyond the year 2000. Even if one assumed the high forecast demand of 3000 tons per year (for nuclear uses) in the year 2000, increasing at 6 percent per year thereafter, cumulative demand would not exceed our present Reserves until 2020.[9] Since thorium-fueled reactors have a high thermal efficiency, do not produce plutonium, and tend to stretch the life of nuclear fuels, they could assume a considerably more important role in the future development of nuclear power than they do at present.

Fusion

Thermonuclear fusion is viewed by a large segment of the scientific community as a very attractive, long-term solution to the world's energy supply problem. The fusion reaction was first demonstrated about forty-five years ago. To date, the only man-made, self-sustaining fusion reaction has been the explosion of the hydrogen bomb. Fusion may use fuels which are essentially in-

exhaustible, and appears to be environmentally safe. A great deal of research is needed, however, before even a working model can be built. Electrical power from fusion may not be commercially available until well after the year 2000.

The fusion process, like that of fission, releases energy by the conversion of heavy nuclei to nuclei of intermediate weight (see Fig. 4–2). The two fundamental differences between fusion and fission are the nature of the products of the reaction and the techniques used to make the two processes occur. For the purpose of this discussion, emphasis will be placed on the basic fusion processes and their fuel requirements.

Three reactions involving isotopes of hydrogen appear to offer the greatest potential for the creation of energy to be used in a thermonuclear fusion power plant (Fig. 4–31). In equations 4–31A and 4–31B, two deuterium nuclei (Hydrogen–2) combine to form either tritium (Hydrogen–3) or helium (Helium–3) and release 4.0 and 3.2 millions of electron volts (MeV) of energy respectively (Fowler). The equation described in 4–31C uses deuterium and tritium to produce ordinary helium (Helium–4). In this deuterium-tritium (D–T) reaction, 17.6 MeV of energy are released, a considerably larger

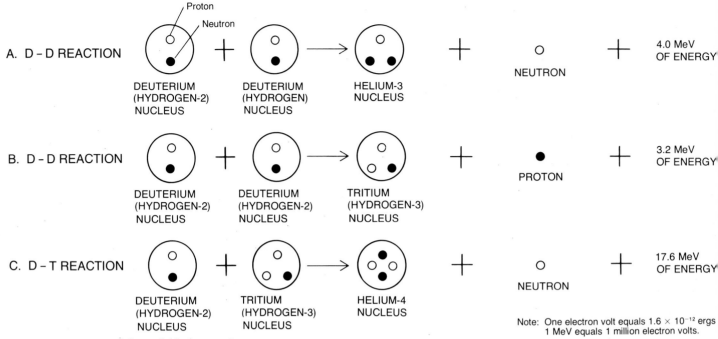

Fig. 4–31   Three potentially useful fusion reactions.

amount than is released in either of the two deuterium-deuterium (D–D) reactions.

The photograph shows the possible configuration of a fusion plant of the deuterium-lithium type. The magnetic confinement machines will all be large, with generating capacities in the thousands of Megawatts range. The major components will be the chamber itself and a surrounding thick blanket of lithium which will absorb the neutrons and convert their energy into heat. The lithium will also be the source of the tritium. Heat extracted from the hot lithium will be used to create steam to turn a steam turbine.

The three fusion reactions described in Fig. 4–31 appear rather simple and direct. Several decades of research, however, have raised a number of major problems that must be solved before useful energy can be obtained from the fusion process. The central problem is how to bring the hydrogen nuclei close enough for fusion to take place. In order to produce this reaction, the two

nuclei, which have the same electrical charge, must approach each other with speeds sufficient to overcome the repulsive force between them (Humle). The only practical way to achieve this is to heat the matter to incredibly high temperatures. The required temperatures are in the order of 40 million degrees Celsius for the D–T reaction and 400 million degrees Celsius for the D–D reaction (Turk, *et al.*). Clearly, containment of these reactions at such temperatures is a problem of immense proportion (see box). Nevertheless, present fusion research indicates that the reactions may be contained through the use of a very strong magnetic field. If and when the scientific feasibility of fusion is demonstrated a large number of engineering problems will inhibit its commercial applications.

Based on our present knowledge, the most promising way of achieving fusion energy will be by means of the D–T reaction. When compared with the D–D reaction, it releases significantly more energy and can be achieved at much lower temperatures and faster rates. For the D–T reaction, the deuterium supply is virtually unlimited: one cubic kilometer of seawater contains an amount of deuterium equivalent to the energy potential of 300 billion tons of coal or 1500 billion barrels of crude oil. The total volume of the oceans is about 1.5 billion cubic kilometers (Hubbert, 1971). On the other hand, tritium exists in only small amounts in nature, and tritium fuel requirements for fusion power must be produced from Lithium-6 by means of a lithium blanket surrounding the reactor vessel. Only 7.4 percent of natural lithium is Lithium-6, limiting still further the potential for producing tritium. It is possible, however, to use the neutrons produced by the D–T reaction to produce even more tritium than is used in the reaction. Ideally we can get both energy and re-

Drawing of a nuclear fusion facility. *Courtesy of American Petroleum Institute.*

Advantages of Fusion

1. If the deuterium-deuterium reaction is perfected, the fuel source will be limitless and cheap.
2. Reaction meltdown, with release of radioactive materials cannot occur.
3. Radioactive wastes produced would be smaller in amount and also less hazardous than those produced by fission reactors.
4. Fuel materials will not be suitable for use in weapons.
5. Because of their safety, fusion plants could be located close to settlements, so their waste heat could be used for space heating.

Disadvantages of Fusion

1. The greatest hazard would be the release of radioactive tritium as gas or in tritiated water. It has a short half-life (twelve years) but as a gas is extremely difficult to contain.
2. Plants will have to be very large and expensive.
3. After ten years of operation, the plant structure itself may be weakened and radioactive and may have to be replaced.
4. Staggering technical problems stand in the way of a working model, let alone an electrical generating plant of commercial scale.

placement of the expensive and scarce portion of the fuel requirements. Nonetheless, the future of a commercial fusion power industry depends to a great extent on the Nation's ability to find, recover, and process lithium resources.

The Nature and Occurrence of Lithium

Lithium is a very light, soft, and ductile metal (see Glossary), found principally in granitic pegmatites, subsurface brines, and salt water. In granitic pegmatites, lithium minerals are usually found in association with sodic and potassic feldspars, quartz, and muscovite (Singleton, et al.). Spodumene is the most plentiful lithium-bearing mineral in pegmatites and typically makes up 20 to 25 percent of the rock mined. The spodumene belt near Kings Mountain, North Carolina is the richest known lithium deposit in the world. In addition to the North Carolina deposits, major occurrences of lithium-bearing spodumene are found in South Dakota (Black Hills), Canada (Bernie Lake, Manitoba), Rhodesia (Bikita Tin Fields), and the Soviet Union. Other lithium-bearing minerals found in granitic pegmatites are lepidolite, petalite, amblygonite, and eucryptite. All successfully mined lithium-enriched pegmatites contain at least 1.0 percent, but normally no more than 2.0 percent Li_2O (Singleton et al.).

Subsurface brines contain more of the world's known lithium resources than do pegmatites, but lithium concentrations in these brines are low, ranging from 0.02 to 0.2 percent. Lithium-rich brine deposits occur in Clayton Valley, Nevada and Salar de Atacama, Chile. In addition, potentially large deposits of lithium occur worldwide in geothermal and oil well brines. The geothermal brines in the Imperial Valley in California contain an estimated million tons of lithium. Major engineering problems, however, inhibit the realization of their potential.

The ocean waters of the world are estimated to contain approximately 300 million tons of lithium, but because of the low concentration, about 0.00002 percent, this is not considered exploitable in the near future. In the United States, the Great Salt Lake in Utah constitutes a significant resource of this type.

WORLD LITHIUM RESOURCES

Figure 4–32 summarizes Identified world lithium resources. Identified world lithium resources amount to almost 6.5 million tons, with Reserves accounting for 18 percent of the total and other resources for 82 percent. The United States houses the vast majority of lithium resources, with 63 percent of the Reserves and 86 percent of the other resources. The United States' share of Identified resources, however, may be smaller than indicated because of the inclusion of more recent data for the United States than for other world regions. Canada has about 9 percent of the Reserves and 4 percent of the other resources. Outside of North America, most of the remaining lithium resources are found in Eastern Europe (including the Soviet Union) and Africa (Rhodesia and Zaire).

UNITED STATES LITHIUM RESOURCES

The U.S. Geological Survey has classified the Nation's lithium resources into three categories: (1) *Identified resources,* which are profitable to mine at present prices and technology; (2) *Identified Low-Grade resources,* which are not economic at today's relatively low prices and the present level of technology but may become so with increased prices and improved technology;

Lithium

Lithium is the lightest of all metals. Bombardment of lithium with neutrons in a nuclear reactor yields a heavy isotope of hydrogen, tritium, a fuel for use in fusion reactions.

Symbol	Li
Atomic number	3
Atomic weight	6.94
Density	0.534 g/cm^3
Melting point	180.5°C
Boiling point	1347°C

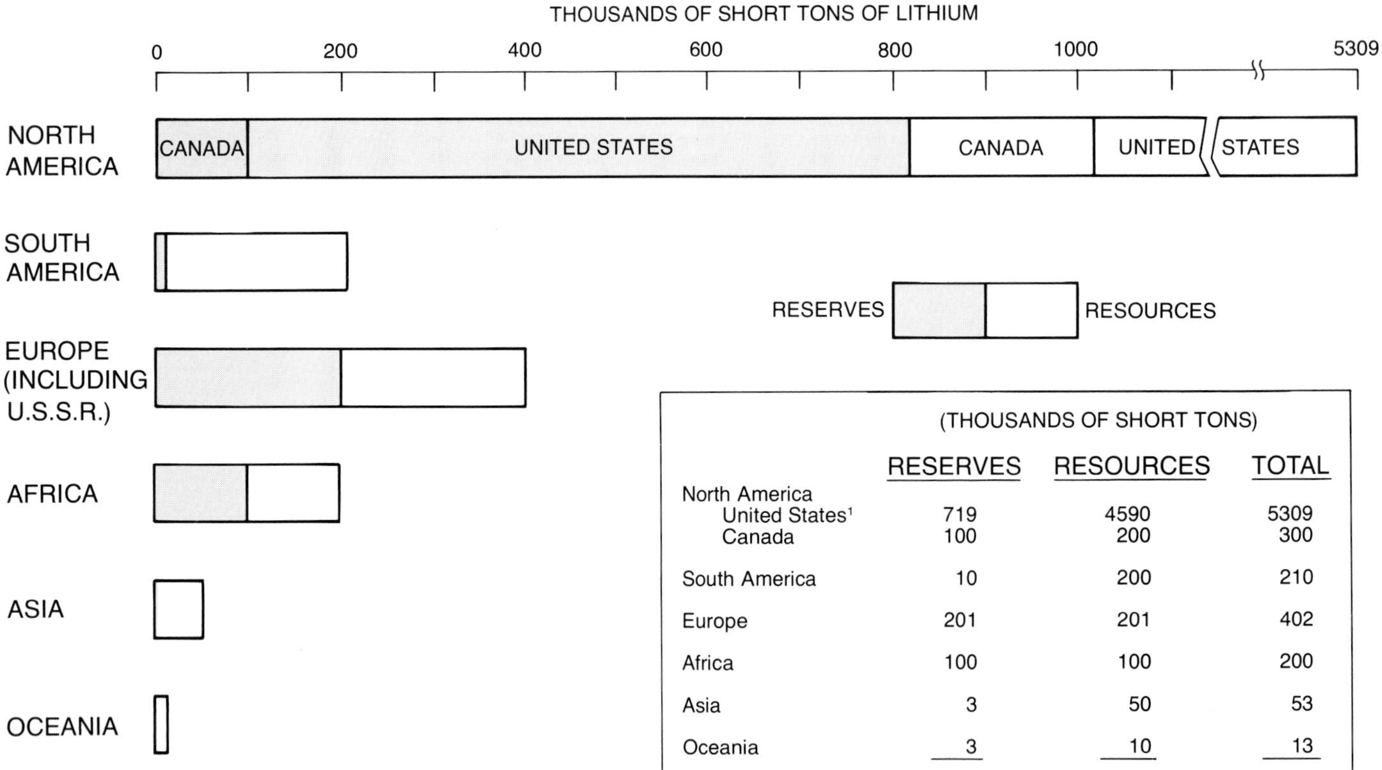

THOUSANDS OF SHORT TONS OF LITHIUM

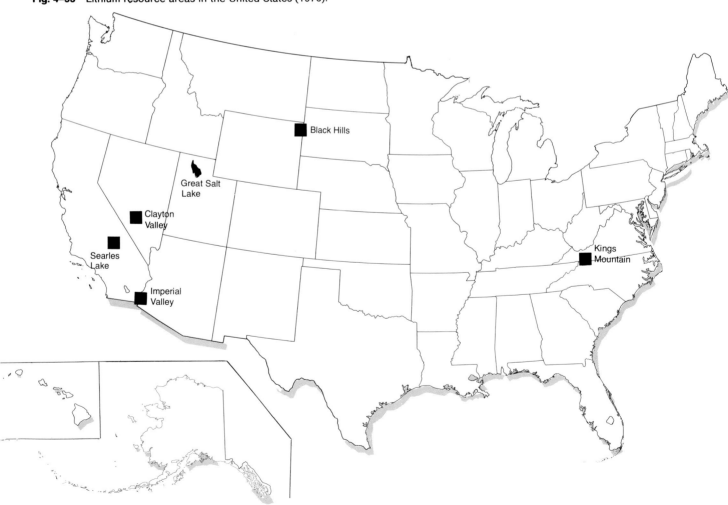

	(THOUSANDS OF SHORT TONS)		
	RESERVES	RESOURCES	TOTAL
North America			
United States[1]	719	4590	5309
Canada	100	200	300
South America	10	200	210
Europe	201	201	402
Africa	100	100	200
Asia	3	50	53
Oceania	3	10	13
World	1136	5351	6487

[1]United States lithium resource data were taken from Vine, U.S.G.S., 1976. Bureau of Mines lists U.S. reserves at 327,000 short tons and other resources at 600,000.

Fig. 4–32 World identified lithium resources (1975).

Fig. 4–33 Lithium resource areas in the United States (1976).

and (3) *Hypothetical* and *Speculative* resources which have yet to be discovered and developed (Vine, 1976).

Lithium resources of the United States are found in five principal locations: (1) Kings Mountain, North Carolina; (2) Clayton Valley, Nevada; (3) Searles Lake, California; (4) Imperial Valley, California; and (5) the Black Hills of South Dakota. Additional resources of significant quantity are known to exist in oilfield brines and borate mine waste at various other locations (Fig. 4–33). As of early 1976, the total United States' lithium resources were estimated at 7.9 million short tons. According to the Geological Survey's classification scheme, Reserves account for 9 percent of the total; Low-Grade resources for 58 percent; and Hypothetical and Speculative resources for 33 percent.

Lithium Reserves are estimated at 719,000 short tons of lithium (see Fig. 4–32) the vast majority of which (85 percent), are located in the spodumene belt of North Carolina. The remainder is found in subsurface brines of Clayton Valley, Nevada (13 percent), and potash-bearing brines at Searles Lake, California (2 percent). Identified Low-Grade resources are estimated to be approximately 4.6 million tons, 90 percent of which are in the spodumene belt of North Carolina, Clayton Valley, Nevada, the Imperial Valley of California, and in oilfield brines. Hypothetical and Speculative resources, estimated by the Geological Survey at 2.6 million tons of lithium, are thought to occur mostly in association with clays and brines at various locations, with about 20 percent of the estimated amount found in pegmatites.

RECOVERABILITY OF LITHIUM RESOURCES

Although the total amount of lithium resources appears to be more than adequate to meet the demands of alternative energy applications, the figures are, in fact, misleading. They fail to forecast amounts which will actually be available to the marketplace (Vine, 1976). In reality, a large part of the lithium resources are inaccessible as a result of economic, technical, and political problems as well as environmental restrictions. In order to allow a reasonable evaluation of the future supply and demand for lithium, the Geological Survey has estimated the percent of lithium recoverable and the probable yield by the year 2000 from various resource locales. Table 4–8 and Fig. 4–34 provide a summary.

From the three categories of resources, it is estimated that 1.3 million tons of lithium will be recovered by the year 2000. Reserves are projected to yield 276,000 tons, or 23 percent, of the lithium. Low-grade resources and Hypothetical and Speculative resources are expected to yield 30 and 47 percent respectively. In summary, more than three-fourths of the lithium recovered by the year 2000 is expected to be extracted from subeconomic and undiscovered resources. If this forecast is correct, its realization will require a great deal of advanced planning on the part of the lithium industry, and a healthy economic climate which provides incentives for exploration and development of undiscovered lithium deposits.

Table 4–8 United States' Lithium Resources, Classified by Geologic Certainty (in millions of short tons of contained lithium)

SOURCES	In-Place	Estimated Percent Recoverable by 2000	Probable Yield
Reserves (Economic)			
Kings Mt., N. C.	.612	35	.222
Clayton Valley, Nev.	.096	50	.048
Searles Lake, Calif.	.011	50	.006
TOTAL	.719		.276
Low-Grade Resources (Subeconomic)			
Kings Mtn., N. C.	1.008	25	.240
Clayton Valley, Nev.	.750	10	.075
Searles Lake, Calif.	.032	10	.003
Imperial Valley Calif.	1.200	1	.012
Oilfield Brines	1.200	1	.012
½ Great Salt Lake	.316	5	.012
Borate Mine Waste	.072	7	.005
Black Hills, S. D.	.012	30	.004
TOTAL	4.590		.363
Hypothetical and Speculative Resources			
Clays	.960	25	.240
Brines	1.200	20	.240
Pegmatites	.420	20	.084
TOTAL	2.580		.564

Source: Vine, James D., "The Lithium Resource Enigma." *Lithium Resources and Requirements by the Year 2000.* (U.S. Geological Survey Professional Paper 1005), Washington: Government Printing Office, 1976.

LITHIUM USES AND PROJECTED REQUIREMENTS

Presently, lithium-bearing minerals are used in the manufacture of a number of products including glass for special uses, glass ceramics, porcelain enamels and refractories. In addition, they are used in the production of lithium metals and chemicals. Domestic production of lithium minerals and chemicals is carried on primarily by two companies, Foote Mineral Company and Lithium Corporation of America. At present, the United States produces and consumes more than one-half of the world's lithium supply. In 1974, approximately 70 percent of the world's lithium production, about 8000 tons, originated in the United States (Singleton, 1975). In the same year, the United States demand was estimated at 4,500 tons. Based on conventional uses only, the Nation's demand for lithium has been projected to increase at an average annual rate of 8.8 percent each year to the

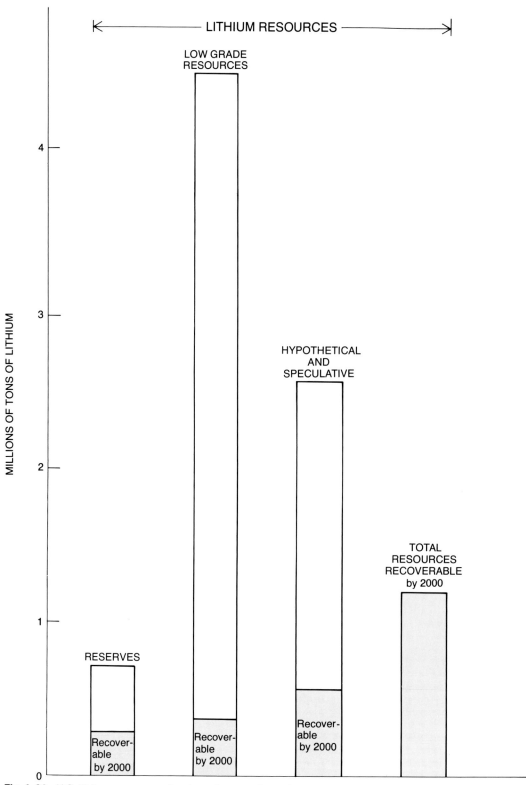

Fig. 4–34   U.S. lithium resources of various degrees of certainty compared with projected cumulative requirements for conventional uses, batteries, and fusion reactors.

Notes: 1. In the year 2000 lithium requirements for energy
storage and electric vehicles would be about equally divided.
2. Base case projection for lithium use in fusion reactors
assumes a continuation of 1976 electrical growth trends
with a smaller growth rate after 1990.
3. Massive shift case assumes that many energy users not
satisfied by conventional fuels will be satisfied by electricity.

LITHIUM REQUIREMENTS

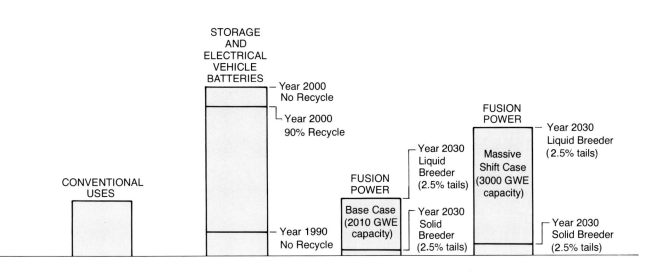

STORAGE
AND
ELECTRICAL
VEHICLE
BATTERIES

— Year 2000
No Recycle

└ Year 2000
90% Recycle

FUSION
POWER

— Year 2030
Liquid
Breeder
(2.5% tails)

FUSION
POWER

— Year 2030
Liquid Breeder
(2.5% tails)

CONVENTIONAL
USES

Base Case
(2010 GWE
capacity)

Massive
Shift Case
(3000 GWE
capacity)

— Year 2030
Solid
Breeder
(2.5% tails)

— Year 1990
No Recycle

— Year 2030
Solid Breeder
(2.5% tails)

year 2000 (Vine, 1976). If this is the case, by the year 2000 conventional uses will consume over 360,000 short tons of lithium, one-third of the countries recoverable resources (Fig. 4–34). Anticipated lithium requirements for energy-related uses are difficult to predict because of many technological uncertainties. If, however, the lithium anode battery is developed for commercial use in electric vehicles and off-peak storage (see Glossary), lithium supplies could be seriously threatened by the year 2000. Furthermore, the commercial development of nuclear fusion in the early decades of the twenty-first century could further complicate the lithium supply and demand situation.

LITHIUM REQUIREMENTS FOR LITHIUM ANODE BATTERIES For the past several years, Argonne National Laboratory has been working on the development of lithium-aluminum/iron-sulfide batteries for use as energy storage devices on electric vehicles (Chilenskas *et al.*, 1976). Argonne anticipates these batteries will come into commercial use by 1985, and that by the year 2000, as much as three percent (3×10^8 Kilowatt hours) of the total United States' energy consumption may be supplied by such batteries. At the same time, as many as 18 million electric vehicles may be powered by the lithium-aluminum/iron-sulfide battery. Based on the Argonne

New batteries of the lithium-metal sulphide type developed at Argonne National Laboratories, Chicago, Illinois. They can store roughly five times the energy of a comparable lead-acid battery. If they are developed commercially, such batteries will demand large quantities of lithium which is a resource that may be needed for production of fusion fuels. *Argonne National Laboratory.*

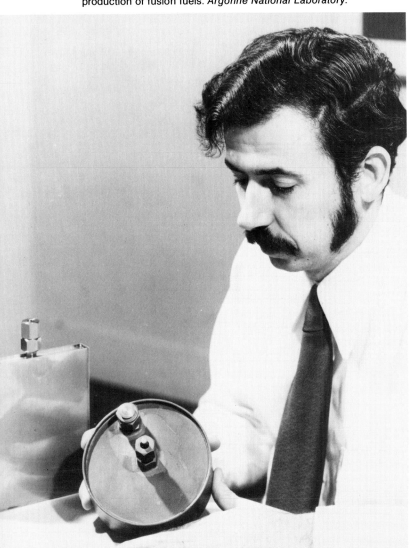

projections, which assume a ten-year battery life, no recycling of the battery itself, and no recycling of the lithium, the maximum cumulative lithium requirement will be over 1.1 million tons up to the year 2000 (Fig. 4–34). This amount can be reduced to .97 million tons with 90 percent recycling of the lithium contained in the spent batteries. Anticipated demand for these same uses, through 1990, is expected to be approximately 156,000 tons.

A comparison of the cumulative demand for lithium use in batteries only, and the cumulative amounts thought to be recoverable by the year 2000 reveals that the supply will be 184,000 tons greater than the demand (Fig. 4–34). When the requirements for conventional uses are also considered, the demand will exceed the supply by more than 170,000 tons, assuming no recycling, and by 32,000 tons assuming 90 percent recycling.

It is clear that the anticipated demand for lithium use in batteries in the 1985 to 2000 period could create a major supply problem. If lithium anode batteries are to be used to the extent projected by Argonne, and if they are to make a significant impact upon petroleum conservation, the lithium industry will have to work at accelerated rates to identify new resources and expand their production facilities to meet the demand.

LITHIUM REQUIREMENTS FOR FUSION POWER The development schedule for fusion power calls for a test reactor to operate in the 1980 to 1981 time period; an experimental power reactor in 1986 to 1988; a small fusion electric plant in 1990 to 1991; and a demonstration power plant by the end of the century. This development plan is, of course, tentative and depends on satisfactory solutions to a number of technical problems. Without adequate funding on the part of both private and public sectors, the research needed for demonstrating the feasibility of commercial fusion power in this century will be impossible. For the sake of providing an assessment of expected lithium requirements for fusion power, however, let us assume that adequate funding will be found for research and development.

Fusion's demand for lithium is highly dependent on the reactor design. For example, in a liquid breeder, where lithium (metallic or salt) is used as both a breeder and coolant, the amount of lithium required, or *inventory*, is very high. On the other hand, in solid breeder reactor designs, the initial lithium fuel needs are less than half those for the liquid breeder, but they must be highly enriched in the Lithium-6 isotope. Furthermore, because of a high burn-up rate in the solid breeder, periodic refueling is necessary and will affect the annual net lithium requirements for fusion power. The Energy Research Development Administration (ERDA) has developed two scenarios which extrapolate cumulative needs to the year 2030. The first scenario is designated the *Base Case* and assumes a continuation of present electric energy growth trends, with the effects of conservation slowing down the growth rate after 1990. It calls for a 2010-Gigawatt (billion watts) electric capacity by the year 2030 and a 270-

Gigawatt fusion capacity. The second scenario, defined as the *Massive Shift Case,* assumes a significant increase in the use of electricity and a decrease in the use of conventional fuels such as oil and gas. The Massive Shift Case also assumes a national electric capacity of 3000 Gigawatts in the year 2030 and a fusion capacity of 614 Gigawatts. In both cases, lithium demand was calculated for two tail fractions, 2.5 and 0.3 percent, for the solid breeder and one tail fraction, 2.5 percent, for the liquid breeder.[10]

Figure 4–34 summarizes the projected lithium requirements for these two scenarios. For the *Base Case,* the cumulative requirements for lithium are as follows: 1) 373,000 short tons using liquid breeders and assuming 2.5 percent Lithium-6 remaining in the tails; 2) 27,000 short tons using solid breeders and assuming 2.5 percent tails; and 3) 14,000 short tons using solid breeders and assuming 0.3 percent tails. Using the same technology, the lithium requirements for the *Massive Shift Case* are 848,000 short tons for liquid breeders, and 62,000 and 44,000 short tons for solid breeders assuming the same two tails fractions.

The results of these extrapolations clearly demonstrate the much larger fuel demands of the liquid breeder concept. Not only does the solid breeder require much smaller initial inventories of fuel but it also allows most of the lithium to be re-used after the Lithium-6 isotope is removed. For example, over 94 percent of the calculated lithium demand may be returned to the market place for use in other industries, assuming a tails fractions of 2.5 percent.

In conclusion, the use of liquid breeders for fusion power could place an intolerable strain on our presently defined supply of lithium resources. Although solid breeders would require far less lithium, the technological uncertainties associated with this concept make it too early to tell which of the two reactor concepts (if either) might be the best choice for commercial fusion power production. In the near term, lithium requirements for use in electric vehicles and off-peak storage batteries are a greater threat to the Nation's lithium supply than are the demands from fusion power. Nevertheless, the availability of lithium is fundamental to the use of fusion as an energy supply option, and a strategy needs to be developed to fully assess the future supply and demand for lithium.

NOTES

[1]The half-lives of U-235 and U-238 are 7.13×10^8 and 4.51×10^9 years respectively. Because of these slow rates of radioactive decay, natural uranium is only mildly radioactive.

[2]According to the U.S. Department of Energy, ore costing up to 15 dollars per pound averages 0.17 percent U_3O_8. The bulk of the Reserves in the up to 30 dollars per pound class range from 0.08 to 0.11 percent U_3O_8, (average 0.10 percent), lie less than 700 feet below the surface, and are found exclusively in sandstone. In many instances low-grade materials are included in the low-cost Reserve amounts because they are co-located with adjacent high-grade ores and it is economic to mine them together.

[3]The exception is a 330-Megawatt high-temperature gas-cooled reactor in operation at Fort St. Vrain, Colorado.

[4]As the chain reaction grows, the fuel rods get hotter and more bubbles are formed. The increase in space occupied by the steam bubbles reduces the moderating effectiveness of the water, reduces the number of slow neutrons, and thus slows the growth of the chain reaction.

[5]Assuming that it would take 3.5 million tons of uranium to fuel 800 converter reactors for a forty-year life, Lapp estimates that the same amount of fuel would supply 800 breeder reactors for 3700 years.

[6]Each additional 1000-Megawatt nuclear generating plant would add 5,300 tons of U_3O_8 to the total requirements over thirty years of operation.

[7]This assumes enrichment of UF_6 to the point at which the rejected materials (tails) contain only 0.2 percent U-235.

[8]There is a 330-MW commercial HTGR operating at Fort St. Vrain, Colorado.

[9]An annual demand of 3,000 tons increasing at 6 percent per year would amount to a cumulative demand of 119,894 tons between the years 2000 and 2020. If one adds the high cumulative demand forecast (see Fig. 4–30) of approximately 18,000 tons up to the year 2000, the total demand through the year 2020 would just about equal our present Reserves of 140,000 tons.

[10]The tails fraction represents the percentage of Lithium-6 remaining in the depleted lithium after fuel processing.

5

Geothermal Heat

The Nature of the Resource

Geothermal heat is one of three types of *primary* energy that flow continuously to the surface of the earth, the other two being solar radiation and tidal energy. Despite its constant flow from the earth's interior, geothermal heat is not defined here as a renewable resource but rather is placed under the broad heading of *nonrenewable* energy sources.

The sources of earth heat are the molten core of the earth and the heat generated by radioactive decay of elements in the crust. Both sources are diminishing very slowly, and the amounts of heat within the earth are virtually unlimited. This immensity and virtual permanence of the earth's heat supply are not so relevant, however, as the size and the renewability of *occurrences that offer accessible high temperatures* for applications such as space heating or electrical generation. Such occurrences are limited in size, and are located only at certain sites. More importantly, these occurrences are not created rapidly by nature. As a result, an intensive program of exploitation could deplete these usable resources, even as the vast reservoir of heat within the earth continues practically undiminished. If depletion is defined as drawing down local temperatures until they are no longer usable, then the resource is not renewed until natural processes restore the original temperatures. For very large occurrences, renewal of usable temperatures in a few decades is a reasonable expectation; for small occurrences, however, renewal is more doubtful; and for resources of the geopressured type (see later section on *exploitable occurrences*) there is no prospect for natural renewal.

World Patterns

At any location, heat flow from the interior to the surface of the earth can be deduced from the temperature gradient, or rate of change of temperature with depth, and the conductivity of the rock or sediments in which temperatures are measured.

Heat flow measurements from over 5,000 sites on continents and ocean floors around the world show a pattern that is best explained by the suggestion that large plates of the earth's crust—each plate often embracing both continent and sea floor—drift slowly across the globe in response to internal forces. Where they collide, they give rise to compressional forces that build mountains; where they separate, they cause tensional forces, rift valleys, and volcanic trends; and where they grind together, they cause earthquakes.

Nero Geothermal well on the island of Hawaii. *Courtesy Department of Energy.*

195

Areas of tension, that is, zones of spreading plates, allow upwelling of molten rock through the crust, causing such volcanic mountain trends as the mid-Atlantic ridge. Not surprisingly, heat flow is high along such trends in the Indian, and Atlantic Oceans. It is especially high in the Pacific west of South America, and where the Pacific ridge system intersects North America along the Gulf of California (Fig. 5–1).

Areas of compression, where one crustal plate is forced under an opposing plate, lead to belts of both low and high heat flow. The plate that is forced down (subducted) introduces a cool mass of rock that weakens the heat flow; but at the same time, some of the plate melts and rises buoyantly to cause volcanoes and high heat flow. Parallel elongate zones of high and low heat flow, suggesting this process, are seen in the vicinity of the Japanese island chain (Fig. 5–1) in which the islands themselves were formed by such upwelling and volcanism.

The association of high heat flow and volcanism confirms a common-sense interpretation of geothermal patterns. Where hot igneous rock is brought from the depths to near surface level by volcanic activity, heat flow anomalies result. Volcanic activity coincides with plate edges where there is communication with the interior. The opposite characteristics, low heat flow and lack of recent volcanism, occur in stable continental areas well removed from the edges of presently active plates. Shield areas of North America, South America, and Africa exemplify this best (Fig. 5–1).

In a few countries, geothermal heat has been used extensively for heating purposes and for generation of electrical power. Table 5–1 shows that, as of 1977, the world's total of 1,400 Megawatts of generating capacity is concentrated in the United States and Italy.

United States Regions: The Background Heat

Within the United States, substantial information on temperature gradients is available in data obtained from oil and gas drill holes and from test holes drilled expressly for geothermal measurement. Analysis of this information points to three broad types of regions in the country and allows for an estimate of the total amounts of heat stored in those regions. The following materials on background heat are an interpretation of an article dealing with this aspect of the resource in the U.S. Geological Survey Circular which is currently the best source of geothermal information (Diment, W. H., *et al.*).

Fig. 5–1 World heat-flow areas, crustal plate edges, and selected major geothermal systems.

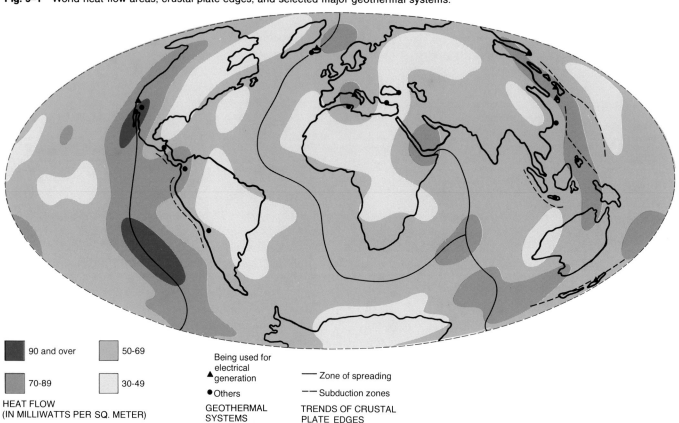

90 and over
70-89
50-69
30-49
HEAT FLOW
(IN MILLIWATTS PER SQ. METER)

▲ Being used for electrical generation
● Others
GEOTHERMAL SYSTEMS

—— Zone of spreading
-- Subduction zones
TRENDS OF CRUSTAL PLATE EDGES

Table 5-1 Installed Geothermal Electric Generating Capacity for Seven Countries, 1968–1977 (in thousand kilowatts)

COUNTRY	1968	1969	1970	1971	1972	1973	1974	1975	1976	1977
El Salvador	0	0	0	0	0	0	0	30	60	60
Iceland	0	2	2	2	2	2	2	2	3	3
Italy	372	395	368	382	382	382	382	398	398	398
Japan	31	31	31	31	31	31	50	54	54	77
Mexico	0	0	4	4	3	75	75	75	75	75
New Zealand	192	192	192	192	192	192	192	192	192	192
United States	84	84	84	203	322	441	441	559	559	599
WORLD	679	704	681	814	932	1,123	1,142	1,310	1,341	1,404

Source: United Nations, 1978

CHARACTER OF HEAT FLOW REGIONS

As explained before, heat flows to the earth's surface from two distinctly different sources: one is in the lower crust or in the *mantle*, the semi–liquid layer that underlies the crust; and the other source is decay of radioactive elements in the crust itself.

The first of these two sources is more important in determining the character of a region. While the base of crust may be deep in some areas, in areas of recent or current volcanic activity, molten rock is much closer to the surface. In such an area, high temperatures, revealed by steep temperature gradients, are encountered at relatively shallow depths and are more accessible.

The second component, heat due to radioactive decay, varies throughout a region according to the character of the crustal rocks. It is measured in heat generation units (HGU) which can range from 1 to 20, but most often are in the vicinity of 4 or 5. One example of extremely high radioactive heat generation is the igneous rocks of the White Mountain Series in New Hampshire, whose HGU values are over 20. The Conway granite in that area is well known as a potential low-grade ore of uranium.

Figure 5–2 shows three types of regions classified according to the deep crustal component of heat flow: 1.) Eastern or Normal type; 2.) Basin and Range, or Hot type; and 3.) Sierra Nevada, or Cold type. The Eastern type typifies the stable continental interior with no recent volcanic activity, which, as noted in Fig. 5–1 occurs near the middle, not the edges, of crustal plates. The Basin and Range type characterizes much of the western part of the country, where volcanic activity is much more recent. The Basin and Range map pattern is modified by two elongate areas of Eastern or Normal type: one coincides with the Wyoming Basins and Colorado Plateau; while that along the Pacific Northwest and the Great Valley of California may be due to an ancient subduction zone like the one recognized near Japan (Fig. 5–1). Associated with the same trend is the single region of abnormally low heat flow, the Sierra Nevadas. Despite the volcanic activity and heat flows along its western edge, the mountain range appears to be the top of a Cold block subducted into the mantle, causing molten rocks to be more remote.

Evidently, the most important characteristic that distinguishes among the three types of regions is the proximity to the surface of hot molten rocks. In the Normal or Cold type they are deep at the base of the crust, and in the Basin and Range (Hot) type are closer to the surface because of recent volcanism. The various *temperature gradients* that accompany the different types are all-important to the possibilities of using earth heat for they determine the drilling depth required in order to reach useful temperatures. Of course, local hot spots are in reality the sites where drilling would be undertaken. Nevertheless, it is helpful to examine the depths dependent on the general character of a region.

Figure 5–3 is a rearrangement of conventional temperature gradient plots, and is designed specifically to show how depths of useful temperatures vary from one type of region to another. Considered in the plots are the effects of varying radioactive heat generation, so that for any chosen HGU value, 0 through 20, the depths necessary to reach temperatures of 90 or 150 degrees Celsius may be estimated.

Assuming, for simplicity, that 5 is the HGU value for local rocks, part A of the figure shows that in the Sierras a depth of about 7 kilometers or 23,000 feet would be necessary in order to reach 90 degree temperatures. In this region, 150 degrees is out of reach since drilling beyond 30,000 feet is impossible with current technology. In the Eastern or Normal type of region (part B) the lower temperature would be reached at 12,000 feet and the higher temperature at 25,000 feet. In the Basin and Range type of region (Fig. 5–3, parts C and D), the higher temperature can be expected at depths less than 17,000 feet,

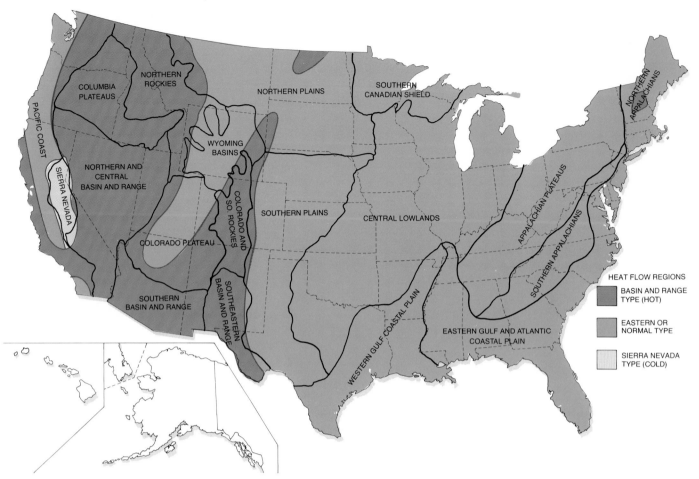

Fig. 5–2 Physiographic provinces, and three types of heat flow regions in the U.S.

and in the more favorable sub-type (D), at less than 14,000 feet. One revealing feature of the charts is the differing slopes or "droop" of the reference lines on parts A through D. In the Cold type of region, a change in the degree of radioactive heat generation is of vital importance. It is much less important in a region favored by high rate of flow from molten rock at depth: that flow becomes the dominant factor as shown by the slight change in depth for changing HGU values. Also noteworthy is the vertical spacing between the two reference lines on the series of plots. The progressively closer spacing on plots A through D reveals steeper temperature gradients in the more favored regions, that is, a shorter vertical (depth) difference between 90 and 150 degree temperatures.

HEAT IN-PLACE IN THE COUNTRY AS A WHOLE

Using information that reveals at what depth temperatures occur (similar to that shown in Fig. 5–3), and data on thermal conductivity of rocks, it is possible to estimate the *heat content* per unit area of a region to a chosen depth. Then the content for a geologic province can be obtained by multiplying by the area of the province. Only the temperatures exceeding mean annual tempera-

tures at the earth's surface (assumed to be 15 degrees Celsius) are used in this calculation in order to distinguish earth heat from heat due to solar radiation.

Table 5–2 shows the seventeen geologic provinces that make up the conterminous United States. These provinces are divided into three region types for the sake of the calculations. Alaska and Hawaii are treated separately because they are geologically so different from the conterminous United States. The table shows estimates of stored heat from surface to 3 kilometers depth, and from 3 to 10 kilometers. The total for each province, and for the country as a whole, therefore, represents stored heat to a depth of 10 kilometers (6.25 miles, or 33,000 feet). This is the practical limit to drilling in the foreseeable future.

The relative contributions of various regions and the provinces they comprise is shown graphically in Fig. 5–4. From this figure it is clear that the Eastern type region contributes greatly to the total of available geothermal heat because of its large area. Provinces of the Basin and Range type, despite their shallow high temperatures, contribute less to the national total because of smaller areas. As noted there, the best estimate of the United

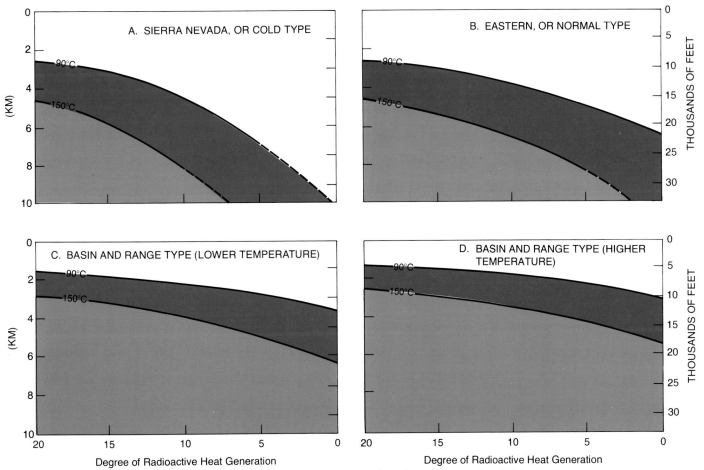

Fig. 5–3 Estimated depths to temperatures adequate for space-heating (90° C) and power generation (150° C) in four types of regions, showing effects of variations in radioactive heat generation.

States' total stored heat is about 800×10^{22} calories, not the 680.2×10^{22} calories which is the sum of province estimates. This revised total is used later in a summation of heat in-place.

Exploitable Occurrences

There are three types of occurrences in which high temperatures in the earth are more accessible: hydrothermal, igneous, and geopressured. The first two types are found in the Basin and Range type of province, and are associated directly or indirectly with recent volcanism that has left molten rock relatively close to the surface. The third type is quite different, and not dependent upon volcanism.

HYDROTHERMAL SYSTEMS

Natural hot springs occur where waters from the surface flow down through fractures or aquifers into deeply buried rock where they are heated. These heated waters then are forced to the surface through buoyancy, and underground pressure. Although hot springs occur through-

out the country, those in which the waters encounter very high temperatures are located in those areas in the West where molten or very hot igneous rocks are unusually near the surface.

Figure 5–5 shows a hypothetical circumstance in which the prime source of heat is the magma chamber, a body of molten rock embedded near the surface as a remnant of volcanic activity. Surface waters descend down an aquifer to the level of very high temperatures, and a fault allows water to rise to the surface. Not all hydrothermal systems depend on molten rock near the surface. Apparently, in some systems, water is carried by faults to great depths where, in a region of the Basin and Range type, sufficiently high temperatures are encountered simply by virtue of the steep temperature gradient (Renner, J. L., *et al.*, p. 52).

In general, hydrothermal systems are extremely attractive for geothermal heat applications because the circulating water serves to bring the high temperatures within reach, at or near the surface. The magma or other "source rocks" may be so deeply buried that drilling to reach them is far more expensive than drilling to the hot

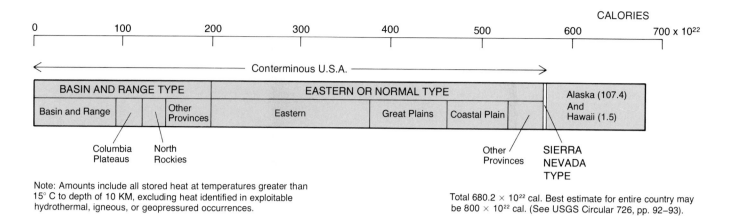

Fig. 5–4 Estimated heat content to depth of 10 KM, showing regions.

Note: Amounts include all stored heat at temperatures greater than 15° C to depth of 10 KM, excluding heat identified in exploitable hydrothermal, igneous, or geopressured occurrences.

Total 680.2 × 10²² cal. Best estimate for entire country may be 800 × 10²² cal. (See USGS Circular 726, pp. 92–93).

water or steam. Most importantly, the circulating waters provide a system by which heat is drawn continuously from a substantial volume of hot rock. The best hydrothermal system can be developed by drilling a number of holes to increase the flow of steam or hot water so that heat exchangers can draw energy for large-scale applications, such as district heating or electrical power generation.

Among hydrothermal systems that have been identified, the most important variables for evaluating the systems are: the temperature of the fluids, and whether the fluid circulating to the surface is water or steam. Although *low temperature* systems, with fluid temperatures below 90 degrees Celsius, are mapped here, their energy potential has not yet been estimated. Those of *intermediate temperatures*, between 90 and 150 degrees Celsius, are assessed for non-electric applications; and *high temperature* systems, over 150 degrees Celsius, are assessed with a view to electrical power generation.

Hydrothermal systems yielding dry steam are rare (only three have been identified in the United States) and are more desirable than those yielding hot water. Steam can be fed directly into turbines, whereas hot water must be routed through heat exchangers in order to boil water and raise steam. Furthermore, hot waters bring minerals to the surface in solution and cause difficulties with corrosion and pollution.

IGNEOUS OR VOLCANIC POTENTIAL SYSTEMS

Although often referred to as "igneous systems," these occurrences are better understood as *potential* systems because they lack the natural plumbing by which to extract heat from hot rock and carry it to the surface.

Figure 5–6 shows a hypothetical occurrence like that presented in Fig. 5–5, but without permeable rocks overlying the magma. Since there are neither aquifers nor fractures to convey fluids, the embedded magma mass is without communication to the surface. Such occurrences are often very large and hold a great store of heat undiminished *because* of the lack of communication.

Table 5–2 Stored Heat (Resource Base) in Nineteen Provinces, Grouped into Three Types of Region

REGION TYPE	PROVINCE	STORED HEAT (CALORIES × 10²²) 0–3 KM.	3–10 KM.	Total
Sierra Nevada	Sierra Nevada	0.31	2.80	3.11
Eastern or Normal	Eastern	17.20	160.40	177.60
	Coastal Plain	6.70	62.40	69.10
	Great Plains	8.30	77.60	85.90
	Wyoming Basins	0.625	5.84	6.47
	Peninsular Range	0.046	0.43	0.49
	Pacific Northwest	0.468	4.37	4.84
	Klamath Mts.	0.254	2.38	2.63
	Great Valley	0.295	2.76	3.06
	Colorado Plateau	1.990	18.60	20.60
Basin and Range	Basin and Range	8.72	83.70	92.40
	Northern Rockies	2.49	23.90	26.40
	Central Rockies	1.26	12.10	13.40
	Southern Rockies	1.13	10.90	12.00
	Columbia Plateaus	2.70	25.90	28.60
	Cascade Range	1.07	10.30	11.40
	San Andreas Fault Zone	1.26	12.10	13.40
TOTAL COTERMINOUS UNITED STATES		54.80	516.90	571.30
Alaska		10.20	97.20	107.40
Hawaii		0.14	1.37	1.50
TOTAL UNITED STATES		65.14	615.47	680.2[1]

[1]This total was revised to 800 × 10²² calories in the source document (U.S.G.S. *Circular 726*, p. 92–93)

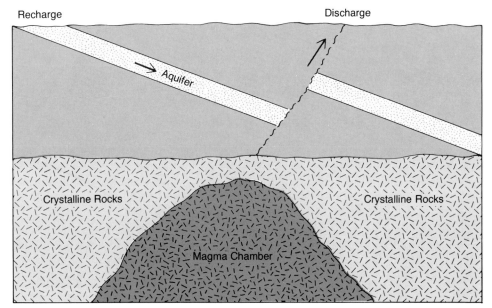

Fig. 5–5 Schematic profile showing elements of a hydrothermal convection system.

Development of such occurrences has not yet been accomplished, first because a number of identified hydrothermal systems await development; and second because geothermal energy has received serious attention only in recent years. Now, some research is being done on artificially induced circulation systems (see final section of this chapter).

GEOPRESSURED RESERVOIRS OF THE GULF COAST
Whereas the foregoing two types of exploitable occurrence depend on the location of hot or molten igneous rocks unusually near the surface, the third type of oc-

currence is in a sedimentary rather than igneous environment. Furthermore, it depends upon deep rather than shallow burial of the critical rocks.

Figure 5–7 represents a cross-section of a thick sequence of young sedimentary rocks in the Gulf coast region. At depths ranging from 6,000 feet (1,800 meters) to 15,000 feet (4,600 meters) abnormally high subsurface pressures have been encountered in sand bodies penetrated by holes drilled for oil and gas exploration. The sands were initially of interest because they were a problem for petroleum engineers who had to contain the pressures when drilling through the sands. They have since

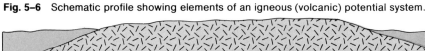

Fig. 5–6 Schematic profile showing elements of an igneous (volcanic) potential system.

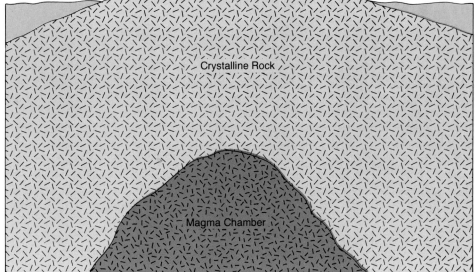

Testing of a geothermal well near Milford, Utah. *Courtesy American Petroleum Institute.*

energy, these waters are hot and also contain dissolved methane (natural gas). Their temperatures, usually in excess of 150 degrees Celsius, are *considerably higher than expected for the depth*, according to the Eastern or Normal type of region in which the Gulf Coast province occurs (Table 5–2). In that type of province, a depth of 5 kilometers (16,000 feet) is consistent with 150 degree Celsius temperatures only if the radioactive heat flow is extremely high, that is, between 15 and 20 heat generation units (HGU) (Fig. 5–3b). With no evidence of highly radioactive rock, the abnormally high temperatures at depth are apparently due to the low thermal conductivity of overlying porous sediments, which constitute an insulating blanket.

Development of this geopressured type of energy will be accomplished by drilling through the isolated sand bodies, or "reservoirs" and allowing the hot methane-laden salt water to rush into the drillhole and up to the surface. The methane will be separated at the surface and used in the same way as any natural gas. The waters are hot enough to raise steam for electrical generation or to be used in industrial processes.

ACCESSIBLE HEAT IN-PLACE IN IDENTIFIED HYDROTHERMAL, VOLCANIC, AND GEO-PRESSURED OCCURRENCES

Just as in the above estimates of background heat for provinces and regions, the amount of heat in-place is only the portion at greater than 15 degrees Celsius, (the temperature that represents average conditions at the surface). Furthermore, it is only that heat which is *accessible* in identified occurrences (Fig. 5–8). For hydrothermal systems this estimate considers heat in rocks down to a depth of 3 kilometers (9,843 feet). For volcanic type the depth limit is 10 kilometers (33,000 feet). For geopressured reservoirs the limit is 6.86 kilometers (22,500 feet). *Accessible heat in-place*, defined this way, is comparable to tonnage of coal that is *mineable* by underground methods (see Chapter 1). The geothermal heat actually *recoverable* at well-head is markedly less than the amounts accessible. Not shown on Fig. 5–8 is the fact that the amount of heat *applied* to some task is less than the amount recovered at well-head because of energy lost in conversion or exchange of heat. The same is true of the heat value in coal or oil: much of it is lost, whether it is burned in a residential furnace or a power plant. For geothermal heat, however, the loss tends to be a larger portion, because the relatively low temperatures make for low conversion efficiencies.

In the materials that follow, amounts of heat in-place and recoverable are tabulated in units of 10^{18} *calories* which is the heat equivalent of 690 million barrels of crude oil, or 154 million short tons of bituminous coal.

HYDROTHERMAL SYSTEMS OF THE HIGH-TEMPERATURE TYPE Figure 5–9 shows identified systems in western states and in Alaska with average reservoir temperatures over 150° Celsius. The remainder of the coun-

been recognized as a source of usable energy (Dickinson, 1953).

The pressures labelled "abnormal" in the figure are greater than "hydrostatic." This means that they exceed the pressure exerted by a column of water reaching from the surface to the depth of the sand body. The inset in Fig. 5–7 shows one interpretation. Pressure in the thick and continuous main sand series is *not* abnormal because formation waters have been able to drain out and allow the sandstones to be compacted by the weight of overlying sediments. In sand bodies isolated by faults, or by enclosing shales, formation waters are trapped so that compaction is prevented. Pressures build as more sediments are deposited at the surface and the weight of overburden increases.

In addition to pressures that will raise water to the surface in artesian fashion and offer usable mechanical

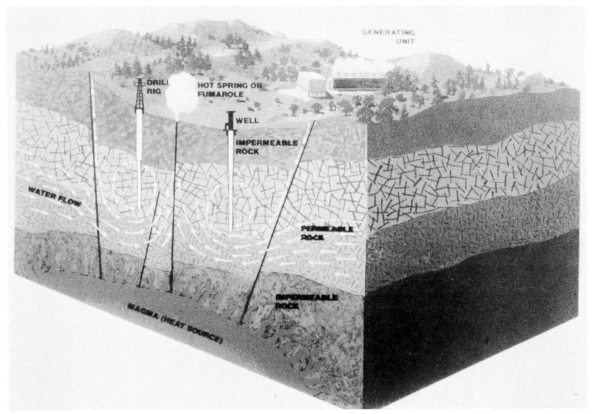

Schematic cross-section at the Geysers, eighty miles north of San Francisco. Show in white are the natural circulation of water, and steam escaping through a fumarole. Drilled wells tap the steam underground and bring it to the surface where it is piped to a power plant. *Courtesy of Pacific Gas and Electric Company, San Francisco, California.*

Fig. 5–7 Schematic profile through onshore Gulf Coast region, showing depths to geopressured zones.

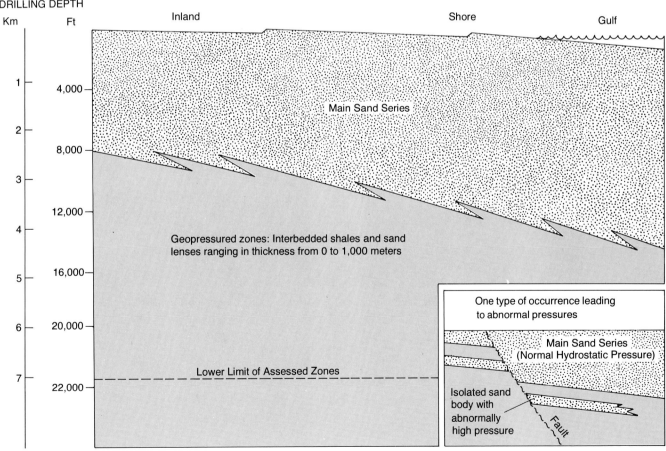

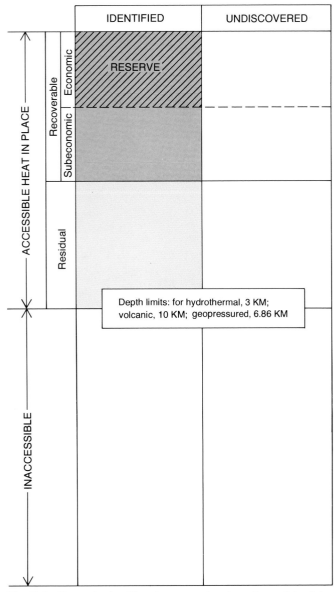

		IDENTIFIED	UNDISCOVERED

Depth limits: for hydrothermal, 3 KM; volcanic, 10 KM; geopressured, 6.86 KM

Fig. 5–8 Conceptual setting for amounts of geothermal heat in place and amounts recoverable.

try is not shown because neither high nor intermediate-temperature systems occur there.

Among the systems mapped, there is some variety in reservoir temperatures, some being barely over 150 degrees Celsius, while many exceed 200 degrees. One reservoir in the Salton Sea area of southern California has temperatures of 340 degrees Celsius (Renner, J. L., *et al.*, p. 10–21). Systems with higher temperatures are the more desirable, both because they tend to increase total heat content, and because they lead to higher efficiencies in application of the energy. Generally, heat in-place is based upon reservoir temperatures, specific heat of the rocks, approximate area of the subsurface reservoir, and a lower depth limit of 3 kilometers, which is the current limit of geothermal drilling.

A good proportion of the systems mapped are fairly large, having stored heat greater than 1×10^{18} calories. There are fewer systems with stored heat greater than 10×10^{18} calories. The largest of them is Yellowstone, with a heat content of 296×10^{18} calories, while Long Valley and Coso Hot Springs in California each contain about 20×10^{18} calories.

Systems with temperatures over 150 degrees Celsius can be used to raise steam for electrical power generation, unlike those of intermediate temperature. Among these high-temperature systems are those dominated by steam, rather than hot water, making them more favorable for power generation. The three identified systems of this type, the Geysers, Mt. Lassen, and the mud volcano system at Yellowstone, are labelled on Fig. 5–9. Two of the three are quite large, Geysers containing 24, and Mt. Lassen containing 10 of the 10^{18} calorie units, while the steam-dominated portion of Yellowstone accounts for only 1.1 units (U.S.G.S. *Circular 790*, p. 41–45). The heat contents of the Mt. Lassen and the Yellowstone systems are included in the state summary of heat in-place, but are not in later assessments of *recoverable* energy because they are in national parks where no exploitation is allowed.

One of the well-heads at the Geysers. Pipes carry steam from wells to fourteen separate power plants that together have the capacity to produce 800 Megawatts of electricity as of Spring, 1980. *Courtesy of Pacific Gas and Electric Company, San Francisco, California.*

When heat content is summarized by state (Fig. 5–10, Table 5–3) it is clear that Wyoming and California dominate. However, none of Wyoming's 296 units of heat will be available for extraction (because of national parks) whereas California's total of 135 includes only 10 units untouchable at Mt. Lassen. Although Oregon, New Mexico, Utah, and Nevada all have very substantial amounts of heat in-place, the amounts are tiny in comparison to those of California and Wyoming.

HYDROTHERMAL SYSTEMS OF THE INTERMEDIATE-TEMPERATURE TYPE   Figure 5–11 shows systems with intermediate reservoir temperatures—between 90 and 150 degrees Celsius. These are more plentiful than the high-temperature type—both in the contiguous states and in Hawaii and Alaska, with a total of 164, as opposed to 56 of the high-temperature type. None of the intermediate-temperature systems mapped has a reservoir temperature under 100 degrees Celsius. Many have temperatures of 145 and 150 degrees, and may in fact be usable for electric power generation. For the most part, however, these systems are suited to tasks such as space heating.

Large systems are less prevalent than small ones. One of the largest is Klamath Falls, Oregon, which has 7×10^{18} calories in-place and is being studied as a source of energy for space heating in Portland. The Bruneau-Grandview system near Boise, Idaho is a giant among hydrothermal systems, covering 1,483 square kilometers and holding an estimated 107×10^{18} calories as accessible heat in-place, or the equivalent of more than 74 billion barrels of crude oil.

The distribution of heat in-place (Fig. 5–12, Table 5–4) is similar to that for the high-temperature systems. Idaho, however, has the overwhelming total of 122.22 units because of the Bruneau-Grandview system referred to above.

LOW-TEMPERATURE GEOTHERMAL WATERS (LESS THAN 90 DEGREES CELSIUS)   Very common throughout the country, even in the East, are warm springs where waters travel to considerable depth, then return to the surface—usually by means of faults or fractures in the rocks. The process is similar to that of hydrothermal systems, but lacks the magma body and the shallow high temperatures associated with it. Warm springs, or *thermal springs* are often defined as those that carry water at temperatures more than 10 degrees Celsius higher than average annual air temperatures of the area. To qualify, water would need to be only 10 degrees Celsius in parts of Alaska, but 30 degrees Celsius in southern Arizona.

Thermal springs occurring at the surface are one indication that low-temperature geothermal waters are accessible at shallow depths. Unusually steep temperature gradients in water wells and oil and gas wells in some areas are another indication—even where there are no surface springs. Together, the thermal springs and well data show that there are quite extensive areas with prospects for encountering low-temperature geothermal waters. Although these waters cannot be used for power generation or directly for space heating, they can be used for drying tobacco or grain crops, or as a source from which heat pumps may draw energy for space heating.

Figure 5–13 maps thermal springs and areas favorable for finding useful geothermal waters *at depths less than 1 kilometer* (3,281 feet), which is thought to be the economic limit. No estimate has been made of the amounts of accessible heat. In western states, the number of thermal springs is quite remarkable, and favorable areas are extensive. In these areas the low-temperature waters augment the region's generous endowment of high- and intermediate-temperature hydrothermal resources and the great energy potential in volcanic occurrences.

IGNEOUS OR VOLCANIC POTENTIAL SYSTEMS   In addition to hydrothermal systems, there are many volcanic hot spots with potential for exploitation. Their potential depends on the existence of molten rock (magma) masses in the upper crust, *some of which are contributing to hydrothermal systems already recognized and assessed*, and others which are without any connection to the surface, as suggested by Fig. 5–6. Since the heat content of such a molten mass is so large, its heat content is virtually unaffected by the heat drawn off by hydrothermal systems.

Only molten rock of a certain character is likely to form storage chambers which are in upper levels of the crust and therefore accessible to drilling and exploitation. Magmas of *basic* composition, that is, those which form basalts, andesites and comparable rocks, have low viscosity and tend to flow directly from the lower crust and exit through pipes and fissures. On the other hand, magmas of more *acidic* composition such as those that form granites, are more viscous and tend to create storage chambers within 10 kilometers of the surface. It is from these chambers that eruptions occur. Coincidentally, the acidic magmas are known to cause explosive volcanic activity, whereas the low-viscosity basic magmas are associated with quiet eruptions and widespread flows such as those in the Columbia plateaus of the state of Washington.

The volcanic hot spots mapped and assessed here are those deduced from surface evidence of acidic eruptions, which provide a hint that a chamber exists below. Critical to estimates of present temperature and heat content is a knowledge of the age of the eruption. The molten rock in the chamber is assumed to have been cooling at a given rate from a temperature of approximately 850 degrees Celsius at the time of eruption until the present. The size of the chamber is estimated on the basis of surface indications of its area, and upon the assumption that its thickness ranges from 2.5 to 10 kilometers, which is the depth limit for the volumes that contribute to heat content calculations. The heat contents reported are ". . . that part of the heat content . . . that still remains in the ground, both within and around the original magma

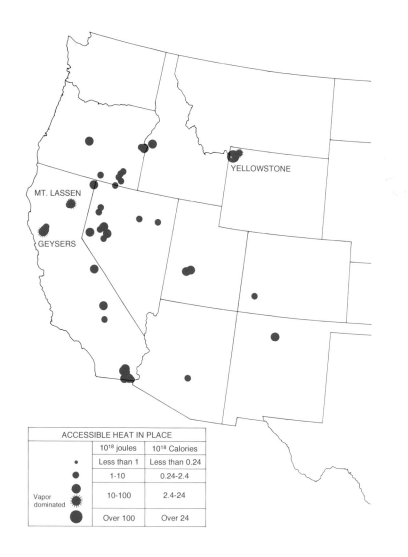

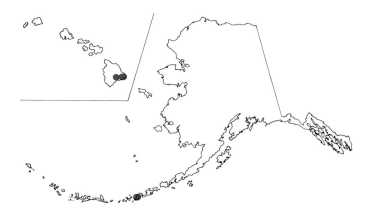

Fig. 5–9 Identified high temperature hydrothermal convection systems, i.e., those with temperatures 150° C or greater.

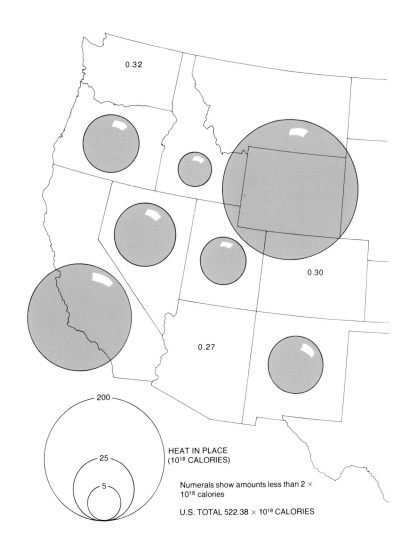

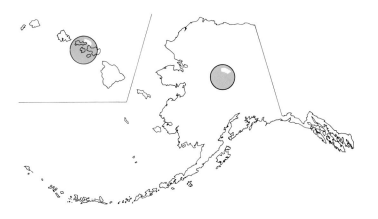

Fig. 5–10 Accessible heat in place in identified high temperature hydrothermal systems, 1978.

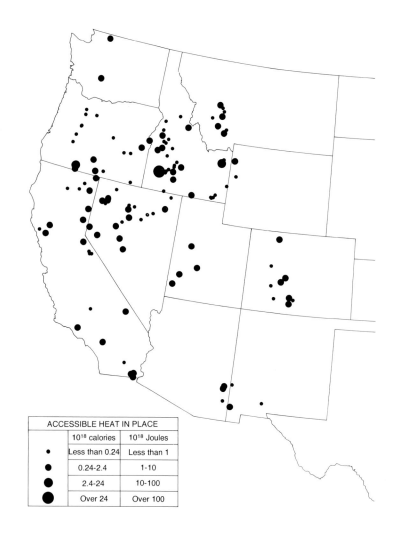

ACCESSIBLE HEAT IN PLACE		
	10^{18} calories	10^{18} Joules
•	Less than 0.24	Less than 1
●	0.24-2.4	1-10
⬤	2.4-24	10-100
⬤	Over 24	Over 100

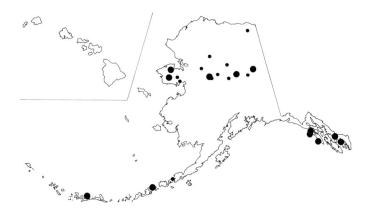

Fig. 5–11 Identified intermediate-temperature hydrothermal systems, i.e., those with temperatures 90°-150° C.

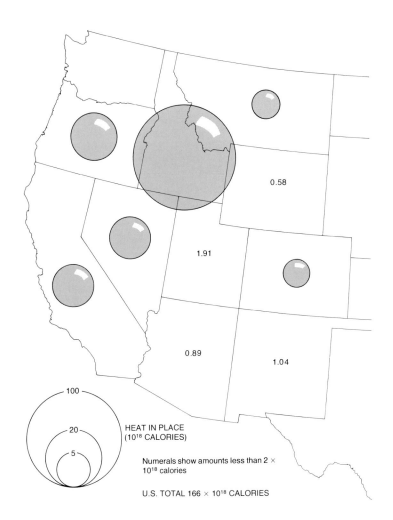

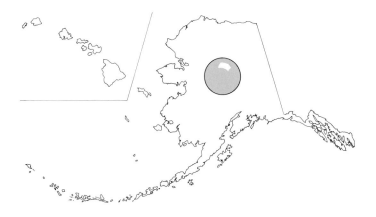

Fig. 5–12 Accessible heat in place in identified intermediate-temperature hydrothermal systems, 1978.

chamber'' (Smith and Shaw 1976, p. 75). Of the total heat content in all occurrences, about half exists in molten or partially molten magma while half is in solid or "dry" hot rock surrounding the magma chamber. An exception to this occurs in Alaska, where practically all the heat in-place is contained in molten or partially molten masses.

Figure 5–14 shows the locations of identified volcanic "systems" in the conterminous United States, Alaska, and Hawaii. Because of the volcanic character of the Aleutian chain, there are eighty-eight identified occurrences in Alaska. Hawaii, with large oceanic volcanoes, is credited with five despite the low-silica character of the magmas. In western states there are sixty-one identified, with nineteen in California and twenty in Oregon. A substantial number of these do support hydrothermal systems (see Table 5–5).

Occurrences which have been studied completely enough to determine an estimate of heat in-place are classified according to size in Fig. 5–14. Because of the great magnitude of heat content in these magma masses, the size classification is quite different from that used on maps of hydrothermal systems. The highest grouping in Fig. 5–14 is for occurrences with greater than $1,000 \times 10^{18}$ calories. The largest are in Yellowstone and the neighboring area in Idaho, where one Idaho occurrence holds estimated heat of greater than $4,000 \times 10^{18}$ calories,

that is, *more than thirty times the size* of the gigantic Bruneau-Grandview hydrothermal system.

The summary of heat content by state (Fig. 5–15, Table 5–5), makes it clear that Wyoming and Idaho have the highest totals by far on the basis of volcanic occurrences of known size. There are a number of identified occurrences of unknown size, however, whose heat content could change the relative resource standing of some states. Alaska, in particular, has a large number of such occurrences in the Aleutian chain.

GEOPRESSURED RESERVOIRS OF THE GULF COAST Sand bodies of abnormally high pressures containing hot water and dissolved methane occur in a zone whose top ranges from approximately 1.8 kilometers (6,000 feet) below the surface at the inland edge of the area in Texas, to as deep as 4.6 kilometers (15,000 feet) off the coast near New Orleans. Sands between these depths and a lower limit of 6.86 kilometers (22,500 feet) have been assessed by the U.S. Geological Survey for an area of 310,000 square kilometers, which extends offshore into federal waters and is limited by the 600 foot depth line that marks the edge of the continental shelf (Fig. 5–16).

Mechanical energy associated with the flow of fluids driven to the surface by high pressures constitutes less than 1 percent of the total accessible fluid resource, so it was ignored in this study. The energy in hot water and dissolved methane was estimated for the study area by using more than 3,500 oil and gas well records from which data on sand thicknesses, sand porosity, fluid pressures, and fluid salinity were extracted. To organize the intricately interbedded sand and shale in which the geopres-

Table 5–3 Accessible Heat In-Place in Identified High-Temperature Hydrothermal Systems (Including Vapor-Dominated Systems)

STATE[1]	MEAN HEAT IN-PLACE[2] (10^{18} Calories)
Alaska	1.95
Arizona	0.27
California[3]	135.59
Colorado	0.30
Hawaii	2.23
Idaho	4.23
Nevada	27.39
New Mexico	20.78
Oregon	21.62
Utah	11.47
Washington	0.32
Wyoming	296.23
UNITED STATES TOTAL	522.38

[1]Including National Parks.
[2]To a depth of 3 kilometers (9,843 feet).
[3]Of the California heat in place 23.9×10^{18} calories are in vapor-dominated systems whose conversion efficiency is lower than that for hot water systems.
Sources: U.S.G.S. *Circular 790*, pp. 44–57.

Table 5–4 Accessible Heat In-Place in Identified Intermediate-Temperature Hydrothermal Systems

STATE	HEAT IN-PLACE[1] (10^{18} Calories)
Alaska	5.57
Arizona	0.89
California	8.13
Colorado	2.31
Idaho	122.22
Montana	2.59
Nevada	8.51
New Mexico	1.04
Oregon	13.05
Utah	1.91
Wyoming	0.58
UNITED STATES TOTAL	166.00

[1]To a depth of 3 kilometers (9,843 feet). Source: U.S.G.S. *Circular 790*, p. 58–85.

Thermal springs

Area favorable for recovery of low
temperature geothermal waters

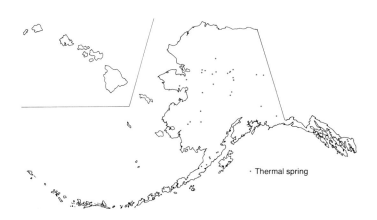

· Thermal spring

Fig. 5–13 Low-temperature geothermal waters: areas of springs and prospects:
A. Western states, Alaska, and Hawaii
B. Central and Eastern states

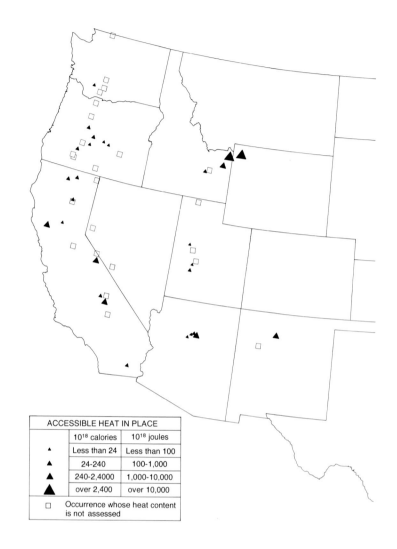

ACCESSIBLE HEAT IN PLACE		
	10^{18} calories	10^{18} joules
▴	Less than 24	Less than 100
▲	24-240	100-1,000
▲	240-2,4000	1,000-10,000
▲	over 2,400	over 10,000
☐	Occurrence whose heat content is not assessed	

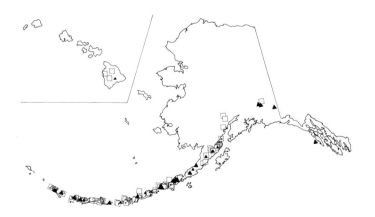

Fig. 5–14   Identified volcanic occurrences.

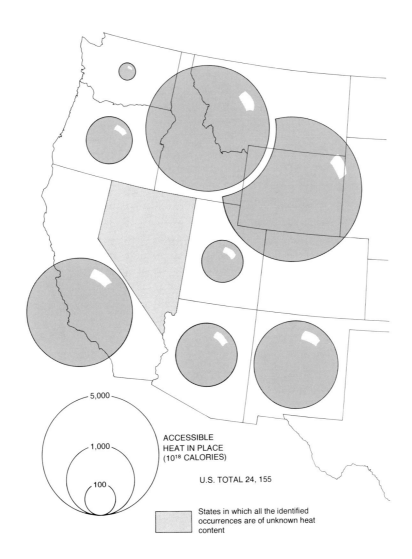

ACCESSIBLE
HEAT IN PLACE
(10^{18} CALORIES)

U.S. TOTAL 24, 155

States in which all the identified
occurrences are of unknown heat
content

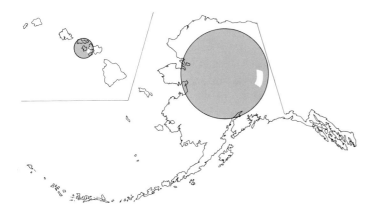

Fig. 5–15 Accessible heat in place in identified volcanic occurrences, 1978.

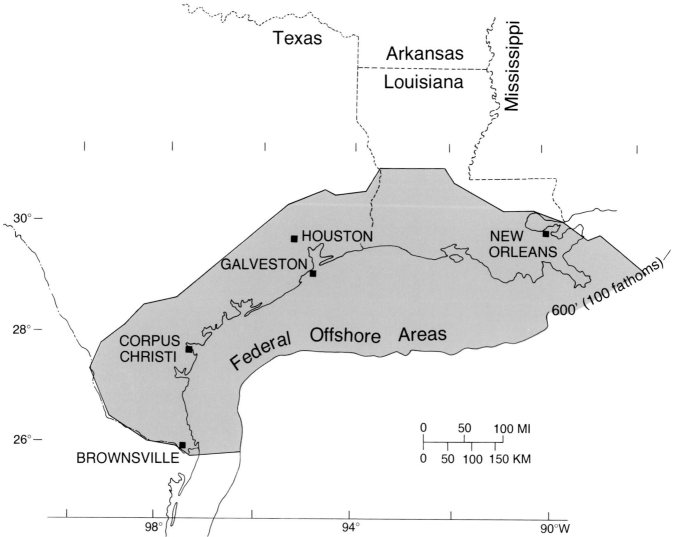

Fig. 5–16 Northern Gulf Coast area studied by U.S. Geological Survey in assessing geopressured reservoirs.

sured sands occur, the volume of sediments studied was hypothetically sliced into fourteen horizontal intervals, each 457 meters (1,500 feet) thick; and average values of rock and fluid characteristics were assigned to each interval. On the basis of a number of simplifying assumptions and calculations, the volumes of hot water, volumes of dissolved methane, and the associated thermal and methane energy in-place were estimated for each depth interval, then summed for Texas, for Louisiana, and for federal areas on the continental shelf (U.S.G.S. *Circular 790*, p. 136–142).

Table 5–6 shows the similarity of the distributions of thermal energy and methane energy, Texas holding almost half the total for each type. Figure 5–17 maps the total accessible energy *in-place*, of which Texas holds 45 percent, Louisiana 17 percent, and federal offshore areas 37 percent. The sizes of spheres in the Gulf Coast area, drawn to the same scale as those for hydrothermal and volcanic heat in-place, show heat amounts in geopres-

sured reservoirs to be very large. Recovery rates, though, are quite low because much of the hot water is in shales which have little permeability.

SUMMARY OF HEAT IN-PLACE IN EXPLOITABLE OCCURRENCES AND IN THE COUNTRY AS A WHOLE   Figure 5–18 and Table 5–7 summarize heat in-place in all *identified* occurrences: hydrothermal, volcanic, and geopressured, ranking states according to their totals derived from the three types. The large amounts of geopressured energy in-place makes the federal offshore areas, Texas, and Louisiana leading areas on this basis. California and Wyoming also show very high totals. Most of Wyoming's energy, however, is located in Yellowstone National Park, and cannot be exploited.

Figure 5–19 summarizes national totals of both *identified* and *undiscovered* energy in-place in exploitable occurrences, along with amounts of background heat in the entire country. The reference cube compares this geothermal heat in-place to another in-place resource: the

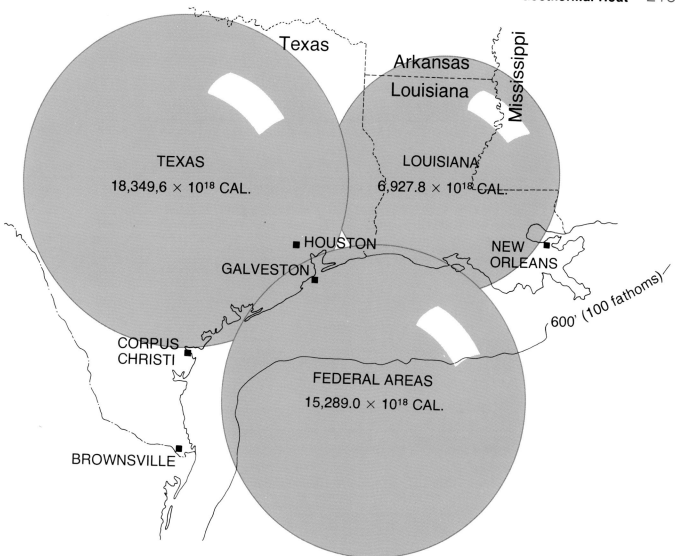

Texas

Arkansas

Louisiana

Mississippi

TEXAS
18,349,6 × 10^{18} CAL.

LOUISIANA
6,927.8 × 10^{18} CAL.

■ HOUSTON

GALVESTON

NEW
ORLEANS

CORPUS
CHRISTI

600' (100 fathoms)

FEDERAL AREAS
15,289.0 × 10^{18} CAL.

BROWNSVILLE

Fig. 5–17   Accessible heat in place in identified geopressured reservoirs of the northern Gulf Coast, 1978.

1,800 billion barrels of potential crude oil in identified Green River oil shales.

The bar graph in Fig. 5–19 shows how estimates of undiscovered heat in-place relate to identified amounts for the three types of usable occurrence. For the hydrothermal type, undiscovered refers to concealed natural systems of circulation. For volcanic occurrences, the undiscovered amounts include identified magma bodies for which no estimate has been made, and others not yet identified. For geopressured reservoirs there are possibilities beyond the Gulf Coast in a number of areas shown in Fig. 5–20. Some of the sedimentary basins shown in three western states, Wyoming, Utah, and Colorado, are the same basins in which prospects exist for unconventional natural gas from tight sands (see chapter on Oil and Gas). On the strength of a very large undiscovered component, the volcanic occurrences appear to hold the greatest potential; geopressured reservoirs are ranked next; and hydrothermal types have the least po-

tential. This observation must be modified by recognizing that the recovery rate for energy in geopressured reservoirs is around 2.5 percent; for volcanic systems it is unknown; but for hydrothermal systems it is about 25 percent.

ENERGY RECOVERABLE FROM EXPLOITABLE OCCURRENCES

Hydrothermal systems have already been exploited in the United States and other countries. The amounts of heat recoverable are therefore estimated on the basis of experience. Exploitation of geopressured reservoirs will likely require a combination of the technologies used for hydrothermal systems and those used in the completion of oil and gas wells. By contrast, volcanic occurrences lack hot fluids which can be raised to the surface, and require extraction techniques not yet developed. For these reasons, estimates of energy recoverable are made only for hydrothermal and geopressured occurrences. If cur-

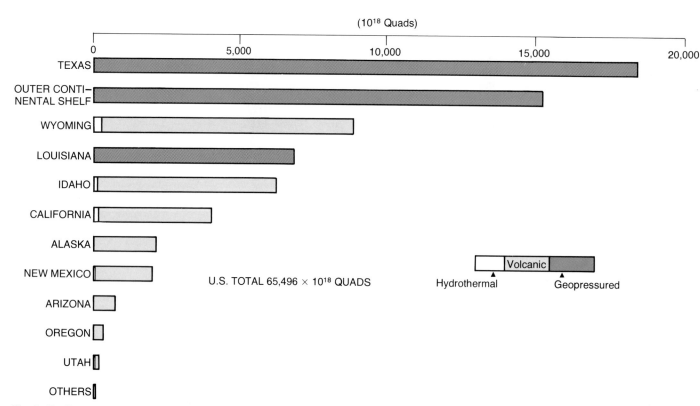

Fig. 5–18 National summary of accessible heat in place in identified hydrothermal, volcanic, and geopressured occurrences showing the roles of leading states, 1978.

Table 5–5 Identified Volcanic Occurrences

| | HEAT IN-PLACE[1] | | NUMBERS OF OCCURRENCES | | |
STATE	(10^{18} Calories)	Total	With Heat Content Estimated	With Heat Content Unknown	With Associated Hydrothermal Systems
Alaska	2,159	88	26	62	3
Arizona	767	4	4	0	0
California	3,859	19	11	8	10
Hawaii	23	5	1	4	1
Idaho	6,147	5	4	1	0
New Mexico	2,013	3	1	2	1
Nevada	0	2	0	2	1
Oregon	345	20	9	11	4
Utah	201	6	3	3	2
Washington	17	7	2	5	2
Wyoming	8,624	1	1	0	1
UNITED STATES TOTAL	24,155	160	62	98	25

Source: U.S.G.S. *Circular 790*, p. 13.

[1]To a depth of 10 kilometers (33,000 feet). For each state the heat in-place is the accessible resource in the occurrences that have been assigned values. Column three shows that many occurrences are not represented; some of those occurrences, especially in Alaska, have magma masses that are deeper than 10 kilometers. The heat in unevaluated known occurrences and undiscovered occurrences may be over 215,000 calories $\times$ 10^{18} (U.S.G.S. *Circular 790*, p. 15, 158).

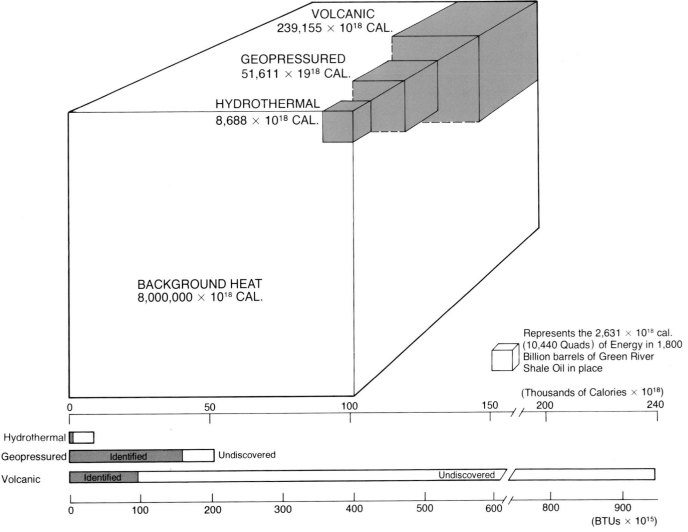

Fig. 5–19 Summary of accessible heat in place in exploitable occurrences both identified and undiscovered and background heat in the whole country to depth of 10 KM., 1978.

rent research leads to usable processes of heat extraction, the large amount of heat in volcanic occurrences may be usable in the future.

Applications of geothermal heat fall into two categories: electrical power generation, using fluids with temperatures over 150 degrees Celsius, and space-heating or industrial uses for water with temperatures lower than 150 degrees. For the sake of a national overview, the first information presented is for well-head heat recoverable without regard to application. This is followed by estimates of the potential for space-heating uses and for electrical power generation.

RECOVERABLE WELL-HEAD HEAT, REGARDLESS OF APPLICATION   Table 5–8 summarizes the accessible heat in-place and heat recoverable from identified hydrothermal and geopressured occurrences, and includes the heat in-place in volcanic occurrences.

Roughly 25 percent of the heat in-place in hydrothermal systems can be recovered at well-head. Recoverable amounts for geopressured reservoirs are only 2.6 percent

Table 5–6 Heat In-Place in Geopressured Fluids of the Northern Gulf of Mexico Basin (in 10^{18} calories)

AREA	HEAT IN-PLACE[1]		TOTAL
	Thermal	Methane	
Texas, onshore and inner continental shelf	11,227.9	7,166.7	18,394.6
Louisiana, onshore and inner continental shelf	4,538.9	2,388.9	6,927.8
Federal areas, outer continental shelf	10,033.4	5,255.6	15,289.0
GULF BASIN TOTAL	25,800.2	14,811.2	40,611.4

Source: U.S.G.S. *Circular 790*, p. 142.
[1]This is the accessible fluid resource base (according to U.S.G.S. *Circular 790*, p. 142) to a depth of 6.86 kilometers (22,500 feet).

Table 5–7 Summary of Accessible Heat In-Place in Identified and Undiscovered Hydrothermal, Volcanic, and Geopressured Occurrences (in 10^{18} calories)

	IDENTIFIED	UNDISCOVERED[1]		TOTAL
		Estimated	Mean	
Hydrothermal				
Over 150° C	522	2800–4900	3,850	4,372
90–150° C	166	3100–5200	4,150	4,316
TOTAL	688		8,000	8,688
Volcanic	24,155	215,000		239,155
Geopressured	40,611[2]	11,000[3]		51,611
TOTAL	65,454	231,900–236,100	234,000	299,454[4]

[1]Taken from U.S.G.S. *Circular 790*, p. 157.
[2]Gulf Basin.
[3]In other geopressured basins. Thermal energy only.

[4]For comparison, the background heat above 15 degrees Celsius and to a depth of 10 kilometers in the United States is estimated at 800×10^{22} calories (U.S.G.S. *Circular 726*, p. 92–93).

Table 5–8 Summary of Heat In-Place and Recoverable Geothermal Energy in Identified Hydrothermal, Volcanic, and Geopressured Occurrences (in 10^{18} calories)

STATE	HYDROTHERMAL[1] Over 150°C In-Place	Over 150°C Well-Head	90–150°C In-Place	90–150°C Well-Head	VOLCANIC[2] In-Place	VOLCANIC Well-Head	GEOPRESSURED RESERVOIRS[3] Thermal In-Place	Thermal Well-Head	Methane In-Place	Methane Well-Head	TOTAL In-Place	TOTAL Well-Head
Alaska	1.95	0.49	5.57	1.39	2,159.0	0	0	0	0	0	2,166.5	1.9
Arizona	0.27	0.07	0.89	0.22	767.0	0	0	0	0	0	768.5	0.3
California	135.59	29.37	8.13	2.03	3,859.0	0	0	0	0	0	4,042.7	31.4
Colorado	0.30	0.07	2.31	0.58	0	0	0	0	0	0	2.6	0.7
Hawaii	2.23	0.49	0	0	23.0	0	0	0	0	0	25.2	0.5
Idaho	4.23	1.06	122.22	30.57	6,145.0	0	0	0	0[4]	0	6,273.5	31.6
Louisiana	0	0	0	0	0	0	4,538.9	117.1	2,388.9	64.5	6,927.8	181.6
Montana	0	0	2.59	0.65	0	0	0	0	0	0	2.6	0.7
Nevada	27.39	6.81	8.51	2.13	0	0	0	0	0	0	35.9	8.9
New Mexico	20.78	5.26	1.04	0.26	2,013.0	0	0	0	0	0	2,034.8	5.5
Oregon	21.62	5.40	13.05	3.26	345.0	0	0	0	0	0	379.7	8.7
Texas	0	0	0	0	0	0	11,227.9	291.5	7,166.7	188.7	18,394.6	480.2
Utah	11.47	2.87	1.91	0.48	201.0	0	0	0	0	0	214.4	3.4
Washington	0.32	0.01	0	0	17.0	0	0	0	0	0	17.3	0.0
Wyoming	296.23	0	0.58	0.15	8,624.0	0	0	0	0	0	8,920.8	0.2
Federal Areas Outer Continental Shelf	0	0	0	0	0	0	10,033.4	258.6	5,255.6	138.6	15,289.0	397.2
UNITED STATES TOTAL	522.38	51.89	166.00	41.72	24,155.0	0.0	25,561.4	667.2	15,050.2	391.8	65,495.6	1,152.6[5]

[1]Well-head heat recoverable is simply 0.25 of accessible heat in-place.
[2]No recoverable amounts can be estimated because methods of extracting heat from hot dry rock have not been developed.
[3]Amounts of energy recoverable depend on the development plan. Plan Two (U.S.G.S. *Circular 726*, pp. 136–137) which assumes drastic reduction of formation pressures and considerable surface subsidence, is the basis for these estimates, which are roughly 2.6 percent of accessible energy in-place. Less ambitious development plans would recover less of the energy in-place: only 0.25 percent, for instance, is recovered in Plan Three. The tabulated amounts, therefore, represent *maximum* recoverable from identified reservoirs of the Gulf of Mexico.
[4]Methane energy recoverable is smaller than amounts of thermal energy recoverable from geopressured reservoirs. However, the more favorable conversion efficiency for methane (as in space heating) renders greater usable energy than do geopressured waters used in electrical generation.
[5]Or 4,573.8 Quads.

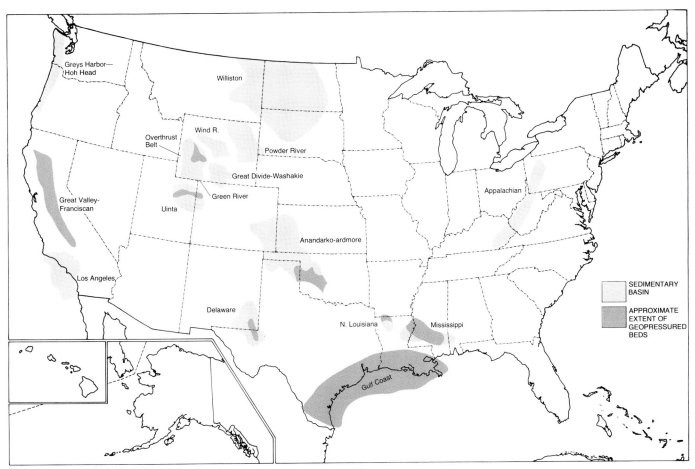

Fig. 5–20 Sedimentary basins having potential for geopressured reservoirs.

of heat and methane energy in-place. Despite the small recovery factor, the energy from geopressured reservoirs in Texas and Louisiana overshadows amounts recoverable from hydrothermal systems in the most favored states, California and Idaho. Those amounts from geopressured reservoirs, however, are *maximum* amounts, and once withdrawn will not be renewable (though injection of fluids from the surface may prolong the life of reservoirs). In total, the recoverable energy amount is quite large: the quantity $1,152.6 \times 10^{18}$ calories translates to 4,573.8 Quads which is of the same magnitude as the energy in all coals recoverable from the Demonstrated Reserve Base (see Chapter 1). What proportion of this well-head heat may be recoverable economically today, hence considered *Reserves,* is not clear. Whereas a government report in 1975 implied that roughly half the hydrothermal well-head heat may be Reserves now or in the very near future, the 1979 report declines to make an estimate (U.S.G.S. *Circulars 726* and *790*).

Because the temperatures of geothermal fluids are well below those attained by burning fuels, the efficiency of application is much lower than for fuels such as coal or oil. It may be misleading, therefore, to translate well-head heat into BTUs and compare it with the energy content of coal and crude oil amounts that are recover-

able. A more sound approach is to estimate how *much of those fuels would be saved or displaced* by the substitution of geothermal heat in specific applications.

SPACE HEATING WITH HYDROTHERMAL SYSTEMS OF INTERMEDIATE TEMPERATURES—(90–150 DEGREES CELSIUS) Figure 5–21 and Table 5–9 assume that intermediate-temperature waters are used for space heating. The well-head heat recoverable will do the heating task of a certain number of gallons of fuel oil. This may be deduced by recognizing that the geothermal heat exchanger will use only 24 percent of the well-head heat, whereas an oil furnace uses roughly 85 percent of the energy content of fuel oil. If the fuel oil is not burned, then it may be assumed that corresponding amounts of crude oil need not be refined. In a number of states, the amount of crude oil saved or displaced in this way exceeds one billion barrels, an amount comparable to reserves in a very large field. In Idaho, for instance, the savings of 26.2 billion barrels of crude oil is almost equivalent to the present proved reserves for the whole country.

ELECTRICAL ENERGY RECOVERABLE FROM HIGH-TEMPERATURE HYDROTHERMAL SYSTEMS (OVER 150 DEGREES CELSIUS) AND FROM GEOPRESSURED RESERVOIRS The potential for electrical power generation exists in both high-temperature hydrothermal systems and in geopressured reservoirs where fluids are generally above 150 degrees Celsius. These two heat sources could also be used for space heating and other low-temperature applications, but electrical generation is assumed here.

Figure 5–22 maps the potential for hydrothermal electric power output, listed in Table 5–10. California shows a potential of almost 14,000 Megawatts for thirty years, that is, the equivalent of fourteen very large nuclear power plants operating at 100 percent capacity for thirty years. Some of this potential is already developed in the Geysers area where installed capacity was nearly 900 Megawatts in 1979. Nevada, New Mexico and Oregon each has the potential for at least two such plants. Amounts of fuel saved can be calculated directly from the average fuel needs of coal-burning and nuclear plants.

Elements of the concept being investigated by the U.S. Department of Energy at the Los Alamos Scientific Laboratory, Los Alamos, New Mexico. Hot rocks are fractured by means of hydraulic techniques used in oil and gas fields. Then, cold water is injected through the longer pipe, is heated as it passes through fractures to reach the second pipe through which water at high temperature and high pressure is withdrawn. The pancake-shaped fractured zone, 1,500 feet in radius, would be formed if the fractures propagate in response to the rapid cooling of hot rock. *Courtesy Department of Energy*

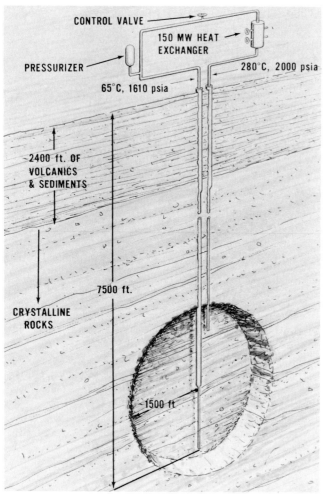

CONTROL VALVE

150 MW HEAT EXCHANGER

PRESSURIZER

280°C, 2000 psia

65°C, 1610 psia

~2400 ft. OF VOLCANICS & SEDIMENTS

7500 ft.

CRYSTALLINE ROCKS

~1500 ft

Table 5–10 shows a total of more than 3 billion tons of coal saved—a very significant amount when compared to the present annual production of the country, which is less than 1 billion tons.

Figure 5–23 and Table 5–11 show the potential electrical output from geopressured reservoirs of the Gulf Coast to be roughly ten times greater than that of identified high-temperature hydrothermal systems throughout the West—provided that the very ambitious development plan (Plan Two, which is explained in Table 5–11) is undertaken, and assuming only thermal energy, not methane, is used for power generation. Of course, the implied fuel savings are correspondingly large; over 32 billion tons of coal would be theoretically displaced during the thirty-year period, or over 2 million tons of U_3O_8 (uranium oxide).

SUMMARY Figure 5–24 and Table 5–12 show that the total potential fuel savings from space-heating and electrical applications of energy in identified hydrothermal and geopressured occurrences is over 1,700 Quads. For comparison, the total energy used in the United States in 1978 was roughly 75 Quads. Included in those summations is the methane which would be produced in exploitation of Gulf Coast geopressured reservoirs. Again, the development plan assumed is the most ambitious one contemplated; and the amount of methane energy, therefore, a maximum estimate.

Table 5–9 Fuel Saved If Well-Head Heat Recoverable from Intermediate-Temperature Hydrothermal Systems is Used for Space Heating

STATE	Well-Head Heat Recoverable[1] (10^{18} Calories)	Fuel Savings[2] (Billions of Barrels of crude oil)
Alaska	1.39	1.19
Arizona	0.22	0.91
California	2.03	1.74
Colorado	0.58	0.50
Idaho	30.57	26.20
Montana	0.65	0.56
Nevada	2.13	1.83
New Mexico	0.26	0.22
Oregon	3.26	2.80
Utah	0.48	0.41
Wyoming	0.15	0.13
UNITED STATES TOTAL	41.72	36.48

[1]Well-head heat recoverable is simply 0.25 of heat in-place to a depth of 3 kilometers. Source: U.S.G.S. *Circular 790*, p. 58–85.

[2]Assume beneficial heat is 0.24 of well-head heat (U.S.G.S. *Circular 790*, p. 26) whereas an oil furnace extracts 0.85 of the energy in fuel oil, which is 3.55×10^5 calories per gallon. Fuel oil saved is translated to *crude oil not refined* by using the average of 100 barrels of crude for every 22 barrels of fuel oil produced (Cook, p. 93).

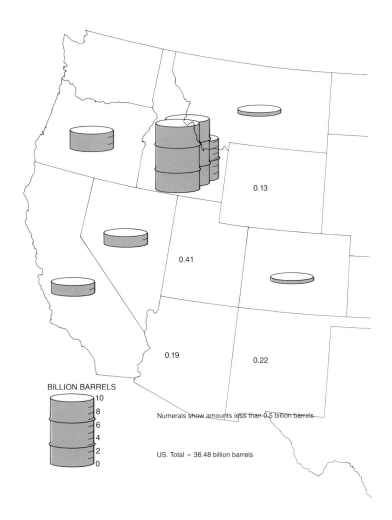

BILLION BARRELS

10
8
6
4
2
0

0.13

0.41

0.19

0.22

Numerals show amounts less than 0.5 billion barrels

U.S. Total = 36.48 billion barrels

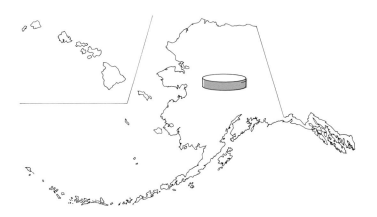

Fig. 5–21 Crude oil saved in western states if heat recovered from identified intermediate-temperature hydrothermal systems (90-150° C) is used for space heating.

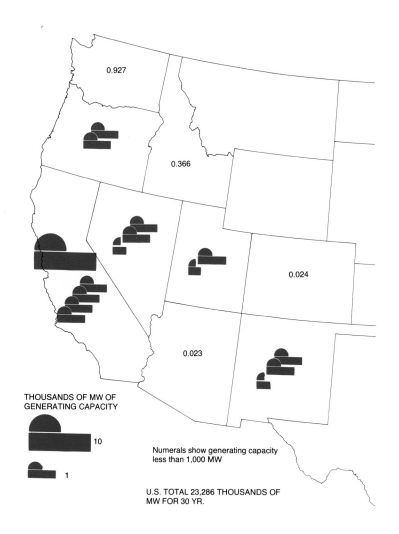

THOUSANDS OF MW OF
GENERATING CAPACITY

10

1

Numerals show generating capacity
less than 1,000 MW

U.S. TOTAL 23,286 THOUSANDS OF
MW FOR 30 YR.

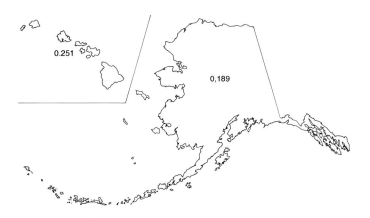

Fig. 5–22 Potential electrical output from identified high-temperature hydrothermal systems, for period of 30 years.

Table 5-10 Fuels Saved through the Use of Electrical Energy Recoverable from Identified High-Temperature Hydrothermal Systems

STATE	Well-Head Heat Recoverable (10^{18} Calories[1])	Electrical Energy (Thousands of Megawatts Electric Output for Period of 30 Years[2])	FUELS SAVED[3] Coal (Millions of Tons)	FUELS SAVED[3] U_3O_8 (Tons)	STATE	Well-Head Heat Recoverable (10^{18} Calories[1])	Electrical Energy (Thousands of Megawatts Electric Output for Period of 30 Years[2])	FUELS SAVED[3] Coal (Millions of Tons)	FUELS SAVED[3] U_3O_8 (Tons)
Alaska	0.49	0.189	26.082	1,770	New Mexico	5.26	2.700	372.600	25,288
Arizona	0.07	0.023	3.174	215	Oregon	5.40	2.031	280.278	19,022
California	29.37	13.816	1,906.608	129,307	Utah	2.87	1.300	179.400	12,176
Colorado	0.07	0.024	3.312	224	Washington	0.01	0.027	3.726	253
Hawaii	0.48	0.251	34.638	2,350	UNITED STATES TOTALS	51.89	23.286[4]	3,213,468	217,999
Idaho	1.06	0.366	50.508	3,427					
Nevada	6.81	2.559	353.142	23,967					

[1]Excluding national parks. Heat recoverable at well-head is estimated to be 0.25 of accessible heat in-place.

[2]Electrical energy is derived by applying recovery factors that vary according to reservoir temperatures and depths (U.S.G.S. *Circular 790*, pp. 23–26). Only the heat to a depth of 3 kilometers (9,843 feet) is considered.

[3]In thirty years of operation a coal-fired plant rated at 1000 Megawatts would burn roughly 90 million tons of bituminous coal (Nuclear Energy Policy Study Group, p. 78); and a light water reactor would need roughly 6090 tons of unenriched U_3O_8 (Sondermayer, p. 1118). If the plants are assumed to run at only 65 percent of capacity, then these fuel requirements are for only 650 Megawatts of output. The fuels needed for each 1000 Megawatts of *output* would be roughly 138 million tons of coal, and 9,366 tons of U_3O_8; these are the figures used to convert geothermal electric output into coal and uranium oxide saved.

[4]In undiscovered hydrothermal systems the potential is estimated at 72 to 127 thousand Megawatts of output for thirty years (U.S.G.S. *Circular 790*, p. 41).

Table 5-11 Electrical Energy Recoverable from Identified Geopressured Reservoirs in Gulf of Mexico Basin, According to Development Plan Two

AREA	Well-Head Heat Recoverable from Thermal Component Only[1] (10^{18} Calories)	Electrical Energy (Thousands of Megawatts for 30 Years[2])	FUELS SAVED[3] Coal (Millions of Tons)	FUELS SAVED[3] U_3O_8 (Tons)
Texas, onshore and inner continental shelf	291.45	103.19	14,240	966,477
Louisiana, onshore and inner continental shelf	117.06	41.46	5,721	388,268
Federal areas, outer continental shelf	258.62	91.59	12,639	857,794
GULF BASIN TOTAL	667.13	236.24	32,600	2,212,539

[1]Dissolved methane is assumed to be not used for electrical generation; and mechanical energy is ignored. Well-head heat recoverable (U.S.G.S. *Circular 790*, p. 144) is to a depth of 6.86 kilometers (22,500 feet).

[2]Conversion efficiency of 0.08 is assumed (U.S.G.S. *Circular 726*, p. 138). The resulting energy is translated to electrical according to the equivalence, 1 MW Century = 7.53×10^{14} Calories (U.S.G.S. *Circular 726* p. 4). Energy is extracted from reservoirs at rates consistent with Development Plan Two which is the most ambitious of three proposed plans, and entails 25,840 wells spread roughly 2 kilometers apart, and considerable surface subsidence (U.S.G.S. *Circular 726*, p. 136–37). The least ambitious of the three plans (Plan Three) would yield only 23,000 Megawatts of output for thirty years (U.S.G.S. *Circular 790*, p. 160).

[3]In thirty years of operation a coal-fired power plant rated at 1000 Megawatts would burn roughly 90 million tons of bituminous coal (Nuclear Energy Policy Study Group, p. 78) and a light water reactor of the same rating would need roughly 6090 tons of unenriched U_3O_8 (Sondermayer, p. 1118). If the plants are assumed to run at 65 percent of capacity, then these fuel requirements are for only 650 Megawatts of output. The fuels needed for each 1,000 Megawatts of output would be roughly 138 million tons of coal, and 9,366 tons of U_3O_8; these are the figures used to convert geothermal electrical output into coal and uranium oxide saved.

Prospects and Development

WESTERN STATES, ALASKA AND HAWAII

Most development has occurred in the western states, Alaska, and Hawaii because these states contain the identified hydrothermal systems whose exploitation is now feasible.

Figure 5–25 shows a number of features that pertain to development and prospects. One is the generalized pattern of heat flows as deduced from well and test-hole information—high rates of heat flow being indicative of good prospects, whether or not there is some surface expression of an exploitable occurrence. Three areas of high heat flow are evident. One is the Imperial Valley in southern California where high-temperature hydrothermal occurrences are presently being exploited. Another is the Snake River Plain in southern Idaho and its extension into Nevada referred to as the Battle Mountain High. Many intermediate-temperature hydrothermal systems

occur in the Snake River Plain, and a number of high- and intermediate temperature hydrothermal systems occur in the Battle River High, apparently due to deep-circulating waters, not to any local igneous intrusions. The third area of high heat flow is located in southern New Mexico where only a few hydrothermal systems of intermediate temperature occur. The most prominent area of low flow is the Sierra Nevada range (see Fig. 5–2).

KNOWN GEOTHERMAL RESOURCE AREAS   Mapped as small blocks in Fig. 5–25 and listed in Table 5–13 are areas in which "the prospects for extraction of geothermal steam or associated resources are good enough to warrant expenditures of money for that purpose (Godwin, *et al.,* p. 8–9). These are called known geothermal resource areas (KGRAs). Unusually high temperature gradients and subsurface temperatures that warrant the KGRA classification are indicated either by measurement in test-holes or by analysis of surface deposits and the composition of hydrothermal waters. KGRAs are shown

Fig. 5–23   Potential electrical output from thermal energy in identified geopressured reservoirs of the northern Gulf Coast.

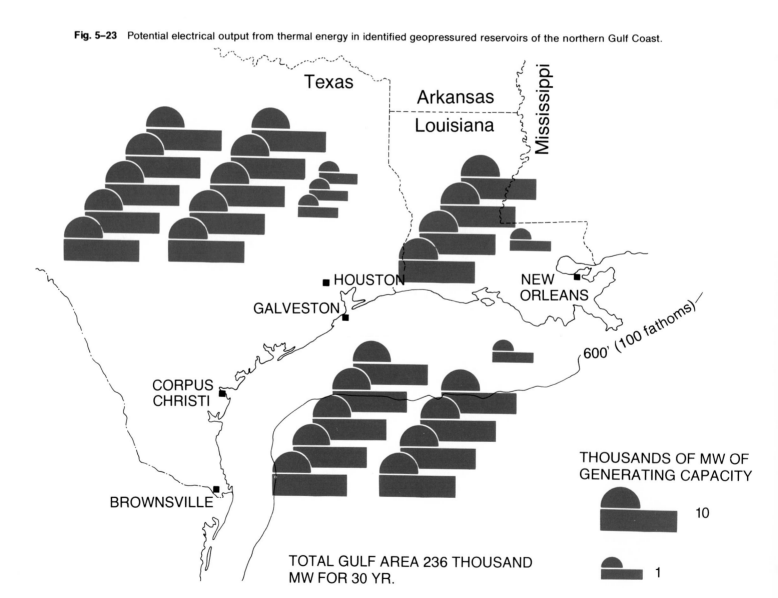

TEXAS

Arkansas

Louisiana

Mississippi

HOUSTON

GALVESTON

NEW ORLEANS

600' (100 fathoms)

CORPUS CHRISTI

BROWNSVILLE

THOUSANDS OF MW OF GENERATING CAPACITY

10

1

TOTAL GULF AREA 236 THOUSAND MW FOR 30 YR.

Table 5–12 National Summary of Potential Fuel Savings through Space Heating and Electrical Applications of Geothermal Energy

| TYPE OF OCCURRENCE | ALTERNATIVES FOR SPACE HEATING | | ALTERNATIVES FOR ELECTRICAL GENERATION | | TOTAL SAVINGS |
| | Conventional Natural Gas | Crude Oil | Coal | U_3O_8 | |
	(Trillions of Cu. Ft.)	(Billions of Barrels)	(Millions of Tons)	(Tons)	Quads, (10^{15} BTU)
Intermediate-Temperature Hydrothermal (space heating)	0	36.48[1]	0	0	211.56[3]
High-Temperature Hydrothermal (electrical)	0	0	3,213	217,999	83.54[4]
Geopressured Reservoirs					
Methane for Space-Heating	617[2]	0	0	0	620.00[4]
Thermal for Electrical	0	0	32,600	2,212,539	847.60
TOTAL					1,762.70

[1]Assuming 100 barrels of crude refined for every 22 barrels of fuel oil produced.
[2]According to the most ambitious development plan, Plan Two (U.S.G.S. *Circular 790*, p. 144).

[3]On the basis of heat recovered from all identified intermediate-temperature systems over an unspecified period.
[4]On the basis of power plants operating for period of thirty years.

Fig. 5–24 Summary of potential electrical output and fuels saved through space heating and electrical generation from identified hydrothermal and geopressured occurrences.

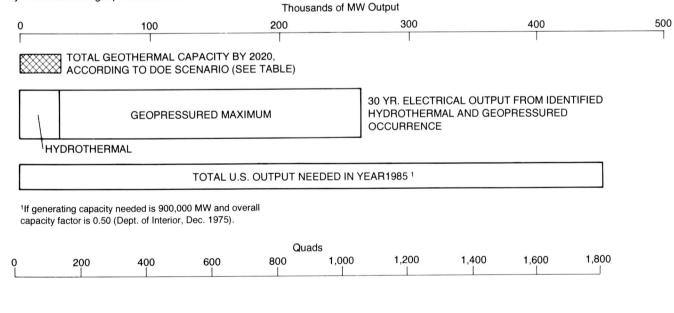

Thousands of MW Output

0   100   200   300   400   500

TOTAL GEOTHERMAL CAPACITY BY 2020, ACCORDING TO DOE SCENARIO (SEE TABLE)

GEOPRESSURED MAXIMUM

30 YR. ELECTRICAL OUTPUT FROM IDENTIFIED HYDROTHERMAL AND GEOPRESSURED OCCURRENCE

HYDROTHERMAL

TOTAL U.S. OUTPUT NEEDED IN YEAR 1985 [1]

[1]If generating capacity needed is 900,000 MW and overall capacity factor is 0.50 (Dept. of Interior, Dec. 1975).

Quads

0   200   400   600   800   1,000   1,200   1,400   1,600   1,800

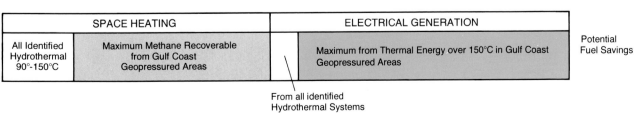

| SPACE HEATING | | ELECTRICAL GENERATION | |
| All Identified Hydrothermal 90°-150°C | Maximum Methane Recoverable from Gulf Coast Geopressured Areas | Maximum from Thermal Energy over 150°C in Gulf Coast Geopressured Areas | Potential Fuel Savings |

From all identified Hydrothermal Systems over 150°C

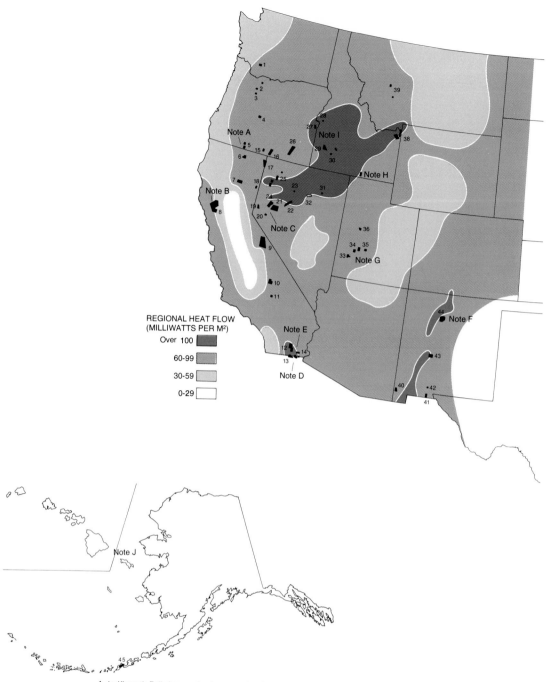

REGIONAL HEAT FLOW
(MILLIWATTS PER M²)

Over 100

60-99

30-59

0-29

A   In Klamath Falls homes, businesses, churches and schools have used intermediate-temperature waters for space heating for years. Plans now call for development of a district heating system to serve center city.

B   In The Geysers area the world's largest geothermal-electric installation makes use of a rare vapor-dominated reservoirs. Generating capacity in 1979 is roughly 900 Megawatts, and is expected to grow to 1128 Megawatts by 1982.

C   At Desert Peak a large hidden high-temperature reservoir was discovered in 1976. Seven miles northwest, at Brady Hot Springs, fluids at 150°C are being used for process heat and vegetable dehydration.

D   Located at Cerro Prieto in Mexico is the first major hot water geothermal-electric installation in the Western Hemisphere. It has operated since 1973, now has a capacity of 75 Megawatts, with plans for an additional 215 Megawatts.

E   The Imperial Valley, like Cerro Prieto to the south, is an area of crusral spreading, and high heat flow. Reservoir temperatures in the sediments that fill the valley range up to 360°C. These installations are under development or planned: at East Mesa, a 10 MW and a 45 MW electric plant; at Heber, a 45 and a 50 MW plant; at Brawley, a 10 MW electric plant, and process heat for sugar refining; at Salton Sea, two 10 MW electric plants.

F   At Valles Caldera, the Department of Energy plans a 50 MW electric demonstration plant, using waters of 275°C. Also a research project is testing methods of fracturing hot dry rock of a volcanic occurrence and inducing an artificial circulation of water.

G   The Roosevelt area is a major geothermal discovery of the 1970s. A 52 MW electric plant will make use of its 265°C waters.

H   At Raft River Valley, water classified as intermediate is actually at 150°C and will be used in a Department of Energy project for electrical generation, aquaculture, food processing, and crop irrigation. The electric pilot plant will have a capacity of 5 MW and will use isobutane rather than water as the working fluid.

I   At Boise, water at 77°C has been used for space heating since 1892. Now a demonstration project heats state and federal buildings, and there are plans for a district heating utility system.

J   Puna area. In 1976, a well drilled to 1,966 meters (6,300 ft.) found water at temperatures up to 358°C. A 5 MW electrical plant was being built in 1979.

Fig. 5–25   Geothermal prospects and development activities in Western states, Alaska, and Hawaii.

Table 5–13 Known Geothermal Resource Areas (KGRAs) and Associated Hydrothermal and Volcanic Occurrences

AREA		TYPES OF OCCURRENCES[2]			
		Hydrothermal		**Volcanic**	
Number[1] (For Map Reference Only)	**Name**	Over 150° C	90– 150° C	Assessed	Not Assessed
1				X	
2	Mt. Hood				X
3			X		
4		X		X	
5	Klamath Falls		X		
6				X	
7	Morgan Springs	*		X	X
8	Geysers	*	X	X	
9	Long Valley	X	X	X	X
10	Coso	X		X	
11		X		X	
12	Salton Sea	X		X	
13	Heber and East Mesa	X			
14			X		
15			X		
16	Crump's Hot Springs	X	X		
17	Surprise Valley	X			X
18			X		
19	Steamboat Springs	X	X		X
20			X		
21	Desert Peak and Brady	X			
22			X		
23	Leach Hot Springs	X			
24		X	X		
25	Pinto Hot Springs	X			
26	Alvord	X			
27	Vale	X			
28	Crane Creek	X			
29	Bruneau		X		
30	Grandview		X		
31	Elko		X		
32	Beowawe	X			
33			X		
34	Roosevelt and Cove Fort	X			
35			X		
36			X		
37	Raft River		X		
38	West Yellowstone		X		
39			X		
40			X		
41			L		
42			X		
43			L		
44	Valles Caldera	X	X	X	
45	Geyser Bright and Hot Springs Cove	X	X		

[1]KGRAs are generalized for presentation on atlas map.
[2]Occurrences deduced from Maps 1 and 2 in U.S.G.S. *Circular 790.* An entry in table denotes that one or more such occurrences exists in or near the KGRA.

* Indicates that vapor-dominated reservoirs occur.
L Indicates that only low-temperature (under 90° C) waters occur in or near the KGRA.

here without the locations of hydrothermal or volcanic occurrences. The character of the prospective areas may be deduced by comparing with Figs. 5–9, 5–11, and 5–14. The coincidence of KGRAs with hydrothermal and volcanic occurrence is noted in Table 5–13 which makes it clear that most KGRAs are based on hydrothermal occurrences.

Leasing activity since 1974 has resulted in greatest acreage leased in the state of Nevada (Table 5–14). The greatest activity, though, has been in the state of California, where thirty-five drilling rigs were active in fiscal year 1978, whereas only two were active in Nevada, and one each in Idaho and Oregon (Interagency Coordinating Council, p. 40–41).

ACTIVITY HIGHLIGHTS Figure 5–25 shows areas where important development is taking place. The growth of electrical generating capacity at the Geysers, for example, reached 900 Megawatts by the end of 1979 and is expected to exceed 1,100 Megawatts by 1982. Also in 1979, a 10-Megawatt electric plant was completed at East Mesa in southern California by Imperial Magma Power Company, while experimental power plants are being built at Puna on the island of Hawaii and at Valles Caldera, New Mexico.

Researchers at Valles Calderas, near Los Alamos, are attempting to devise methods of inducing fracture permeability in hot dry rock. Success in this venture could open up the use of very large amounts of accessible heat in volcanic (hot dry rock) occurrences. In one test, fractures were successfully introduced in rock of tempera-

Table 5–14 Summary of Competitive Geothermal Lease Sales in Known Geothermal Resource Areas (KGRAs) as of December 31, 1978

STATE	Total Number of Acres Leased in Sales 1974–1978
Nevada	154,931
New Mexico	89,400
Utah	88,211
Oregon	68,873
California	40,654
Idaho	27,448
Colorado	8,036
Arizona and Montana	0[1]
TOTAL	474,553

[1]No bids for leases offered.
Source: Interagency Coordinating Council, p. 1–11.

Table 5–15 Installed Generating Capacity by 2020 at Various Western and Gulf Sites[1] (in Megawatts electric)

	Installed Geothermal-Electric Generating Capacity by Year 2020	Identified Potential Capacity Derived from Tables 5–10 and 5–11
Region I Sites		
Alvord, Oreg.	300	
Baker Hot Springs, Wash.	50	
Brawley, Calif.	1,000	
Coso Hot Springs, Calif.	600	
East Mesa, Calif.	100	
Geysers, Calif. (hot water)	1,000	
Geysers, Calif. (steam)	2,170	
Glass Mt., Calif.	50	
Heber, Calif.	1,000	
Lassen, Calif.	100	
Mono-Long Valley, Calif.	250	
Mount Hood, Oreg.	50	
Puna, Hawaii	920	
Salton Sea, Calif.	2,000	
Surprise Valley, Calif.	2,000	
Vale Hot Springs, Oreg.	800	
REGION TOTAL	12,390	15,858
Region II Sites		
Brady Hot Springs, Nev.	1,000	
Beowave, Nev.	1,000	
Bruneau-Grandview, Idaho	3,150	
Chandler, Ariz.	230	
Cove, Fort-Sulphurdale, Utah	1,500	
Leach, Nev.	1,500	
Raft River, Idaho	800	
Roosevelt Hot Springs, Utah	1,000	
Safford, Ariz.	100	
Steamboat Springs, Nev.	200	
Thermo, Utah	500	
Valles Caldera, N. Mex.	1,500	
Weiser-Crane Creek, Idaho	1,000	
West Yellowstone, Mont.	50	
REGION TOTAL	13,390	16,100
Region III Sites		
Acadia Parish, La.	350	
Brazoria, Tex.	2,225	
Calcasieu Parish, La.	350	
Cameron Parish, La.	500	
Corpus Christi, Tex.	1,650	
Kenedy County, Tex.	300	
Matagorda County, Tex.	500	
REGION TOTAL	5,875	236,240
WESTERN AND GULF TOTAL	31,655	268,198

[1]According to Interagency Geothermal Coordinating Council scenario.
Source: Interagency Coordinating Council, p. 45–58.

tures near 185 degrees Celsius, and water was circulated for seventy-five days at a heat extraction rate of 5 Megawatts (thermal). Plans are underway to enlarge the system and to extract enough heat to generate 50 Megawatts electric (Cummings, Ronald G., *et al.*).

GULF COAST AREA OF GEOPRESSURED RESERVOIRS

Figure 5–26 shows areas of greatest potential for development, some of which are close to large cities and might ultimately supply them with electricity. Three drill-holes of interest are shown on that map, one of which is the first hole drilled expressly for testing geo-pressured resources. The area is covered with thousands of oil and gas holes, 3,000 of which were used in assessing the resource potential, and two of which have been selected for geopressured reservoir testing.

FEDERAL GOALS FOR ELECTRICAL DEVELOPMENT IN WESTERN STATES AND THE GULF COAST

Figure 5–27 shows sites for development of electrical generating capacity in accord with the goals of the U.S. Department of Energy. By the year 2020 a total of over 30,000 Megawatts of geothermal generating capacity is envisioned, most of it from hydrothermal occurrences in western states, while a smaller capacity is anticipated from geopressured reservoirs of the Gulf Coast. Table 5–15 lists the sites, and the projected generating capacity at each. It also shows the potential geothermal electrical generating capacity for each of the planning regions—derived by summing the state potentials from Tables 5–10 and 5–11 for identified hydrothermal and geopressured reservoirs respectively. Apparently the Department of Energy aims at developing a very large part of the identified potential in Regions I and II (hydrothermal potential) but is not contemplating development of geopres-

Fig. 5–26 Prospects and activity in the northern Gulf Coast area of geopressured reservoirs.

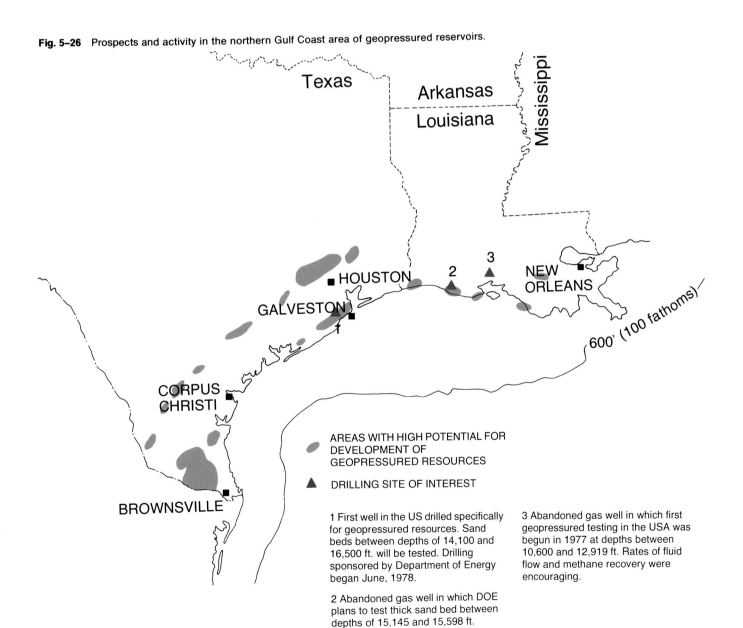

AREAS WITH HIGH POTENTIAL FOR DEVELOPMENT OF GEOPRESSURED RESOURCES

▲ DRILLING SITE OF INTEREST

1 First well in the US drilled specifically for geopressured resources. Sand beds between depths of 14,100 and 16,500 ft. will be tested. Drilling sponsored by Department of Energy began June, 1978.

2 Abandoned gas well in which DOE plans to test thick sand bed between depths of 15,145 and 15,598 ft.

3 Abandoned gas well in which first geopressured testing in the USA was begun in 1977 at depths between 10,600 and 12,919 ft. Rates of fluid flow and methane recovery were encouraging.

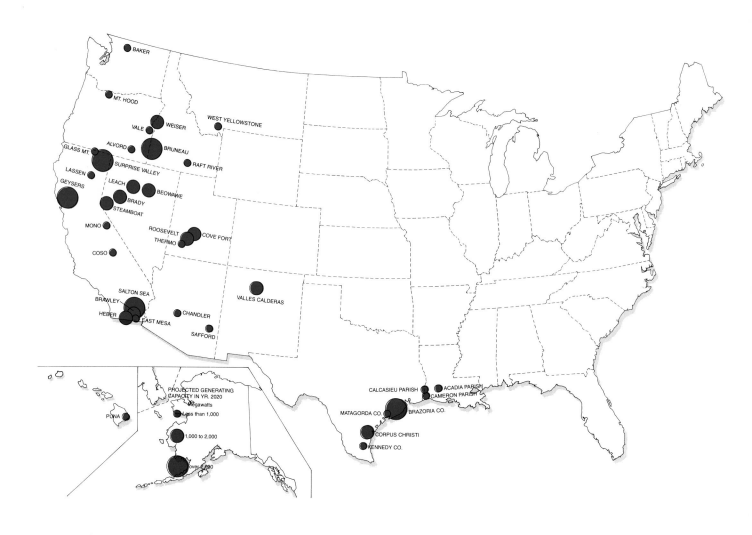

Fig. 5–27 Sites for geothermal electric generation according to Department of Energy Interagency Coordinating Committee, 1978, showing four regions used by DOE in development planning.

sured reservoirs on a scale consistent with the Plan Two estimate of 236,000 Megawatts capacity.

EAST COAST LOW-TEMPERATURE GEOTHERMAL WATERS

While some development drilling for these waters is taking place across the country, the activity on the East Coast is especially interesting because the potential is so close to densely populated areas with great demand for low-temperature energy. The Department of Energy has undertaken a program of shallow test holes to explore promising areas as far south as Georgia (see Fig. 5–13B). Figure 5–28 shows locations of forty-nine holes drilled during 1978 as part of a continuing program.

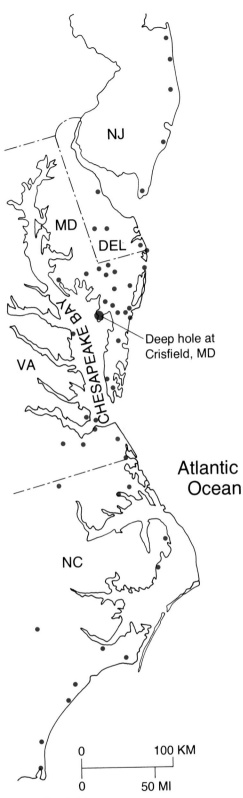

Deep hole at Crisfield, MD

Fig. 5–28 Forty-nine test holes drilled by Department of Energy in summer of 1978 to explore low-temperature geothermal waters on the East Coast.

A geothermal project site in Imperial Valley, California. *Courtesy American Petroleum Institute.*

PART
TWO

Renewable Resources

Chapters 6 through 10 deal with solar radiation, wind, hydropower (and tidal), ocean thermal gradients, and biomass. Only hydroelectric power can be represented by relatively firm numbers that express the total potential and the proportion of it that has been developed already. The total potential of the other renewable sources is much less certain and for solar radiation and biomass materials, estimates of maximum potential are virtually impossible. Instead, the use of these resources is estimated by determining how much energy might be delivered (and how much fuel energy saved) if certain assumptions are made about fuel prices and other economic factors.

Some renewable sources—wind, hydropower, and ocean thermal gradients—are regarded here as essentially producers of electrical energy, and are represented in units of generating capacity and units of output. Solar radiation and biomass resources have varied applications in which their contributions are best expressed as amounts of energy (in Quadrillion British Thermal Units) delivered, and amounts of mineral fuel energy displaced.

Solar Radiation

Since the Arab oil embargo and price hike in 1974, American consumers of petroleum products face not only soaring prices but also the very real threat of supplies being interrupted without notice. To deal with the high price of gasoline, many are driving less and switching to small vehicles. The response to expensive and uncertain supplies of fuel oil has been a sudden and growing interest in the use of solar energy for heating buildings (space heating). Coincidentally, the federal government has supported research on a number of solar-related technologies, including solar heating. The various applications of solar energy will gain momentum as fuel prices continue to rise, so that by the year 2000 solar energy may contribute 20 to 25 percent of the Nation's total need for raw (primary) energy (Stobaugh and Yergin, p. 211).

As the individual citizen, the utility company, or the society as a whole, contemplate the use of new solar technologies, they are forced to adopt a new attitude toward energy costs. The costs of an installation to gather the "free" solar energy must be endured initially, and then recouped as the burning of expensive fuels is averted in the years that follow. The economic attractiveness of any such investment is affected not only by the costs of fuels saved, but also by institutional factors such as tax credits which effectively lower the costs of the installation for an individual or a business.

The various forms of solar energy are reviewed in *Introduction to Renewable Resources* (see Part One of this book). The principles are expressed more concretely in Table 6–1 in order to provide a clear setting for the materials in this and the following chapters.

Part A of Table 6–1 distinguishes solar radiation itself from the physical and biological effects that depend on it. Radiation can be used for either its thermal or its photovoltaic effects. With low-temperature devices it can heat buildings. In addition, it can generate electricity either by high-temperature thermal devices or by photovoltaic installations. The three major physical effects that depend on the sun—wind, hydro energy, and ocean thermal gradients—all can be used for generating electricity. The biological phenomena that are based upon photosynthetic conversion of radiation by green plants are greatly varied and can be used to produce a rich variety of fuels. Although some of those fuels could be used to generate electricity, they are usually regarded as substitutes for coal, oil, and natural gas, in non-electric applications.

Part B of the table identifies the three major applications that are recognized by the Department of Energy: 1) thermal applications, such as space heating, 2) electrical generation, and 3) production of alternate fuels. In the right-hand column the resources drawn upon for those

The photovoltaic concentrator array at the U.S. Department of Energy's Sandia Laboratories. *Courtesy Department of Energy.*

Table 6–1 Solar and Solar Related Phenomena, and Their Applications

A Arranged According to Origins

RADIATION			SOLAR-DEPENDENT PHYSICAL EFFECTS	SOLAR-DEPENDENT BIOLOGICAL EFFECTS
Thermal		**Photo-Voltaic**		
Low-Temperature	High-Temperature			
Hot water and space heating	Industrial	Electrical generation	Wind electric	Photosynthesis, supporting all plant and animal life
Agricultural and industrial	Electrical generation		Waves	
			Ocean currents	Biomass: organic wastes and energy crops yielding alternate fuels
			Hydroelectric energy	
			Ocean thermal electric	

B Arranged According to Applications

APPLICATION OR PRODUCT	RESOURCE USED	CHAPTER
Hot water and space heating (and cooling)	Radiation	6
Agricultural and industrial process heat		6
Electrical generation	Radiation Thermal Photo-voltaic	6
	Wind	7
	Hydropower	8
	Ocean thermal	9
Alternate fuels: solid, liquid, and gaseous	Biomass: wastes and crops	10

applications are arranged to coincide with the order of the chapters in Part Two.

This chapter is concerned only with solar radiation used directly for its thermal and photovoltaic effects. It is arranged according to applications, as in Table 6–1B. After a brief discussion of radiation and its measurement, hot water heating is reviewed. This is followed by a more lengthy section dealing with solar space heating and the climatic and economic factors that determine its contribution to heating needs and its economic feasibility in different parts of the United States. Then electrical power generation by thermal and photovoltaic means is reviewed. Finally, some estimates of the future contributions of these solar technologies are presented.

The main purpose of this chapter is to promote an understanding of the resource, its distribution, and the geographic factors that condition its use. Secondarily, the status of the different applications is clarified, and some government projects are mapped. The sections covering different applications vary in length because for some applications there is less information available on the locational aspects, whereas for others (such as space heating) there exist some very useful studies whose principles are applied in this chapter.

The Resource and Its Distribution

Solar radiation reaching the earth is measured by instruments called pyranometers, and is expressed in units of energy received per square unit of surface area. *Langleys,* that is, calories per square centimeter, is the unit most commonly used. It can be translated easily into BTUs (British Thermal Units) per square foot of surface area. Most of the available data on radiation is for the amounts received on a horizontal surface. In order to estimate the amounts received by a surface that is tilted to intercept the radiation more directly, such as that of a solar collector, this amount must be converted by calculations (see Liu and Jordan). Measuring radiation by the amounts falling onto a given surface makes it very convenient to determine how large a collecting device must be built if it is to gather enough energy for some specific task.[1]

WORLD DISTRIBUTION OF RADIATION

Two factors control how much radiation reaches any part of the earth's surface: the intensity of radiation, and its duration. Intensity depends on the angle at which the rays strike the surface (more direct rays, more intense radiation) and also upon the clarity of the atmosphere. Duration depends on length of the daylight period. Figure 6–1 shows that the annual total of radiation received is arranged roughly according to latitude. High latitudes receive less because the radiation received there is less direct; while the low latitudes receive more because of more direct radiation. The maximum for the year is not along the equator, as would be dictated by geometrical considerations alone, but in the vicinity of Latitudes 20 to 25 North and South, where clear skies in desert areas admit more radiation than the cloudy skies near the equator. A number of radiation maxima are obvious, the largest in area being those of North Africa, Southwest Asia, and Australia, all straddling their respective Tropics. Farthest poleward in location are the maxima in southern Argentina and its counterpart in the southwestern United States.

RADIATION IN THE UNITED STATES

On maps of radiation (Figs. 6–2A to 6–2D, and Fig. 6–3) the maximum evident on the world map shows concentrations in southern California, Nevada, Arizona, and New Mexico. In these areas, the clear skies of desert and semi-desert climates give the area markedly greater radiation than areas of comparable latitude in the southeastern part of the country. The *gradient* or slope from higher values in the South to lower in the North is most pronounced in January, when low sun and short days conspire to bring much less radiation to northern areas. In July, the long days in higher latitudes bring so much radiation that the north-south gradient is replaced by a west to east gradient that is due to more clear skies in the west. Some areas in the North, such as Boise, Idaho, therefore, receive greater radiation than areas in the South, such as Louisiana.

It is important to understand that the amounts of radiation mapped here are the sum of direct and indirect or diffuse radiation, which occurs when cloudy skies scatter the incoming rays. In the Southwest, where skies

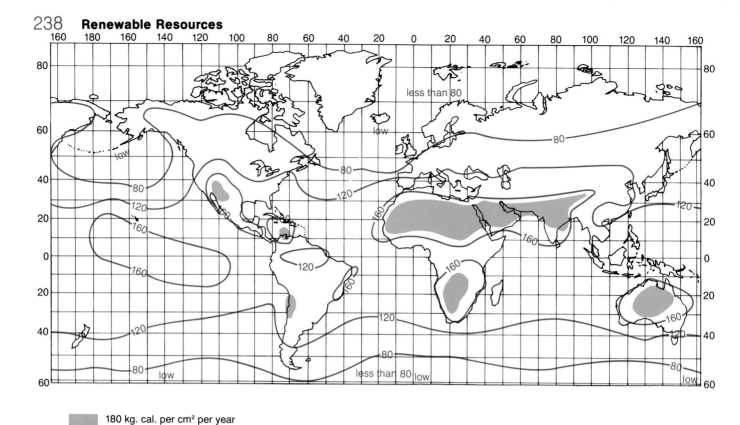

180 kg. cal. per cm² per year

Fig. 6–1 Total solar radiation (direct and diffuse) at the surface of the earth during the year. Measured in thousands of calories (kg. cal.).

are usually clear, the total radiation is not only high, but is dominated by direct radiation. Only this direct form can be *focused* by devices such as parabolic reflectors or lenses to achieve the high temperatures essential for solar-thermal electric power generation and some other applications. These areas of clear skies are definitely more favorable for such high-temperature applications.[2]

Because mapped values are long-term averages, the actual insolation (or amount of solar radiation) received on any day will deviate, sometimes greatly, from the daily value suggested by the monthly map. In addition, the reliability of the data on which radiation maps are based may be questionable (Bennet; Durrenberger and Brazel). Radiation measurement is a challenging and recent undertaking. In 1930 there were only 32 insolation stations in the world, and although there are now 107 in the United States alone, their periods of record differ from station to station, some of which have substantial periods of missing data. Therefore, mapped values should be regarded as estimates of long-term averages, based on necessarily imperfect instrumentation.

Hot Water Heating

Hot water heating and space heating are *low-temperature* thermal applications of solar radiation. As such, they are very logical ways of using the sun's energy because solar radiation is converted to low-temperature heat very efficiently. A dark surface can absorb as heat 85 to 90 percent of the radiation falling on it. In contrast, photoelectric cells convert only about 15 percent of the

radiant energy to electrical energy; and photosynthesis is only 1 percent efficient. Substituting solar radiation for fuels in low-temperature applications is wise because it matches the energy source to the task. Fuel oil and natural gas are capable of producing very high temperatures and are wasted on applications such as hot water or space heating.

Heating of domestic (or commercial) hot water is a solar application even more logical than space heating because for domestic hot water the need (or the "load") is constant, whereas for heating there is a need only during the winter months. This means that for domestic hot water an installation properly sized will be fully used throughout the year and will more quickly lead to fuel savings that will repay the initial investment.

The collector and other elements of a hot water heating system are sketched in Fig. 6–4. Optimum collector size, and hence the cost of the system, depends on a number of factors, including: 1) the demand and the implied size of the storage tank, 2) the radiation expected at the location in question, and 3) the price of fuel that would otherwise be burned to heat water.

A rough guide to the sizing, without regard for optimizing the system in the face of specific fuel costs, is as follows. Assuming there is a backup system (using a fuel), the size of tank need not consider storage for cloudy days, and can be arrived at simply by per capita use. On the basis of 20 gallons per person per day, a family of four could use a system of 80 to 100 gallons capacity. While this factor of need is essentially constant throughout the country, the radiation varies greatly and dictates that a

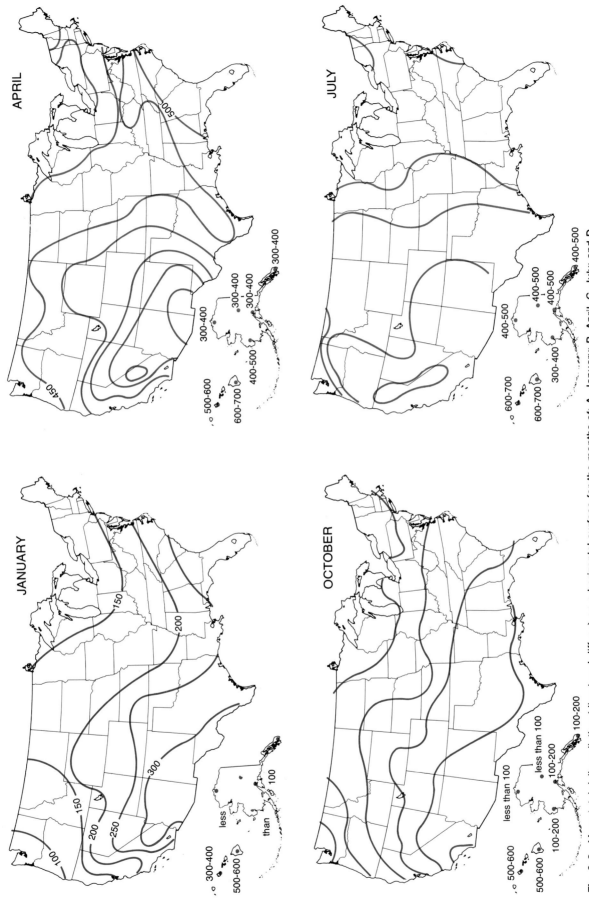

Fig. 6–2 Mean total daily radiation (direct and diffuse) on a horizontal surface for the months of: A. January; B. April, C. July; and D. October. Measured in langleys (gm. cal. per square cm.)

Fig. 6–3 Mean total daily radiation (direct and diffuse) on a horizontal surface, in the U.S., based on annual records. Measured in langleys (gm. cal. per sq. cm.).

A commercial application of solar hot water heating. The 144 panels are on the roof of the Red Star Industrial Service Laundry in Fresno, California. This is the largest installation of its type in the country as of 1978. *Courtesy Department of Energy.*

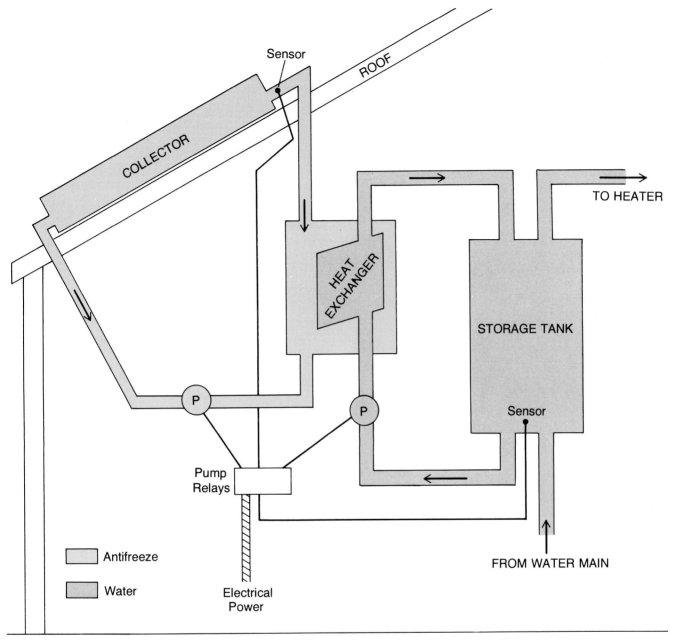

Fig. 6–4   Solar domestic hot water system.

larger collector is needed where fewer BTUs are received per square foot of surface area. Figure 6–5, based on zones implicit in a map of mean daily radiation from annual records (Fig. 6–3) suggests how collector size must vary in order to provide enough energy for a 100-gallon system in various parts of the country. In the northern zone, collector area in square feet should be roughly 1.5 times the number of gallons required, while in the southern zone the multiplier is only 0.5.[3]

A more precise estimate of the system size that is best for a given location in light of the local cost of fuels saved can be obtained through a service initiated by the federal Energy Research and Development Administration (now Department of Energy). If the potential buyer submits

certain information—water needs, location, cost of relevant fuel at the location, the specifications and the cost per square foot of the installation under consideration, and the type of financing—he will receive guidance on the sizing of the system and the savings expected.[4]

The economic attractiveness of solar water heating is shown in Fig. 6–6. This map plots the results of SOL-COST analysis at six locations using 1978 installation costs, and assuming electricity (at local rates) as the alternative for heating water. For this analysis, a system with 80 square feet of collector was used at each of the locations. It provided only 48.3 percent of the need at Fargo, but 72 percent of the need at Dallas. Net savings due to fuel costs averted in these hypothetical installa-

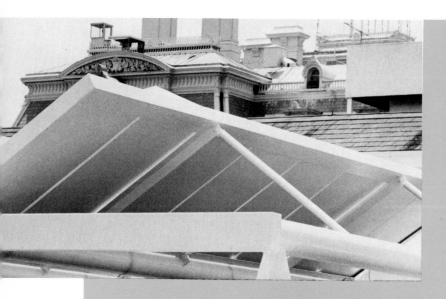

the kitchen of the Staff Mess, which is located in the basement. As is normal in such installations, the solar system preheats water before it enters the regular hot water heating system, thereby reducing the energy required. The amount of energy saved depends on the weather, incoming water temperature, and water use.

The energy provided by the White House solar system will replace approximately 1,000 dollars per year of steam, at 1979 prices. As energy costs increase in the future, so will savings.

Technical details are as follows:

Collectors	32 panels in 4 rows of 8
	Total collector area: 611 square feet
	Orientation: due south, 33° tilt
Storage	600 gallons
Heating Fluid	50 gallons ethylene glycol solution
Flow Rate	10 gallons per minute
System Output	75–130 thousand BTUs per year

The White House Solar System

The White House solar hot water heating system was completed at the end of April 1979 on the roof of the West Wing of the White House. The solar panels (collectors) can be seen from Pennsylvania Avenue. The primary user of hot water in the West Wing is

Fig. 6–5   Three simple radiation zones in the U.S., and the implied areas of collectors for domestic hot water heating.

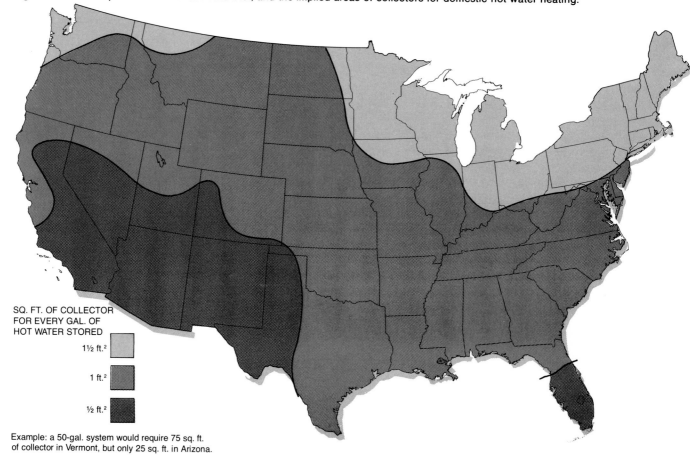

SQ. FT. OF COLLECTOR FOR EVERY GAL. OF HOT WATER STORED

1½ ft.²

1 ft.²

½ ft.²

Example: a 50-gal. system would require 75 sq. ft. of collector in Vermont, but only 25 sq. ft. in Arizona.

tions is enough to fully pay for the installations in the periods noted. The investment in these installations is paid back in less than ten years, indicating a sound investment, even without benefit of government tax rebate.

The boxed insert shows the solar installation at the White House which began operating on June 20, 1979, and provides a schematic of how the solar hot water heater operates. The photograph shows how the collectors look on the roof of the White House itself.

Space Heating and Cooling

Application of solar energy to space heating and cooling is more interesting and complex than its use for hot water heating. The need for either heating or cooling varies throughout the country, and is also seasonal; the season of greater need and season of greater supply coincide differently in the cases of heating and cooling; and much more energy is used for heating and cooling than for water heating, making the opportunities for fuel savings far greater.

THE NEED FOR HEATING AND COOLING

The distribution of the *need* is fundamental to any study of heating and cooling, regardless of the energy source. The occurrence of temperatures that demand heating or cooling is expressed in maps of "degree-days" (Figs. 6–7 and 6–8).

Heating degree-days (Fig. 6–7) are the total for a whole year, of the daily needs, each of which is derived this way: (65 degrees Fahrenheit) minus (daily mean temperature). For example, a day with a mean temperature of 45 degrees Fahrenheit would contribute 20 degree days to the annual total. In this way the annual or seasonal need for heating is captured in a single number which may be used by fuel suppliers to anticipate customers' needs and by meteorologists to express the severity of a month or a winter. The pattern, considerably simplified in Fig. 6–8, is dominated by greatest values in northern areas and by an intrusion southward of greater values in highland areas, especially the Rocky Mountains and the Sierra Nevada.

Cooling degree-days are derived by a similar procedure which uses the "base" of 65 degrees Fahrenheit, and sums the daily differences between that and the daily mean, that is: (daily mean temperature) minus (65 degrees Fahrenheit). When these values are mapped (Fig. 6–8) it is not surprising that greater values occur in the South. The pattern is modified, however, by greater values extending northward in the middle of the country where the lack of highland or ocean moderation causes higher summer temperatures.

Fig. 6–6 Years to payback for a solar domestic hot water system at six locations according to SOLCOST analysis, Spring, 1978.

Note: At every location, installation cost assumed is $2,170. Alternative is electric hot water heating using power at local 1978 electricity rates escalated at 7% per year.

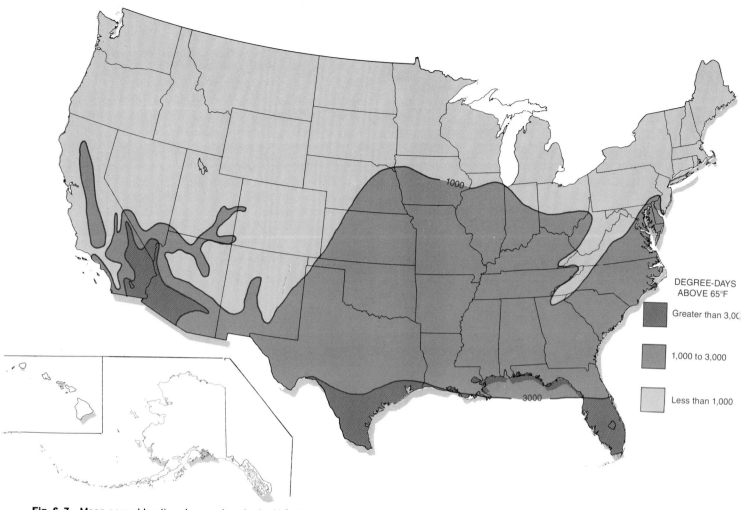

Fig. 6–7 Mean annual heating degree-days in the U.S. Base 65° F (19° C).

DEGREE-DAYS
ABOVE 65°F

Greater than 3,00

1,000 to 3,000

Less than 1,000

Fig. 6–8 Mean annual cooling degree-days in the U.S. Base 65° F (19° C).

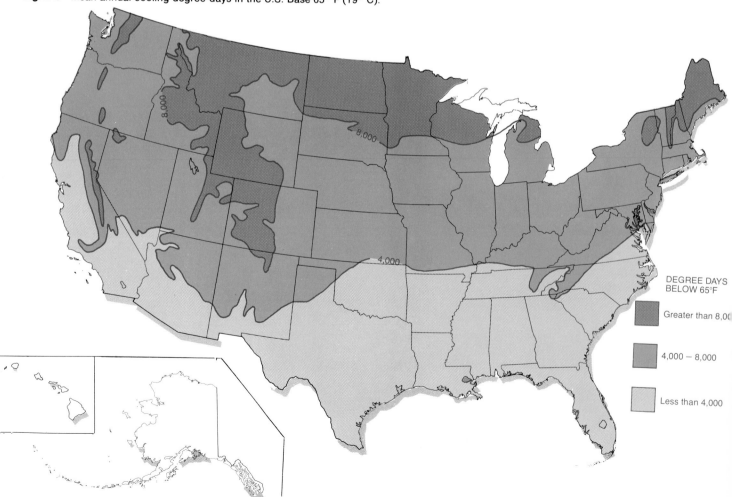

DEGREE DAYS
BELOW 65°F

Greater than 8,00

4,000 – 8,000

Less than 4,000

THE NEEDS AND THE SOLAR SUPPLY

How the supply of energy from the sun matches and satisfies the needs for heating and cooling is one of the most important geographic phenomena of our time. Responding to the research opportunity offered by the federal government in the past few years, a number of engineers have tackled the problem. Provided here is the result of one such study by the TRW Systems Application Center in their "Phase Zero" report on impacts of solar heating and cooling.

HEATING DEMAND AND SUPPLY   In winter, it is self evident that areas of greatest heating need are at the same time deficient in solar radiation. Actual conditions within the country, however, are more complex than might be expected, because of certain combinations of demand and supply. Figure 6–9 (TRW Systems Group, p. 2–2) shows two degree-day values, 2,500 and 5,000 degree days, that divide the country into three belts with regard to heating need. At the same time, two radiation values, 250 Langley and 350 Langley, slice the country into three belts with regard to supply of solar energy *during the heating season*. The combination of the two sets of three belts leads to nine theoretical region types summarized in the table (Fig. 6–9).

Type 1, in the extreme southeastern and southwestern parts of the country, is where the heating need is least and the supply of solar energy is greatest. At the other extreme is Type 9 in the North, where heating need is greatest and solar supply is least. Understandably, Type 7 is non-existent, because it combines very little heating need with very low radiation values. Type 8 is unique to the Northwest where ocean moderation causes a mild winter of intermediate heating need, and at the same time cloudy skies in the winter reduce radiation to the lowest category. Type 6 is unique to the Northern Plains where heating need is great and the solar supply is greater than the latitude would suggest. Both these areas will appear again in the more thorough analysis of heating supply and demand in a later section.

COOLING DEMAND AND SUPPLY   Using an approach similar to that for heating demand and supply, three zones of cooling demand taken from a cooling degree-days map may be combined with three zones of greater to lesser solar radiation received during the cooling season. Again the theoretically possible combinations are reflected in nine types in Fig. 6–10.

In summer the cooling need is well-matched to the supply. Logically, the areas of high temperatures tend to

Fig. 6–9 Heating demand, as reflected in heating degree-days, combined with mean daily solar radiation during the heating season.

	Heating DEGREE DAYS		
	0–2500	2500–5000	5000–9000
Mean Daily Radiation (ly) 350–450	1	2	3
250–350	4	5	6
175–250	7	8	9

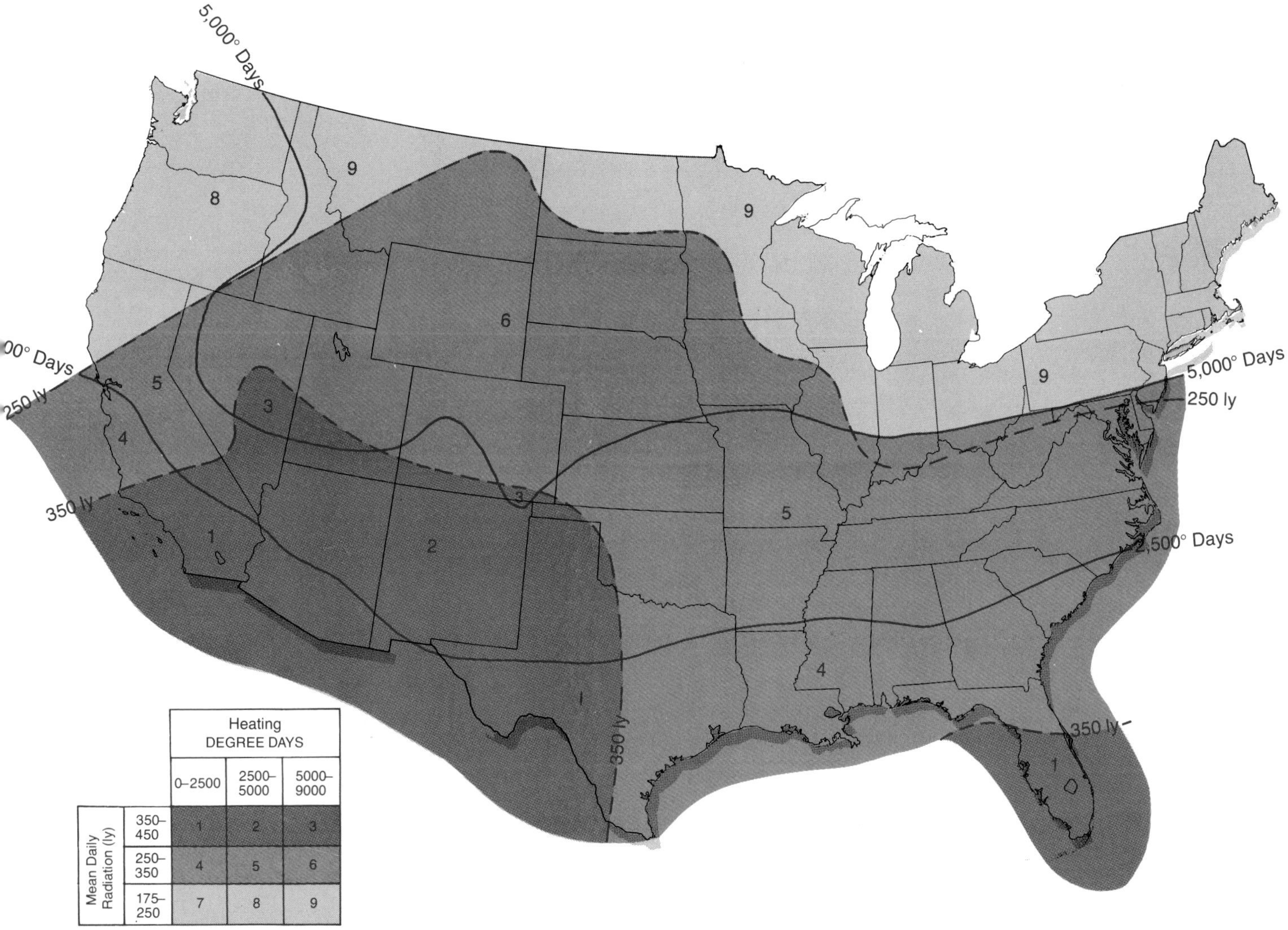

Regions numbered according to combinations of demand (degree days, base 65 F) and supply (mean daily radiation, in langleys, during period of November-April)

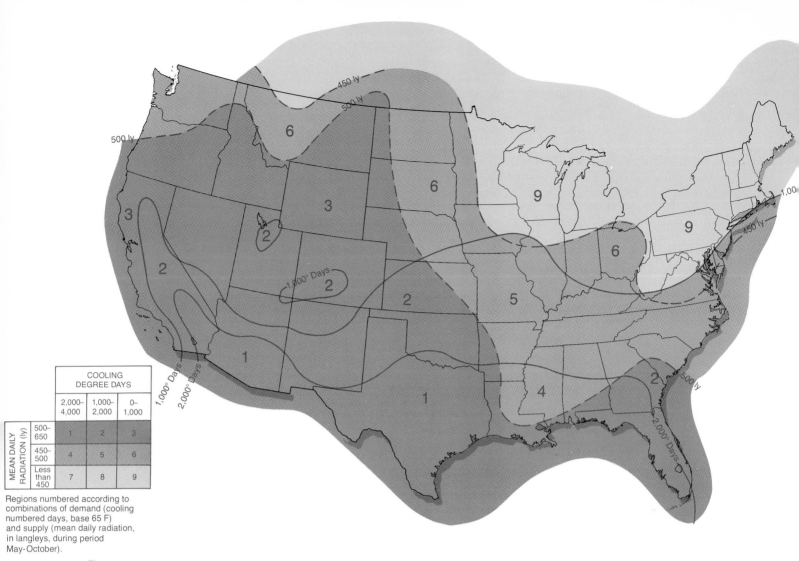

COOLING DEGREE DAYS			
	2,000– 4,000	1,000– 2,000	0– 1,000
500– 650	1	2	3
450– 500	4	5	6
Less than 450	7	8	9

MEAN DAILY RADIATION (ly)

Regions numbered according to combinations of demand (cooling numbered days, base 65 F) and supply (mean daily radiation, in langleys, during period May-October).

Fig. 6–10 Cooling demand, as reflected by cooling degree-days, combined with mean daily solar radiation during the cooling season.

be areas of great radiation (Type 1) and areas of lower temperatures tend to be areas of lower radiation (Type 9). Clear skies in the West allow the condition of greatest radiation to sweep into northern areas, while in the East the high radiation values are restricted to southern latitudes. Once again, not all theoretical types actually exist, Types 7 and 8 being the unlikely combinations of relatively great cooling demand and lowest radiation.

Although the use of heat for cooling is not new (for example, in the 1940s refrigerators were gas-fired) the equipment for solar-fueled air conditioning is nevertheless unsatisfactory. Solar-driven air conditioning requires high-temperature operation which renders collectors inefficient. Concentrating collectors are suited to the task, and have been used successfully, but the technology is regarded as immature, and for that reason no thorough geographic analysis of cooling feasibility in various parts of the country is available.

SOLAR SPACE HEATING, AND THE EFFECTS OF CLIMATIC VARIATION

Solar energy may be applied to heating buildings in a number of different ways. *Passive* solar heating refers to the use of architectural design that properly orients the

building, and makes deliberate use of south-facing surfaces either to admit or to absorb radiation. The structure itself is used as the solar collector and storage device, and can attain a large part of its heating needs this way. *Active* solar heating systems entail a separate collector, a storage device, and controls linked to pumps or fans that draw heat from storage when it is available (see Box). Systems which are dominantly passive, but make use of fans to move air are often called ''hybrids.'' In severe climates, both passive and active systems are backed up by a fuel-burning heater of some kind.

PASSIVE SOLAR HEATING

Passive designed buildings can achieve great fuel savings with relatively small investment and little or no complicated apparatus. It can be argued, in fact, that *every* building should be oriented and built with solar radiation and other climatic factors as paramount considerations in design.

Buildings can intercept and store solar energy in three basic ways: direct gain, indirect gain, or isolated gain (Fig. 6–11). A building may employ more than one of the three strategies. In *direct gain*, the sun enters through a south-facing window and strikes ''thermal masses'' de-

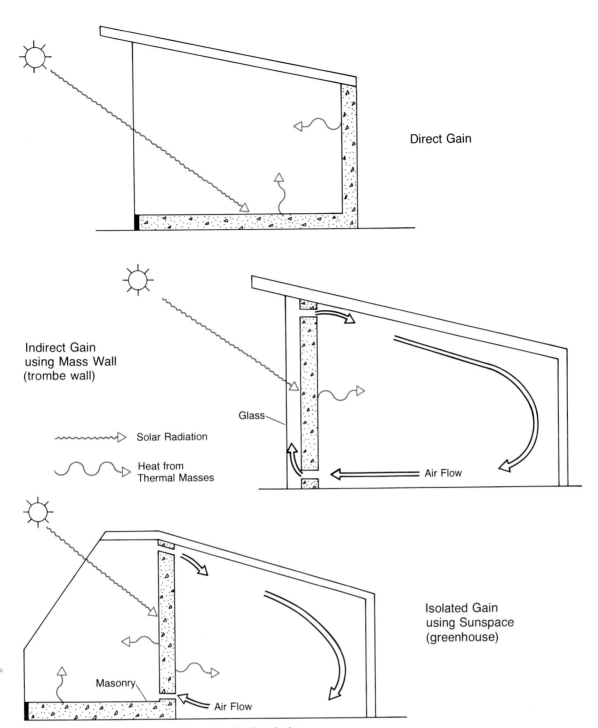

Direct Gain

Indirect Gain
using Mass Wall
(trombe wall)

〜〜〜〜▷ Solar Radiation

〜〜〜▷ Heat from
Thermal Masses

Glass

Air Flow

Isolated Gain
using Sunspace
(greenhouse)

Masonry

Air Flow

Fig. 6–11 The three types of passive solar heating design.

Direct Gain. Radiation penetrates south-facing windows and is absorbed as heat by masonry floor and wall. These thermal masses store heat then release it overnight.

Indirect Gain. A massive south wall is covered with glass, creating an air space through which air circulates. Cool air from the building enters near the floor, is warmed, and flows out at the top. Vents are closed overnight to prevent circulation of air. The glass reduces radiation loss from the wall overnight when it is giving up its stored heat to the adjacent room.

Isolated Gain. A "sunspace" room that may be used as a greenhouse which collects heat and circulates warm air during the day. Excess heat may be stored in a rock bed (not shown) and drawn upon at night. Glass areas may be covered at night to reduce heat loss. If the sunspace is located downsloped from the building, heat will flow very effectively upward into the building, while cool air drains down, in a convective circulation called *thermosiphon.*

Fig. 6–12 Climate regions used by American Institute of Architects Research Corporation in proposing guidelines for energy-conserving and passively heated homes.

Region 4, essentially areas over 5,000 feet (1,524 m) elevation, is shaded only to simplify the patterns

liberately placed to absorb radiation. These usually are a heavy concrete floor, and sometimes masonry walls, fireplaces, or water stored in drums. Thermal masses store the heat, then release it at night. *Indirect gain* is typified by a glazed masonry wall. *Isolated gain* uses a south room as a collector and a heavy masonry wall to absorb heat and protect the dwelling space from wide swings of temperature (see Box).

CLIMATIC FACTORS AND PASSIVE DESIGN   Since passive design requires that the structure itself be used to gather and store heat, the cost of solar heating features and the energy gained through them cannot be estimated accurately. Active systems lend themselves better to a geographic analysis of optimal design and economic feasibility. Nevertheless some useful observations can be made about how building design should be adjusted to various climates in the country. Some estimates can also be made of how much the sun contributes to the building's heating needs.

Figure 6–12 shows sixteen generalized climatic regions in the contiguous states, developed by the American Institute of Architects (A.I.A.) in cooperation with the National Climatic Center, and published as *Regional Guidelines for Building Passive Energy Conserving*

Homes. For each region, this publication notes the proportions of each year that are either too hot or too cool, and also the extent to which passive design alone can reduce those periods of discomfort. Most important, the book offers down-to-earth suggestions on building design and site landscaping to take advantage of the climate's assets and to minimize its liabilities. In Region 1A, for instance, typified by Hartford, Connecticut, the climate is too hot (over 80 degrees Fahrenheit) for 13 percent of the year and is too cool (under 68 degrees Fahrenheit) for 75 percent of the year. Through passive design alone, the A.I.A. estimates the periods of discomfort can be reduced to 0 percent for "too hot" and 63 percent for "too cool." Similar estimates for other regions show significant contributions from passive design.

In fact, the A.I.A. "percent-of-the-year" approach tends to greatly understate the role of passive heating in winter months. A more useful indicator of a building's effectiveness is the proportion of heating (or cooling) needs that are supplied by the sun. Table 6–2 lists performance estimates from a few of the existing passive solar buildings and shows that very large portions of the winter heating needs are gained from the sun—a typical proportion being 75 percent.

Table 6-2 Selected Buildings of Passive Solar Design, and the Proportion of Heating Needs Gained from the Sun

BUILDING	HEATING DEGREE-DAYS	SOLAR PROPORTION OF HEATING ENERGY NEEDS (Percent)	HEATING STRATEGY
Sea Ranch Sundown, 100 miles north of San Francisco, California	2,969	95	Direct gain
Kelbaugh residence, Princeton, New Jersey	5,100	75	Indirect gain (mass wall)
Whitcomb residence Los Alamos, New Mexico	6,500	30	Indirect gain (water mass wall)
The Rural Center, Northern California	7,520	100[1]	Indirect gain (roof pond)
Erwin residence Nacogdoches, Texas	1,500	75-90	Isolated gain (sunspace)
P. Davis residence, Albuquerque, New Mexico	4,348	75	Isolated gain (thermosiphon)
Crowther residence Denver, Colorado	5,500	65	Indirect gain

[1]Tolerating 55° F indoors.
Source: HUD, November, 1978.

ACTIVE SYSTEMS, THEIR COMPONENTS, AND THE METHOD OF ANTICIPATING PERFORMANCE

The schematic, Fig. 6–13, shows how an active solar system works. The collector feeds heat through an exchanger into a storage tank from which heat is drawn for domestic hot water and for the major task of heating the building. After a sunny period hot water will be available in storage, and the system will draw upon it for heating. Overnight, or during a cloudy period, stored heat may be exhausted, requiring the system to draw heat from the auxiliary furnace. Active systems may, in fact, be used in many localities in combination with *heat pumps* to aid those heat-transfer devices when outdoor temperatures are too low for their efficient operation (Gilmore).[5] The following analyses, however, assume active solar heating is backed up by a fuel-burning heater. Collectors use either hot air or hot water as the medium for moving and storing heat. While the following materials are derived from studies that specify hot water, systems using hot air are not very different in their economic feasibility (Duffie, Beckman, and Decker, p. 2).

Predicting the performance of a solar space-heating system in a variety of climates is complicated, because both heat demand and radiation per square unit of collector vary throughout the country. An important goal in such a study is an optimal design—one in which the collector and associated storage are large enough to meet the winter's demands, but not so large as to make the installation uneconomic. It would be possible to build a collector and storage large enough to warm the building on the coldest day of the year. The system would be unnecessarily large, however, for heating on most winter days. Of course, rising fuel prices will encourage the development of larger solar installations which allow greater fuel savings, and which would then offset the cost of installation and maintenance.

Evaluating the performance of existing active space-heating systems can provide some guidance for the design of new systems. This will provide only a starting point, however, and not principles that have application in any area of the country. What is needed is a procedure for *simulating* the performance of a solar heating system as it responds to the temperatures and the radiation that characterize a number of sites across the country. The following are the main factors in such an analysis.

DEMAND ESTIMATES A building's demand for heat can be expressed by a single figure called a *heat loss factor*. This figure is arrived at by conventional heating and ventilation formulas which consider areas of floors, walls, and windows, along with their materials and the degree of insulation. Heat loss for a building of a certain size may be expressed as *heat needed per degree-day*. Thus, a given mean temperature for a day implies a certain heat demanded by the building. The question of what climatic data best *represents* a locality for the sake of such simulation work is vital, and a number of different studies have used different approaches to this problem.

SUPPLY ESTIMATES Solar radiation records, on which maps such as those in Figs. 6–2 and 6–3 are based, provide the amount of solar energy per square unit of *horizontal surface*. For a tilted collector, this amount must be augmented to account for the more favorable attitude. With the amended figure, and with knowledge of the collector's tendency to reflect some radiation and to radiate heat to the environment, an estimate can be made of how much heat will be gathered and stored for every square unit of collector. Various collector sizes may be tried in a simulation to see how the heat gained matches the heat needs for the building in question.

SOME IMPORTANT MEASURES RELATED TO SYSTEM PERFORMANCE As the hypothetical heating system performs in response to climatic data that drive the simulation, a number of quantities can be noted.

1. *Solar Fraction or Solar Dependency*. This is the proportion of the building's heating load that is supplied by the sun. Usually it is reported as an average for the heating season, though it will be a greater proportion in the mild portions of the season and a smaller proportion in severely cold weeks. Solar fraction will rise as collector area increases; but, as suggested above, it is not necessarily economical to strive for very high solar fraction, since the increased cost of a larger solar installation may not be offset by fuel savings.

2. *Fuel Savings*. For every heat unit supplied by the solar system, some fuel is not burned. The total value of such fuels is the fuel savings. These may be considered *gross* fuel savings or *gross savings* to avoid confusion with an economic criterion, *savings,* which is used elsewhere.

3. *Critical Fuel Cost*. Peculiar to a location and its climate, this is the price level to which competing fuels must rise before a solar installation of specified cost is likely to be attractive economically.

4. *Optimum Collector Area*. As progressively larger collectors and associated storage are used, either in practice or in simulation, the (gross) fuel savings increase along with the cost of the system. The collector size which appears to maximize the *net savings* is the optimum.

DESIGNS FOR VARIOUS CLIMATES

There have not been many comprehensive studies of how the many variables in hypothetical solar heating systems interact in various parts of the country. This is largely because of the massive task of calculation required to follow a system through a heating season, to experiment with alternative designs, and finally to express the results in a meaningful way.

Fig. 6–13 An active solar hot water and space-heating system.

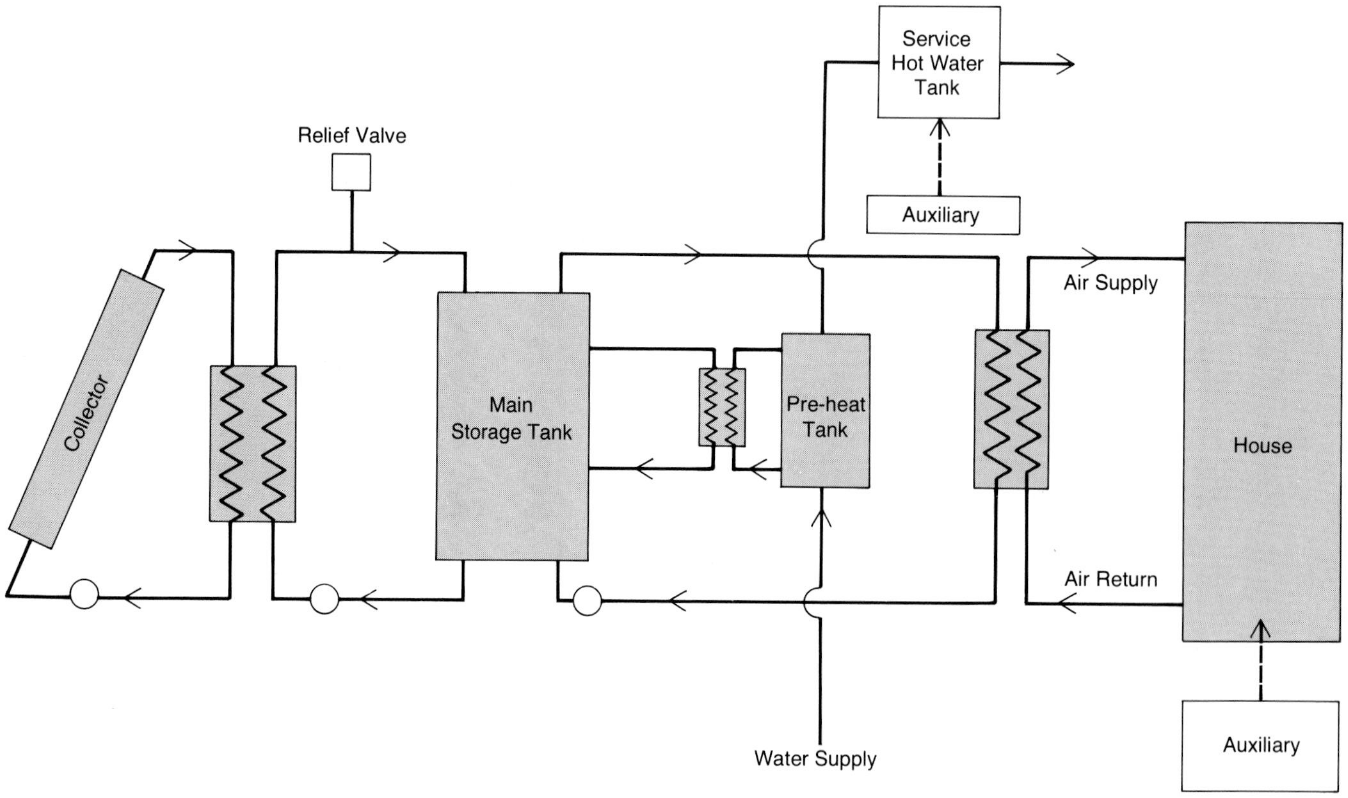

Components of an Active Solar Space-Heating Unit

Collector The flat plate collector is most commonly used. It consists of a rectangular box, 3 feet by 7 feet, and 8 inches deep, covered by glass or plastic. The bottom of the box is blackened to absorb solar radiation. The heat that is generated in the absorber is carried away by water, either in pipes or by trickling water across the surface, or by air. In more advanced systems, a vacuum is created inside the box in order to reduce energy losses; the windows are often double-layered, or specially coated to admit incoming radiation but trap outgoing heat radiation.

Storage The storage unit stores heat for night and for cloudy days. The common storage medium is water in an insulated tank. If the heating system circulates air, crushed rocks or pebbles are used to store the heat. Water can store more heat per pound than the rocks, and therefore the storage for water systems is more compact. Eutectic salts (such as sodium sulfate decahydrate) when perfected will offer excellent compact storage for air systems. When the salt melts (at a temperature near 90 degrees Fahrenheit) it stores a great deal of energy as the *latent heat of fusion.* When the salt solidifies it releases this heat. One pound of the salt stores 108 BTUs when it melts, then 20 BTUs more as its temperature rises 30 degrees. To store the same amount of heat in a 30-degree rise of temperature it takes 4 pounds of water or 20 pounds of rock.

Two earlier studies made an impact (Löf and Tybout, 1970 and 1974) but the most important recent analysis was made by Professors Duffie, Beckman, and associates at the Solar Energy Laboratory, University of Wisconsin, Madison. While processing results from a large number of performance simulations based on *half-hourly* radiation and temperature data, they discovered an empirical relationship that makes possible the estimation of system performance from *monthly data* (Klein, Beckman, and Duffie). A full explanation of their procedures will not be attempted here; their paper, however, clarifies how to modify local radiation data to consider the tilted collector surface, and shows how that information along with collector characteristics and heating load for a particular building—all derived from *monthly data*—may be entered into a simple graph, now widely known as the *F-Chart,* in order to learn the *solar fraction* for a collector of a given size. The solar fraction implies a certain amount of fuel saved, and that can be balanced against installation and other costs. A series of ever-larger systems (collectors) can then be tested in the simulation in order to arrive at the optimum size.

The F-Chart shortcut has made possible the hypothetical testing of various systems at a large number of stations—leading, therefore, to a national picture of variations due to climatic differences. Most studies of solar system feasibility made since the F-Chart appeared in 1976 have made use of it, though often coupling it with an economic analysis different from that used by Duffie and associates. In fact, the procedure is used by SOL-COST, the analysis service now offered by the Department of Energy through its service contractors.

THE ECONOMIC CRITERION To determine economically optimal system size Duffie and associates used the single value, *average annual savings,* which is the total fuel savings in a year minus the annual cost of the system. Fuel savings is simply the amount of fuel which would have otherwise been used multiplied by its price (which is assumed to escalate through time). The annual cost of the system is calculated to be the total installation cost multiplied by a single factor called the *annual payment rate.* Figure 6–14 shows that a payment rate of 10 percent, which is the one used in generating the results mapped here, could represent a loan at 5.57 percent for 15 years, at 7.75 percent for 20 years, or at 10.31 percent for 30 years.

ANNUAL SAVINGS AND FUEL PRICES Figure 6–15 shows (for Madison, Wisconsin) how annual savings increase with larger collector area—up to a point—and

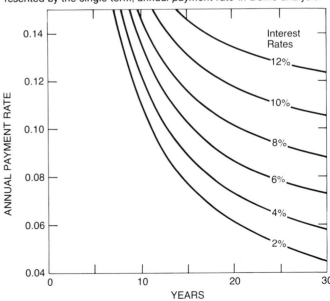

Fig. 6–14 Varying length of loan and interest rate that may be represented by the single term, *annual payment rate* in Duffie analysis.

how the savings are greater and smaller according to the price of fuel saved.

In any study of a system at a particular site, local fuel prices are used. For the sake of the national study, how-ever, *fuel prices were held constant* at future prices for all locations. The price used was $10 per million BTUs. Considering the conversion losses for gas and oil fur-naces, this price for BTUs delivered in the heating sys-tem corresponds to 3.4 cents per Kilowatt-hour for elec-tricity, 98 cents per gallon for fuel oil, and $7.25 per thousand cubic feet for natural gas.

By late 1979 or early 1980 the price for fuels which Duffie assumed had been exceeded by electrical rates, and almost matched by fuel oil prices. Only natural gas prices remained well below the assumed price per million BTUs.[6]

An important principle is evident on Fig. 6–15: for higher fuel prices, the optimal collector size is larger. This means that any system built to optimal size at to-day's fuel prices will be undersized as fuel prices rise. Some studies have built in an annual escalation of fuel prices, but the approach in this study was to choose a single value that represented average price of fuel energy delivered over the period under consideration.

Labeled in Fig. 6–15 is the *critical fuel cost*. This is the fuel price (price and cost are used interchangeably here) that allows a system to just break even. At fuel prices lower than critical, even the optimized system yields annual savings that are negative.

GEOGRAPHIC VARIATIONS IN DESIGN AND PERFOR-MANCE FACTORS

• *Collector area.* Figure 6–15 indicates that at the $10 price for fuel, savings are maximized with a collector of 30 square meters (323 square feet) in Madison, Wiscon-sin. Similar analysis at eighty-six other stations led to op-timal collector areas throughout the United States (Fig.

Installation Costs and Other Assumptions in the University of Wisconsin Study

The following list presents the specific costs as-sumed in the study at the Solar Energy Laboratory at the University of Wisconsin (Duffie, Beckman, and Dekker).

Fixed Costs   These are costs of the parts of the in-stallation that are independent of collector size, and are inevitable in every system; they include piping, pumps, controls and valves, exchangers, and the labor to install them. The amount used was $1,000 (which by 1978 standards is low).

Variable Costs   Costs that vary with the size of a system were as follows:

1. **Collector.** The type of collector used was a flat plate, with a single glass cover and a selective coating to reduce long-wave radia-tion. The price used was $110 per square meter ($10.22 per square foot) which was lower than the $15 to $20 per square foot being quoted in the summer of 1978, and anticipated prices falling as a result of mass production.

2. **Storage cost** was estimated at $15 for every square meter of collector ($1.39 per square foot of collector area)—roughly the cost being quoted in 1978.

Other Assumptions of the Study   The hypothetical building to be heated had a floor area of 150 square meters (1,615 square feet) and a wall area of 120 square meters (292 square feet). It was insulated to meet ASHRAE[7] standards, which are different in different parts of the na-tion; therefore no single heat-loss factor can be stated for the building. Domestic hot water heating load was assumed to be 30 litres (79 gallons) per day which had to be heated from an inlet temperature of 11 degrees Celsius (53 de-grees Fahrenheit). Operation and maintenance costs were ignored, and the study assumed no increase in property tax due to the installation, no income tax credit for interest paid on the loan, and no government investment tax credit or other incentive to the buyer.

Fig. 6–15 Annual savings in an active solar heating system being maximized at different collector sizes for different fuel costs per mil-lion BTU's of delivered heat. Cost of solar system held constant.

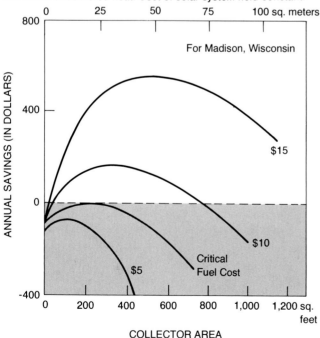

COLLECTOR AREA

6–16). (The square symbols on that map are *not* made with areas proportional to collector areas). Largest collectors (over 400 square feet in area) are appropriate in the Northern Plains; moderately large (300 to 400 square feet) are indicated in northern and northeastern states; and small systems are indicated in the extreme South. Conversion to metric is simplified by Fig. 6–17.

An estimate of relative installation costs can be made from Fig. 6–16, using the fixed cost of $1,000 plus $11.61 per square foot for collector and storage combined (the odd numbers are due to conversion from metric units in the study by Duffie and associates). For example, where a collector size of 400 square feet is indicated, the installation would cost $1,000 plus (400 × $11.61), or $1,000 plus $4,644 for a total of $5,644. Where a system with collector of half that size is indicated, the installation would cost $1,000 plus (200 × $11.61) for a total of $3,322.[8]

• *Solar Fraction.* At each location, the series of trials with ever-larger collectors led to an increase in solar fraction, as shown for Madison in Fig. 6–18. When the savings equation indicates optimal collector size, the corresponding solar fraction is identified; at Madison, for in-

stance, it is 50 percent. At the other locations it is the value mapped in Fig. 6–19.

Plains and Mountain states and extreme southern areas show fractions of greater than 60 and 70 percent, while systems in the Pacific Northwest and Middle Atlantic states carry only 20 to 40 percent of the season's heating load.

• *Annual Savings.* The value of fuels saved, diminished by the annual cost of the installation, yields annual savings—the quantity maximized in successive trials of system size, and the major criterion used here to show relative economic feasibility. For optimal systems based on assumed system cost and assumed fuel cost, annual savings are the amounts mapped on Fig. 6–20.

The savings pattern shows a very clear maximum in the Northern Plains and Rocky Mountain states where a simplified approach to heating need and solar supply (Fig. 6–9) revealed a region combining substantial heating need and radiation amounts exceptionally high for the latitude (Type 6). In addition, savings would be relatively high in Maine. *The substantial heating need* is a key factor, because part of the installation costs are for the "fixed" items which do not produce solar BTUs. In an

Fig. 6–16 Optimal flat plate collector areas in various parts of the country if competing fuel delivers heat at $10 per million BTU's and if solar installation costs are those assumed (see text).

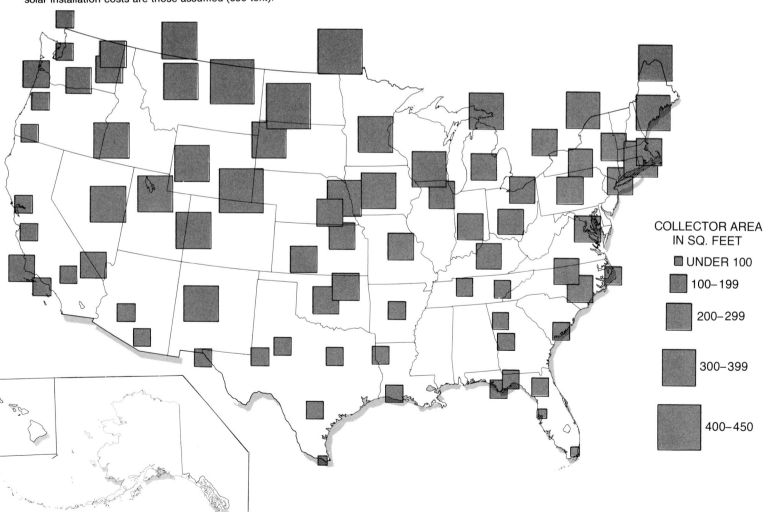

COLLECTOR AREA
IN SQ. FEET

☐ UNDER 100

▪ 100–199

◼ 200–299

◼ 300–399

◼ 400–450

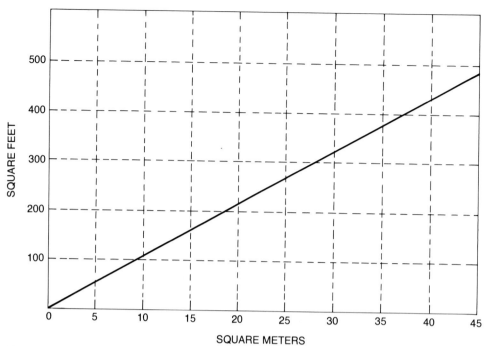

Fig. 6-17 Conversion of square feet to square meters.

installation with a small collector (for a small heating need), these fixed costs are large in relation to the amounts of fuel saved.

Savings in Fig. 6–20 are calculated on the basis of collectors costing 110 dollars per square meter, and competing fuels delivering energy at 10 dollars per million BTUs. Figure 6–21 shows how savings would rise if fuel costs increased to 15 dollars per million BTUs, and also shows how savings would be "shifted down" if collectors were to cost 150 dollars per square meter rather than 110 dollars.

Fig. 6–18 Solar fraction of heating load varying with collector area at one location, Madison, Wisconsin.

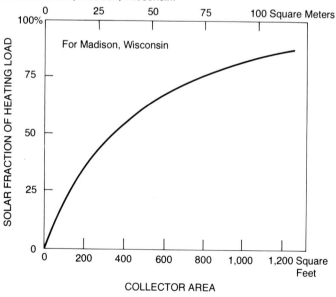

• *Critical Fuel Costs.* It was pointed out in Fig. 6–15 that fuel costs must be at a certain level or else even an optimized system will not yield positive savings. The level is far from uniform across the country. In fact, it varies with temperature and radiation factors as do the optimal collector areas and the solar fraction. Calculations for each weather station, trying a series of fuel prices in turn, can lead to the critical fuel cost that applies at each station. A shortcut procedure used at the University of Wisconsin (Duffie, Beckman, and Brandemuehl) allows the critical costs to be estimated in a procedure that will accommodate any chosen combination of fixed and variable installation costs. Our analysis used their procedure, and it assumed fixed costs of $2,000 and variable costs of $230 per square meter of collector which represent the national average installation costs in the summer of 1978.[9] These costs are for installed systems, and therefore include labor, some design fees, and a markup for the installer. A do-it-yourself installation would lower the costs dramatically.

Figure 6–21 maps the critical fuel costs for fifty-nine stations. The values for these and for thirty-six other stations in Table 6–3 all are expressed in dollars per million BTUs of energy delivered (competing with solar energy delivered through the active heating system). Those costs may be translated into familiar fuel prices in cents per gallon for fuel oil, dollars per thousand cubic feet for natural gas, or cents per Kilowatt-hour for electricity by procedures explained in a note.[10]

The values shown in Fig. 6–21 are dated because they depend on 1978 installation costs. Furthermore, those installation costs were assumed to apply everywhere in the country, whereas in fact there may be considerable vari-

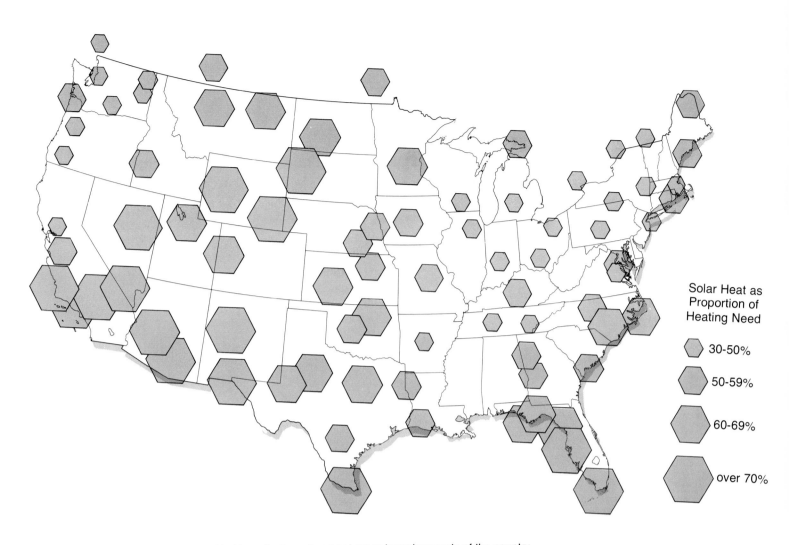

Fig. 6–19 Solar fraction of heating load, as provided by collectors of optimal areas in various parts of the country.

Fig. 6–20 Annual savings (fuels savings minus costs) from systems with optimal collector areas in various parts of the country. Competing fuel delivers heat at $10 per million BTU's.

Solar Heat as Proportion of Heating Need

- 30-50%
- 50-59%
- 60-69%
- over 70%

Dollar Savings

- less than 100
- 100-199
- 200-299
- over 300

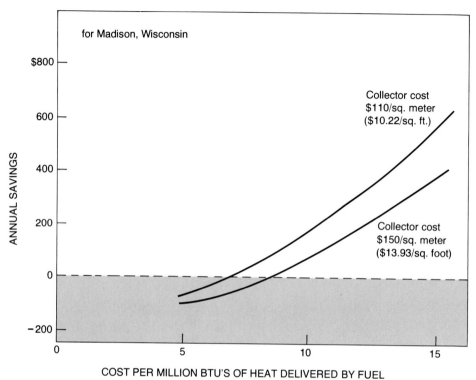

Fig. 6–20-A Annual savings at various fuel prices, and for higher and lower collector costs.

Solar collector panels on the roof of the National Parks visitors center at Mount Rushmore, South Dakota. Solar energy provides 50 percent of the heating needs and 40 percent of the cooling needs of the building. *Courtesy Department of Energy.*

ation in costs in different regions. Nevertheless the *relative values* on that map are extremely useful indicators of the relative feasibility of active systems. Lowest values in the Great Plains and Mountain states indicate that fuel prices will not have to be as high there as in other areas in order to make a solar system economically feasible. In coastal Washington and in a narrow eastern corridor from the Gulf of Mexico to the Great Lakes the systems will not be economic unless much higher fuel prices occur.

Examination of the variables in the Denver area clarifies the application of critical fuel costs and also reveals recent changes in both installation and fuel costs.

Table 6–4 shows that the critical fuel cost in the summer of 1978 was $6.00 per million BTUs, which translates to 51 to 59 cents per gallon for fuel oil (depending on furnace efficiency), roughly 2 cents per Kilowatt-hour for electricity, and $4.35 per thousand cubic feet of natural gas. At that time, the critical electricity cost was exceeded by the electrical rates in Denver, but neither fuel oil nor natural gas was expensive enough to meet the critical level.

Since 1978, energy prices, especially the price of fuel oil, have increased so sharply that their crossing of the critical threshold would seem a certainty. Unfortunately, installation costs have risen at the same time and have

Table 6–3 Critical Fuel Costs at Ninety-Five Locations According to 1978 Installation Costs for Active Systems

STATION	Critical Fuel Cost[1] in $/$10^6$ BTU	STATION	Critical Fuel Cost[1] in $/$10^6$ BTU	STATION	Critical Fuel Cost[1] in $/$10^6$ BTU
Albany, N.Y.	8.62	Fresno, Cal.	8.36	Norfolk, Va.	8.75
Albuquerque, N.M.	6.27	Glasgow, Mont.	6.27	Omaha, Neb.	7.31
Amarillo, Tex.	6.66	Gr. Junction, Colo.	6.66	Oak Ridge, Tenn.	9.40
Ames, Iowa	8.10	Hartford, Conn.	8.36	Oklahoma City, Okla.	7.58
Annapolis, Md.	9.01	Indianapolis, In.	9.40	Parkersburg, W. Va.	9.40
Atlanta, Ga.	8.75	Ithaca, N.Y.	9.67	Peoria, Ill.	8.62
Atlantic City, N.J.	8.23	Jackson, Miss.	9.40	Phoenix, Ariz.	7.58
Baltimore, Md.	8.88	Jacksonville, Fla.	9.27	Philadelphia, Pa.	8.75
Billings, Mont.	6.92	Kansas City, Mo.	8.36	Pittsburgh, Pa.	8.75
Birmingham, Ala.	9.01	Lake Charles, La.	9.27	Pocatella, Idaho	6.66
Bismark, N.D.	6.66	Lander, Wyo.	5.62	Portland, Maine	7.44
Boise, Idaho	7.58	Lansing, Mich.	9.01	Portland, Ore.	10.32
Boston, Mass.	9.80	Laramie, Wyo.	5.75	Pullman, Wash.	7.44
Brownsville, Tex.	9.80	Las Vegas, Nev.	7.05	Raleigh, N.C.	8.23
Cape Hatteras, N.C.	7.84	Lexington, Ky.	8.10	Rapid City, S.D.	6.27
Caribou, Me.	7.44	Lincoln, Neb.	7.84	Richland, Wash.	7.84
Charleston, S.C.	8.88	Little Rock, Ark.	9.27	Richmond, Va.	8.88
Charlotte, N.C.	8.36	Los Angeles, Cal.	7.31	St. Louis, Mo.	8.75
Chicago, Ill.	8.75	Louisville, Ky.	9.27	Salt Lake City, Utah	7.58
Cleveland, Ohio	9.67	Lynn, Mass.	10.06	San Antonio, Tex.	8.88
Columbia, Mo.	8.36	Macon, Ga.	8.75	San Francisco, Cal.	6.79
Columbus, Ohio	9.53	Madison, Wis.	8.10	Seattle, Wash.	10.19
Corvallis, Ore.	9.14	Manhattan, Kan.	8.23	Shreveport, La.	9.53
Dallas, Tex.	9.01	Medford, Ore.	8.49	Spokane, Wash.	7.84
Denver, Colo.	5.75	Memphis, Tenn.	9.01	State College, Pa.	9.53
Des Moines, Iowa	8.23	Miami, Fla.	9.27	Summit, Mont.	7.31
Detroit, Mich.	9.27	Milwaukee, Wis.	8.23	Tallahassee, Fla.	8.49
Dodge City, Kan.	6.66	Minneapolis, Minn.	8.36	Trenton, N.J.	8.62
Duluth, Minn.	7.44	Nashville, Tenn.	9.53	Tuscon, Ariz.	7.44
Ely, Nev.	5.35	New Orleans, La.	10.58	Tulsa, Okla.	8.75
Fargo, N.D.	7.97	Newport, R.I.	8.36	Wash., D.C.	9.53
Ft. Wayne, Ind.	8.62	New York, N.Y.	10.71		

[1]Calculated from information in Duffie, Beckman, and Brandemuehl (see text) assuming fixed cost of $2,000, variable cost of $230 per square meter.

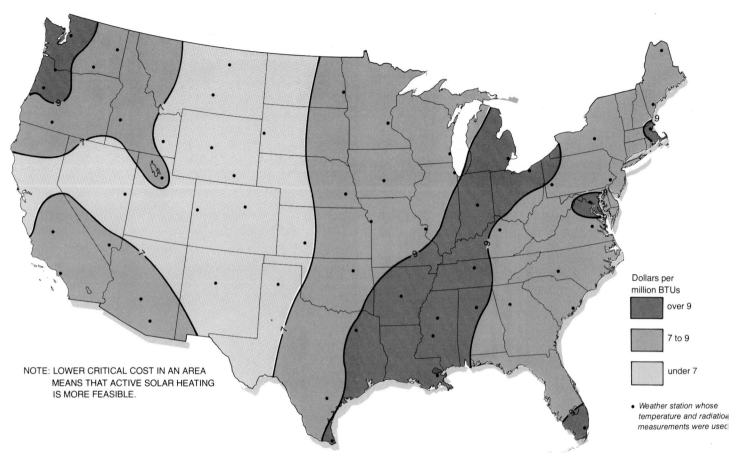

Fig. 6–21 Critical fuel costs for 59 stations, expressed in dollars per million BTU's of energy delivered.

Table 6–4 Critical Fuel Prices for Active Space-Heating Systems in the Denver Area versus Actual Fuel Prices, 1978 to 1980

	CRITICAL FUEL PRICE	FUEL OIL ($/Gal.)			ELECTRICITY ($/KWhr)		NATURAL GAS ($/Mcf)	
		Critical Prices if Furnace Efficiency Is		Actual Price[1]	Critical Price	Actual Price[2]	Critical Price	Actual Price[3]
DATE	($/Million BTU)	60%	70%					
Summer 1978	$6.00[4]	0.51	0.59	0.45	0.02	0.03	4.35	1.83
Early 1980	$9.90[4]	0.84	0.98	0.94	0.3	0.05	7.17	3.56

[1]From Department of Energy, *Energy Data Report: Heating Oil Prices and Margins,* (monthly reports) April, 1978 through January, 1980. Prices for "upper 10 percent" were used.

[2]From U.S. Department of Labor, Bureau of Labor Statistics, *News: Consumer Prices, Energy,* December, 1979. Average price per Kilowatt-hour was used.

[3]From U.S. Department of Labor, Bureau of Labor Statistics, *News: Consumer Prices, Energy,* December, 1979. Average price per therm of utility (piped) gas was used and translated into price per thousand cubic feet.

[4]Calculated by reference to methodology and tabulation provided by Duffie *et al.* (Dufffie, Beckman, and Brandemuehl, 1978). For summer of 1978 calculation, fixed costs were assumed to be $2,000 and variable costs $230 per square meter. For early 1980 calculation, fixed costs were assumed to be $2,000 but variable costs assumed to have risen to $484 per square meter ($45 per square foot). Guidance on current installation costs was provided by Mr. Loren Lantz (Lantz, 1978 and 1980).

roughly kept pace with the fuel prices. Figure 6–22, which deals only with fuel oil, shows the actual price of oil climbing from 45 cents in the summer of 1978 to 94 cents per gallon in January, 1980. If our assumption of 1980 installation costs is valid, then the critical fuel oil cost has become 84 to 98 cents. The professionally installed solar heating system competes with fuel oil only if the oil furnace is inefficient in its use of fuel. Electrical rates in Denver in 1980 greatly exceed the critical, making the solar clearly feasible in competition with electrical heating (Table 6–4), but the natural gas prices are still far too low to compete with. It should be emphasized that this study of costs is *not* definitive, because less expensive installations are possible and because tax incentives can greatly reduce the effective costs of installation—especially if a state tax credit may be combined with the federal tax credit.

FACTORS PROMOTING THE FEASIBILITY OF SOLAR HEATING SYSTEMS

Three separate factors would tend to make active solar heating systems competitive: falling costs of installation combined with rising fuel prices, tax breaks that would reduce the costs of installation, and low-cost loans for the solar investment.

FALLING INSTALLATION COSTS   It is often assumed

Fig. 6–22   Fuel oil prices, 1978–1980, showing comparison with estimated critical fuel costs (for two furnace efficiencies) at Denver, Colorado.

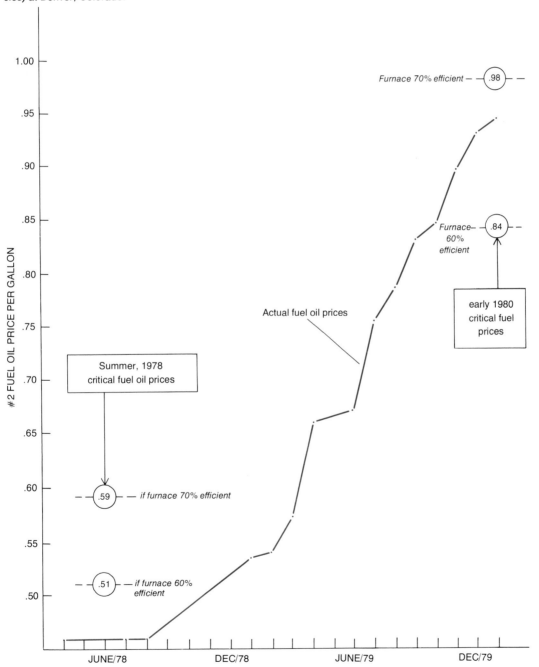

that mass production of solar collectors will markedly reduce their cost. Recent experience throws some doubt on that prospect (see Table 6–4) but fuel prices may rise quickly enough to make even expensive installations economically attractive.

TAX BREAKS FOR THE HOME OWNER   The Federal Investment Tax Credit is a proportion of an equipment purchase that can be used to reduce the computed income tax liability. It has been used heretofore by businessmen to ease the purchase of equipment. Part of the National Energy Bill (passed October, 1978) will allow home owners and businessmen to apply this tax credit to solar hot water and space-heating equipment. For home owners, the terms allow 30 percent of the first $2,000 and 20 percent of the next $8,000 on qualifying solar or wind apparatus for a maximum credit of $2,200. For businessmen, a simple 10 percent of qualifying costs is allowed. Some states provide additional tax incentives that are very substantial.

A recent Department of Energy study demonstrated how the economic feasibility of solar systems would be enhanced by just the federal tax incentive at four locations in the country. Table 6–5 illustrates how solar systems would be competitive with electrical resistance heating at Grand Junction, Colorado and Los Angeles, California, and competitive with heat pump and fuel oil only at Grand Junction. *Without* the federal income tax credit the solar system is only marginally competitive with resistance heating at two locations. Apparently the tax rebate received in the spring would, for the systems assumed, be large enough to return to the owner the amount spent on downpayment on the solar system (see Bezdek, and also Bezdek, Hirshberg, and Babcock).

LOW-COST LOANS   The interest paid on a loan for a solar installation on either a new or an old building is important to the total installation cost. An analysis for the Joint Economic Committee of the U.S. Congress studied the effects of both interest rates and future fuel costs on the feasibility of solar heating (Schulze, Shaul, Balcomb, *et al.*). Differing interest rates were found to be of crucial importance to either slowing or promoting the spread of solar heating in various parts of the country. It was apparent that either rising fuel prices or lowered interest rates would promote solar installations; but, since higher fuel prices have greater impact on low-income families, the study recommends government intervention to provide low-interest loans.

ECONOMIC ANALYSIS FOR THE PROSPECTIVE SOLAR BUYER

The foregoing maps and principles are for a nationwide view of the varying potential for active heating systems in various parts of the country. Most relevant to the prospective buyer are the local installation costs and local fuel prices. Although future fuel prices cannot be predicted any better for specific sites than for regions, the installation costs can be quite firmly defined, and may be very different from those assumed above.

To serve the home-buyer or contractor, the Department of Energy has seized upon the analytical potential of the F-Chart, and has made available a quick custom-made analysis through the SOLCOST service which applies to residential and commercial systems, and to hot-water as well as combined hot-water and space-heating systems.[11] The analysis uses F-Chart procedures to optimize the system in the light of life cycle economic standards (years to payback, and so on) and reports the optimum system size and the economic future that may be expected. Assumptions about future fuel costs are inevitable in this, as in any such analysis; and since the size of the system and its anticipated fuel savings depend on those prices, the optimal sizing and economic future can never be as firm as the prediction may suggest.

A sample of the user's *input* to the analysis and a typical *output* for a proposed system in Denver are shown in Table 6–6. The tabulation of year-by-year savings and costs reflects the SOLCOST approach to economic analysis. Rather than *average savings,* as used by Duffie, this analysis notes the annual value of benefits and costs and then discerns at what year certain milestones or events are reached. The prospective buyer can apply his

Table 6–5 Economic Feasibility in 1978 of Solar Combined Water- and Space-Heating Systems (Flat-Plate Collector) for Single-family Dwellings

| CITY | SYSTEMS COMPETING WITH SOLAR HEATING | | | |
	Electrical Resistance	Heat Pump	Fuel Oil	Natural Gas
Boston, Mass.	Marginal	X	X	X
Washington, D.C.	Marginal	X	X	X
Grand Junction, Colo.	Competitive	Competitive	Competitive	X
Los Angeles, Calif.	Competitive	X	X	X

Criteria for Feasibility. The solar installation, tested by simulation of its performance (see Booz, Allen & Hamilton) is *marginally competitive* if only one of the following criteria is met, and is *competitive* if two of the three are met.

1. *Years to Positive Cash Flow* (must be less than three). The years elapsed until fuels savings exceed the extra expenses of the solar system after taxes.
2. *Years to Recover Down Payment* (must be less than five). Years for accumulated savings to offset initial cash payments and early cash flow losses after taxes.
3. *Payback Period* (must be less than ten). Years before accumulated savings repay the full costs of the solar system, or equal the remaining principal on loan for system.

Note: Federal Income Tax Credit is assumed.

Source: Bezdek, Hirshberg, and Babcock, p. 1218.

own yardstick, or that of a financial advisor, in order to decide "feasibility."

In order to effectively cover the country, radiation and temperature data for 124 cities are stored for reference in SOLCOST analysis (Table 6–7). The user selects the city nearest his location, then defines carefully the characteristics of the building and the collector being considered, as well as financial terms affecting his investment. Very often, a prospective buyer will consult with a heating engineer when specifying the building's characteristics and the expected performance of the collector.

THE NATIONAL SOLAR HEATING AND COOLING DEMONSTRATION PROGRAM

Under the Solar Heating and Cooling Demonstration Act of 1974, the U.S. Department of Energy entertains proposals and awards grants for construction of buildings using solar heating or cooling. Many new federal buildings are included in the program.

Residential projects are managed by the Department of Housing and Urban Development (HUD) which has sponsored a series of five *cycles* in which proposals are received and grants awarded. Table 6–8 shows the numbers of projects funded in Cycles 1, 2, and 3; and Figure 6–23 plots the locations of those residential projects. Applicants are mostly developers and construction companies, though grants have been awarded to municipal housing authorities, and to the Blackfoot Indian tribe for ten townhouses in Cutbank, Montana.

Commercial projects, managed by the Department of Energy, are mapped on Fig. 6–24, showing the accumulated number of projects in each state as of November 1, 1978. Every state except Alaska is represented in the commercial program, but the greatest number of projects have been undertaken in Louisiana, California, Texas, and Washington, D.C. Three types of projects are grouped here as "commercial." Privately owned commercial buildings account for most projects in most states. Others include non-residential federal buildings, and residential federal building units (which in Louisiana account for forty of the forty-two mapped).

THE NATIONAL SOLAR DATA NETWORK A large number of buildings in the demonstration program are, or will be, equipped with instrumentation and monitored for the sake of gathering and distributing information on performance of various types of installations in various climates.[12]

Figure 6–25 plots locations of some of these instrumented buildings. There are eighty-four of the *active space-heating type,* half of which are residential, and half commercial. (Their approximate locations are mapped, while their names and city or town locations are listed in Table 6–9). A sketchy analysis of some of the buildings appears in the Department of Energy (D.O.E.) publication SOLAR/0025-79/42, July, 1979. The six *passive heating* buildings being monitored are in a variety of climates (DOE/0022-79/39, July, 1979). A total of twenty-

one buildings using *solar cooling* are in the monitoring network, but cooling season data is currently available on only the four that are mapped (DOE SOLAR/0023-79/40). There are also domestic hot-water heating installations, some of which are in buildings being monitored for heating or cooling operations. (The sixty-seven domestic hot-water installations with monitoring, operational in July 1979, are listed in DOE SOLAR/0024-79/41, but their locations are obscured).

A preliminary economic analysis of ten monitored installations of various types, including domestic hot water, active heating, passive heating, and active cooling, has been completed. It uses data gathered over a period of four to twelve months at the various installations. The report concludes that only the passive space-heating project has a payback period of twenty years or less (DOE SOLAR/0823-79/01).

Industrial and Agricultural Process Heat

Separate from the Solar Heating and Cooling Demonstration Program is a D.O.E. program to promote the use of solar radiation for heat in a number of industrial and agricultural processes. As with space heating, the temperatures easily attained from radiation are quite suitable for a number of industrial and agricultural tasks which now consume large amounts of fuels. Locations for D.O.E.-funded industrial heat projects, as of late 1978, are shown on Fig. 6–26, while the projects are described in Table 6–10. Agricultural projects are mapped in Fig. 6–27, and described in Table 6–11.

Electrical Generation

Generation of electrical power by direct use of the sun can be accomplished by two quite different means— *thermal* systems, and *photovoltaic* systems.

THERMAL SYSTEMS

Thermal systems are essentially turbines fueled by the sun's heat rather than by conventional fuel such as coal, oil, or natural gas. They operate either independently or in association with a fuel-burning steam plant.

The idea that solar heat can be used to drive an engine is not new. In fact, sun-powered steam engines were used in the 1920s for pumping water. Because high temperatures are required, the sun's energy must be concentrated by some sort of focusing device, such as a trough with a tube along its axis, parabolic mirrors, or lenses.

Two modes of collecting and concentrating solar radiation for power generation have been recently proposed. The *solar farm* approach uses lenses to focus energy on miles of selectively coated pipes containing some liquid which would transfer heat to boilers (Meinel and Meinel). The *power tower,* on the other hand, is a "solar furnace" system in which mirrors focus the sun onto a central receiver near the boiler itself. In this kind of system there is no need for long-distance transport of hot fluids: it is the sun's rays that are transported to the central point.

Table 6–6   An Example of SOLCOST Use for Residential Homeowner

INPUT

Input Parameter	User Input
Solar system type	Code 1 (space-heating)
Fuel type for reference heating system	Code 2 (electricity)
Fuel type for solar auxiliary heating system	Code 2 (electricity)
Collector type	Code 3 (water, single-cover)
Collector tilt angle	55 degrees
Collector azimuth angle	+10 degrees
Site location	Denver
Building heat loss coefficient	8.3 BTU/Sq. Ft./Deg.-Day
Building floor area	1,950 sq. ft.
Solar system fixed initial cost	$1,000
Solar collector installed cost/sq. ft.	$12.00
Loan interest rate	9%
Loan term	20 years
Loan down payment	22%
Property tax rate	2%
Income·tax rate	30%
Inflation of maintenance, insurance, property taxes	4%
Present electricity cost ($/Kw-hr)	$.035
Electricity cost escalation per year	10%

Parabolic solar collector being tested at The U.S. Department of Energy's SANDIA Laboratories, Albuquerque, New Mexico. As the sun's rays are focused onto the glass tube along the mirror's axis, fluids circulated through the tube can be heated to high temperatures. *Courtesy Department of Energy.*

Table 6-6 (continued)

OUTPUT: CASH FLOW SUMMARY

Year	(A) Fuel/ Utility Savings	(B) Mainte-nance + Insur-ance	(C) Property Tax	(D) Annual Interest	(E) Tax Savings	(F) Loan Payment	(G) Net Cash Flow
							$-1500
1	$ 500	$ 70	$ 135	$ 468	$ 181	$ 570	-94
2	550	73	140	459	180	570	-53
3	605	76	146	449	178	570	-8
4	665	79	152	438	177	570	42
5	732	82	158	426	175	570	98
6	805	85	164	413	173	570	159
7	886	89	171	399	171	570	228
8	974	92	178	384	168	570	303
9	1,072	96	185	367	166	570	387
10	1,179	100	192	349	162	570	480
11	1,297	104	200	329	159	570	582
12	1,427	108	208	307	155	570	696
13	1,569	112	216	284	150	570	821
14	1,726	117	225	258	145	570	960
15	1,899	121	234	230	139	570	1,113
16	2,089	126	243	199	133	570	1,283
17	2,297	131	253	166	126	570	1,470
18	2,527	136	263	130	118	570	1,676
19	2,780	142	273	90	109	570	1,904
20	3,058	147	284	47	99	570	2,156
TOTAL	$28,637	$2,086	$4,020	$6,192	$3,064	$11,400	$12,703

COLLECTOR SIZE OPTIMIZATION
Collector type: Flat-plate 1 glass selective.
Best solar collector size for tilt angle of 55 degrees is 400 sq. ft.
Solar costs: $1,000 fixed + $4,800 collector + $900 storage.
Input conventional system costs: 0.
Initial solar investment: $6,700. Down payment: $1,500.
Financial scenario: Residence.

Payback time for fuel savings to equal total investment: 8.9 years.
Payback time for net cash flow to equal down payment: 9.9 years.
Rate of return on cash flow: 16.3 percent.
Annual portion of load provided by solar: 72.0 percent.
Annual energy savings with solar system: 91.3 million BTU.
Tax savings: Income tax rate × (C + D)
Net cash flow: A − B − C + E − F.

Parabolic reflecting collectors producing high-temperature fluids used to drive a engine for pumping irrigation water at Willard, New Mexico. The project is operated by Sandia Laboratories for the U.S. Department of Energy. *Courtesy Department of Energy.*

A. CYCLE 1, SPRING, 1976

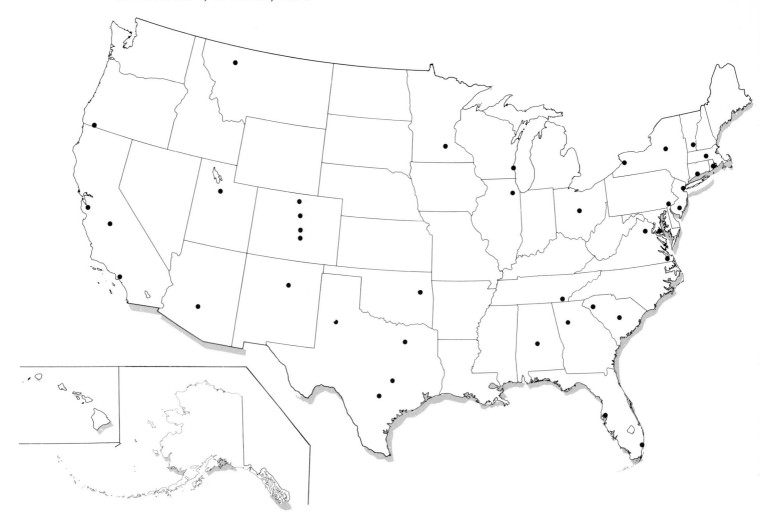

B. CYCLE 2, FALL, 1976

Fig. 6–23 Project sites for residential Solar Heating and Cooling Demonstration Program:
A. Cycle 1, Spring, 1976   B. Cycle 2, Fall, 1976   C. Cycle 3, 1977   D. Cycle 4, 1978

C. CYCLE 3, 1977

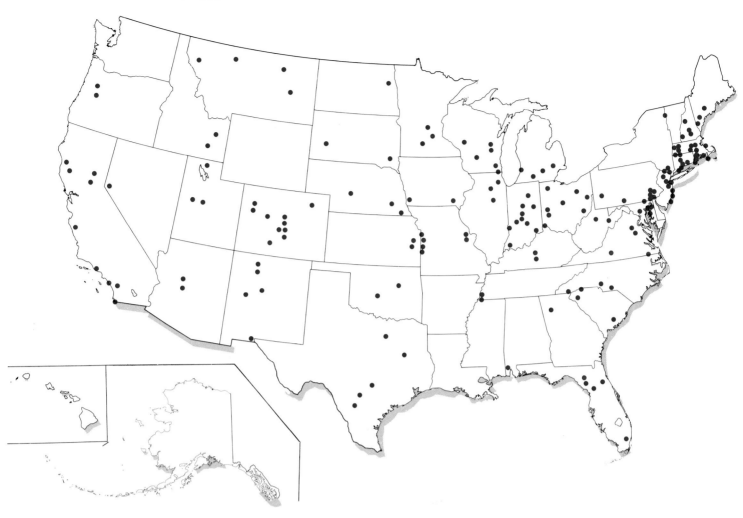

D. CYCLE 4, 1978

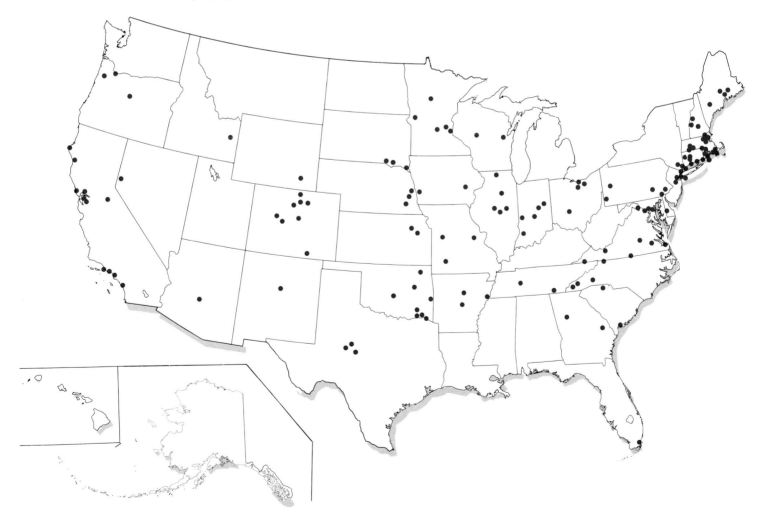

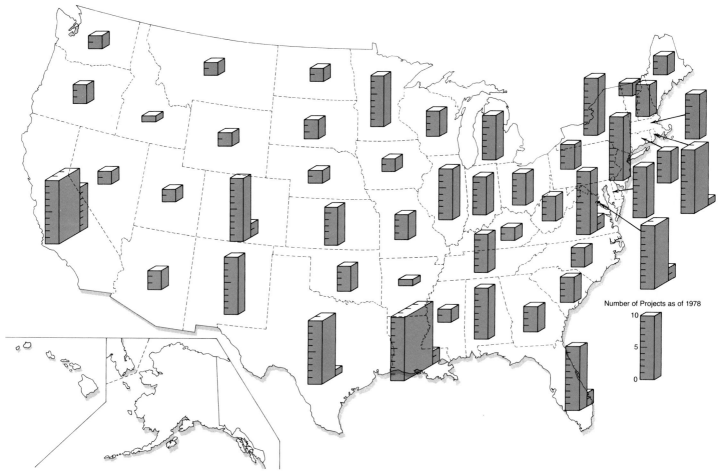

Number of Projects as of 1978

10

5

0

Fig. 6–24 Commercial projects in the National Solar heating and Cooling Demonstration Program.

Fig. 6–25 Some sites in the National Solar Data Network, i.e., monitored projects in the Solar Heating and Cooling Demonstration Program. Performance data and analysis are available from the Department of Energy.

Greenmoss,
Waitsfield, Vt.

Kalwall Corp.,
Manchester, N. H.

Spence-Urban,
West Des Moines, Iowa

Living Systems,
Davis, Calif.

Colorado Sunworks,
Longmont, Colo.

Irvine School,
Irvine, Calif.

Hullco,
Prescott, Ariz.

Alabama Power Co.,
Montevallo, Ala.

Dallas Recreation Center,
Dallas, Tex.

Reedy Creek Utilities
Lk. Buena Vista, Fla.

MONITORED PROJECTS

Active Space-heating
Total of 84 being monitored, half of
which are residential and half commercial.
See tabulation for site names.

Passive Space-heating
All residential, except for Kalwall project in
N. H. Others are or will be in the program.

Active Cooling
Cooling season data available in 1979 for
only four of the 21 solar cooling projects
on the program.

In addition, there are 67 domestic
hot water projects being monitored.

Fig. 6–26 Industrial projects in the Solar Energy for Agricultural and Industrial Process Heat Program as of 1978.

Fig. 6–27 Agricultural projects in the Solar Energy for Agricultural and Industrial Process Heat Program as of 1978.

Table 6-7 Cities for Which Radiation and Temperature Data are Stored for Reference in SOLCOST Analysis of Solar Heating Feasibility in 1978

Alabama
Birmingham
Mobile
Montgomery

Arizona
Phoenix
Yuma

Arkansas
Little Rock

California
Eureka
Fresno
Los Angeles
Red Bluff
Sacramento
San Diego
San Francisco

Colorado
Denver
Grand Junction

Connecticut
Hartford

District of Columbia
Washington

Florida
Apalachicola
Jacksonville
Key West
Miami
Tampa

Georgia
Atlanta

Hawaii
Hilo
Honolulu

Idaho
Boise
Pocotello

Illinois
Cairo
Chicago
Springfield

Indiana
Evansville
Fort Wayne
Indianapolis

Iowa
Des Moines
Dubuque
Sioux City

Kansas
Concordia
Dodge City
Wichita

Kentucky
Louisville

Louisiana
New Orleans
Shreveport

Maine
Eastport

Massachusetts
Boston

Michigan
Alpena
Detroit
Grand Rapids
Marquette
Sault Ste. Marie

Minnesota
Duluth
Minneapolis

Mississippi
Vicksburg

Missouri
Kansas City
St. Louis
Springfield

Montana
Havre City
Helena
Kalispell

Nebraska
Lincoln
North Platte

Nevada
Ely
Las Vegas
Reno
Winnemucca

New Hampshire
Concord

New Jersey
Atlantic City

New Mexico
Albuquerque
Roswell

New York
Albany
Binghampton
Buffalo
Canton
New York
Syracuse

North Carolina
Asheville
Raleigh

North Dakota
Bismark
Devils Lake

North Dakota
Fargo
Williston

Ohio
Cincinnatti
Cleveland
Columbus

Oklahoma
Oklahoma City

Oregon
Baker
Portland
Roseburg

Pennsylvania
Harrisburg
Philadelphia
Pittsburg

Rhode Island
Block Island

South Carolina
Charleston
Columbia

South Dakota
Huron
Rapid City

Tennessee
Knoxville
Memphis
Nashville

Texas
Abilene
Amarillo
Austin
Brownsville
Del Rio
El Paso
Fort Worth
Galveston
San Antonio

Utah
Salt Lake City

Vermont
Burlington

Virginia
Norfolk
Richmond

Washington
Seattle
Spokane
Walla Walla
Yakina

West Virginia
Elkins
Parkersburg

Wisconsin
Green Bay
Madison

Wyoming
Cheyenne
Lander
Sheridan
Yellowstone Park

Puerto Rico
San Juan

THE RADIATION RESOURCE FOR SUCH SYSTEMS Whether the system is of the solar farm or the power tower type, it can use *only radiation which can be focused* by a concentrating device. In other words, it can only use *direct* radiation received through a clear sky, and not *diffuse* radiation that filters through clouds and haze. Solar radiation values presented earlier in this chapter for twelve months and for the annual average were for the *total* radiation received, that is, the sum of direct and diffuse components. Furthermore, it is the total radiation falling onto a horizontal surface, and thus may be termed *total-horizontal*. Such radiation values are not appropriate for assessing solar-thermal electric potential partly because they include the diffuse component, but also because they do not consider the enhancement that occurs when a collecting device is tilted toward and follows the sun to receive radiation at right angles to the collecting plane.

A study, out of SANDIA Laboratories, provides a guide to the radiation most relevant to solar-thermal

Table 6-8 The Five Cycles of the Residential Solar Heating and Cooling Demonstration Program

CYCLE	Year of Awards	Number of Projects
1	Spring 1976	55
2	Fall 1976	102
3	Fall 1977	169
4	Spring 1978	239[1]
4A	Fall 1978	6,612[2]
5	Spring 1979	?

[1]Single-family units.
[2]Multi-family units.
Sources: Department of Housing and Urban Development, *Selling the Solar Home*, April, 1978. Department of Housing and Urban Development, July, 1979.

Table 6–9 Active Space-Heating Installations Being Monitored in the National Solar Heating and Cooling Demonstration Program

LOCATION[1]	Project Name	Installation Type	LOCATION	Project Name	Installation Type
Alabama			**Missouri**		
Montevallo	Alabama Power Company	Liquid, flat-plate	Gladstone	Bond Construction	Liquid, flat-plate
Huntsville	Chester West	Liquid, flat-plate	**Montana**		
Arizona			Billings	Billings Shipping	Liquid, flat-plate
Tempe	Tempe Union High School	Liquid, flat-plate	Big Fork	Design Construction	Liquid, flat plate
California			**Nebraska**		
Irvine	Irvine School	Liquid, evacuated tube	Lincoln	Lincoln Housing	Liquid, flat-plate
Tahoe Paradise	Lake Valley Firehouse	Liquid, flat-plate	**New Mexico**		
Santa Rosa	Montecito Pines	Liquid, flat-plate	Albuquerque	Albuquerque Western	Liquid, concentrating collector
Escondido	Ortiz & Reill #5	Liquid, flat-plate			
Escondido	Ortiz & Reill	Liquid, flat-plate	Albuquerque	Homes by Marilynn	Liquid, flat-plate
Colorado			Albuquerque	Kirtland AFB	Liquid, flat-plate
Boulder	Boulder Post Office	Liquid, flat-plate	Albuquerque	Public Service of New Mexico #7	Air, flat-plate
Denver	Perl-Mack	Liquid, flat-plate			
Westminster	Waverly Homes #13	Air, flat-plate	Albuquerque	Public Service of New Mexico #21	Air, flat-plate
Westminster	Waverly Homes #14	Air, flat-plate			
Connecticut			**New York**		
New Haven	Frank Chapman	Air, flat-plate	Malta	Stewart-Teele-Mitchell	Liquid, flat-plate
Hamden	Hamden Apartments	Liquid, flat-plate	Chester	Suntech Homes	Air, flat-plate
Stamford	Lutz-Sotire	Liquid, flat-plate	**North Carolina**		
Florida			Charlotte	Charlotte Medical Center	Liquid, flat-plate
Winter Springs	Florida Gas Company	Liquid, flat-plate	Salisbury	Salisbury House	Liquid, trickle down
Gainesville	Matt Cannon	Liquid, flat-plate			
Lake Buena Vista	Reedy Creek Utilities	Liquid, concentrating collector	**North Dakota**		
			Bismark	Bismark Power Co.	Liquid, flat-plate
Yulee	State of Florida	Liquid, concentrating collector	**Ohio**		
			Canton	Alpha Construction	Air, flat-plate
			Columbus	Columbia Gas	Liquid, concentrating collector
Georgia					
Shenandoah	Shenandoah Community Center	Liquid, flat-plate	Columbus	Davidson-Phillips	Air, flat-plate
Atlanta	Towns Elementary School	Liquid, flat-plate	Troy	Troy-Miami Library	Liquid, evacuated tube
Indiana			**Oregon**		
Greenwood	Moulder Corp.	Air, flat-plate	Stayton	Landura Corp.	Liquid, flat-plate
Iowa			Coos Bay	Twin City Builders	Liquid, flat-plate
West Branch	Scattergood School	Air, flat-plate	**Pennsylvania**		
Des Moines	Spence-Urban	Liquid, flat-plate	Westchester	Bell of Pennsylvania	Liquid, flat-plate
Kansas			Washington	Walnut Ridge	Air, flat-plate
Kansas City	Du-Cat Investments	Air, flat-plate	**Rhode Island**		
Shawnee	John Byram	Air, flat-plate	Jamestown	M. F. Smith Associates	Liquid, flat-plate
Topeka	Kaw Valley State Bank	Liquid, evacuated tube	**South Carolina**		
			Greenwood	Blakedale Prof. Center	Liquid, flat-plate
Kentucky			St. Matthews	Phillips Kauric #1	Liquid, flat-plate
Louisville	Rademaker Corp.	Liquid and air, flat-plate	St. Matthews	Phillips-Kauric #1	Liquid, flat-plate
			Clemson	RHRU Clemson	Air, flat-plate
Louisiana			**South Dakota**		
Baton Rouge	LSU Field House	Liquid, flat-plate	Aberdeen	First Baptist Church	Air, flat-plate
Maryland			Rapid City	Harney Lumber Co. #87	Liquid, flat-plate
Damascus	Damascus Land	Liquid, evacuated tube	Rapid City	Harney Lumber Co. #88	Liquid, flat-plate
			Keystone	S. D. School of Mines	Liquid, flat-plate
Columbia	J. D. Evans #A	Liquid, flat-plate	**Texas**		
Columbia	J. D. Evans #B	Liquid, flat-plate	Dallas	Brad Popkin	Air, flat-plate
Cockeysville	Padonia Elementary School	Liquid, concentrating collector	Austin	College Houses	Liquid, concentrating collector
Baltimore	Upton Center	Liquid, flat-plate	Dallas	Dallas Recreation Center	Liquid, flat-plate
Massachussetts			Austin	Radian Corp.	Liquid, concentrating collector
Concord	Concord Light Plant	Air, flat-plate			
Needham	Saddle Hill Trust #36	Liquid, flat-plate	San Antonio	Randolph AFB	Liquid, flat-plate
Burlington	Technology Property	Liquid, flat-plate	San Antonio	Trinity University	Liquid, concentrating collector
Minnesota					
Blue Earth	Telex Communication	Liquid flat-plate			

Table 6-9 (Continued)

LOCATION[1] Project Name		Installation Type	LOCATION	Project Name	Installation Type
Virginia			**Wisconsin**		
Leesburg	Loudoun County Schools	Liquid, flat-plate	Howard's Grove	Howard's Grove School	Air, flat-plate
Virginia Beach	Sir Galahad	Liquid, flat-plate	Milwaukee	Zein Mechanical	Air, flat-plate
Lynchburg	T. D. Moseley	Liquid, flat-plate	**West Virginia**		
Washington			Charles Town	Page-Jackson Elementary	Liquid, flat-plate
Richland	Olympic Engineering	Liquid, flat-plate	**Wisconsin**		
Seattle	Washington Nat. Gas	Air, flat-plate	Howard's Grove	Howard's Grove School	Air, flat-plate
			Milwaukee	Zein Mechanical	Air, flat-plate

[1]See accompanying map for approximate locations.

Source: U.S. Department of Energy, *Thermal Performance Analysis of Space-Heating Systems in the National Solar Data Network,* SOLAR/0025–79/42, July, 1979.

A commercial application of solar hot water for space heating, at the East Norristown branch of the First Pennsylvania Bank near Philadelphia. *Courtesy Department of Energy.*

Table 6–10 Industrial Heat Projects in the Solar Heating and Cooling Demonstration Program in 1978

LOCATION	Contractor	Application	Type of Heat
Alabama			
Decatur	Teledyne Brown Engineering Huntsville, Alabama	Soybean drying	Hot air
Fairfax	Honeywell, Inc. Minneapolis, Minnesota	Textile drying	Steam
California			
Fresno	California Polytechnic State University Foundation	Drying of prunes and raisins	Hot air
Gilroy	Trident Engineering Assoc. Inc. Annapolis, Maryland	Onion and garlic dehydration	Hot air
Pasadena	Jacobs-Del Solar Systems, Inc. Pasadena, California	Commercial dry cleaning and laundry	Steam
Sacramento	Acurex Corporation Mt. View, California	Washing food cans	Hot water
Florida			
Bradenton	General Electric Company Philadelphia, Pennsylvania	Orange juice pasteurizing	Steam
Mississippi			
Canton	Lockheed Missiles & Space Company, Inc. Huntsville, Alabama	Kiln drying of lumber	Hot air
Pennsylvania			
Harrisburg	AAI Corporation Baltimore, Maryland	Curing of concrete blocks	Hot water
South Carolina			
La France	General Electric Company Philadelphia, Pennsylvania	Textile drying	Hot water
Texas			
Sherman	Acurex Corporation Mt. View, California	Gauze bleaching	Steam

Source: DOE/CS-0053, September, 1978, p. 109.

Power tower solar thermal electric test facility at U.S. Department of Energy Sandia Labs, Albuquerque, New Mexico. Completed late in 1977, the plant is rated at 5 Megawatt electric. The tower is twenty stories tall, and each of the square "heliostats" (mirrors) holds 25 mirrors that are four feet on a side. *Courtesy Department of Energy.*

Table 6–11 Agricultural Projects in the Solar Heating and Cooling Demonstration Program in 1978

LOCATION	Contractor	Application
Alabama		
Auburn	Auburn University	Heating broiler houses
Huntsville	Lockheed Missiles & Space Company, Inc.	Crop drying
Arizona		
Tuscon	University of Arizona	Heating and cooling of dairy facilities
Tuscon	University of Arizona	Food processing
California		
Brentwood	Suntek Research Association Corte Madera, California	Commercial greenhouse
Colorado		
Ft. Collins	Colorado State University	Heating of greenhouses
Florida		
Gainesville	University of Florida	Grain drying
Gainesville	University of Florida	Food processing
Lake Alfred	Agricultural Research and Education Center	Food processing
Winter Haven	U. S. Citrus and Subtropical Products Laboratory	Food processing
Georgia		
Athens	Southeast Poultry Research Laboratory	Heating of chick brooding house
Atlanta	Georgia Institute of Technology	Drying of crops other than grain
Byron	Southeastern Fruit and Tree Nut Research Laboratory	Heating of greenhouses
Tifton	Georgia Coastal Plain Experiment Station	Drying of crops other than grain
Tifton	Georgia Coastal Plain Experiment Station	Tobacco curing
Illinois		
Urbana	University of Illinois	Hay drying
Urbana	University of Illinois	Grain drying
Indiana		
West Lafayette	Purdue University	Ventilating air in swine building
West Lafayette	Purdue University	Food processing
West Lafayette	Purdue University	Grain drying
West Lafayette	Purdue University	Heating of greenhouses
Iowa		
Ames	Iowa State University	Grain drying
Ames	SEA-FR, NC Region	Grain drying
Kansas		
Manhattan	Kansas State University	Heating swine confinement building
Manhattan	Kansas State University	Grain drying
Manhattan	Kansas State University	Grain drying
Kentucky		
Lexington	University of Kentucky	Tobacco curing
Lexington	University of Kentucky	Grain drying
Maryland		
Beltsville	Genetics and Management Laboratory	Livestock shelters
Massachusetts		
Amherst	University of Massachusetts	Food processing
Tewksbury	Daystar Corporation Burlington, Massachusetts	Commercial greenhouse
Michigan		
East Lansing	Michigan State University	Heating poultry house
East Lansing	Michigan State University	Food processing
Minnesota		
St. Paul	University of Minnesota	Turkey house ventilating
Nebraska		
Lincoln	University of Nebraska	Heating of swine house
Lincoln	University of Nebraska	Computer model development
New Jersey		
New Brunswick	Rutgers University	Heating greenhouse
Allentown	Rutgers University	Commercial greenhouse
New York		
Ithaca	Cornell University	Heating greenhouses

Table 6–11 (Continued)

LOCATION	Contractor	Application
North Carolina		
Raleigh	North Carolina State University	Tobacco curing and peanut drying
Raleigh	North Carolina State University	Heating greenhouses
Ohio		
Wooster	Ohio Agricultural Research and Development Center	Computer simulation of grain drying
Wooster	Ohio Agricultural Research and Development Center	Heating greenhouses
Springfield	Lockheed Missiles & Space Company, Inc. Huntsville, Alabama	Commercial greenhouses
Oklahoma		
Stillwater	Oklahoma State University	Peanut drying
Pennsylvania		
University Park	Pennsylvania State University	Heating greenhouses
South Carolina		
Clemson	Rural Housing Research Unit, USDA/SEA	Greenhouse cropping
South Dakota		
Brookings	South Dakota State University	Heating swine buildings
Tennessee		
Knoxville	University of Tennessee	Hay drying
Texas		
Beaumont	Texas A & M University Research Ext. Center	Rice drying
College Station	Texas A & M Research Foundation	Heating greenhouses
El Paso	Solargenics, Inc. Chatsworth, California	Commercial greenhouses
Virginia		
Blacksburg	Virginia Polytechnic Institute and State University	Solar-assisted heat pump for heating and cooling swine houses
Washington		
Pullman	Washington State University	Hop drying
Pullman	Washington State University	Food processing
Seattle	Ecotope Group	Heating of greenhouses
Wisconsin		
Madison	University of Wisconsin	Food processing

Source: DOE/CS-0053, September, 1978, p. 105–106.

electrical generation. It estimates, for twenty-six stations throughout the country, the values of direct radiation received on a surface at right angles to the radiation, a quantity referred to as *direct-normal* radiation (Boes, *et al.*). For the sake of better-quality records only five years of records at only twenty-six recording stations were employed. Direct radiation was estimated on the basis of recorded figures for *percent possible sunshine,* that is, the ratio of total radiation received at the surface to the radiation theoretically received at the edge of the atmosphere.

That study presents twelve monthly maps of estimated daily totals for both direct-normal and total-horizontal radiation. In Table 6–12 the daily mean totals of direct normal radiation (based on monthly data) are listed for reference. Figure 6–28 represents the regional variations by estimated daily totals based on the annual mean (as calculated by the present authors) from data in the Boes study.

The *pattern* in Fig. 6–28 is very similar to that of daily total-horizontal radiation in Langleys (Fig. 6–3). It is the *values* that are uniquely useful since they express the energy usable in concentrating devices: while the mapped values are in Kilowatt-hours per square meter of collecting surface (daily) they may be converted to annual Kilowatt-hours per square meter simply by multiplying by a factor of 365. Such values are a guide to the possible annual output of a solar-fired generating device if the collector areas and system efficiencies are specified.

APPLICATION OF THE MAPPED RESOURCE If a 10-Megawatt electric power plant were to be built in the most favorable part of the country, how much collector surface, and land area, would be required?

If the capacity factor of the plant were 40 percent, then the annual output of the plant rated at 10,000 Kilowatts (10 Megawatts) would be $(10,000) \times (8,760) \times (0.40) = 3.5 \times 10^7$ Kilowatt-hours.

Given the favorable insolation of roughly 3,000 Kilowatt-hours per square meter per year in a location such as southern Arizona, the number of square meters of collector surface required would be $\frac{3.5 \times 10^7}{3.0 \times 10^3} = 1.167 \times 10^4$, or 11,670 square meters; but because the overall efficiency of converting the solar energy into

Table 6–12 Daily Total of Direct Radiation on Surface Normal to the Sun (mean values for each month, and annual mean units are kilowatt-hours per square meter of surface)

LOCATION	J	F	M	A	M	J	J	A	S	O	N	D	ANNUAL
Albuquerque, N.M.	6.6	6.9	7.8	8.9	9.9	10.6	9.2	9.2	7.9	7.8	6.6	6.1	8.1
Appalachicola, Fla.	4.7	5.2	5.5	6.6	7.1	6.9	6.5	5.9	6.1	6.1	5.4	4.6	5.9
Bismark, N.D.	4.3	5.4	5.9	7.1	7.3	8.8	8.7	8.1	6.7	5.7	3.9	3.5	6.3
Blue Hill, Mass.	3.6	3.9	4.8	4.9	5.7	5.6	5.0	5.4	4.9	4.2	3.1	3.3	4.5
Boston, Mass.	3.2	3.6	4.8	4.8	5.9	5.9	5.6	5.5	5.0	3.9	2.8	2.9	4.9
Brownsville, Tex.	3.3	4.0	3.9	4.8	6.1	6.4	7.4	6.6	5.3	4.9	3.8	3.0	5.0
Cape Hatteras, N.C.	5.0	5.0	5.9	7.4	8.4	7.5	8.2	7.1	7.1	5.7	4.9	4.5	6.4
Caribou, Me.	3.9	4.8	6.2	5.6	5.8	6.0	6.1	5.9	4.6	3.4	2.3	3.1	4.8
Charleston, S.C.	3.9	4.3	4.8	6.5	6.4	5.7	5.4	5.4	5.2	5.3	4.3	4.0	5.1
Columbia, Mo.	3.9	4.2	4.3	5.4	6.8	6.8	6.7	7.4	5.7	5.2	4.1	3.4	5.3
Dodge City, Kan.	6.0	4.8	5.9	6.8	7.5	7.9	8.0	7.9	6.8	6.9	5.8	5.0	6.6
El Paso, Tex.	6.6	7.5	8.6	9.6	10.2	9.8	8.7	8.7	7.7	7.7	6.4	6.1	8.1
Ely, Nev.	5.5	5.4	7.2	8.5	8.8	10.1	8.8	8.6	8.4	7.4	6.0	5.6	7.5
Fort Worth, Tex.	4.8	4.8	5.5	5.8	6.8	7.2	7.4	7.4	6.1	5.6	4.7	4.1	5.8
Great Falls, Mont.	2.8	4.2	5.6	5.7	6.7	8.3	8.2	7.2	5.9	4.5	3.0	2.4	5.4
Lake Charles, La.	3.7	3.6	4.7	5.2	6.3	5.9	5.6	5.1	5.0	4.7	4.0	3.3	4.7
Madison, Wis.	4.0	4.9	5.9	6.1	6.8	8.0	7.8	7.4	5.8	4.7	3.4	3.8	5.7
Medford, Ore.	1.6	2.6	4.1	6.0	6.8	8.7	9.8	8.4	6.6	4.4	2.4	1.4	5.2
Miami, Fla.	5.1	6.4	5.8	6.6	6.3	5.8	6.0	5.8	5.2	5.8	5.2	5.2	5.8
Nashville, Tenn.	3.1	3.6	4.0	5.6	6.3	6.1	5.8	5.6	5.2	4.7	3.6	2.9	4.7
New York, N.Y.	2.8	3.3	4.6	4.6	4.9	5.6	4.7	4.3	5.3	4.0	2.6	2.8	4.1
Omaha, Neb.	4.6	5.1	5.4	6.3	6.5	7.0	7.2	7.3	5.8	5.4	4.1	3.8	5.7
Phoenix, Ariz.	5.8	6.5	7.9	9.3	10.0	9.3	8.6	7.7	7.4	6.9	5.8	5.1	8.1
Santa Maria, Calif.	5.3	5.0	7.1	8.5	9.1	9.2	8.5	7.9	7.2	7.0	5.7	5.1	7.1
Seattle, Wash.	1.5	2.0	3.7	4.8	6.0	7.6	8.6	6.1	4.6	3.0	2.0	1.2	4.3
Washington, D.C.	3.8	4.1	5.0	5.7	5.6	6.4	5.5	5.5	5.2	4.7	3.6	3.6	4.9

Source: Boes, *et al.*

electrical is only about 15 percent, the collector area must be increased by a factor of $\frac{1.0}{0.15}$ or 6.667 resulting in collector surface area of around 77,800 square meters. This is consistent with the 63,000 to 92,000 square meters of *collector area* estimated by three contractors; but the actual *field area* needed to encompass collector array and receiver tower is around 3 million square meters according to the designs proposed by McDonnell-Douglas, Honeywell, and Martin-Marietta (Central Receiver Thermal Power System, p. 14). While this converts to nearly 75 acres, the ERDA booklet just cited assumes that roughly 100 acres (0.156 square miles) may be taken as the area required for a rated 10-Megawatt installation *whose actual output is roughly 4.22 Megawatts.*

It is possible to estimate the land areas needed to replace the output of a large fuel-burning plant of a rated capacity of 1,000 Megawatts. Operating at 50 percent of capacity, on the average, such a plant would have an output of only 500 Megawatts. Using a simple proportion, the area needed would be $\frac{(500)(0.156)}{4.22} = 18.12$ square miles.

FEDERALLY SPONSORED RESEARCH AND DEVELOPMENT The Department of Energy is presently sponsoring a project for the power tower or central receiver type of solar-thermal generating plant (see ERDA, Central Receiver Solar Thermal Power System). Systems like those in the schematic, Fig. 6–29 are to be developed through four stages: a Test Facility, Pilot Plants, Demonstration Plants, and Commercial Plants according to a schedule outlined in the publication above.

As of 1978, a 5-Megawatt Test Facility exists at SANDIA Laboratories in Albuquerque. A design has been chosen for the first Pilot Plant, as well as a site at Barstow in the Mojave Desert in Southern California (Fig. 6–28).

The Mitre Corporation (working for the Department of Energy) is exploring the idea that solar thermal plants be used to augment existing fuel-burning power plants in favorable areas throughout the Southwest. This applica-

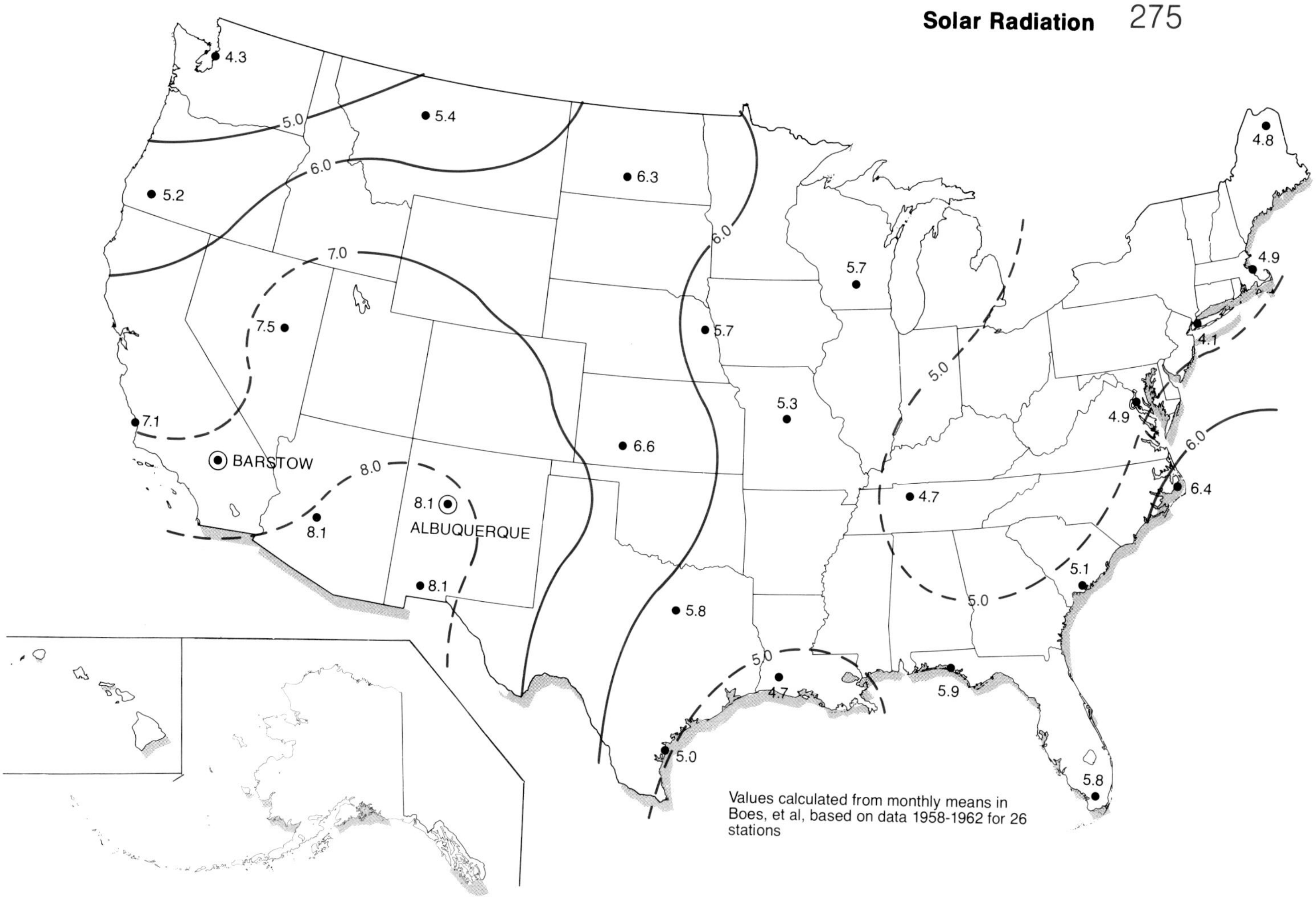

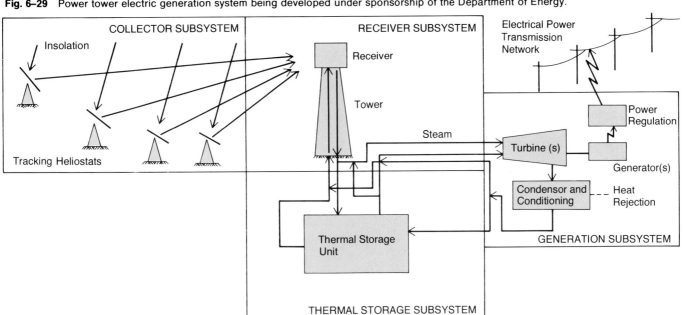

Fig. 6–28 Daily total of direct solar radiation normal to the collecting surface (direct-normal radiation). Annual mean values in KW hr. per square meter of surface.

Fig. 6–29 Power tower electric generation system being developed under sponsorship of the Department of Energy.

tion, called *repowering* by the Mitre Corporation, places the solar installation where the demand already exists, and avoids the building of new generating equipment since the solar system provides heat for an existing turbine (Curto and Nikodem).

PHOTOVOLTAIC SYSTEMS

In photovoltaic systems solar radiation is converted directly into electricity. Although there are several ways in which this is accomplished, solar cells, or photovoltaic cells, appear to be the most promising technology. Developed in the 1950s, these solar cells were used on orbiting space satellites and as communication and signalling devices in remote parts of the world. Despite these applications, much research and development is required before photovoltaic systems can make a significant contribution to our energy needs.

An array of photovoltaic cells. Each is 2 to 3 inches in diameter and has a peak output of roughly 0.25 watts. *Courtesy Department of Energy.*

Photovoltaic cells are made by the same principles required in the manufacture of transistors and integrated circuits. The primary requirement in transistors and solar cells is the semi-conductor. Photons, or the particles which make up a beam of light, can shake electrons loose from the atoms they strike. The solar cell combines this property of light with the usual properties of a semi-conductor (see Box). The main ingredient of the semi-conductor is silicon. In the semi-conductor, pure silicon is selectively "contaminated" at high temperatures with different gaseous impurities which serve to define the electrical boundaries between the element cells (see Box). Although silicon is one of the most abundant elements on the earth (the major constituent of sand), the solar cell requires an absolutely pure silicon crystal in order to be efficient in converting light to electricity. Thus the cost of a single solar cell is very high because of the high cost of processing the silicon into pure crystalline form.

The same solar cells used for years on spacecraft and in remote installations on earth are quite adaptable to individual buildings. In general, the size of a typical roof area where solar cells could be installed, would provide enough area to fulfill the electrical needs of a typical dwelling. Two major factors stand in the way of widespread residential use of photovoltaics: the cost of solar cells and the need for cheap reliable storage for electricity within the building.

Presently, photovoltaic installations are most competitive in areas where the cost of stringing utility lines is very high. An excellent example of this is the village of Schuchuli, Arizona, on the Papago Indian Reservation. There, an array of photovoltaic cells, sponsored by the Department of Energy, provides power for pumping water, lighting, and small appliances for the population of 100 persons. Cost of delivered power on a life cycle basis

How Solar Cells Are Made

A semi-conductor consists of a junction between two dissimilar materials. The p-region is transparent and can absorb light. Photons enter this region and release electrons from the silicon atoms. The electrons flow in one direction across the junction and the holes move in the other direction. The movement of electrons sets up a potential difference (voltage) and current will flow—electrons toward the positive terminal, and holes toward the negative terminal.

A typical solar cell is 2 centimeters by 2 centimeters, and its power output is around 0.5 Watts. As a result, many cells must be connected together to create a significant power output.

Source: U.S. Department of Energy, June, 1978.

is $1.76 per Kilowatt-hour, whereas power from the nearest private utility would cost $1.91 per Kilowatt-hour, *plus* hundreds of thousands of dollars for transmission lines. The cost for photovoltaics has dropped from $22 per Watt in 1977 to roughly $7 per Watt of installed capacity in 1979 (*New York Times,* June 24, 1979). For comparison, the investment in a large nuclear or coal-fired plant is roughly $1.00 per installed Kilowatt of generating capacity (Nuclear Energy Policy Group, p. 116, 123). Apparently photovoltaic power installations are not ready to displace nuclear or coal-fired utility plants; but they are now very practical and may be the least expensive source of electrical power in remote areas—especially in the Southwest where radiation is so abundant throughout the year.

Because solar cells, like flat-plate solar heat collectors, can make use of both *direct* and *diffuse* radiation, the two may be combined in future installations in order to provide both heat and electricity. For the same reason, *no special resource maps* are needed here: the total radiation as mapped in Fig. 6–3, when viewed with 15 percent conversion efficiencies in mind, can lead to estimates of cell array areas needed for a given generating capacity.

The Future Role of Solar Energy

It is possible to estimate the contribution of solar energy on the basis of three factors: 1) the projected total demand in various sectors of the economy; 2) the expected rise in fuel prices; and 3) the government incentives which will make solar energy competitive. Since fuel prices and the suitability of regions to solar applications vary throughout the country, solar energy will play different roles in different parts of the country. To be consistent, one projection is used here as the major source of information. Its views of three applications of *solar radiation itself* are shown immediately below, and similar projections (from the same source) are at the ends of chapters dealing with wind, ocean thermal gradients, and biomass, all of which are energy sources dependent upon radiation.

The projection reported here was made in 1977 by the METREK division of the Mitre Corporation under contract to the Energy Research and Development Administration (ERDA), now the U.S. Department of Energy (see METREK, August, 1978). Two scenarios are employed throughout the analysis: one based on policies specified in the National Energy Plan as proposed in April, 1977; the other based on recent trends in energy prices, policies and technology. The National Energy Plan scenario presented here consistently leads to greater solar contributions than the recent trends scenario because of various government incentives. Both projections are inaccurate to the extent that Presidential and Congressional initiatives and new technical information have altered policy and have made inappropriate some assumptions in the mathematical model used to generate the two cardinal outputs: the extent of market penetration (proportion of need served by solar) and the amounts of solar energy expected to be delivered.

THE FUTURE OF HOT-WATER HEATING, SPACE HEATING AND SPACE COOLING

In these applications, referred to by METREK as the "buildings sector," solar energy is allowed to compete only with electrical energy (the role of heat pumps is not specified). In the regional and national totals shown below for the year 2000, residential installations are expected to be about three times the number of commercial installations; retrofit are roughly 1.5 times the number of new installations; and about 75 percent of all installations are expected to be for hot-water heating.

Figure 6–30 assumes sixteen regions, each of which has some climatic "uniformity" and is represented by only one climatic station. The map shows projected *market penetration* percentages, the highest of which is 13.2 percent for Region 5, represented by Omaha, Nebraska, and the lowest is 0.5 percent for Region 13 represented by Cape Hatteras. These proportions when applied to demand estimates lead to the *projected energy delivered* which is greatest in Region 2 where a high penetration rate combines with a large population and consequently, a large demand. The regional energy amounts and the national total of 0.53 Quads are all estimates of *annual* amounts delivered.

THE FUTURE OF AGRICULTURAL AND INDUSTRIAL PROCESS HEAT

Air or water heated by the sun can be used for a variety of agricultural and industrial applications. Hot air is needed to dry tobacco, fruit crops, and lumber. Hot water is needed for drying textiles, for laundries, and for food processing operations. Altogether, heat supplied for such processes is referred to as agricultural and industrial process heat. As in domestic applications, the extent to which solar energy will displace fuels depends upon fuel prices.

Fuel oil, natural gas, and coal are the alternatives considered in this analysis. Low-temperature solar applications are expected to displace fuels early in the scenario, but high-temperature applications—using concentrating collectors—constitute the majority of solar installations by the year 2000.

Figure 6–31 shows the market penetration of solar energy for process heat to be generally much higher than for the buildings sector. In addition, its distribution is uniform throughout the nine economic regions: the highest penetration is 14.5 percent in the Pacific region, while the lowest, 7.2 percent, is in the East North Central. In estimated amounts of solar energy delivered, the West South Central region clearly leads, by virtue of a 12.2 percent

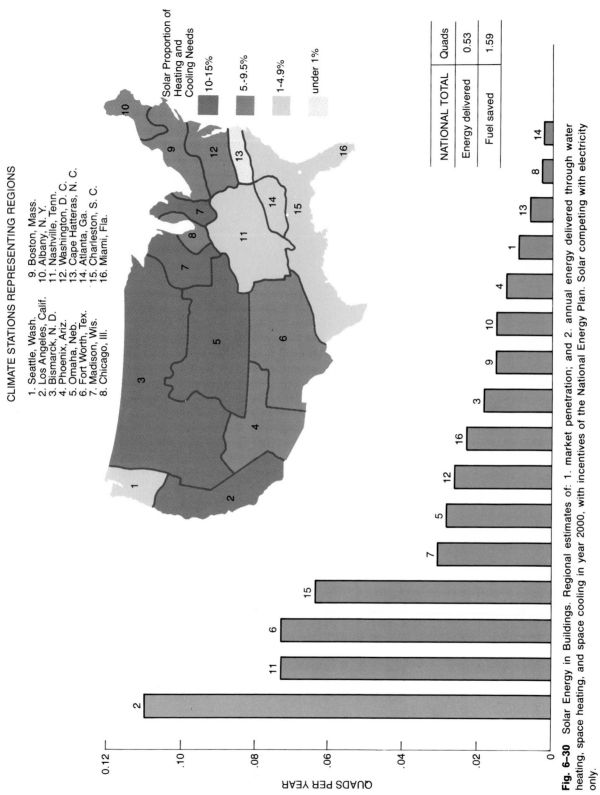

Fig. 6-30 Solar Energy in Buildings. Regional estimates of: 1. market penetration; and 2. annual energy delivered through water heating, space heating, and space cooling in year 2000, with incentives of the National Energy Plan. Solar competing with electricity only.

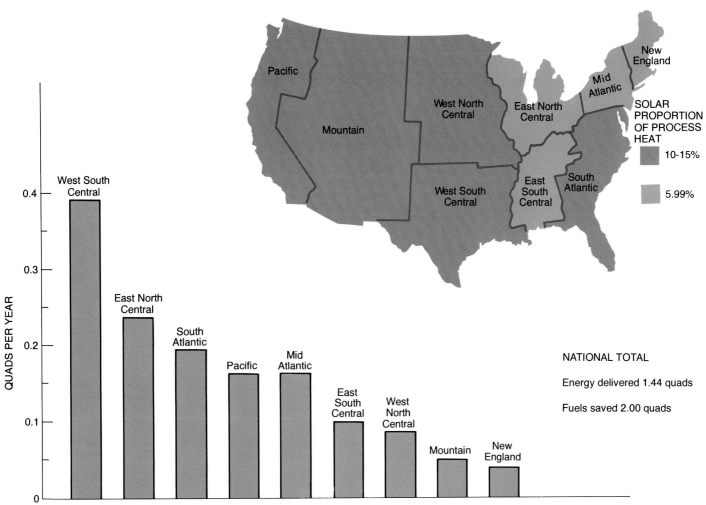

Fig. 6–31 Solar Energy for Process Heat. Regional estimates of: 1. market penetration, and 2. annual energy delivered as process heat in year 2000, with incentives of National Energy Plan.

penetration and a substantial total demand expected. The expected national total of energy delivered annually, 1.44 Quads, is almost three times that anticipated in the buildings sector.

THE FUTURE OF ELECTRICAL GENERATION BY UTILITIES

The four solar-related contributors to future electrical generation are radiation (thermal and photovoltaic), wind power, ocean thermal gradients, and biomass. Hydroelectric power is excluded simply because it is rather thoroughly developed. Projected electrical contributions from the four new sources and their combined market penetration are shown in the final chapter; here, only the projected thermal and photovoltaic output are considered.

Electrical generation from radiation is considered in the METREK projection only in the economic regions which embrace substantial areas of desert, West South Central, Pacific, and Mountain (Fig. 6–32). Expected output is roughly the same in the first two regions, and less than half that amount in the Mountain region where total demand is smaller. In all three cases, solar-thermal generation is expected to overshadow photovoltaic for utility application and will be used largely as fuel savers, that is, not as independent installations but auxiliary to fuel-burning power plants. The (annual) national total electrical output from thermal and photovoltaic together *in the year 2020* is thought to be 378 billion Kilowatt-hours, or 1.3 Quads. This is roughly the same as solar energy to be delivered as process heat in the year 2000 (Fig. 6–31), but the fuel saved will be greater because electrical generation rejects more fuel energy in conversion than does the direct burning of fuels for process heat.

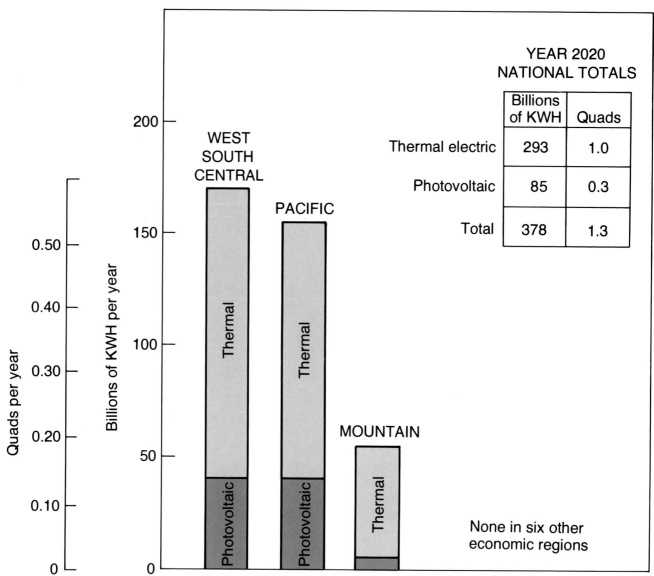

YEAR 2020 NATIONAL TOTALS		
	Billions of KWH	Quads
Thermal electric	293	1.0
Photovoltaic	85	0.3
Total	378	1.3

Fig. 6–32   Solar Electric. Estimated annual output in economic regions from thermal and photovoltaic installations in year 2020, with incentives of the National Energy Plan.

NOTES

[1]In a location such as New York or Philadelphia where 350 Langleys are received on an average day, a surface of one square meter (10.76 square feet) would receive 3.5 million calories, or 13,888 BTUs during one day. If *all* that radiation could be converted to usable heat and none lost to the environment this energy would raise the temperature of 100 pounds of water from 0 to 138 degrees Fahrenheit. It would provide, roughly, the daily domestic hot water needs for one occupant of a home. In fact, much of the radiation striking a solar collector is reflected and lost. Furthermore, some of the heat produced is radiated and convected away from the warm collector to the cool air surrounding it. Because of such energy losses a collector may deliver as heat only 50 to 70 percent of the solar energy it intercepts and would therefore be rated as 50 to 70 percent efficient.

[2]The extent to which direct radiation occurs in various parts of the Nation during each month of the year is revealed in a recent report from SANDIA Laboratories (Boes, *et al.*) Not presented in this brief review of the resource is information on *percent of possible sunshine* (an indication of cloudiness) or on the monthly or annual hours of sunshine,

both of which can be found in the *National Atlas of the United States of America* (U.S. Geological Survey).

Since total radiation receipts measured in calories per square centimeter embody the effects of both intensity of radiation *and* the length of day, that quantity is the most important single measure. Radiation at a number of specific cities may be found in *Local Climatological Data* (NOAA). Unpublished data from original station records on an hourly and half-hourly basis may be obtained through arrangement with the National Climatic Center, Federal Building, Asheville, North Carolina.

[3]These sizing principles are suggested by a Philadelphia engineer (Ridenour) who applied the 0.5 to 1.5 multipliers to states, not regions. Applying the numbers to zones based on mean radiation in Fig. 6–6 is an interpretation made by the authors.

[4]The service is called SOLCOST, and can be obtained through the following contractor: International Business Services Inc., Solar Group, 1010 Vermont Ave., Washington, D.C. 20005. Telephone (202) 628-1450.

[5]Working like a reversed air conditioner, a heat pump can move heat

"uphill" from cool outdoor air to a warm room. When outdoor temperatures fall below 30 degrees Fahrenheit the efficiency of a heat pump drops drastically. If warm water were provided by a solar collector, that water would be a source from which the pump could very efficiently extract heat for the house.

[6]The national average for electrical rates in December, 1979 was 5.2 cents per Kilowatt hour, and that for natural gas was $3.75 per thousand cubic feet (Bureau of Labor Statistics, December, 1979). The national average price for fuel oil in January 1980 ranged from 86 to 95 cents per gallon (DOE Energy Data Report, January, 1980).

[7]ASHRAE stands for American Society of Heating Refrigeration and Ventilating Engineers.

[8]These amounts are based on the *assumed* fixed and variable costs which are lower than those of 1978; furthermore, the optimum size of the collector is itself dependent on those assumed fixed and variable costs. Any current design analysis, using current fixed and variable costs and best estimate of fuel costs, would likely arrive at a slightly different system size and a total cost different from that deduced from Fig. 6–16. (How to obtain a current analysis is explained later.)

[9]These are virtually identical to assumptions made in a recent SOLCOST illustration of solar feasibility (*Changing Times,* April, 1978), and consistent with the estimate by a SOLCOST contractor who suggested fixed costs of $2,000 to $3,000, and variable costs of $20 to $30 (Lantz, Loren). The variable costs depend heavily on collector type, those used here being for the single-cover selective-surface collector. Collectors having other characteristics, especially those used for swimming pool heating, are much cheaper.

[10]Translation of (critical) costs per million BTUs of energy *delivered* into familiar prices for the fuel or for electricity:

Electric Rates Since there are 3,413 BTUs in every Kilowatt-hour, the dollar price per million BTUs should be multiplied by 0.003413 to get cents per Kilowatt-hour. If resistance heating is assumed, there is no need to consider the matter of efficiency because all the electrical energy used is converted to heat.

Fuel Oil Since each gallon of Number 3 heating oil yields 14.1×10^4 BTU, or 0.141×10^6 BTU, then the cost per gallon is 0.141 × the dollar cost per million BTU. Since conversion efficiency in a furnace is only 60 to 70 percent for fuel oil, and the relevant price for competition with solar heat is the price per *delivered* BTU, the fuel price itself must be diminished by 0.60 or 0.70. The critical cost in dollars per million BTUs must therefore be multiplied by (0.141) (0.60), that is, by 0.0846, in order to get the critical "market" price of fuel oil (assuming 60 percent efficiency).

Natural Gas Since each thousand cubic feet of natural gas yields 1.035×10^6 BTU, the cost per thousand cubic feet would be (1.035) (dollar cost per 10^6 BTU). But again, conversion efficiency of 70 percent must be considered, so the critical fuel cost in dollars per 10^6 BTU must be multiplied by (1.035) (0.70), that is, by 0.7245, to get the critical "market" price in dollars per thousand cubic feet.

[11]The input may be mailed, or it may be submitted through national time-sharing computer networks. In either case, advice may be obtained from the following office:

International Business Services Inc., Solar Group
1010 Vermont Avenue, Washington, D. C. 20005
Telephone: (202) 628-1450

Another analysis procedure for determining local feasibility of solar heating is now available. It is called RSVP and can be used with the aid of the *National Solar Heating and Cooling Information Center.* Call toll-free 800-523-2929.

[12]The purpose of the Data Network will be accomplished if those interested in monthly performance or analysis reports will request documents for buildings of interest. The current D.O.E. catalog is: *Availability of Solar Energy Reports from the National Solar Data Program,* (SOLAR/0020-79/37, May, 1979). This and other D.O.E. SOLAR titles referred to may be requested from the U.S. Department of Energy, Technical Information Center, P.O. Box 62, Oak Ridge, Tennessee 37830.

Wind Power

Windmills first appeared around 200 BC in Persia, and possibly as long ago as 1700 BC in Babylon and China. The use of windmills persisted in Europe during the Middle Ages, and by the fourteenth century the Dutch had used them extensively for pumping water during the reclamation of Rhine delta polders. In the sixteenth century windmills drove sawmills and paper mills in Europe, and by the middle of the nineteenth century there were an estimated 9,000 being used for various purposes in the Netherlands. With the introduction of the steam engine, reliance on wind diminished. In the Netherlands, for example, the number of operating windmills dropped from 9,000 to 2,500 by the year 1900, and to less than 1,000 by the year 1960 (Eldridge, p. 10, Reed, Maydew, and Blackwell, p. 1).

In the United States, large numbers of small windmills have been used for pumping water, for powering sawmills, and for generating electricity in later decades. In addition to numerous small electric generators at rural sites, one giant windmill of 175-foot blade diameter, the Smith-Putnam machine near Rutland, Vermont, fed power into the network of the Central Vermont Public Service Company during the period 1941 to 1945. An estimate of wind's contribution to total United States' energy needs (Fig. 7–1) shows the sudden decline around 1860 when steam engines became widely used. Another drop occurred around 1940 as the Rural Electrification Administration introduced to many areas cheap electrical power generated by fossil-fueled and hydroelectric plants.

Ironically, wind power was being abandoned just at the beginning of a post-World War II period which saw an immense increase in absolute and per capita use of energy in various forms—especially electrical. With the present fuel shortage and rising electric rates, wind power is being re-discovered. Already, there is a resurgence in the use of small generators and a growing interest in the possibility that wind can make a substantial contribution to the Nation's electrical utility needs.

The Character of Wind

Wind offers ready-made mechanical energy because nature has converted the sun's energy into moving masses of air. As an energy resource, wind has many advantages: it is renewable, non-polluting, and in some areas is quite reliable. Unfortunately, it has daily and seasonal periods of lesser intensity. Like the sun's radiation, it is rather diffuse, so that large installations are needed to gather enough energy to generate massive amounts of electrical power.

Winds flow in response to differences in atmospheric

←

This *vertical axis* wind turbine, built by the Department of Energy's Sandia Laboratories, Albuquerque, New Mexico, is seven stories tall, and is rated at 30 Kilowatts of generating capacity in winds of 22 miles per hour. Conceived by G.J.M. Darrieus in France in the 1920's, this eggbeater machine can use either two or three blades. The blades are airfoils which move *into the wind*. Turbines of this type have a number of attractive features. The machine does not have to swivel to face upwind. The installation is very stable because the heavy generator assembly is at the base, not on top of a tower. Furthermore, the lack of conventional tower means considerable savings in materials. *Courtesy Department of Energy.*

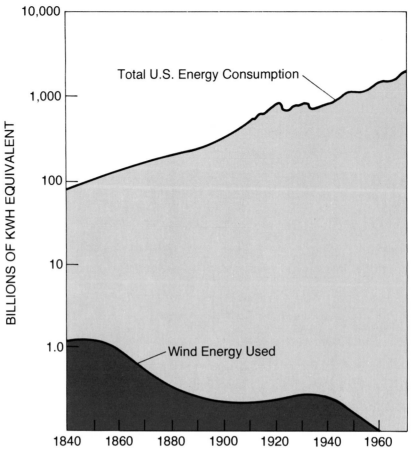

Fig. 7-1   Contribution of wind to U. S. energy needs, 1840–1970.

pressure; and those differences are due to the unequal heating of the earth's surface by the sun. For most of the atmosphere, in either Northern or Southern Hemisphere, the major winds are from west to east, with waves that lend the flows a northern direction in some areas and a southern direction in others. Winds at or near the earth's surface can be explained by maps of surface atmospheric pressures. Figure 7–2 shows the world pattern of high and low atmospheric pressures to be essentially latitudinal, with equatorial lows bordered by a series of subtropical highs in the vicinity of the thirtieth parallels. In the Southern Hemisphere these highs give way to an unbroken belt of low pressure in latitudes beyond the fortieth parallel. In the Northern Hemisphere, continental masses prevent such a coherent belt of low pressure.

Most relevant is the *rate of change* of pressure, that is, the pressure gradient as revealed by the spacing of isobars (Fig. 7–2). In the winter hemisphere the highs and lows are more profound, and the slope or pressure gradient that drives the wind is steeper. This effect is more pronounced in the northern hemisphere because it is exaggerated by continent-ocean contrasts. Winds in North America show two characteristics which can be predicted on the basis of Fig. 7–2. Relatively strong winds are encountered in both Pacific and Atlantic coastal areas where the map shows steep gradients; and

stronger winds occur in the winter season as suggested by steeper gradients in January than July.

Within the United States, an assessment of wind power requires the understanding and mapping of *Power Available* from the wind, and the application of certain assumptions to deduce *Power Extractable*.

POWER AVAILABLE CALCULATION

Since wind energy is the kinetic energy of moving air, it is proportional to air density, wind speed, the volume of air moving, and hence proportional to the cross-sectional area of flow. Thus, Power Available, P_A (per unit of time, t) may be derived as follows (Reed, p. 6):

$$P_A = \frac{\text{kinetic energy}}{t} = \frac{m\,V^2}{2t} = 1/2\rho A V^3$$

in which

ρ = air density
m = mass of air
R = flow rate
V = wind speed
A = cross-sectional area perpendicular to flow direction

It is useful to reduce the expression to one that applies to a cross-sectional area of one unit, such as a square meter. For sea level density of air, the expression becomes:

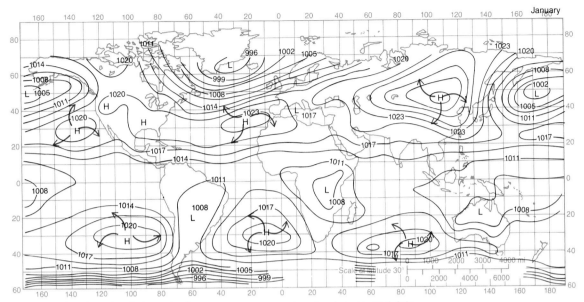

Fig. 7–2 World sea level atmospheric pressures, mean values for January and July.

$P_A = 0.05472 \ V^3$, if V is measured in miles per hour (mph)

or $0.08355 \ V^3$, if V is measured in knots

or $0.6125 \ V^3$, if V is measured in meters per second

It is significant that power is proportional to the *cube* of wind velocity. In all cases the Power Available is expressed in Watts of power for every square meter of area perpendicular to the wind flow (Fig. 7–3). Analogous formulas could be derived to express power in other units such as Watts per square foot.

METEOROLOGICAL DATA

Because Power Available is proportional to the cube of wind speed (Fig. 7–4), detailed wind data and careful analysis are necessary for a fair assessment of power. The average or mean wind speed at a site throughout a given period is not adequate if the total power in the wind is to be calculated. An average speed (V) of 15 mph used in the above formula would yield power of 185 Watts per square meter; and if the wind in fact blew steadily at that speed the calculated power would be representative. But if winds blow at 10 mph half the time, and 20 mph half the time, the sum of the power calculated from the two speeds would be 245 Watts per square meter (Reed, p. 2).

It is apparent that Power Available must be calculated from data that reveals the *frequency of winds of various speeds*. Fortunately, wind data gathered by observers in the network of U.S. Weather Bureau stations (now National Oceanographic and Aeronautical Administration or N.O.A.A.) is organized by wind speed categories. In addition, upper-air balloon observations can also be assigned to speed categories.

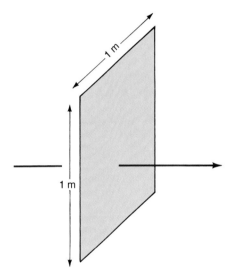

Fig. 7–3 One square meter perpendicular to wind flow, as assumed in expressing Power Available in Watts per square meter.

In response to federal research contract opportunities, three separate national assessments of wind power in the United States have been conducted in the past few years—by SANDIA Labs, by Lockheed-California, and by General Electric. Most recently, an overview of the three assessments was prepared by Batelle Pacific Northwest Laboratories. As will be shown, the wind power estimated for different regions of the country varies considerably from study to study because of different data employed and different treatments of the data.

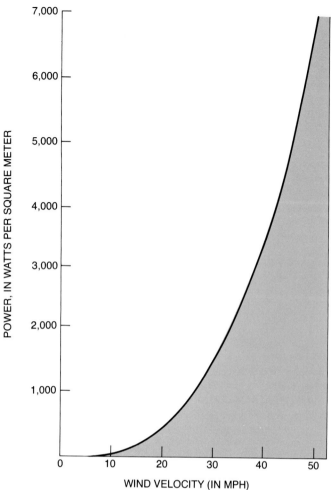

Fig. 7–4 Theoretical wind Power Available at various wind velocities.

THE USE OF SURFACE WIND DATA Central to any national assessment of wind power is wind speed data collected at hundreds of stations throughout the country. The nature of that data and the general procedure for deducing available wind power are outlined here.

The pioneering study employing large numbers of station records was conducted by Jack Reed of SANDIA Laboratories (Reed, 1975). In preparation for that analysis, wind speed data for 758 stations were assembled from records at the National Climatic Center, Asheville, N. C.[1] Table 7–1 shows the organization of data by wind speed categories, the tabulated numbers in the eight columns expressing the percentage of time (during the month) that speeds were in the range shown by the column heading. Monthly power available was calculated according to the formula above, using the *median value* in each range as the velocity factor, and the frequency value to weight the power derived when summing the contributions from each speed category (procedure outlined in Table 7–2). Annual power is not the total of all months, but is computed by following the steps in Table 7–2, using the annual percent frequencies (mean of the twelve monthly frequencies) for each speed category. Similarly, seasonal Power Available can be deduced by using an average frequency for a group of months: for instance, for the winter period, November through February, the frequency of winds in the 1 to 3 knot range would be 5.3 percent.

A similar procedure was used in the Lockheed and General Electric studies (see Table 7–3 for summary of study characteristics) but with variations as to which station records were used and whether median speed or

Table 7–1 Frequencies of Winds in Ten Velocity Categories, Cheyenne Airport (percent)[1]

	VELOCITY RANGES (KNOTS)										POWER
PERIOD	1–3	4–6	7–10	11–16	17–21	22–27	28–33	34–40	41–47	48–55	(Watts/m²)
Jan.	5.5	10.7	24.7	27.2	15.8	10.7	3.7	1.1	0.1	0.0	433.1
Feb.	4.7	10.7	24.5	26.5	16.3	11.4	4.5	0.9	0.1	0.0	453.5
Mar.	5.1	11.9	25.6	27.4	14.5	9.5	4.0	1.3	0.2	0.0	434.6
Apr.	5.3	12.0	25.1	28.3	15.1	8.1	3.2	1.2	0.2	0.0	399.8
May	6.0	13.7	30.8	31.3	11.6	4.5	1.3	0.2	0.0	0.0	242.6
June	6.2	16.0	35.3	30.5	8.4	2.4	0.5	0.1	0.0	0.0	176.2
July	8.5	19.4	38.4	26.3	5.7	1.2	0.1	0.0	0.0	0.0	125.6
Aug.	8.2	18.8	38.0	26.6	6.2	1.3	0.2	0.0	0.0	0.0	132.4
Sept.	7.2	17.8	35.2	28.7	7.9	2.1	0.3	0.0	0.0	0.0	157.1
Oct.	6.0	16.0	35.6	27.5	8.9	3.7	1.3	0.4	0.0	0.0	228.7
Nov.	5.7	11.7	26.8	26.8	15.2	3.1	3.1	1.2	0.0	0.0	402.6
Dec.	5.3	11.1	24.0	26.6	15.6	11.3	4.2	1.0	0.1	0.1	463.9
ANNUAL	6.1	14.1	30.4	27.8	11.8	6.2	2.2	0.6	0.0	0.0	302.7

[1]Not shown is that on annual basis winds are calm 0.7 percent of the time.

Source: Reed, p. 152.

Table 7-2 Calculation of Power Available in the Month of January at Cheyenne Airport

SPEED CATE-GORY (Knots)	Medi-an of Cate-gory (V)	P$_A$, i.e., 0.08355 × V³	Per-cent Fre-quency for Cate-gory	P$_A$X% FRE-QUENCY (Watts/sq. meter)
1–3	2.0	0.67	5.5	.037
4–6	5.0	10.44	10.7	1.117
7–10	8.5	51.31	24.7	12.674
11–16	13.5	205.56	27.2	55.912
17–21	19.0	573.07	15.8	90.545
22–27	24.5	1,228.70	10.7	131.471
28–33	30.5	2,370.53	3.7	87.710
34–40	37.0	4,232.06	1.1	46.550
41–47	44.0	7,117.12	0.1	7.117
48–55	51.5	11,412.17	0.0	0.0
POWER AVAILABLE FOR MONTH				433.135[1]

[1]See Table 7–1.

some other value was employed to represent a speed category.

UPPER AIR DATA In order to approximate the wind potential in highland areas where weather station data is extremely sparse, the velocities of upper-air winds as revealed by balloon soundings is useful. Speeds are mapped in a publication used by all three of the above-mentioned studies for this purpose (Crutcher).

OFFSHORE WIND INFORMATION Coastal stations provide some indication of wind conditions in the near offshore areas. A most useful supplement is actual wind measurements made at sea, and generalized into values applied to "Marsden Squares" (rectangular areas one degree of latitude by one degree of longitude). Of the three studies, only the one conducted by General Electric made use of such data.

Power Available

The General Electric study includes a very thorough examination of Power Extractable based on its maps of Power Available, (shown here in Figs. 7–5 and 7–6). To avoid confusion, maps of Power Available from the other three studies are not presented here, but they will be of interest to anyone who wants to pursue the subject more completely (Reed, 1975; Lockheed-California, 1976; Elliot, 1977).

The G.E. study (*General Electric Final Report Executive Summary*, 1977) arrived at three *regimes* which are

considered favorable and are thus distinguished from the remainder of the country where winds are considered inadequate (Figs. 7–5 and 7–6). Available Power at these three regimes at surface and 50-meter levels is identified by two measures: the first is Watts per square meter, as explained in a previous section; the second is Megawatt-hours per square meter per year which is nothing more than the Watts per square meter value multiplied by 8,760 hours in a year. A quick conversion from one measure to another uses the equivalence: one Megawatt-hour per square meter per year equals 114 Watts per square meter. The odd values in Watts per square meter for each region are due to the fact that G.E.'s maps labelled the regions according to the Watt-hours measures which were converted for presentation here.

The two maps show that favorable regimes are essentially on the coast, in the highlands, or on the Great Plains. Furthermore, High regimes are restricted to Pacific and Atlantic coastal areas, and to highlands such as the Pacific Coastal Range, the Sierras, the Northern Rockies, and the Adirondacks. Favorable regimes are much more restricted at the surface than at the 50-meter (164 foot) height, which is intended to represent roughly the *propeller hub height* of large wind machines of the horizontal axis type.

Only the annual average condition is represented on these maps, because it suits the G.E. method of calculating Extractable Power.[2] Some perspective is provided by Fig. 7–7 which approximates the seasonal variations in the three regimes by use of only a few stations to represent each. The tendency for maximum power to occur in winter (or early spring), as expected on the basis of world maps, is very clear—especially in the High regime. To support this rather general information, monthly values plotted for seven stations in two different parts of the country show a clear maximum in winter season—the values in winter being in some cases two to three times the values of those in summer (Fig. 7–8). Implications for design of wind energy systems independent of other energy sources are obvious. A rough indication of how the season of maximum varies throughout the country is provided in Fig. 7–9, in which the transitions shown are unavoidably more sharp and sudden than the actual change in seasonality. In general, a maximum in winter or spring is prevalent, with a consistent tendency toward spring maximum in the Great Plains. In northern and highland areas, maximum power occurs, fortunately, in the season of greatest power demand; in the South, however, where greatest electrical demand is in summer for air conditioning, the winter and spring maximum is less fortuitous.

COMPARISON WITH SANDIA AND LOCKHEED STUDIES

Using Table 7–4, the G.E. values mapped in Figs. 7–5 and 7–6 may be compared with other estimates of Power Available. Apparently G.E. values are relatively high

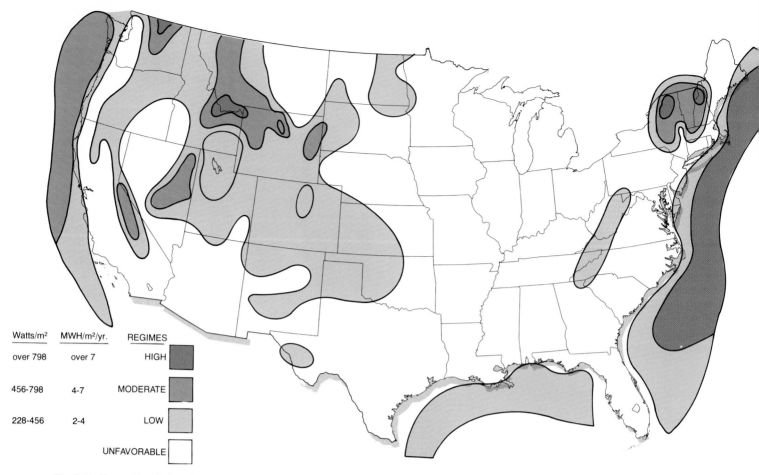

Watts/m²	MWH/m²/yr.	REGIMES
over 798	over 7	HIGH
456-798	4-7	MODERATE
228-456	2-4	LOW
		UNFAVORABLE

Fig. 7–5 Favorable wind regimes at surface level in the conterminous 48 states and offshore, showing Power Available values.

Fig. 7–6 Favorable wind regimes at 50 meter height in the conterminous 48 states and offshore, showing Power Available values.

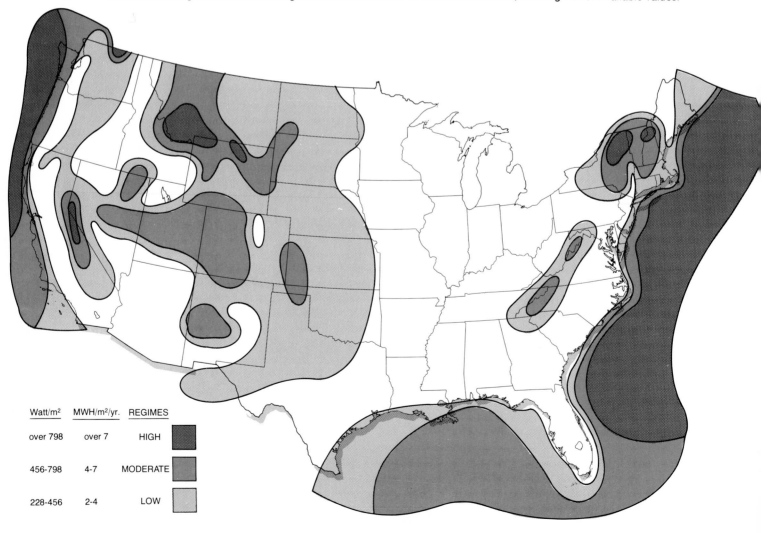

Watt/m²	MWH/m²/yr.	REGIMES
over 798	over 7	HIGH
456-798	4-7	MODERATE
228-456	2-4	LOW

only in the Central Gulf Coast and the Western New Mexico Mountains. Their higher values for the coastal region may be due to the better data employed, and the higher mountain values due to the assumption that mountain winds are equivalent to those in the "free circulation," that is, the upper-air winds.

Power Extractable

While the foregoing analysis provides a useful translation of wind velocities into wind Power Available, the results indicate only areas relatively more attractive or less attractive. What can be accomplished with the power, or conversely how many windmills would be required in certain areas to do a given job, are questions that remain to be answered.

Two approaches to the question of Power Extractable seem to be necessary in light of present energy conversion technology: here they are called the *idealized*, and the *electrical* approaches.

IDEALIZED POWER EXTRACTABLE

This approach, used here for the sake of understanding fundamentals, assumes that virtually all the Power

Table 7–3 Criteria Used in the National Wind Energy Assessments by SANDIA, General Electric and Lockheed

ITEM	Sandia	General Electric	Lockheed	COMMENTS
Surface wind data used	SANDIA wind speed frequency distributions	SANDIA wind speed frequency distributions	SANDIA wind speed frequency distributions	Many additional surface wind summaries and data sets are available
Number of stations used for surface analysis	758 (total available)	Not given. Primarily first order stations	478 and selected stations with limited data	
Method of estimating wind power	V^3 (median of wind speed class)	V^3 (median of wind speed class)	V^3 (fitted cumulative frequency distribution by spline function and evaluated every decile)	Median of class overestimates wind power by 7%, on the average
Accounted for variation in anemometer heights?	No	No. Heights assumed to be 10 m	Specified where known. Assumed at 10 m where varied over period or not known	Anemometer heights vary from 6–60 m. Standard height became 6 m in 1960s
Accounted for density variations with station elevation	No	No	Yes	Use of sea level density overestimates wind power by 8.5–9% per km increase in elevation
Low-level rawinsonde data, e.g., 150 m level	Not used	Estimated wind power at 150 m from frequency distribution for 65 stations. Annual estimate only for 00Z	Not used	Use of this data underestimated wind power by 30–60%
Vertical extrapolation of wind speed	Suggests 1/7 power law. Different methods discussed	Extrapolated downward from 150 m to 50 m using empirical procedure	Extrapolated upward from 10 m to 50 m and 100 m using 0.2 power law	Extrapolating upward to 50 m using 0.2 power law gives 32% more energy than 1/7 power law
Upper air wind data used	Climatological maps by Crutcher	Climatological maps by Crutcher	Climatological maps by Crutcher	
Accounted for upper air density?	No. Sea level density used	Yes	Yes	Density at 3 km is <75% of sea level density
Method of estimating mountain top wind power	850 mb power assumed typical of Eastern mountains; 700 mb power typical of Western mountains	Assumed $V_m = V_f$ thus $P_m = P_f$ m = mountain, f = free air	Assumed $V_m = 1/2\ V_f$ thus $P_m = 1/8\ P_f$ m = mountain, f = free air	Relation of mountain top power to free air power is highly variable, ranging from <1/8 to >1
Coastal and off-shore wind data used	Coastal stations on SANDIA tape	Ship wind data for Marsden squares and Navy Marine areas Isodyns (kmh/m²/yr) for	Coastal stations on SANDIA tape, and a few additional stations	Many coastal stations are not representative of exposed sites
Types of national analyses made	Surface and upper air isodyns (watts/m²), values smoothed objectively by geometric averaging. Annual and seasonal averages for surface; seasonal for upper air	surface, 150 m level, upper air, and 50 m level. Subjective type analysis. Annual averages only.	No isodyns. Power density values plotted on U.S. map for each station. Wind power classes represented by color variations. Annual and seasonal maps of power for 10, 50, 100 m	Discrepancies are found in the energy patterns and estimates

Source: Taken from the Batelle study (Elliot, 1977).

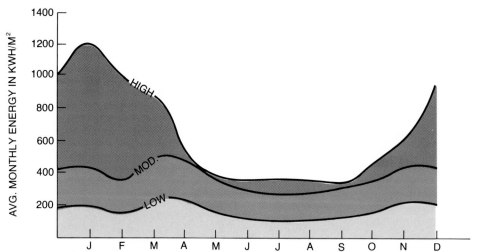

Fig. 7-7 Seasonal variation in wind energy according to General Electric model of winds in three regimes at 50 meter height.

Available, *is in fact available* for extraction. Because Power Available, as derived and mapped in the foregoing discussion, is the sum of contributions from winds of various speeds, it is truly available only if a wind energy conversion device will operate at low, medium, high, and very high speeds. This is not consistent with generation of alternating electrical current (AC), but is compatible with generation of direct current (DC) power, and with the execution of mechanical tasks such as pumping water uphill, or producing hot water through turbulence. Assuming such tasks are relevant, the mapped Power Avail-

able at a selected site must be subjected to the following considerations: 1) the efficiencies of energy conversion, and 2) the size of energy conversion devices.

EFFICIENCIES OF ENERGY CONVERSION   The primary consideration here is how much of the wind's mechanical or kinetic energy can be translated into rotary power on the shaft of a windmill or turbine. Not all the power in the windstream can be extracted, for if it were, the wind would stop flowing at the extraction device. The problem of how much power can be extracted *without* interrupting the flow of air is susceptible to theoretical analysis which

Fig. 7-8   Seasonal variation in Power Available at 10 meter height in two regions of the U.S.

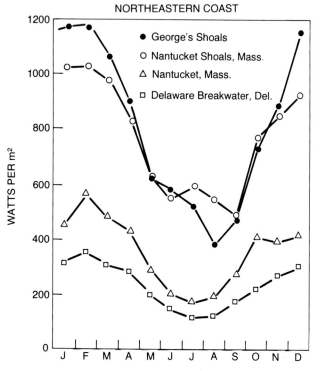

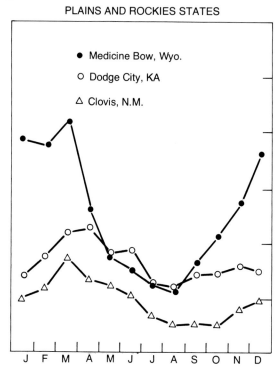

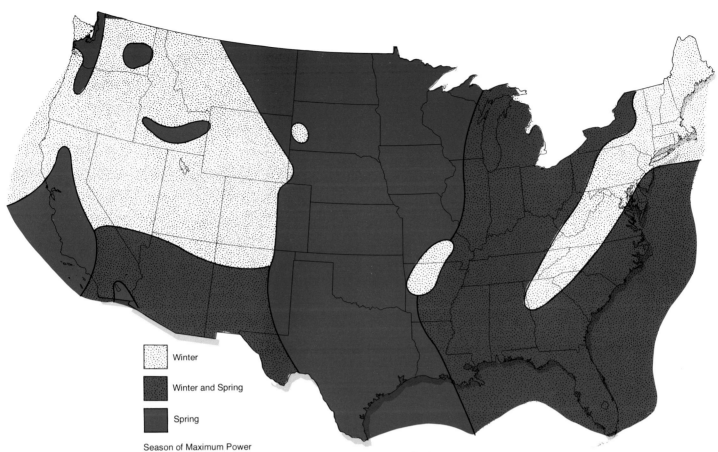

Fig. 7–9 U. S. areas where wind power is maximized in Winter, Winter–Spring, and Spring.

Winter

Winter and Spring

Spring

Season of Maximum Power

yields a maximum efficiency of 0.59, often called the *power coefficient* or the Betz coefficient (Betz, 1927). As suggested in Fig. 7–10 the maximum efficiency for various types of machines is achieved at various combinations of blade speed and wind speed. On that graph, the most relevant entries are for the "high-speed two-blade type" (horizontal axis machine) and the Darrieus rotor, which is the vertical axis or eggbeater-style machine. Since maximum efficiencies are around 0.35 and 0.45 respectively, the use of 0.40 conversion efficiency is realistic. (Conversion from mechanical to electrical power entails another loss which may be expressed as an efficiency of 0.85 to 0.90, but will be ignored in this idealized analysis).

SIZE OF THE CONVERSION DEVICE   Since Power Available has been expressed in *Watts per square unit of swept area*, the Power Extractable is proportional to the area swept by the "blades" of the wind machine, and a machine with a blade radius twice that of a smaller one will intercept *four times* the power. There is another effect of machine size. If the blade radius is great, the device must be so tall that part of the swept area is at a height greater than that at which wind velocity was measured and will encounter winds faster than those measured. If performance of a large machine is estimated by

reference to wind measured at or near the ground, this factor for "height effects" should be considered.[3]

It follows that if the goal is converting mapped Power Available to Idealized Power Extractable, the following formula may be applied (Reed, Maydew, and Blackwell, p. 31).

$$P_E = (0.40) \times (\pi R^2) \times (R/10)^{3/7} \times (P_A)$$

Power Extractable = Efficiency of 40%, Area Swept, Height Effects Factor, Power Available

The expression may be reduced to:

$$P_E = 0.469 \, R^{17/7} \, (P_A)$$ so that only the blade radius of the machine and the Power Available need be known.

The simpler expression,

$$P_E = (0.40) \quad (\pi R^2) \quad (P_A)$$

will be used here because consideration of height effects is inappropriate when Power Available is gleaned from a

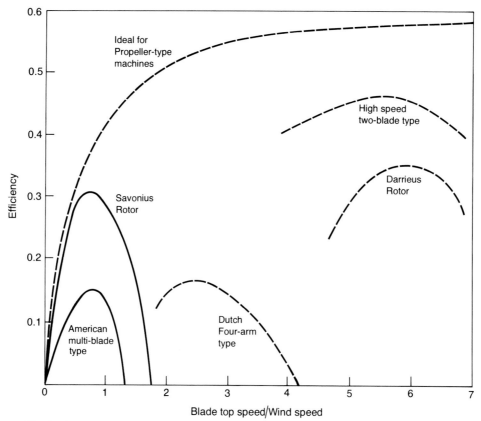

Fig. 7–10 Efficiencies of different types of wind energy conversion devices, showing importance of the ratio of blade tip speed to wind speed.

generalized map, and particularly because the concept will be illustrated with power deduced from the map showing power at 50 meters, which is the hub height for large machines.

Figure 7–11 plots the results of a series of calculations according to that formula, that is, simply multiplying power per square meter by the machine's swept area, and reducing the amount by the factor of 40 percent to consider extraction efficiency. For Power Available of 500 Watts per square meter, which prevails as an annual average in a Moderate regime such as in Western Colorado (Fig. 7–6) it is apparent that a giant machine with a blade radius of 40 meters could capture approximately 1,000 Kilowatts of power. At this point, the analysis might conclude that, based on the capacity of one such machine, it would take 1,000 of them to equal the generating capacity of a 1,000-Megawatt coal-fired plant, but, for reasons pointed out below, this idealized analysis is not relevant to electrical power generation.

ELECTRICAL POWER EXTRACTABLE

WIND CHARACTERISTICS AT THE SITE The main factor that undermines the idealized method of calculating Power Extractable is the fact that wind machines, especially those now built in the United States to produce alternating current, do not make use of all the Power Available in the winds because they do not respond to winds of all speeds. First, a large machine will not start when winds are under 5 to 8 mph, and any Power Available cal-

Table 7–4 Annual Average Wind Power at 50 Meters Height in Selected Regions According to Three National Assessments (in watts/sq. meter)

REGION	SANDIA[1]	POWER General Electric	Lockheed[2]
Texas Panhandle	450–650	250–450	400–990
Eastern Kansas	300–400	200–300	180–990
Iowa	250–350	150–200	290–673
Eastern Dakotas	350–450	200–250	300–812
Southern Wyoming	400–850	250–500	523–1030
Lake Michigan	200–250	150–200	210–840
Indiana	200–250	100–200	200–565
Western New York	200–350	150–300	140–1000
Atlantic Coast	250–300	400–600	190–904
Central Gulf Coast	150–200	300–450	170–220
Texas Coast	300–350	250–300	490–570
Oregon Coast	200–300	400–600	800–1610
Western New Mexico Mountains		400–700	190–340
Arizona Mountains		100–200	100–430

[1]SANDIA values extrapolated from 10 meter height to 50 meters, using law of 1/7 power.
[2]High and low values are extreme because Lockheed maps show no isolines (isodynes) but only spot locations.
Source: Elliot, 1977.

culated on the basis of light winds will be false for this reason. More important, given the impact of the V³ factor in deriving Power Available, are the higher winds. These may account for a very large part of the calculated Power Available; but their power will not be used by horizontal axis machines which feather their blades (adjust blade pitch) in order to maintain a constant rpm on the shaft.

These considerations demand that the winds of a particular site, or a wind regime, be described by the very information that is used (then lost) in calculating total Power Available at a site, namely, the frequency of winds of certain speeds. Table 7–1, along with Figs. 7–12 and 7–13, show how that information may be used to better grasp the potential for electrical generation at a particular site.

Annual percent frequencies in the form shown in Table 7–1 can be manipulated to make a plot such as in Fig. 7–12, which shows proportions of the year during which winds *exceeding* certain speeds may be expected to occur. For any chosen series of wind speeds, then, these frequencies may be converted to hours per year (as in Table 7–5), and the speeds converted to their implied power available per square meter. The derived plot of *power duration* (Fig. 7–13A) can then be constructed using the hours at certain Power Available values. This is extremely useful information, for the *expected hours* at various values of Power Available are crucial to wind machine output and design. One further graph (Fig. 7–13B)

is derived from Fig. 7–13A by noting the hours at various power (wind speed) values, expressing the product as Kilowatt-hours, and plotting the results opposite wind speeds. This plot of *power density* reveals in which wind speeds the greatest amount of power may be expected. The speeds that do contain the most power are those of sufficiently high velocity and sufficiently high frequency to garner the highest product: they are neither the highest velocities, because those are rare, nor the most frequent, because those winds are light. In the Cheyenne, Wyoming case illustrated in Fig. 7–13B it is winds of 20 mph that contain most power in the average year.

Hypothetically, a wind turbine might be designed for peak output in winds of 20 mph velocity. If it were, then Fig. 7–13A would reveal the number of hours during the year the winds may be expected at that velocity, and therefore, the number of hours the turbine would be operating at peak output. The shaded rectangle represents the product of peak power (234 Watts per square meter) and the hours at that power (1095). The stippled area represents hours of operation at speeds of greater than 5 mph but less than 20 mph. If blades feather so that winds of greater velocity than 20 mph are treated as if they were 20 mph, there are no more hours of operation to account for. Since power, or energy, is now expressed in Kilowatt-hours per square meter per year, the expected output from machines with larger and smaller rotors might be deduced.

Fig. 7–11 Idealized Power Extractable from wind at sites of different Power Available for a single wind machine of four sizes, ignoring height effects.

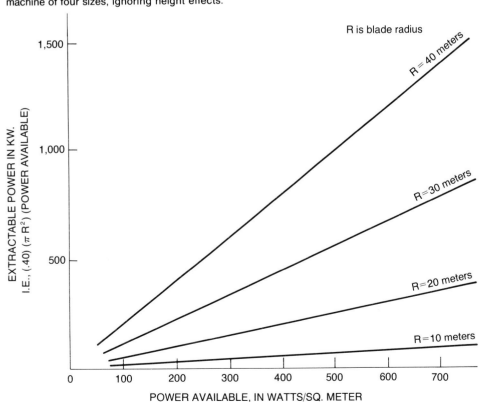

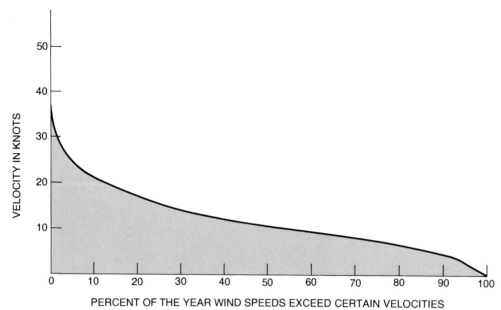

Fig. 7–12 Cumulative percent frequency plot of surface winds at Cheyenne Airport, Wyoming.

THE GENERAL ELECTRIC APPROACH TO POWER EX-TRACTABLE FOR UTILITIES   Although the wind characteristics as described by graphics such as Figs. 7–12 and 7–13 are fundamental to the design of wind turbines, there are various approaches to a practical and economical turbine design. Lockheed's analysis of costs for utility applications led them to recommend rotor diameters of nearly 350 feet (107 meters) at most locations, with various transmissions and generator capacities to suit variations in wind speed (Lockheed-California Co., 1976). The General Electric approach employs generators of the same capacity in various locations, and varies rotor diameters in order to keep shaft rpm constant, and to gather enough energy to power the standard dynamo in three different wind regimes.

In order to assess the potential of the three regimes,

High, Moderate, and Low, the G.E. study chose for each regime one or more stations whose wind characteristics were thought to be representative of that regime. Economical wind turbines were then arrived at by procedures that take into account the need for efficient blade tip speed during operation in various wind speeds and seek to find the minimum investment per Kilowatt of output.

Table 7–6 shows the features of turbine units designed for three median wind speeds frequently occurring in each of the three regimes. In each of the three cases, the "nameplate" capacity of the turbines chosen for electrical utility application is 1,500 Kilowatts (that is, 1.5 Megawatts). Rotor speeds diminish from High to Low regimes, and the rotor diameter needed to achieve the desired rpm is smallest in the High regimes, largest in the Low regime (Fig. 7–14). The *rated wind velocity,* that is,

Table 7–5   Frequency of Wind Speeds, and the Power Associated with Each Surface Data from Cheyenne, Wyoming

	SPEEDS (VELOCITIES) IN KNOTS							
	5	10	15	20	25	30	35	40
Percent frequency, taken from Fig. 7–12	86	52	25.2	12.5	4.5	1.7	0.6	0.2
Hours per year on the basis of percent frequency	7,534	4,555	2,234	1,095	394	149	53	18
Velocity cubed	125	1,000	3,375	8,000	15,625	27,000	42,875	64,000
Power at each speed, in kilowatts	3.6	29.24	98.69	233.9	456.9	789.5	1,253.7	1,871.4
Thousands of kilowatt-hours per year (KW × 8760 × % Frequency)	27.1	133.2	220.5	256.1	180.0	117.6	66.4	33.6

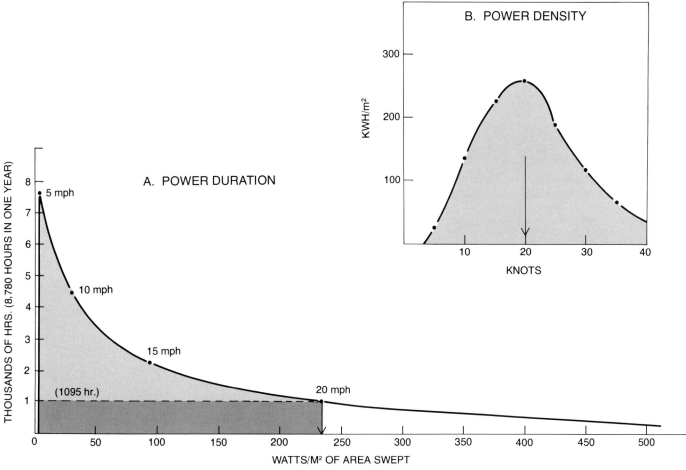

Fig. 7–13 Wind characteristics at Cheyenne Airport, Wyoming: A. Power duration   B. Power density

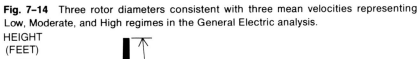

Fig. 7–14 Three rotor diameters consistent with three mean velocities representing Low, Moderate, and High regimes in the General Electric analysis.

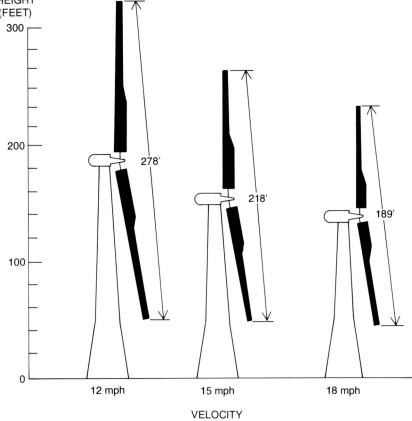

the velocity at which peak power of 1,500 Kilowatts is realized, is always higher than the *design velocity* which represents either mean or median wind speed for a site (both terms are used in the G.E. report). The *capacity factor* is extremely important because, on the basis of velocity duration curves (Fig. 7–12), it expresses the proportion of the time during which winds blow at the rated velocity and at speeds less than that. Differing velocity duration curves characteristic of higher and lower mean wind sites lead to highest capacity factor (0.50) at the 18-mph site, and lowest (0.40) at the 12-mph site.

In estimating the output of the 1,500-Kilowatt units located in each of the three regimes, the study recognized that the wind speeds designed for, namely 18, 15, and 12 mph do not occur with the same frequency at all sites in the regimes. A more conservative capacity factor for each regime was arrived at, therefore, that better represents velocity duration in the regime—in every case smaller than the factor applied to the three design cases.[4]

The revised capacity factors (Table 7–7) are critical to the remainder of the analysis, for they determine the estimated annual output of the standard 1,500-Kilowatt machines in each of the three regimes. Any analysis of 1) how many machines are needed to accomplish a given task, or 2) the maximum electrical energy extractable in an area depends on those estimates.

THE QUESTION OF PACKING DENSITY   Equally important is the matter of how closely the wind machines may be planted in an area without deleterious effects on the wind field. Lacking experience with this, industry researchers estimate that a spacing of 10 to 15 rotor diameters may be necessary. Lockheed assumed spacing of ten times the diameter while General Electric chose the more conservative spacing of fifteen times the diameter, which results in fewer machines planted per unit area. Considering the three rotor sizes in the three regimes, this rule of thumb leads to the following densities: High re-

Table 7–7 Operating Characteristics of 1500-Kilowatt Units in Three Wind Regimes

| | REGIME | | |
	High	Moderate	Low
Annual energy output (KWh × 10⁶)	5.01	4.33	2.92
Regime capacity factor considering variety of winds	0.38	0.33	0.22

gime, 3.01 per square mile; Moderate, 1.851 per square mile; and Low, 1.306 machines per square mile.[5]

INSTALLATION REQUIRED TO MATCH CAPACITY OF A 1,000-MEGAWATT ELECTRIC PLANT   An intriguing question is the nature and size of a wind-electric installation which generates as much power as the competition, that is, a very large fuel-burning steam-electric plant. The comparison is not entirely appropriate, since wind power installations would not be employed in such a concentrated or independent fashion. Nevertheless, the wind installation(s) may be construed as scattered through an area that would otherwise be served by the steam-electric plant.

Assuming the conventional power plant operates in the average year at 50 percent capacity, the actual output to be matched by wind installation is not 1,000 Megawatts, but 500 Megawatts. In the most favorable, that is, the High wind regime, the average capacity factor of 38 percent (Table 7–7) suggests that for an output of 500 Megawatts, the installed capacity must be about 1300 Megawatts. The space requirements for such an installed capacity can be found in a G.E. plot which bypasses the

Table 7–6   Characteristics of 1,500-Kilowatt Units Designed for Three Median Wind Speeds

| | MEDIAN (DESIGN) SPEEDS | | |
	18 mph	15 mph	12 mph
Rotor diameter	190 ft.[1]	219 ft.[1]	278 ft.[1]
Rotor speed	40 rpm	31.5 rpm	20.6 rpm
Rated wind velocity	22.7 mph	20.3 mph	16.8 mph
Design annual output	6.62 KWh × 10⁶	5.85 KWh × 10⁶	5.29 KWh × 10⁶
Capacity factor, *i.e.* proportion of time the winds flow at rated velocity	0.50	0.45	0.40

[1]Optimum rotor diameters to maintain steady rotor rpm in different winds are not known precisely. Elsewhere, these diameters, 218 feet, 278 feet, and 331 feet are used for High, Medium, and Low wind regimes when defining packing density.

number of machines and directly shows the square miles needed (Fig. 7–15). If the conservative spacing of fifteen rotor diameters is assumed, the need is for 300 square miles; but if spacing of ten rotor diameters is assumed the space required drops to about 140 square miles. Thus, the best of the wind cases demands far more land area than the solar-thermal electric plant of the power-tower or central receiver type, which requires only 18 square miles for a comparable output (see Chapter 6 on solar energy). This suggests that research should be directed toward devices or structures that focus and concentrate the wind flow as well as dynamos that make use of the power residing in the higher winds at a given site, rather than spilling the winds that exceed the design speed.

It must be recognized that, as with solar-electric power installations, the area is committed only once, and thereafter produces clean energy without fuel. The coal or nuclear plant, in contrast, requires continual mining which may disturb as much as 15 to 20 square miles during the life of the plant.

Net energy balance for wind-electric systems according to both Lockheed and G.E. studies is not immediately attractive when compared to fuel-burning plants. General Electric estimates the *energy payback period,* that is, the time for energy generated to compensate for the energy expended in materials for the installation, is 0.29 to 0.89 years for wind machines in the three regimes, whereas for a 700-Megawatt coal plant the period is 0.076 years, and for a 1150-Megawatt nuclear plant it is 0.109 yrs. To put it more favorable, the wind machines pay back the energy cost of their materials within 3½ to 11

months, depending on regime, and from then on consume virtually no energy. Fuel-burning plants pay back their material energy costs much sooner, but continue to consume energy. Obviously this "net energy" analysis is incomplete and should be improved by a *life cycle* approach that considers all energy expended and produced during the installation stage and the expected life of the installations. The results of such analysis should supercede any analysis of dollar economics in deciding whether electric power from the wind is desirable.

MAXIMUM POTENTIAL OF THE THREE REGIMES Corollary to the question of how many machines are needed to accomplish a given generating task is the question of how great is the total potential for electrical generation. Some stimulating estimates of the power output from large numbers of machines in offshore areas and in the Great Plains have been made in the recent past (Heronemus); but no thorough study of the actual number of machines that *could* be planted in various parts of the country had been made until the Lockheed and General Electric studies referred to here. The following account is entirely based on the General Electric study which did not attempt to estimate installations in offshore areas, but focused only on land areas in the contiguous states (General Electric, *Final Report*).

With a view to electrical utility application of wind power, the vital assumptions have already been stated.

1. The annual output of the selected 1,500-Kilowatt machines in each of the three regimes is as noted in Table 7–7.

Fig. 7–15 Land area requirements for large installed generating capacities in three wind regimes, with two assumptions about dynamo spacing.

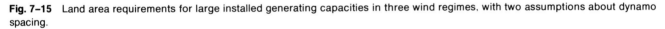

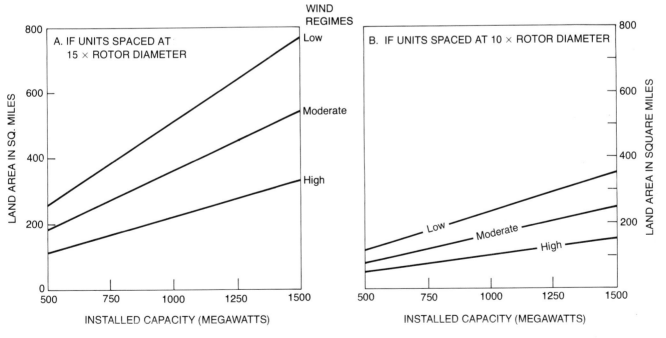

2. The number of machines that can be planted per square mile in each of the three regimes, considering the three rotor diameters, is noted above in the section on *Packing Density*.

It is necessary, therefore, only to estimate the *land areas available* in each of the three regimes, in order to arrive at the maximum number of machines possible in each regime, and hence in the forty-eight states as a whole. This maximum number of machines and the generating output that follows are considered the *saturation case*. The procedure used to arrive at available land areas is as follows:

1. In each regime, land area was diminished by any institutional lands, military bases, national parks, national monuments, and Indian reservations.
2. In each regime all areas were assigned to one of a number of land use categories, such as Irrigated Land, Swamp, or Urban Areas, by reference to land use maps in the *National Atlas of the United States of America* (U.S.G.S., p. 158).
3. For each land use type a factor was assigned to denote the proportion of land which could be used for wind machines (see Table 7–8).
4. By reference to Landform classes in the *National Atlas* (U.S.G.S., p. 61) slope factors were applied to reduce the available land to that with suitably gentle slopes.
5. The resulting "net" land areas were then filled according to the packing density appropriate to the regime, and the resulting number of 1,500-Kilowatt units subjected to the regime's capacity factor to yield total Kilowatts, which multiplied by 8,760 hours yields Kilowatt-hours per year.

Results of the land use analysis are summarized in Table 7–9, which shows for each regime the total land area, amounts not available for land use and slope reasons, and the net available land in square miles and as a

Table 7–8 Land Use Factors Used by General Electric to Determine Land Available for Wind Machines

LAND USE CATEGORY	PROPORTION OF AREA CONSIDERED AVAILABLE FOR WIND MACHINES
Alpine meadows, mountain peaks	1.0
Swamp	1.0
Marshland	1.0
Moist tundra and muskeg	1.0
Forest and woodland, mostly ungrazed	0.9
Sub-humid grassland	0.9
Desert shrubland, mostly ungrazed	0.9
Desert shrubland, grazed	0.8
Open woodland, grazed	0.8
Forest and woodland, grazed	0.8
Forest and woodland with some cropland	0.8
Cropland with pasture, woodland and forest	0.5
Cropland with grazing land	0.2
Mostly cropland	0.2
Urban areas	0.01
Irrigated land	0.0

proportion of the total. Altogether 18 percent of the land was lost to institutional uses, with the least impact of this factor occurring in the High regime which is largely mountainous. Similarly, the overall loss to competing land use is 26 percent, with the greatest impact in the Low regime which coincidentally is of lower elevation and more heavily farmed. The factor of slope deleted 42 percent of the land overall, and 70 percent in the High wind regime. The last column in Table 7–9 shows that of

Table 7–9   Net Land Areas Available for Wind Power Systems (in thousands of square miles)

WIND REGIME	Total Land Area	Institutional Lands	Not Available (Other Land Use)	Not Available (Poor Topography)	Net Available Land	Fraction of Regime Area (Percent)	Distribution of Available Area (Percent)
Low	1,104	201.5	325.5	420.4	156.6	14.2	74
Moderate	415	85.6	79.7	195.8	53.9	12.0	25
High	55	2.2	10.8	39.0	3.0	5.5	1
TOTAL	1,574	289.3	416.0	654.7	213.5	13.5[1]	100

[1]Mean.

Source: General Electric. *Wind Energy Mission Analysis, Final Report*, February 18, 1977.

all the remaining land area 74 percent is in the Low regime, 25 percent in the Moderate, and only *1 percent in the High wind regime*. Apparently the question of whether sloping land can be employed is critical to a fuller exploitation of areas where wind potential is greatest.

Table 7–10 summarizes the results of hypothetical planting of 1,500-Kilowatt capacity wind machines in the available land in the three regimes, noting the number of units at saturation and their output in Watts and in Watt-hours per year. The relative capacity (or output) of the three regimes is roughly as suggested by relative net land areas in Table 7–9. The more dense packing of machines in the High regime, however, allows 3 percent of the total capacity to be installed there despite its having only 1 percent of the available land area. This more effective use of available land in the higher wind regimes is indicated by the *Efficiency Factor* column in Table 7–10, which shows 14.90 millions of Kilowatt-hours per square mile in the High regime, versus only 3.78 per square mile in the Low. The overwhelming majority of the generating capacity resides in the Low wind regime, with about half that amount in the Moderate. The total capacity of 470.3 Gigawatts, that is, 470,300 Megawatts, is only for land areas, excluding therefore, the considerable potential in Pacific, Atlantic and Gulf offshore areas.

IMPACT OF DIFFERENT ASSUMPTIONS   In such an assessment of national potential output at saturation, certain assumptions have a great impact on the resulting numbers. It is clear, for instance, what a drastic effect the slope requirements have upon the numbers of machines in the High regime, much of which is mountainous.

The matter of packing density, which follows from the estimated spacing required between machines is critical. Figure 7–15 shows how the factor can affect land areas

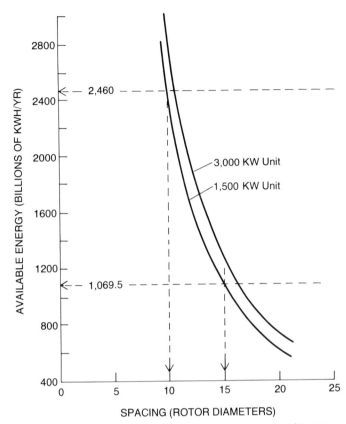

Fig. 7–16 Effects of dynamo spacing and dynamo capacity upon potential wind energy output in the conterminous states at saturation.

Table 7–10 Total Wind Energy in Each Regime and Available Wind Energy at Saturation (Electric Utilities Only)

WIND REGIME	Total Land Area in Regime (Sq. Miles × 1,000)	Total Energy in Regime (Billions KWh/Yr.)	Available Area for WECS (Sq. Miles × 1,000)	Energy Output (Billions KWh/Yr.)	Required Number of 1,500 KW Units	Installed Electrical Capacity (Gigawatts)	Efficiency Factor (Millions KWh/mi²)	Proportion of Total Capacity
Low	1,104	4,168	156.6	591.8	204,800	307.2	3.78	0.65
Moderate	415	3,331	53.9	433.0	99,800	149.7	8.03	0.32
High	55	827	3.0	44.7	8,960	13.4	14.90	0.03
TOTAL	1,574	8,326	213.5	1,069.5	313,500	470.3	5.01[1]	

[1]Mean.

Source: General Electric. *Wind Energy Mission Analysis, Final Report,* February 18, 1977.

required for certain generating capacities; and, logically, the effect on total capacity at saturation is profound. Figure 7–16 shows how spacing of ten times rotor diameter could boost the national total at saturation from 1,069.5 to 2,460 billion of Kilowatt hours per year. Installing machines of 3,000 instead of 1,500-Kilowatt rating has a much smaller impact.

Not demonstrated here are the effects of choosing *land use factors* other than those noted in Table 7–8.

IMPLEMENTATION RATES AND THE IMPACT ON NATIONAL UTILITY NEEDS

Figure 7–17 shows, for the electric utility sector only, three possible futures for the use of wind power, in which the numbers of 1,500-Kilowatt units installed increases at rates consistent with Rapid, Medium, or Slow implementation cases.[6] The land area of all favorable regimes is theoretically saturated by the year 2015, 2025, and 2035 according to the Rapid, Medium, and Slow cases respectively.

The estimated generating capacity implied by the numbers of machines, and how that capacity relates to national needs is dealt with in Fig. 7–18. Wind generating capacity *at saturation*, 470 Gigawatts (taken from Table 7–10), would be just under one-half of the country's estimated needs in the year 1990; but a substantial number of wind machines cannot be installed that soon. By the year 2000 at Rapid implementation approximately 250 Gigawatts of wind generating capacity might be installed, by which time the country's electrical needs, increasing at between 5 and 6 percent, would have reached 1,650 Gigawatts. At most, therefore, wind power could contribute 15 percent of the projected 1,650-Gigawatt need. If the Nation's total generating need were to remain constant at the 1980 level of just over 600 Gigawatts, the 470-Gigawatt capacity of wind machines could account for over two-thirds, assuming saturation.

More relevant than installed generating capacity is the *expected output* from electrical plants, and in this regard, wind power from the machines assumed in the foregoing analysis is less productive than most other generating installations. Fuel-burning plants put out roughly 45 percent of their rated capacity, nuclear plants about 40 percent, and hydropower plants about 50 percent; but the capacity factor for the foregoing wind machines is only 22 to 38 percent in Low to High wind regimes respectively (Table 7–7). The expected annual output from all wind machines in the United States *at saturation* is 1,069.5 billion Kilowatt-hours (Table 7–10) while the national need in the year 1980 is 2,769 billion Kilowatt-hours (U.S. Department of the Interior, December, 1975). Apparently, wind machines could provide roughly 38 percent of the needed 1980 output if all favorable land regimes were saturated according to the General Electric assumptions. An even larger contribution of wind power can be anticipated if any of the following are realized:

1) more machines are packed per square mile by virtue of new designs.
2) the offshore potential is used, especially in New England.

Fig. 7–17   Numbers of 1,500 KW wind dynamos in total land area, according to three different implementation rates. Saturation of available areas is reached around the year 2015, in the case of rapid implementation.

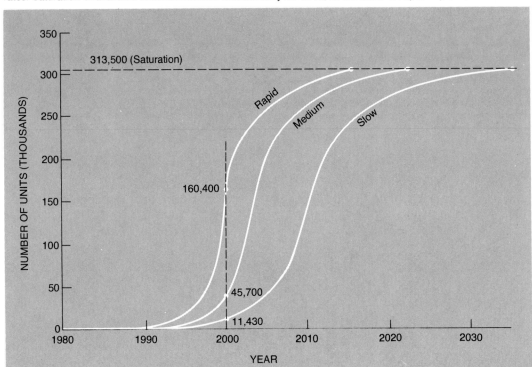

3) the needed output is reduced through conservation, and by using electricity only for those tasks that cannot be accomplished by any other means.

ECONOMIC FEASIBILITY AND NUMBERS OF MACHINES PRODUCED

General Electric uses the concept of *break-even costs*, that is, the dollar cost per installed Kilowatt of wind-generating capacity that would make wind power economical for a utility company, assuming 1975 conditions of fuel costs and electric rates as represented by four regions of the country surveyed (General Electric, *Final Report*, p. 5–16).

Included in any consideration of the installed costs is the matter of how per unit costs may be expected to drop with mass production, the rate of drop in per unit cost being expressed in "learning curves" (Fig. 7–19). The 0.90 learning curve applies if doubling of cumulative amounts produced leads to per unit costs that are 90 percent of the pre-doubling cost. Similarly, if doubling of cumulative production reduces per unit costs to only 85

percent of their former level the falling costs are described by the 0.85 learning curve.

In Fig. 7–20, falling costs according to both 0.90 and 0.85 learning curves determine the *numbers of units required* in order to reach break-even cost. Because of the greater capacity factors in higher wind regimes, break-even costs can be higher (see table in Fig. 7–20) and the required numbers of units for break-even correspondingly lower. The logarithmic vertical scale reveals the numbers of units required, and the dates noted at the head of each bar show the years in which those numbers of units might be installed according to the Rapid implementation case only. Thus, in the High wind regime the units required would be reached by the year 1988 or 1993, depending on the learning curve that is chosen; in the Moderate regime break-even would be achieved in 1991 or in 2000; and in the Low regime breakeven is reached in 1998 only by virtue of the more favorable learning curve. At the 0.90 learning curve, per unit costs do not fall rapidly enough, so the required number of units far exceeds saturation for the Low regime, and break-even is not achieved there with the assumptions of Fig. 7–20.

EFFECT OF RISING FUEL COSTS

As with application of solar energy to space heating, economic feasibility of using wind power depends not only on falling costs of the system, but also upon the costs of fuels which would be burned as an alternative. As fuel costs rise, the wind systems in competition can tolerate higher and higher installation costs (see break-even costs in Table 7–11) and it follows that the number of units mass-produced in order to reach those costs is smaller. Table 7–11 shows how those required numbers of units vary in the three regimes for three different fuel price escalation assumptions. Apparently a rise in fuel prices is a very potent factor, because as fuel price doubles (up by 100 percent) the break-even costs rise by a factor of 1.55. A much more dramatic effect is seen in the numbers of units required to reach break-even costs. As fuel prices double, the number of units required drops by a factor of 18, using the 0.90 learning curve, and by a factor of 6.5 using the 0.85 learning curve.

POSSIBLE FUEL SAVINGS

On the basis of electrical generating capacity growing in accordance with a specified implementation case, it is possible to estimate the amounts of fuel saved as wind generation displaces generation by fuel-burning plants. Using the Rapid implementation case as an example (Fig. 7–18), growing cumulative amounts of fuels saved to the year 2,000 may be projected as in Fig. 7–21, based on estimates by G.E. of the actual fuel-burning generators which would be supplanted by wind in various regions of the country. Estimated amounts of fuels that may be saved by the year 2000 according to the Rapid implementation case, are shown in Fig. 7–22.[7]

Fig. 7-18 Estimated U.S. generating needs and installed utility wind generating capacity according to three implementation rates.

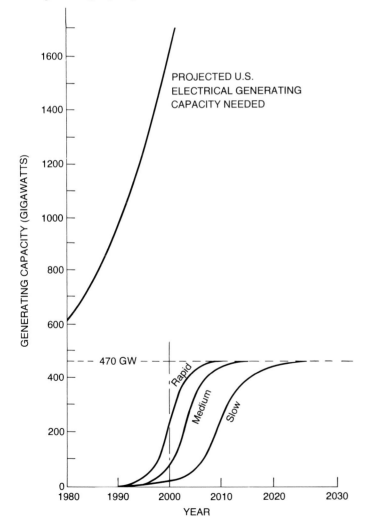

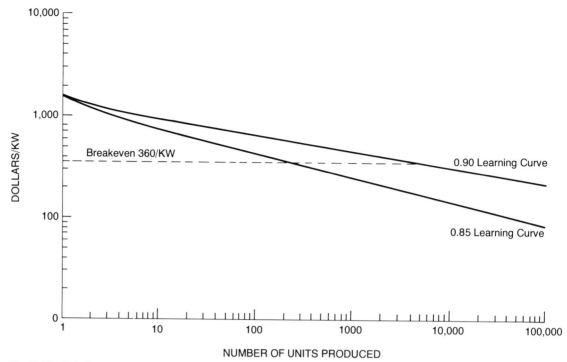

Fig. 7–19 Relationship between costs of a single 1,500 KW wind unit and the number of units produced, according to two learning curves in High Wind regime.

REGIONAL DIFFERENCES IN WIND IMPLEMENTATION

One projection completed for the federal government estimates the contributions of various solar-related generating methods: solar radiation (thermal and photovoltaic methods) wind, ocean thermal gradients, and biomass (see METREK, 1978). An overview of electrical output expected from these renewable sources is presented in the final chapter: here, only the wind electrical output is shown for comparison with General Electric's forecast.

The METREK study, working with assumptions based on the National Energy Plan of 1976, foresees a national wind generating capacity of 63 Gigawatts by the year 2000 and 143 Gigawatts by the year 2020, which is consistent with G.E.'s Medium implementation rate (Fig. 7–18). Regional contributions to the total, as seen by METREK for 2020, are rather surprising in light of the more and less favorable wind regimes in the country as outlined earlier. Figure 7–23 shows the largest output would be in the South Atlantic region: presumably some offshore wind in the northern part of the region would be utilized, along with sites in the Appalachian highlands. The other two leading regions, East North Central and Middle Atlantic, are not regions with favorable winds, except for lake shores in the first region. They must owe their large projected output to very high demand for power in these regions and to the fact that other renewable energy sources do not compete successfully. The annual output for the country, 2.26 Quads, is more than twice that of solar-thermal electrical generation projected in the same analysis.

Table 7–11 Fuel Prices and the Break-Even Point for Utility Wind Generators

A Numbers of Wind Generating Units Needed to Reach Break-even Costs

LEARNING CURVE	WIND REGIME	NUMBERS OF UNITS NEEDED TO REACH BREAK-EVEN IF FUELS INCREASE RATE IS:		
		0%	50%	100%
0.90	High	5,832	1,536	320
	Moderate	53,519	14,022	2,934
	Low	8,596,479[1]	2,261,986[1]	470,938[1]
0.85	High	180	76	28
	Moderate	753	316	115
	Low	20,232	8,515	3,080

B Break-even costs

WIND REGIME	BREAK-EVEN COSTS ($ PER KW) IF FUEL PRICE INCREASE IS:		
	0%	50%	100%
High	360	441	560
Moderate	310	380	482
Low	200	245	311

[1]These exceed the saturation number in Low regime.
Source: General Electric, *Final Report*, Feb. 18, 1977

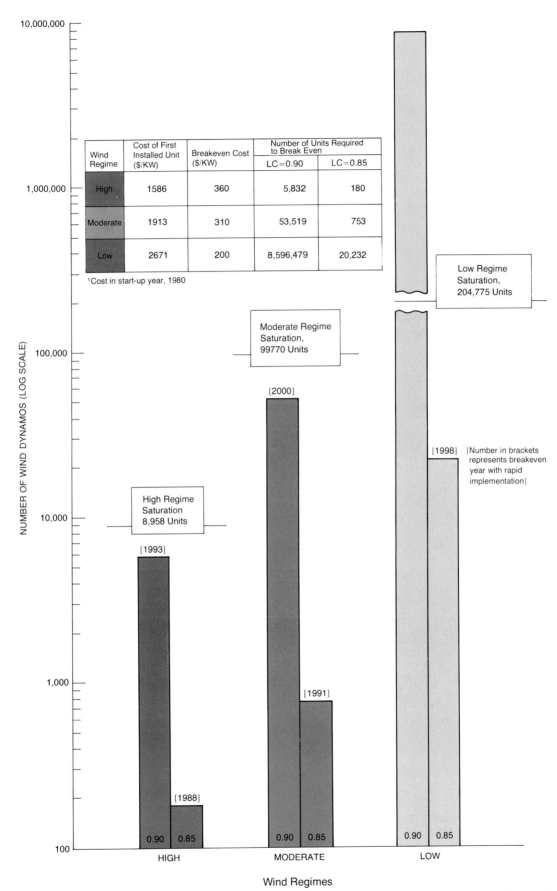

Wind Regime	Cost of First Installed Unit ($/KW)	Breakeven Cost ($/KW)	Number of Units Required to Break Even	
			LC=0.90	LC=0.85
High	1586	360	5,832	180
Moderate	1913	310	53,519	753
Low	2671	200	8,596,479	20,232

¹Cost in start-up year, 1980

Low Regime Saturation, 204,775 Units

Moderate Regime Saturation, 99770 Units

[2000]

[Number in brackets represents breakeven year with rapid implementation]

[1998]

High Regime Saturation 8,958 Units

[1993]

[1991]

[1988]

NUMBER OF WIND DYNAMOS (LOG SCALE)

10,000,000

1,000,000

100,000

10,000

1,000

100

0.90   0.85   0.90   0.85   0.90   0.85

HIGH   MODERATE   LOW

Wind Regimes

Fig. 7–20 Numbers of wind dynamos needed to reach breakeven costs in three wind regimes, showing breakeven year in rapid implementation case. Left bar assumes 0.90 learning curve, while right bar assumes greater cost reductions of 0.85 learning curve.

VARIOUS OTHER APPLICATIONS

While the foregoing analysis has focussed on utility application, and has therefore depended in its geographical aspect upon the rotor requirements for the 1,500-Kilowatt dynamo chosen by G. E., there are other applications of wind-generated electricity which have in common an independence from utility networks, specifically in residential, industrial, agricultural, and remote community uses.

General Electric has chosen wind machines suitable for these applications for each of the three wind regimes and has specified rotor diameters—all of which is con-

sidered not essential here since the geographic dimension of extractable wind power is clearly expressed in the analysis of utility power. A comprehensive summary of various applications showing numbers of units and the resulting installed capacity and annual output values for the whole country is provided in Table 7–12: the two parts of the table are *for saturation,* in the first case, and *for the year 2000* at Rapid implementation rates in the second case.

On the basis of market potential it was determined that the tabulated numbers of units would be installed. In both the "saturation" and the "year 2000" cases, the numbers of residential machines are about three times the number of those run by utilities; but the annual energy output from utility machines is about five times that of all the residential machines. The other applications are relatively small, agriculture being conspicuously larger than the single industry chosen (paper mills) or the remote community application.

Recent Developments

The federal Energy Research and Development Administration (ERDA) now part of the Department of Energy (D.O.E.) has since 1975 administered the Federal Wind Energy Program whose objective is to accelerate development and utilization of both large and small wind systems. Applications being studied involve *small systems* for farm and rural uses, *100-Kilowatt scale systems*

Fig. 7–21   Estimated cumulative fuel savings, in billions of barrels of oil equivalent, from wind dynamo additions to U. S. utilities. Rapid implementation case.

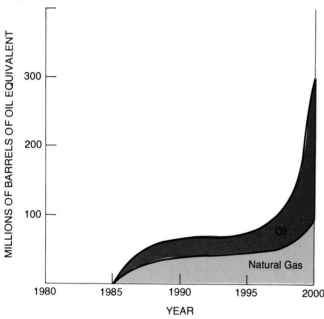

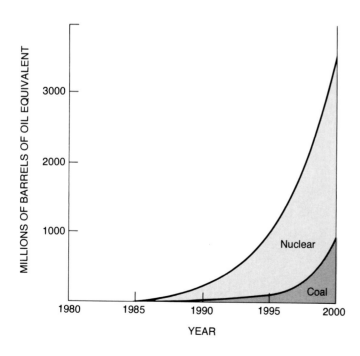

Fig. 7–22   Estimated annual fuel savings in the year 2000 due to wind dynamo additions to U. S. utilities. Rapid implementation case.

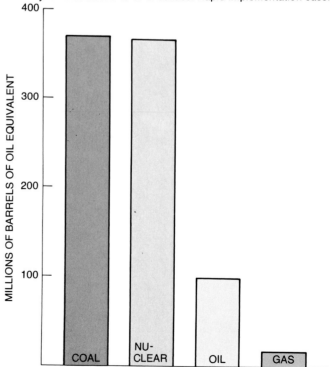

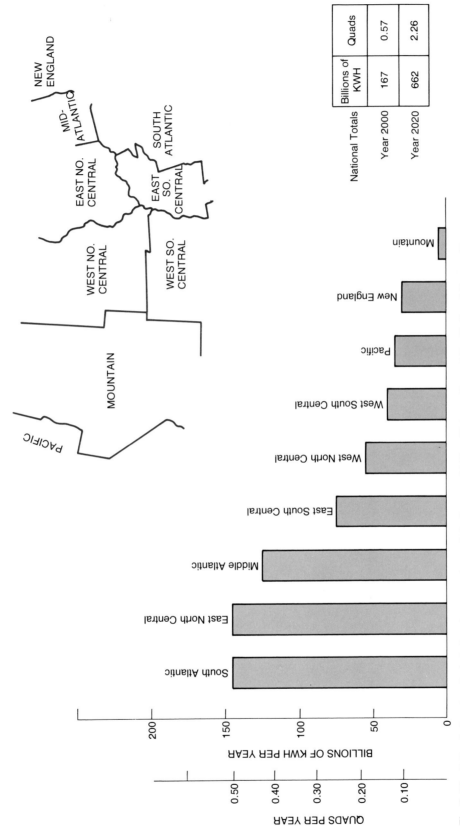

National Totals

	Billions of KWH	Quads
Year 2000	167	0.57
Year 2020	662	2.26

Fig. 7-23 Estimated regional wind-electric output (annual) in the year 2020, with incentives of the National Energy Plan.

The first of the utility-scale wind turbines sponsored by the federal government was this machine installed in 1977 near Sandusky, Ohio. With blade diameter of 125 feet, this turbine is rated at 100 Kilowatts in winds of 18 miles per hour. *Courtesy Department of Energy.*

Table 7–12 (Part 1) Potential at Saturation for Various Applications

APPLICATION	Number of Units	Size of Units (KW)	Installed Capacity (Giga-watts)	Annual Energy Output (Billions KWh)
Electric utility	313,503	1,500	470.3	1,069.5
Residential	9,300,000	10	93.0	207.0
Paper industry	1,000	1,500	1.9	2.9
Agriculture	780,000	35	27.0	43.0
Remote communities	1,000	1,500	1.9	2.9
TOTAL	10,395,503		594.1	1,325.3

for irrigation and small industry uses, *Megawatt-scale systems* for utility uses, and *Multi-unit systems* which are arrays of Megawatt-scale systems employed for a substantial proportion of a utility's generating capacity.

Small systems are being studied now at the ERDA test center at Rocky Flats, west of Denver, Colorado, and numerous studies are being conducted on novel conversion devices, system technology, wind characteristics, and other aspects of the program (see Federal Wind Energy Program: *Summary Report,* January 1, 1978).

The most exciting developments are in the Megawatt-scale systems for utility application as envisioned in the foregoing analysis of national potential. Large machines are being developed in a series with three broad categories, Mod-Zero, Mod-One, and Mod-Two. The Mod-Zero, rated at 100 Kilowatts, and with a blade diameter of 125 feet, has operated at the Plumbrook site near Sandusky, Ohio since 1976. Three machines termed Mod-Zero-A, with 125-foot blades but 200-Kilowatt generators, constitute the next phase: One at Clayton, New Mexico was completed early in 1978; the second, at Culebra Island, Puerto Rico, was completed in July, 1978; and the third, at Block Island, was under construction in 1978 (see Fig. 7–24).

The third phase currently entails a single Mod-One machine, with blade diameter of 200 feet and a rating of 2,000 Kilowatts (2 Megawatts) completed in July, 1979 near Boone, North Carolina.[8] With a diameter of 200 feet, this machine will be the largest ever built, exceeding by 25 feet the diameter of the well-known Smith-Putnam machine which generated utility power near Rutland, Vermont for a few years in the 1940s.

The fourth phase involves a Mod-Two machine now visualized as 300 feet in diameter and rated at 2,500 Kilowatts (2.5 Megawatts) in a relatively light wind regime at a site not yet chosen. Figure 7–24 shows the seventeen sites, selected by ERDA, from sixty-five proposed, for the large machines to follow the Plumbrook Mod-Zero initiative. The candidate sites were selected to

A single large wind turbine capable of supplying five to 15 percent of Block Island's electricity was dedicated in June, 1979. According to the Department of Energy, a commercial wind turbine of this size would save Block Island $30,000 a year in fuel costs. *Photo by Dick Peabody. Courtesy Department of Energy.*

Table 7–12 (Part 2)   Potential in Year 2000 for Various Applications—Rapid Implementation

APPLICATION	Number of Units	Size of Units (KW)	Installed Capacity (Gigawatts)	Annual Energy Output (Billions KWh)
Electric utility	160,000	1,500	240.6	581.3
Residential	4,600,000	10	46.0	103.0
Paper industry	1,000	1,500	1.9	2.9
Agriculture	360,000	35	13.0	21.0
Remote communities	1,000	1,500	.8	1.45
TOTAL	5,121,900		302.3	709.65

Source: General Electric, *Wind Energy Mission Analysis, Final Report,* Feb. 18, 1977.

Fig. 7–24 The seventeen candidate field test sites selected by ERDA for large wind turbine generators, showing sites now occupied by Mod-Zero, Mod-Zero-A, and Mod-One machines.

Table 7–13 Candidate Sites Chosen from Those Proposed for Large Machines, to Follow the Plumbrook, Ohio, Mod-Zero Experiment

SITE	ORGANIZATION	SITE SELECTED FOR
Clayton, N. M.	Town of Clayton	Mod-Zero-A
Block Island, R. I.	Block Island Power Co.	Mod-Zero-A
Culebra Island, P. R.	Puerto Rico Water Resources	Mod-Zero-A
Boone, N. C.	Blue Ridge Electric Membership Corp.	Mod-One
Amarillo, Texas	Southwestern Public Service Co.	
Augsburger Mt., Wash.	Bonneville Power Administration	
Boardman, Ore.	Portland General Electric Co.	
Cold Bay, Alaska	Alaska Bussell Electric Co.	
Huron, S. D.	East River Power Cooperative	
Kaena Point, Hawaii	Hawaiian Electric Co.	
Ludington, Mich.	MERRA	
Point Arena, Calif.	Pacific Gas and Electric Co.	
Montauk Point, N. Y.	Long Island Lighting Co.	
Holyoke, Mass.	City of Holyoke Gas and Electric	
Kingsley Dam, Neb.	Central Nebraska Public Power and Irrigation	
Russell, Kan.	City of Russell	
San Gorgonio Pass, Calif.	Southern California Edison Co.	

Source: ERDA *Press Release No. 77–97*, June 8, 1977.

include a variety of climates, power distribution networks, and companies able to provide useful data to ERDA. The utility companies which proposed the sites finally selected are listed in Table 7–13.

BEYOND THE FEDERAL PROGRAM

In addition to the numerous small machines being installed by individual pioneers across the country, there are at least two sizable ventures of larger scale that are not supported by federal programs, but rather by electric companies.

One is the installation of a 132-foot diameter machine in 1979 near Palm Springs, California, for Southern California Edison. This is a Schachle, three-bladed horizontal axis machine rated at 3 Megawatts in winds of 40 mph. If successful, this may be followed by a series of 300 machines for the same utility company, an order that would initiate mass production of wind machines, and bring about the all-important economies of large-scale production.

Another California venture is on the horizon. U.S. Windpower Inc. of Massachusetts has an agreement with the California State Department of Water Resources to build twenty machines, rated at 20 Kilowatts each, in a valley about 80 miles south of San Francisco. The machines are expected to produce by June, 1981 (*Boston Globe*, April 21, 1980).

An artist's rendering of the "largest windmill in history" to be designed and built by Boeing Engineering and Construction for a test program being conducted by the Department of Energy and the National Aeronautics and Space Administration. Designed to operate at sites with a mean wind speed of 14 miles per hour, it will drive a generator to produce 2500 kilowatts of electricity. *Courtesy Department of Energy.*

NOTES

[1] Indispensible to subsequent wind power studies, the data on magnetic tape is now known as "the SANDIA tape."

[2] Studies by SANDIA, Lockheed-California, and Batelle all include maps of seasonal as well as annual average Power Available.

[3] The factor for height effects dramatically increases calculated Power Extractable for blades of large radius: the value of $(R/10)^{3/7}$ is only 0.428 for a blade radius of 10 meters, but is 1.811 for a radius of 40 meters.

[4] More understanding of capacity factor can be gained from records of Rapid City, South Dakota, considered by G. E. to represent the Low regime, with design speed of 12 mph. Records for that station (Reed, 1975, p. 133) show winds exceeding 12 mph for 64 percent of the year at the surface. If conversion efficiency of 0.40 is assumed, the resulting "composite" capacity factor would be (0.64) (0.40) = 0.256, which is consistent with the factor used by G. E. for the Low regime (Table 7–7).

[5] These are consistent with rotor diameters of 218, 278, and 331 feet, respectively, which are larger than those in Table 7–6 and thus lead to a more conservative estimate of number of machines per square mile.

[6] Shape of curves is based on the Fischer-Prye method of predicting market penetration of new products (G. E. *Final Report*, P. 5–1).

[7] Early P.M. peak load is supplied by oil-fired or gas-fired turbines at many installations. In areas where winds peak during those same hours, wind dynamos will be able to provide that peak power and save oil and gas. Elsewhere, wind generators will be used for base load, so great savings of coal and nuclear fuels may be expected.

[8] According to George Tennyson, Program Manager, D.O.E., Wind Systems Branch, personal communication, Aug., 1978. In earlier reports on the Wind Energy Program this machine was to be rated at 1,500 Kilowatts, just like the machines assumed in the foregoing national assessment of potential.

Hydroelectric Power

Hydroelectric power is produced from the natural movement or flow of water, as in ocean tides, or in rivers. The power of water can be harnessed by taking advantage of its natural fall from one level to another, or by creating this fall with dams. The falling water can do work in water wheels or modern turbines, and can then be converted into other energy forms, such as electrical.

The hydrologic cycle explains how solar energy is converted to mechanical energy. Water is continually evaporated from the oceans by the sun. The wind then transports it, sometimes thousands of miles, from tropical to mid-latitude regions. This moisture is then dropped as precipitation on both oceans and continents. Finally, the water runs off the land back to sea level. In this way, solar radiation is converted into the kinetic, or mechanical energy, of a flowing fluid.

The use of falling or swiftly running water to produce energy dates back to ancient Chinese and Egyptian civilizations, where the falling water was used to turn water wheels. During the early period of the Industrial Revolution, mechanical work was generated by this type of water power in both North America and Europe.

Prior to electricity, power from water wheels could be transported only for very short distances, and its use was severely restricted. The development of the electrical generator and transmission lines greatly increased the ability to transmit power over a wider area. In spite of these advances, however, hydroelectricity is still generally consumed in the region of its production. Its limited use is due partly to problems of transmission and partly to competition from cheaper forms of energy.

Hydroelectricity is considered to be a renewable resource, although in some circumstances, a dam may over a long period of time (100 years) collect enough silt to fill the reservoir and render it useless. Furthermore, there are several environmental factors that condition the use of hydropower.

First of all, reservoirs flood large areas which could be used for other purposes, and which often include irreplaceable wild lands. Fish migrations are restricted by dams, and fish health is affected by changes in water temperature and introduction of excess nitrogen at spillways. Water supplies may be affected by reservoir evaporation and by water release practices (Healy, 1974). Although considerable potential remains unused in the country and more hydroelectric plants will be built in the years ahead, environmental pressures may restrict the growth of large hydropower developments. The potential which is contained in small streams and at existing dams not utilized for hydro power is considerable, and its development would have a minimal impact on the environment.

Hoover Dam and Lake Mead on the Colorado River, Arizona and Nevada.

World Hydroelectric Power Resources

In 1975, the total world hydroelectric power potential was estimated at more than 212 million Megawatts. Only 13.6 percent of this total has been developed (Department of Interior, 1976). In 1974, Asia contained over 30 percent of the total world capacity, followed by Africa and North America with 19 and 15 percent respectively (Fig. 8–1). The most intensive development of hydroelectricity has occurred in Europe and North America. These two regions have developed 48 and 27 percent of their respective capacities and over 60 percent of the world's developed capacity. On the other hand, Asia, Africa, and South America have over 60 percent of the world's total hydroelectric capacity, but only 20 percent of the total developed capacity. The Soviet Union holds about 10 percent of both world total and world developed capacity.

Presently, hydroelectric power production accounts for only 5 percent of total world energy production. Between now and 1990, projected production will increase at an average rate of 1.7 percent per year (Department of Interior, 1976). If the projection holds true, hydroelectric power's share of total energy production will have decreased to about 4.5 percent by 1990. In the foreseeable future, hydropower seems destined to play a minor role in the development of energy resources in most parts of the world. In some underdeveloped countries, however, where fuel resources are scarce, and hydroelectric capacity great, hydropower will likely be of increasing importance.

United States Hydroelectric Power Resources

There are two basic types of hydroelectric power plants in the United States: *conventional* and *pumped storage*. Conventional hydroelectric plants produce power from natural stream flow, and are limited by the hydrostatic head created by the dam and reservoir (Fig. 8–2). Pumped storage installations use the same generation principles as conventional systems, but produce power during peak load periods by using water pumped to an upper reservoir during off-peak periods.

Very often the pumped storage facility is supplemental to a fuel-burning power plant. To provide for peak loads, one alternative is to build a larger power plant. This is undesirable because the expensive generators would be fully used only during the time of peak loads. The pumped storage alternative makes that expansion unnecessary by storing excess energy (as raised water) when it is not needed, then delivering it as electrical energy at the time of peak demand (Fig. 8–2). The pumping facility is sometimes located at an existing hydroelectric dam. When it is not associated with any natural hydrostatic head, it is considered a "pure" pumped storage installation. Figure 8–2 illustrates the "pure" type. When demand is low, an electric motor pumps water from lower pond to upper pond. When demand is high, the pump serves as a turbine, and the motor serves as a generator. Pumped storage generating capacity should *not* be regarded as an additional energy source to be added to that of the conventional hydroelectric potential because

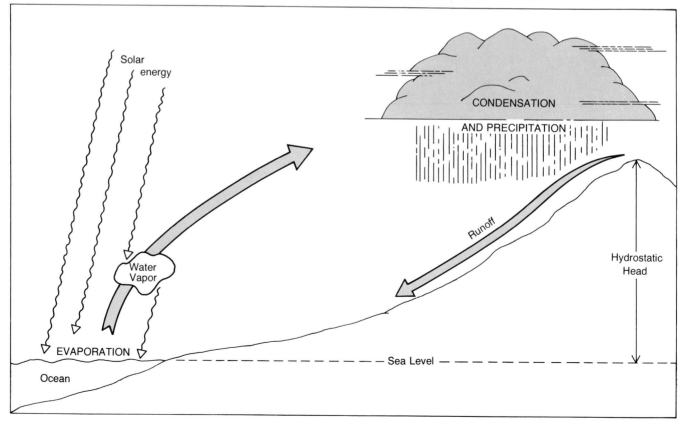

Simplified view of the hydrologic cycle in which water is raised by solar energy and can perform work as it falls back to sea level.

pumped storage depends on external electrical energy for the pumping. Pumped storage is included in this chapter for the sake of a more complete review of hydroelectric generating capacity in the country.

TRENDS IN DEVELOPMENT OF HYDROELECTRIC POWER RESOURCES

Installed conventional capacity in the United States has grown from 3.7 million Kilowatts in 1921 to over 59 million Kilowatts in 1978 (Federal Power Commission, 1976 and 1978). Since 1960, however, the rate of growth has slowed (Fig. 8–3). The declining rate of growth in conventional hydroelectric capacity can be explained in part by the increasing concern over the ecological and social consequences of proposed hydropower projects. In addition, hydroelectric plants have not been able to compete profitably with steam-electric generating plants because the initial capital outlay is much greater. In the past, conventional hydroelectric facilities have provided as much as a third of the Nation's electric generating capacity, but today they generate only 11 percent of the total. Future growth in hydro capacity will be slow unless shortage of fossil fuels, increase in fuel prices, and the high cost of developing other energy sources make hydroelectric power a relatively attractive alternative.

The development of pumped storage hydropower facilities did not begin until the early 1960s. Pumped storage capacity has increased from .4 million Kilowatts in 1964 to 9.7 million Kilowatts in 1978 (Fig. 8–3), and is expected to grow more rapidly in future years, especially in areas with large peak load demands.

CONVENTIONAL HYDROELECTRIC POWER CAPACITY IN THE UNITED STATES

In 1978, the potential conventional hydroelectric power capacity (at dams or sites with a capacity of 5,000

Fig. 8–1 Seven World regions: developed and undeveloped hydroelectric capacity.

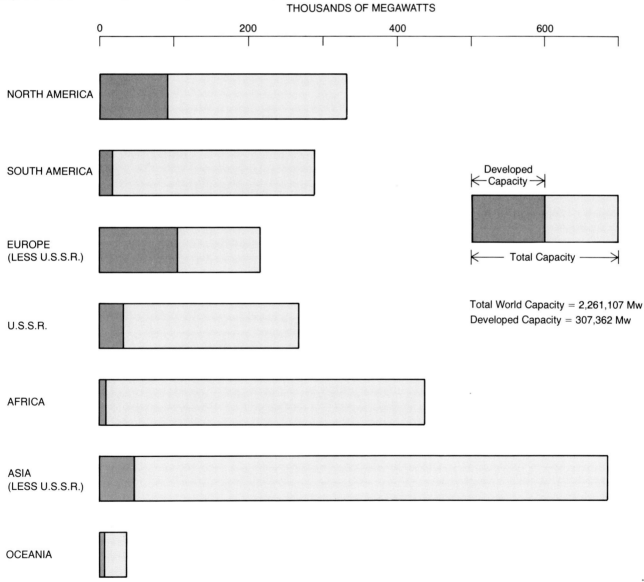

THOUSANDS OF MEGAWATTS

Total World Capacity = 2,261,107 Mw
Developed Capacity = 307,362 Mw

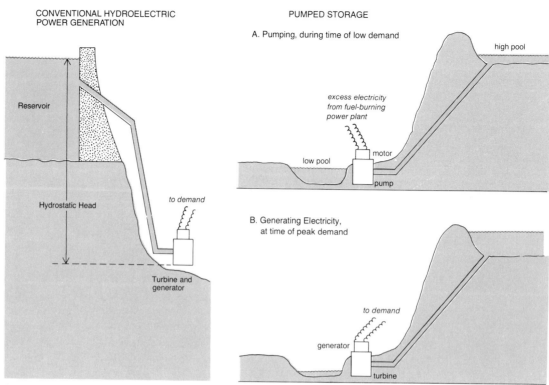

CONVENTIONAL HYDROELECTRIC
POWER GENERATION

PUMPED STORAGE

A. Pumping, during time of low demand

Reservoir

Hydrostatic Head

to demand

Turbine and
generator

*excess electricity
from fuel-burning
power plant*

high pool

low pool

motor

pump

B. Generating Electricity,
at time of peak demand

to demand

generator

turbine

Fig. 8–2   The pumped storage plant, showing two phases of operation.

Fig. 8–3   Growth in U.S. developed hydroelectric capacity, 1921–1978, showing conventional and pumped storage types.

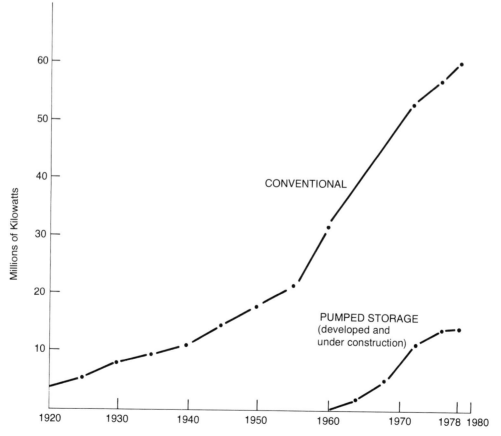

Millions of Kilowatts

60

50

40

30

20

10

CONVENTIONAL

PUMPED STORAGE
(developed and
under construction)

1920   1930   1940   1950   1960   1970   1978   1980

Kilowatts or greater) was estimated at 168.6 million Kilowatts. Fifty-nine million Kilowatts has been developed and the remaining 109 million is classified as undeveloped potential (Federal Energy Regulatory Commission, 1978). Of this 59 million Kilowatts (59 thousand Megawatts), 2,800 Megawatts is located at the Grand Coulee plant on the Columbia River in Washington state. This largest facility in the country is now being expanded. Other conventional hydroelectric plants with installed generating capacity greater than 1,500 Megawatts are the John Day plant on the Columbia River in Oregon, and the R. Moses plant on the Niagara River in New York.

Although it appears that nearly two-thirds of the total conventional hydroelectric power capacity remains undeveloped, these statistics can be misleading. Taking into consideration economic, social, and environmental factors, the number of acceptable undeveloped sites on large rivers is probably small. In addition, nearly 30 percent of the undeveloped potential is located in Alaska where demand is very small.

The developed hydroelectric power capacity represents roughly 11 percent of the Nation's total generating capacity. In the unlikely event of *full* development of undeveloped potential (including Alaska), the 169 million Kilowatts would constitute 26 percent of the national generating capacity needed in the year 1980 and only 9 percent of the capacity needed in the year 2000, according to one projection (Dupree and Corsentino, p. 35). It is clear that unless the national need for generating capacity is reduced, conventional hydroelectric power alone cannot provide a large part of it.

DISTRIBUTION OF HYDROELECTRIC POWER CAPACITY BY DRAINAGE, BY ECONOMIC REGION, AND BY STATE

The Federal Power Commission has delineated fourteen major drainage basins[1] for the purpose of assessing the regional distribution of hydroelectric resources (Fig. 8–4). The North Pacific stands above the rest of these basins in both developed and undeveloped hydroelectric capacity (Fig. 8–4 and Table 8–1), having over 24 million Kilowatts of developed hydroelectricity and an estimated 36.5 million Kilowatts of undeveloped capacity. The South Pacific drainage is the only other area with more than 10 percent of the total developed capacity. Alaska ranks second to the North Pacific drainage in undeveloped capacity, housing more than 30 percent of the total. Together these two drainages have almost two-thirds of the Nation's undeveloped capacity (Fig. 8–4). In the eastern

Table 8–1 Conventional Hydroelectric Power in the United States, by Major Drainages—1978

MAJOR DRAINAGE	CAPACITY (Megawatts)			PROPORTION DEVELOPED (Percent)
	Developed	Undeveloped	Total Potential	
North Atlantic	2,710.2	7,385.8	10,096.0	27
South Atlantic	3,409.2	3,905.9	7,315.1	47
Eastern Gulf	2,370.5	2,606.7	4,977.2	48
Ohio	5,229.5	3,657.1	8,886.6	58
Great Lakes-St. Lawrence	4,602.9	1,313.4	5,316.3	75
Hudson Bay	13.6	0	13.6	100
Upper Mississippi	616.0	563.2	1,173.2	52
Missouri	3,397.1	4,631.1	8,028.3	42
Lower Mississippi	2,044.5	2,038.7	4,083.2	50
Western Gulf	458.2	978.8	1,437.0	32
Colorado	3,206.3	2,191.5	5,397.8	59
Great Basin	353.7	300.1	653.8	54
North Pacific	24,320.7	36,517.5	60,838.2	40
South Pacific	6,936.9	9,856.5	16,793.3	41
Alaska	127.2	33,459.8	33,587.0	0.5
Hawaii	18.1	35.0	53.1	34
UNITED STATES TOTAL	59,208.7	109,441.2	168,649.9	35

Source: Federal Energy and Regulatory Commission, 1978.

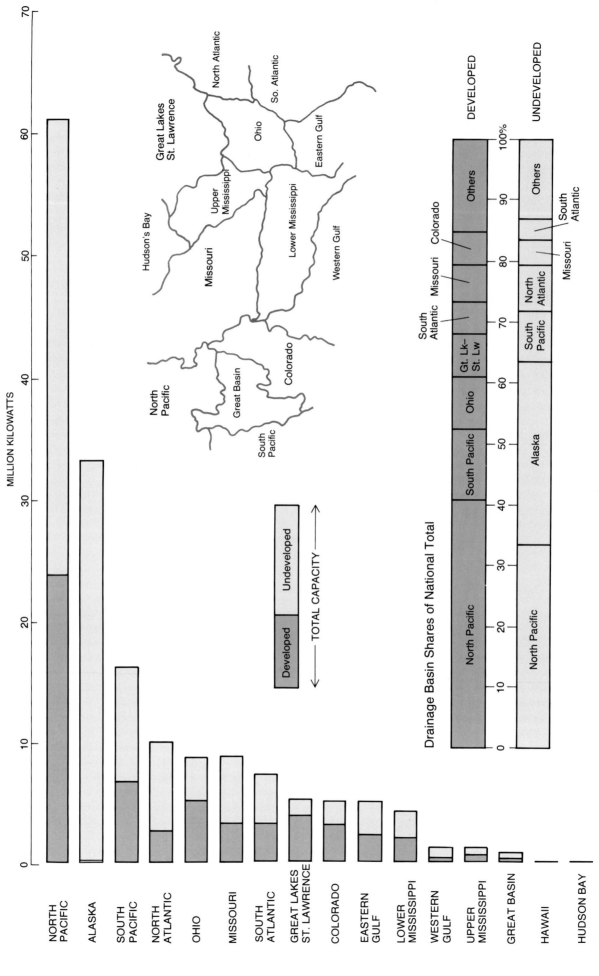

Fig. 8–4 Major drainage basins: developed and undeveloped conventional hydroelectric capacity as of January, 1978.

United States, the Ohio and Great Lakes-St. Lawrence drainage basins are the most intensively developed.

To allow comparison with population, economic and resource data, hydroelectric resources have also been mapped according to eleven geographic regions as delineated by the Bureau of the Census (Fig. 8–5). The Pacific region dominates all others with over 37 percent of the total hydroelectric potential and 49 percent of the developed potential (Fig. 8–5 and Table 8–2). The Mountain and South Atlantic regions rank second and third in terms of developed potential with 12 and 10 percent respectively of the United States' total. The Pacific, Mountain, and Alaska regions hold over 70 percent of both the total developed and potential capacity.

Five states, Washington, Alaska, California, Oregon, and Idaho, have almost two-thirds of the total potential hydroelectric capacity of the country (Fig. 8–6 and Table 8–3). These same states have approximately 50 percent of the developed hydropower capacity and 70 percent of the undeveloped potential. Washington and California are the most intensively developed states, containing almost 40 percent of the Nation's hydroelectric capacity. Alaska is the least developed with only 127,000 Kilowatts. In the eastern United States, the most intensive hydropower development has occurred in Tennessee and Alabama where over 8 percent of the Nation's developed capacity is located. The principal agency involved in developing hydroelectric power in these two states is the Tennessee Valley Authority (T.V.A.).

CONVENTIONAL HYDROELECTRIC POWER UNDER CONSTRUCTION, PLANNED, OR PROJECTED

In 1976, approximately 170 new hydroelectric plants or additions were under construction, planned, or projected in the United States. When completed, these plants and additions will have an installed capacity of 8.2 million Kilowatts. Planned and projected installations call for a total capacity of 2 and 12 million Kilowatts each (Table 8–4). If all of these new facilities are eventually built, they would add over 22 million Kilowatts to the present United States' hydroelectric capacity.

Over 95 percent of new plants or additions now under construction are located in the Mountain and Pacific regions of the country, Washington and California having approximately 90 percent (Fig. 8–7). In the eastern part of the country construction activity is confined to Alabama, Georgia, Kentucky, North Carolina, and Tennessee.

Approximately 70 percent of the planned capacity[2] to be completed by 1985 will be located in the Mountain and Pacific regions, with Montana, Washington, and Idaho accounting for over 50 percent. Outside of the western United States, the major facilities planned will be located in Georgia and Alabama, holding about one-fourth of the total planned capacity.

Projected hydroelectric plants or additions are defined as potential developments not under construction or included in reports of the Regional Electric Reliability Council. These projected plants have obtained licenses or

Table 8–2 Conventional Hydroelectric Power Capacity in the United States by Economic Region—1978

GEOGRAPHIC DIVISION	CAPACITY (Megawatts)			PROPORTION DEVELOPED (Percent)
	Developed	Undeveloped	Total Potential	
New England	1,497.4	3,354.3	4,851.6	31
Middle Atlantic	4,237.1	4,488.6	8,725.8	47
East North Central	953.7	1,340.3	2,293.9	42
West North Central	2,799.4	2,092.9	4,892.4	57
South Atlantic	5,801.5	7,145.4	12,946.9	45
East South Central	5,552.5	2,897.0	8,449.6	66
West South Central	2,255.1	2,353.9	4,609.0	49
Mountain	7,080.4	18,271.9	25,352.2	28
Pacific	28,886.0	34,002.3	62,888.3	46
Alaska	127.2	33,459.8	33,587.0	0.5
Hawaii	18.1	35.0	53.1	34
UNITED STATES TOTAL	59,208.7	109,441.2	168,649.9	35%

Source: Federal Energy Regulatory Commission, 1978.

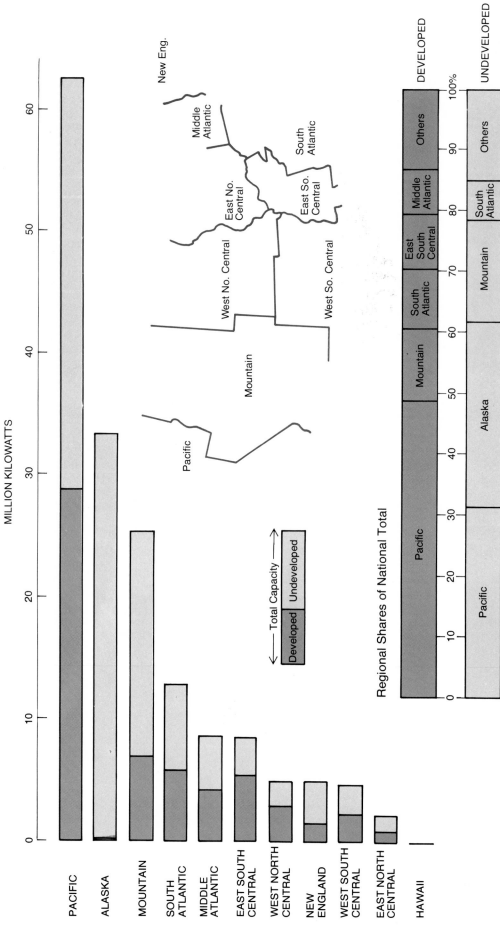

Fig. 8-5   Economic regions: developed and undeveloped conventional hydroelectric capacity as of January, 1978.

Fontana Dam (built 1942 to 1944) is the Tennessee Valley Authority's highest, towering 480 feet. It is the highest concrete dam east of the Rocky Mountains. Fontana is on the Little Tennessee River in North Carolina and borders the Great Smoky Mountains National Park. Power installation: 238,500 kw in three units. *Courtesy Tennessee Valley Authority.*

permit status from the Federal Power Commission and have been authorized or recommended for federal construction, or they have structural provisions for plant additions. The Pacific states, California, Oregon, Washington, and Alaska are scheduled to house almost one-half of the projected plants and additions. Another 17 percent are to be located in the Mountain states. In the eastern and midwestern states, sizeable capacities are projected for Maine, Georgia, and South Dakota.

PUMPED STORAGE HYDROELECTRIC CAPACITY

As of 1978, the total developed pumped storage hydroelectric capacity was 16.0 million Kilowatts (Table 8–5), 75 percent of which was in pure pumped storage plants, and the remainder in combined conventional and pumped storage plants. Between 1972 and 1978 the total pumped storage capacity increased by 6.1 million Kilowatts, or nearly 150 percent. By the end of this century it is possible that the pumped storage hydroelectric capacity will be four times that of 1978.

DISTRIBUTION OF PUMPED STORAGE HYDROELECTRIC CAPACITY BY STATE   In 1978, eighteen states held all of the developed pumped storage capacity of the United States (Fig. 8–8). Almost three-fourths of the total developed capacity is located in states east of the Mississippi where the population is dense and the demand for peak load power is large. The largest pumped storage hydroelectric facility in the United States is located at

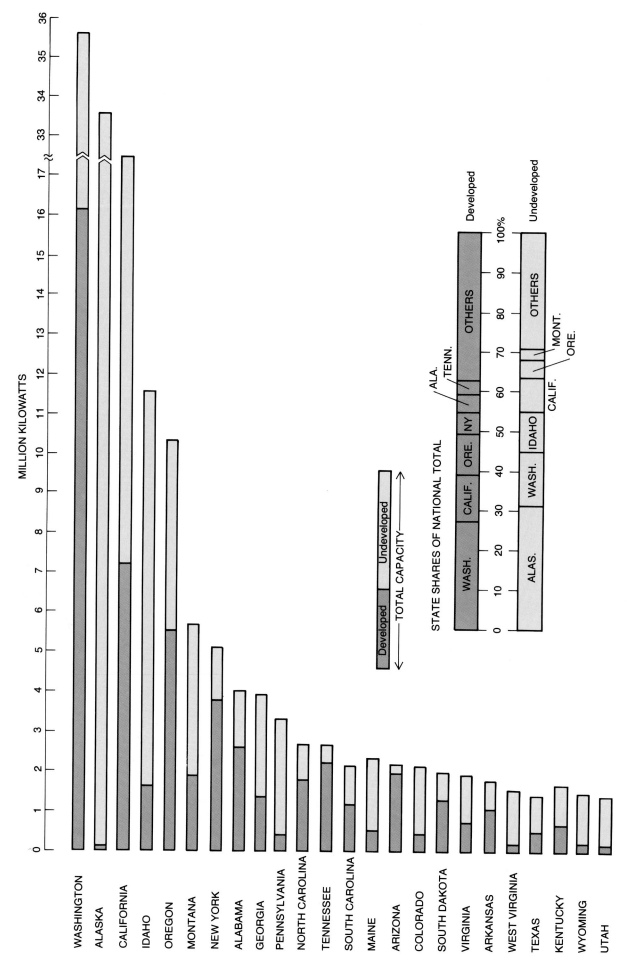

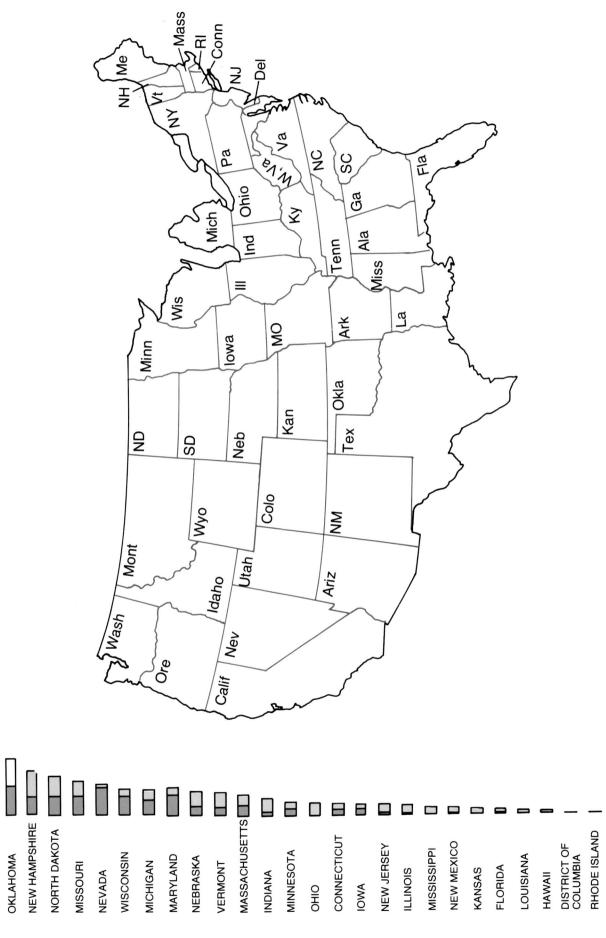

Fig. 8–6 Developed and undeveloped conventional hydroelectric capacity as of January, 1978.

Ludington, Michigan. This facility has a capacity of almost 2,000 Megawatts. Two other pumped storage plants in the eastern United States have capacities of 1000 Megawatts: the Northfield Mountain Plant on the Connecticut River in Massachusetts and the Blenheim-Gilboa Plant on the Schohaire Creek in upstate New York.

Five states, Massachusetts, Michigan, New York, California, and Pennsylvania, have more than 70 percent of the Nation's developed pumped storage capacity. Michigan, alone, has slightly more than 20 percent of the total. In the western part of the United States, California is the leader in developed pumped storage capacity with over 12 percent of the total. In the Midwest, Oklahoma and Missouri have a combined pumped storage capacity equal to 8 percent of the country's total.

PUMPED STORAGE CAPACITY UNDER CONSTRUCTION, PLANNED, AND PROJECTED Presently, there is an estimated 27 million Kilowatts of pumped storage hydroelectric capacity under construction, planned, or projected (Table 8–6). Plants and additions under construction account for 15 percent of the total, and planned and projected installations, 23 and 61 percent respectively. Should all of these installations be developed within the next two decades, the total pumped storage capacity would amount to over 37 million Kilowatts.

The total capacity of these three categories of future potential—under construction, planned and projected —is shown for the affected states in Fig. 8–8, without distinguishing the three components. It appears that six states—Virginia, California, New York, North Carolina,

Fig. 8–7 Economic regions: conventional hydroelectric capacity developed as of January, 1979, and development anticipated in 1976.

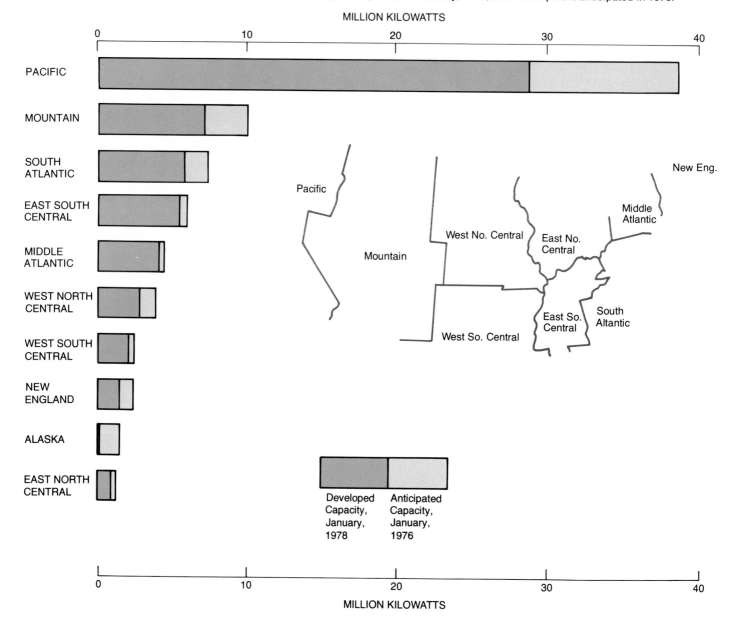

MILLION KILOWATTS

Developed Capacity, January, 1978

Anticipated Capacity, January, 1976

MILLION KILOWATTS

Table 8-3 Conventional Hydroelectric Capacity in the United States by State and Economic Region—1978

REGION	CAPACITY (Megawatts)			PROPORTION DEVELOPED (Percent)
	Developed	Undeveloped	Total Potential	
New England				
Maine	526.5	1,764.4	2,290.9	23
New Hampshire	418.0	685.1	1,103.1	38
Vermont	198.1	375.3	573.4	35
Massachusetts	224.2	279.8	504.0	45
Rhode Island	1.5	0	1.5	10
Connecticut	129.1	169.1	298.2	43
Middle Atlantic				
New York	3,785.2	1,366.6	5,151.8	73
New Jersey	3.6	266.0	269.6	1
Pennsylvania	448.4	2,847.0	3,295.4	14
East North Central				
Ohio	2.5	300.1	302.5	1
Indiana	93.5	316.2	409.7	23
Illinois	33.1	214.2	247.3	13
Michigan	377.7	283.2	660.9	57
Wisconsin	446.9	226.5	673.5	66
West North Central				
Minnesota	170.0	136.3	306.3	56
Iowa	133.6	146.4	280.0	48
Missouri	436.2	346.0	809.2	54
North Dakota	430.0	407.0	837.0	51
South Dakota	1,391.7	543.0	1,934.7	72
Nebraska	236.3	348.7	585.0	40
Kansas	1.8	138.5	140.3	1
South Atlantic				
Delaware	0	0	0	0
Maryland	493.7	163.0	656.7	75
District of Columbia	0	0	0	0
Virginia	718.7	1,207.4	1,926.1	37
West Virginia	207.2	1,341.5	1,548.7	13
North Carolina	1,791.6	903.0	2,694.6	66
South Carolina	1,169.5	954.2	2,123.7	55
Georgia	1,390.7	2,421.2	3,919.9	35
Florida	30.2	47.0	77.2	39
East South Central				
Tennessee[1]	2,186.1	434.5	2,620.6	83
Kentucky[1]	685.6	986.4	1,672.0	41
Alabama	2,591.5	1,267.9	4,003.4	65
Mississippi	0	200.6	200.6	0
West South Central				
Arkansas	1,047.8	727.7	1,775.5	59
Louisiana	0	76.0	76.0	0
Oklahoma	703.4	635.8	1,339.2	53
Texas	504.0	914.4	1,418.4	36

[1]As of January, 1976.

Table 8–3 (Continued)

REGION	CAPACITY (Megawatts)			PROPORTION DEVELOPED (Percent)
	Developed	Undeveloped	Total Potential	
Mountain				
Montana	1,931.8	3,219.3	5,685.1	34
Idaho	1,626.0	9,912.4	11,538.4	14
Wyoming	220.2	1,260.5	1,480.7	15
Colorado	433.4	1,626.4	2,059.8	21
New Mexico	24.3	154.2	178.5	14
Arizona	1,963.7	181.2	2,144.9	91
Utah	199.3	1,206.2	1,405.5	14
Nevada	681.8	41.8	723.6	94
Pacific				
Washington	16,159.5	14,533.3	30,692.8	45
Oregon	5,503.4	4,805.4	10,308.8	53
California	7,223.1	8,966.9	16,190.0	45
Alaska	127.2	33,459.8	33,587.0	0.4
Hawaii	18.2	35.0	53.2	34
UNITED STATES TOTAL	59,208.7	109,441.2	168,649.9	35

Source: Federal Energy Regulatory Commission, 1978.

TVA's Kentucky Dam (built 1938 to 1944) is particularly effective in regulating floods on the lower Ohio and Mississippi Rivers. Located on the Tennessee River near its mouth in Kentucky, the dam creates a reservoir 184 miles long with a volume of more than 6,000,000 acre-feet, two-thirds of which is useful controlled storage. The hydroelectric plant has a generating capacity of 175,000 kilowatts. *Courtesy Tennessee Valley Authority.*

Tennessee, and South Carolina—account for over 70 percent of the anticipated pumped capacity. Tennessee has the largest capacity *under construction* with almost 36 percent of the total. California ranks second with almost 20 percent, followed by South Carolina, Arizona, and Georgia. As a region, the southeastern states hold nearly 60 percent of the capacity of pumped storage plants and additions under construction.

On completion, *planned* pumped storage installations would add 6.4 million Kilowatts to the total United States' capacity.[3] Virginia is scheduled to have 33 percent of the total planned capacity, while nearly 63 percent is to be equally divided among the states of California, Nebraska, New York, and West Virginia.

The total capacity of *projected* pumped storage plants and additions is 16.8 million Kilowatts. Virginia is to house 34 percent of the projected capacity followed by North Carolina and New York with 15 and 12 percent re-

Table 8–4 Conventional Hydroelectric Plants or Additions under Construction, Planned, or Projected—1976 (in megawatts)

STATE	CAPACITY			
	Under Construction	Planned	Projected	Total
Alabama	135.0	140.0		275.0
Alaska	2.1		1,366.9	1,369.0
Arizona	3.0		3.0	6.0
Arkansas			125.0	125.0
California	830.9	135.0	986.0	1,951.9
Colorado	28.0	9.0	128.0	165.0
Georgia	108.0	350.0	650.0	1,108.0
Idaho		317.0	1,145.0	1,462.0
Kentucky	61.0		178.5	239.5
Maine		44.7	830.0	874.7
Missouri	27.0			27.0
Montana	324.0	470.0	575.0	1,369.0
Nebraska			33.3	33.3
New Jersey			70.0	70.0
North Carolina	13.5			13.5
North Dakota			212.0	212.0
Ohio		40.0	180.5	220.5
Oklahoma			25.0	25.0
Oregon	49.0		1,328.0	1,377.0
Pennsylvania			120.0	120.0
South Carolina			84.0	84.0
South Dakota			650.0	650.0
Tennessee	10.4		3.0	13.4
Texas		32.0	213.0	245.0
Utah		133.5		133.5
Virginia			338.0	338.0
Washington	6,522.2	357.5	2,261.2	9,140.9
West Virginia			180.0	180.0
Wisconsin	28.6			28.6
Wyoming		12.6	150.2	162.8
Miscellaneous sites smaller than 25MW			385.1	385.1
TOTAL	8,242.7	2,013.4	12,108.0	22,364.1

Source: Federal Power Commission, 1976.

spectively. Almost 80 percent is located in states east of the Mississippi River and 13 percent in the West Coast states of California and Washington.

RIVER SEGMENTS PRECLUDED FROM HYDROELECTRIC DEVELOPMENT

Federal law precludes development of hydroelectric power in many parts of the United States. Power development in national parks and national monuments, for example, is prohibited unless specifically authorized by Congress (Federal Power Commission, 1976). Furthermore, the Wild and Scenic Rivers Act of 1968 and subsequent amendments designate segments of sixteen rivers as parts of a National Wild and Scenic Rivers system thereby excluding hydroelectric development (Fig. 8–9). The undeveloped conventional hydroelectric capacity of these river segments amounts to 4.7 million Kilowatts at thirty-one sites and 1.4 million Kilowatts of

known pumped storage potential (Federal Power Commission, 1976).[4]

In recent years, special congressional acts have been passed which prohibit or restrict the licensing authority of the Federal Power Commission on certain river segments. These stream segments (Fig. 8–9) have a potential capacity of 4.4 million Kilowatts of conventional hydroelectricity and 1.6 million Kilowatts of pumped storage potential; thus, the total conventional hydroelectric potential withdrawn from development amounts to 9.1 million Kilowatts.

In addition to river segments excluded by legislation, the Wild and Scenic Rivers Act also lists certain rivers to be *studied* for inclusion into the national system. Generally, these river segments are under moratorium for hydroelectric development pending completion of the studies (Fig. 8–9). As of January 1976, there were forty-nine designated study rivers with an estimated total conventional hydroelectric capacity of 11.2 million Kilowatts. This capacity *was included* in the foregoing maps and tabulations of conventional hydroelectric power potential.

Finally, the Wilderness Act of 1964 bars incompatible land and water uses in areas designated by Congress as

Table 8–5 Developed and Anticipated Pumped Storage Capacity by State (in megawatts)

STATE	Developed 1978	Anticipated 1976	Total Capacity
	CAPACITY		
Virginia	132.0	7,894.0	8,026.0
California	1,462.4	2,904.5	4,366.9
New York	1,240.0	3,000.0	4,240.0
North Carolina	59.5	2,500.0	2,559.5
South Carolina	612.0	1,518.4	2,130.4
Michigan	1,978.8	0	1,978.8
Massachusetts	1,446.0	0	1,446.0
Tennessee	0	1,530.0	1,530.0
Pennsylvania	1,196.0	300.0	1,496.0
New Jersey	386.8	1,000.0	1,386.8
West Virginia	0	1,350.0	1,350.0
Washington	100.0	1,200.0	1,300.0
Georgia	250.0	1,191.0	1,441.0
Nebraska	0	1,016.8	1,016.8
Texas	11.2	730.0	741.2
Arizona	148.5	505.4	653.9
Missouri	408.0	191.0	599.0
Colorado	308.5	200.0	508.5
Oklahoma	260.0	260.0	520.0
Vermont	0	80.0	80.0
Arkansas	28.0	0	28.0
Connecticut	7.0	0	7.0
UNITED STATES TOTAL	10,034.7	27,371.1	37,405.8

Sources: Federal Power Commission, 1976; and Federal Energy Regulatory Commission, 1978.

Table 8–6 Pumped Storage Hydroelectric Plants or Additions Under Construction, Planned or Projected—1976 (in megawatts)

STATE	Under Construction	Planned	Projected	Total
	CAPACITY			
Arizona	505.4	0	0	505.4
California	850	1,050	1,235	3,135
Colorado	100	100	0	200
Georgia	466	0	725	1,191
Missouri	191	0	0	191
Nebraska	0	1,000	16.8	1,016.8
New Jersey	0	0	1,000	1,000
New York	0	1,000	2,000	3,000
North Carolina	0	0	2,500	2,500
Oklahoma	0	0	260	260
Pennsylvania	0	0	300	300
South Carolina	518.4	0	1,000	1,518.4
Tennessee	1,530	0	0	1,530
Texas	0	0	730	730
Vermont	0	0	80	80
Virginia	104	2,100	5,690	7,894
Washington	0	200	1,000	1,200
West Virginia	0	1,000	350	1,350
TOTAL	4,264.8	6,450	16,886.8	27,601.6

Source: Federal Power Commission, 1976

"wilderness areas." The President, however, may authorize certain activities in these areas deemed to be in the national interest. Again, some power sites in wilderness areas were included in the foregoing assessment of conventional hydroelectric power.

UNITED STATES' HYDROELECTRIC POTENTIAL AT EXISTING DAMS

During the past decade, rising fuel costs, rapidly escalating construction costs for thermal generating facilities, and increased public concern over the safety of nuclear plants, have not only made it necessary to seek energy alternatives, but also to re-examine old energy options. In his energy message of April 20, 1977, President Carter indicated that new or additional hydroelectric generating capacity at existing dams could be installed at a cost less than the cost of building new coal or nuclear capacity (Army Corps of Engineers, July 1977). Furthermore, at that time it was thought that the installation of additional generating capacity at existing hydropower dams could add as much as 14,000 Megawatts to the Nation's generating potential.

Fig. 8–8 Pumped storage hydroelectric generating capacity in 22 states—amounts anticipated in 1976, and amounts developed as of January, 1978.

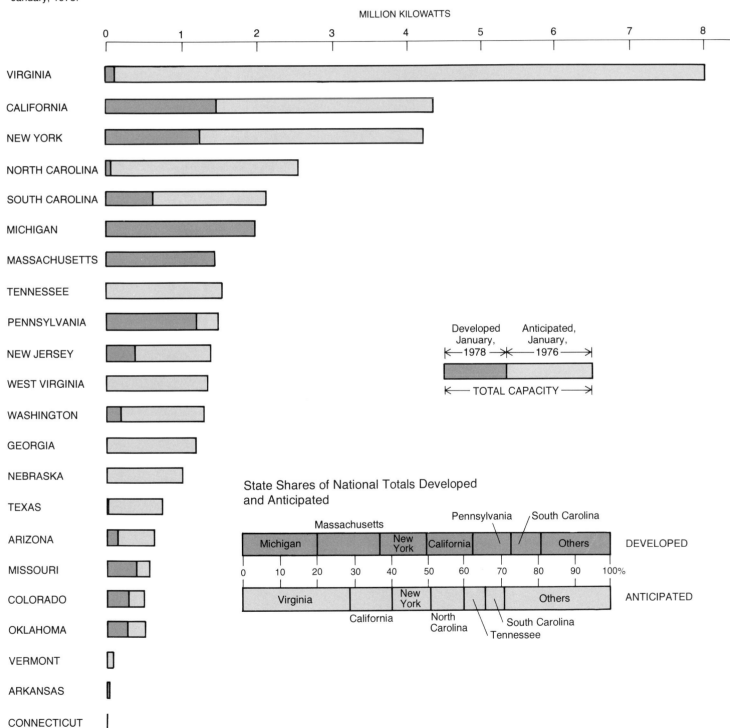

Fig. 8–9 Areas not available for hydroelectric development: river segments covered by the Wild and Scenic Rivers Act and special act.

Fig. 8–10 Amounts of undeveloped generating capacity that can be installed at existing dams (by economic region).

The Army Corps of Engineers is currently conducting a detailed assessment of the hydroelectric resources of the United States, including the potential for new or additional generating capacity at existing dams. In July 1979, the Corps published a series of detailed regional reports providing a preliminary estimate of additional hydroelectric potential at existing dams. The discussion to follow uses these preliminary estimates of *additional capacity at existing dams* together with the Federal Energy Regulatory Commission's *undeveloped capacity* (used in the previous section) to reassess the total undeveloped hydroelectric potential of the United States. In the following analysis, *total undeveloped capacity* is the undeveloped capacity reported by the Federal Energy Regulatory Commission *plus* the additional capacity at existing dams reported by the Army Corps of Engineers.

STATE AND REGIONAL ASSESSMENT   Preliminary estimates indicate that over 5,400 existing dams have the potential of producing an additional 94,000 Megawatts of capacity. This additional capacity could be achieved in two ways: 1) by installing turbines at dams where none

previously existed or, 2) adding additional turbines at dams where they already exist. Added to the 109,441 Megawatts at undeveloped sites reported by the Federal Energy Regulatory Commission, the total undeveloped hydroelectric potential of the United States is over 204,000 Megawatts (Table 8–7). It is interesting that the estimate of additional capacity at existing dams is over 1.5 times the developed capacity. Moreover, 46 percent of the total undeveloped hydroelectric capacity of the country is estimated to be at existing dams.

Three states—Washington, Oregon, and New York —have 42 percent of the total hydroelectric potential at existing dams; 34 percent of the potential at undeveloped sites; and 32 percent of the total undeveloped capacity (Table 8–7). The Pacific, East South Central, and Middle Atlantic have 68 percent of the additional capacity at existing dams. The Pacific region alone is estimated to have over 33,000 Megawatts of additional capacity. The East South Central and Middle Atlantic regions have 16,794 and 14,299 Megawatts respectively (Fig. 8–10). These same three regions plus the Mountain

Fig. 8–11   Undeveloped conventional hydroelectric generating potential, showing proportion at existing dams (by economic region).

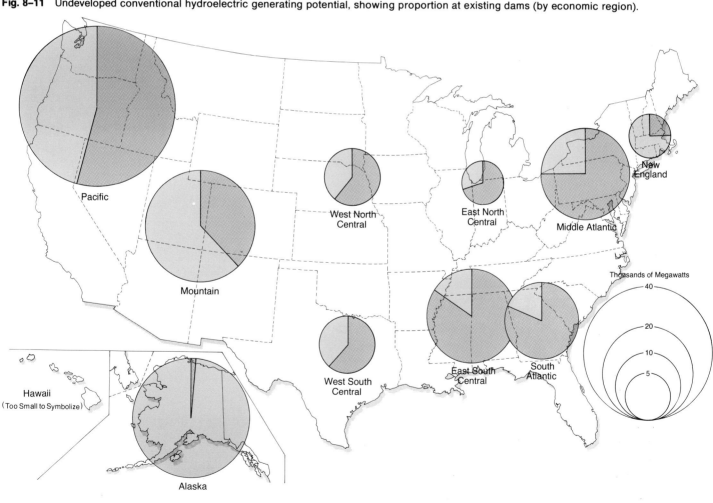

Table 8–7 Summary of Conventional Hydroelectric Capacity by State and Economic Region, Showing Amounts Developed, Undeveloped (Undammed Sites) and Additional Capacity Possible at Existing Dams (in megawatts)

REGION AND STATE	Developed as of Jan. 1, 1978[1]	Additional Capacity At Existing Dams[2]	Additional as a Proportion of Developed	Undeveloped As of Jan. 1, 1978[1]	Total Undeveloped (Cols. 2 & 4)	Additional as a Proportion of Total Undeveloped	Total Potential Capacity
New England							
Connecticut	129.0	88.0	.68	169.7	345.7	.25	474.7
Maine	526.5	369.0	.70	1,764.4	2,133.4	.17	2,659.9
Massachusetts	224.2	115.0	.51	279.8	394.8	.29	619.0
New Hampshire	418.0	261.0	.62	685.1	946.1	.28	1,364.1
Rhode Island	1.5	40.0	26.66	0.0	40.0	1.00	41.5
Vermont	198.1	134.0	.68	375.3	509.3	.26	707.4
TOTAL	1,497.3	1,007.0	.67	3,274.3	4,369.3	.23	5,866.6
Middle Atlantic							
New York	3,785.2	12,458.0	3.29	1,366.6	13,824.6	.90	17,609.8
New Jersey	3.5	40.0	11.42	266.0	306.0	.13	309.5
Pennsylvania	448.4	1,731.0	3.86	2,847.6	4,578.6	.38	5,027.0
TOTAL	4,237.1	14,229.0	3.36	4,480.2	18,709.2	.76	22,946.3
East North Central							
Ohio	2.5	314.0	125.60	300.0	614.0	.51	616.5
Indiana	93.5	96.0	1.03	316.3	412.3	.23	505.8
Illinois	33.1	730.0	22.05	214.2	944.2	.77	977.3
Michigan	377.7	1,133.0	2.99	283.2	1,416.2	.80	1,793.9
Wisconsin	446.9	812.0	1.82	226.5	1,038.5	.78	1,485.4
TOTAL	953.7	3,085.0	3.24	1,340.2	4,425.2	.70	5,378.9
West North Central							
Iowa	133.6	1,117.0	8.36	146.4	1,263.4	.88	1,397.0
Kansas	1.9	220.0	115.79	138.5	358.5	.61	360.4
Minnesota	170.0	989.0	5.82	136.3	1,125.3	.88	1,295.3
Missouri	436.2	1,368.0	3.16	346.0	1,714.0	.80	2,150.2
Nebraska	236.3	94.0	.39	348.7	442.7	.21	679.0
North Dakota	430.0	324.0	.75	407.0	731.0	.44	1,161.0
South Dakota	1,391.7	420.0	.30	543.0	964.0	.44	2,354.7
TOTAL	2,799.7	4,532.0	1.62	2,065.9	6,597.9	.69	9,397.6
South Atlantic							
Florida	30.2	45.0	1.49	47.0	92.0	.96	122.2
Georgia	1,390.7	406.0	.29	2,421.2	2,827.2	.17	4,217.9
Maryland	493.7	532.0	1.08	163.0	695.0	3.26	1,188.7
North Carolina	1,791.6	653.0	.36	903.0	1,556.0	.72	3,347.6
South Carolina	1,169.5	628.0	.54	954.2	1,582.2	.66	2,751.7
Virginia	718.7	497.0	.69	1,207.5	1,704.5	.41	2,423.2
West Virginia	207.2	2,969.0	14.33	1,341.5	4,310.5	2.21	4,517.7
TOTAL	5,801.6	5,730.0	.99	7,037.4	12,767.4	.81	18,569.0
East South Central							
Alabama	2,591.5	4,121.0	1.59	1,267.9	5,388.9	.77	7,980.4
Kentucky	2,961.0	9,271.0	3.13	1,268.2	10,539.2	.70	13,500.2
Mississippi	0.0	133.0		200.0	333.0	.40	333.0
Tennessee	2,185.9	3,269.0	1.49	434.5	3,703.5	.88	5,889.4
TOTAL	7,738.4	16,794.0	2.17	3,170.6	19,964.6	.84	27,703.0

[1]According to Federal Energy Regulatory Commission.   [2]According to U.S. Army Corps of Engineers, July, 1979.

Table 8–7 (Continued)

REGION AND STATE	Developed as of Jan. 1, 1978[1]	Additional Capacity At Existing Dams[2]	Additional as a Proportion of Developed	Undeveloped As of Jan. 1, 1978[1]	Total Undeveloped (Cols. 2 & 4)	Additional as a Proportion of Total Undeveloped	Total Potential Capacity
West South Central							
Arkansas	1,047.8	2,886.0	2.75	727.7	3,613.7	.80	4,661.5
Louisiana	0.0	291.0		76.0	367.0	.79	367.0
Oklahoma	703.4	1,630.0	2.74	635.8	2,265.8	.72	2,969.2
Texas	503.9	372.0	.74	914.4	1,286.4	.29	1,790.3
TOTAL	2,255.1	5,179.0	2.30	2,353.9	7,532.9	.69	9,788.0
Mountain							
Arizona	1,963.7	156.0	.08	181.2	337.2	.46	2,300.9
Colorado	433.4	1,593.0	3.67	1,626.4	3,219.4	.49	3,652.8
Idaho	1,625.9	5,172.0	3.18	9,912.4	15,084.4	.34	16,710.3
Montana	1,931.8	2,332.0	1.21	3,219.3	5,551.3	.42	7,483.1
Nevada	681.8	46.0	.07	41.8	87.8	.52	769.6
New Mexico	24.3	236.0	11.76	154.2	440.2	.65	464.5
Utah	199.3	348.0	1.75	1,206.2	1,554.2	.22	1,753.5
Wyoming	220.2	487.0	2.21	1,260.5	1,747.5	.28	1,967.7
TOTAL	7,080.4	10,420.0	1.47	17,602.0	28,022.0	.37	35,102.4
Pacific							
California	7,223.1	5,447.0	.75	8,966.9	14,413.9	.38	21,637.0
Oregon	5,503.4	14,190.0	2.58	4,805.4	18,995.4	.75	24,498.8
Washington	16,159.5	13,482.0	.83	19,533.3	33,015.3	.48	44,174.8
TOTAL	28,886.0	33,119.0	1.15	28,305.6	61,424.6	.54	90,310.6
Alaska	127.2	418.0	3.29	33,459.8	33,877.8	.01	67,464.8
Hawaii	18.2	31.0	1.70	35.0	66.0	.47	84.2
UNITED STATES TOTAL	59,208.7	94,636	1.60	109,441.2	204,077.2	.46	263,285.9

region have 63 percent of the hydropower potential at undeveloped sites.

In the East South Central region, 84 percent of the total undeveloped capacity is the additional capacity at existing dams. Undoubtedly, this high ratio is due to the existence of Tennessee Valley Authority projects in the region (Fig. 8–11). Other regions in which the additional capacity at existing dams accounts for more than three-fourths of the total undeveloped capacity are the South Atlantic with 81 percent, and the Middle Atlantic with 76 percent. On the other hand, the New England region has the smallest absolute amount of additional capacity at existing dams, just over 1000 Megawatts, which accounts for only 23 percent of the total undeveloped capacity of the region. In the Pacific Region where over 30 percent of the total United States' undeveloped capacity is located, 54 percent is found at existing dams.

STATE AND REGIONAL ASSESSMENT BY SCALE OF SITES The estimates by the Army Corps of Engineers of existing and additional hydropower at existing dams have been grouped into three categories based on Megawatt capacity. These include *small-scale* (.05 to 15 Megawatt); *intermediate-scale* (15 to 25 Megawatt); and *large-scale* (greater than 25 Megawatt). For the United States, 5 percent of the existing hydropower capacity is at small-scale sites; 2 percent at intermediate-scale sites; and 93 percent at large-scale sites (Table 8–8). The estimates for additional capacity at existing dams are similar, with 6 percent of the total assigned to small-scale sites; 3 percent to intermediate-scale sites; and 91 percent to large-scale sites. Regionally, the amounts of additional capacity at small-scale dam sites varies from 89 percent in New England to 1 percent in the East South Central (Fig. 8–12). New England's additional capacity at small dams accounts for less than 1 percent of the total additional capacity of the United States.

Of the approximately 5400 dams with potential for additional generating capacity, 33 percent are in New Eng-

Nickajack Dam was built by the TVA on the Tennessee River west of Chattanooga. The Nickajack project maintains the unbroken series of mainstream reservoirs for navigation, flood control, and power production on the Tennessee. *Courtesy Tennessee Valley Authority.*

Fig. 8–12 Undeveloped generating potential at existing dams, showing proportions at sites of small, intermediate, and large potential (by economic region).

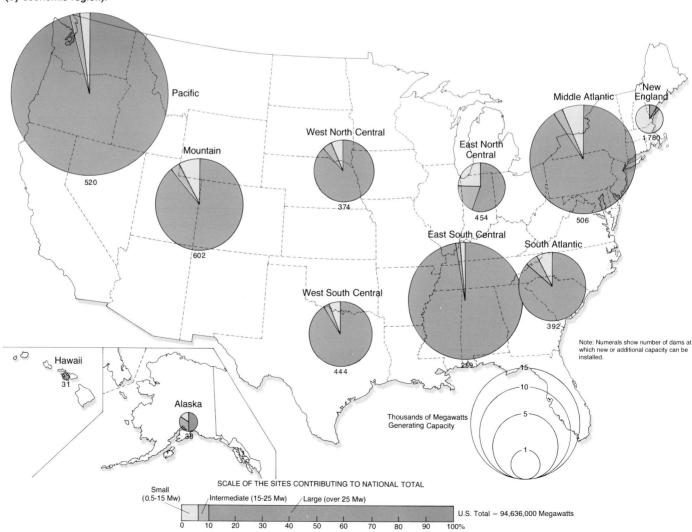

Table 8-8 Installed Generating Capacity at Existing Dams as of 1979, and Potential Additional Capacity

REGION AND STATE	Small (.05–15MW) Existing	Small (.05–15MW) Additional	Intermediate (15–25MW) Existing	Intermediate (15–25MW) Additional	Large (over 25MW) Existing	Large (over 25MW) Additional	CAPACITY TOTALS Existing	CAPACITY TOTALS Additional	NUMBER OF DAMS WITH POTENTIAL FOR ADDITIONAL CAPACITY
New England									
Connecticut	36	88	0	0	68	0	103	88	205
Maine	147	284	58	20	148	64	354	369	472
Massachusetts	73	115	33	0	131	0	237	115	301
New Hampshire	74	238	31	23	281	0	386	261	542
Rhode Island	2	40	0	0	0	0	2	40	105
Vermont	106	134	16	0	74	0	197	134	155
TOTAL	438	899	138	43	702	64	1,279	1,007	1,780
Middle Atlantic									
New York	422	657	216	309	3,103	11,491	3,741	12,458	306
New Jersey	6	21	0	23	0	0	6	40	37
Pennsylvania	0	158	0	107	403	1,466	403	1,731	163
TOTAL	428	836	216	439	3,506	12,957	4,150	14,229	506
East North Central									
Ohio	0	105	0	153	0	56	0	314	77
Indiana	28	58	0	37	0	0	28	76	32
Illinois	100	52	0	145	32	533	132	730	54
Michigan	283	303	52	121	151	709	486	1,133	146
Wisconsin	220	219	112	205	98	387	429	812	145
TOTAL	631	737	164	661	281	1,685	1,079	3,085	454
West North Central									
Minnesota	91	63	0	100	67	825	158	989	114
Iowa	7	28	0	21	128	1,068	135	1,117	38
Missouri	5	22	16	45	577	1,301	598	1,368	42
North Dakota	0	21	0	0	430	303	430	324	45
South Dakota	17	22	0	0	1,483	397	1,500	420	26
Nebraska	16	37	54	21	66	37	136	94	41
Kansas	2	61	0	18	0	141	2	220	68
TOTAL	138	254	70	205	2,751	4,072	2,959	4,531	374
South Atlantic									
Maryland	2	18	0	19	474	496	476	532	20
Virginia	53	94	0	137	633	266	686	497	85
West Virginia	46	18	0	23	102	2,929	148	2,969	36
North Carolina	72	162	103	86	1,762	405	1,937	653	131
South Carolina	88	61	76	54	1,368	513	1,532	628	65
Georgia	20	79	106	23	1,924	304	2,050	406	68
Florida	0	45	0	0	30	0	30	45	17
TOTAL	281	477	285	342	6,293	4,913	6,859	5,730	392

land (Fig. 8–12). No other region has more than 11 percent of the total number. Almost two-thirds of the dams are in states and regions east of the Mississippi.

POTENTIAL FUEL SAVINGS If the estimates of additional hydropower capacity at existing dams prove accurate, the full development of additional hydroelectric potential at existing dams would eliminate the need for 95 1000-Megawatt nuclear or coal-fired electric generating facilities. Considering that hydropower capacity is renewable annually, the development of the full potential at

Table 8–8 (Continued)

REGION AND STATE	SCALE OF SITE POTENTIAL						CAPACITY TOTALS		NUMBER OF DAMS WITH POTENTIAL FOR ADDITIONAL CAPACITY
	Small (.05–15MW)		Intermediate (15–25MW)		Large (over 25MW)				
	Existing	Additional	Existing	Additional	Existing	Additional	Existing	Additional	
East South Central									
Tennessee	11	47	39	80	2,046	3,142	2,096	3,269	49
Kentucky	0	64	0	48	636	9,159	636	9,271	84
Alabama	2	70	0	41	2,269	4,010	2,271	4,121	73
Mississippi	0	20	0	16	0	97	0	133	53
TOTAL	13	201	39	185	4,951	16,408	5,003	16,794	259
West South Central									
Arkansas	11	51	0	67	1,069	2,768	1,080	2,886	105
Louisiana	0	38	0	0	81	253	81	291	23
Oklahoma	0	49	0	87	1,029	1,494	1,029	1,630	115
Texas	52	165	45	22	225	185	321	372	201
TOTAL	63	303	45	176	2,404	4,700	2,511	5,179	44
Mountain									
Montana	29	140	17	43	2,372	2,148	2,418	2,332	88
Idaho	131	140	16	101	2,301	4,931	2,448	5,172	109
Wyoming	19	71	56	63	152	352	227	487	65
Colorado	49	229	22	39	330	1,325	401	1,593	173
New Mexico	0	55	24	24	0	207	24	286	31
Arizona	32	34	0	0	1,374	122	1,406	156	30
Utah	52	135	0	66	138	147	190	348	84
Nevada	9	28	0	18	668	0	677	46	22
TOTAL	321	832	135	354	7,335	9,232	7,791	10,420	602
Pacific									
Washington	157	185	46	130	17,172	13,167	17,374	13,482	124
Oregon	105	231	157	349	6,591	13,609	6,853	14,170	130
California	298	365	171	242	7,167	4,840	7,636	5,447	266
TOTAL	560	781	374	721	30,930	31,616	31,863	33,099	520
Alaska	37	86	15	120	77	212	129	418	38
Hawaii	19	12	0	19	0	0	19	31	12
UNITED STATES TOTAL	2,957	5,455	1,517	3,320	59,230	85,859	63,702	94,636	4,981

Note: Some of these dams may have generating installations that can be enlarged, and some may have none.

Source: U.S. Army Corps of Engineers, July 1979.

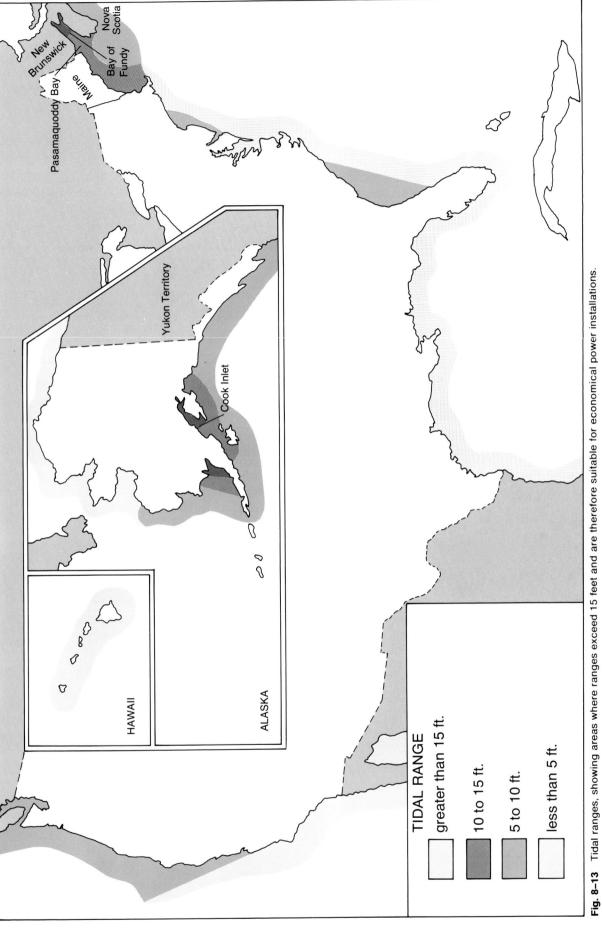

Fig. 8–13 Tidal ranges, showing areas where ranges exceed 15 feet and are therefore suitable for economical power installations.

existing dams could save the United States the equivalent of 578,000 tons of uranium (U_3O_8) or 8.5 billion tons of coal over a thirty-year period.[5] Moreover, there would be minimal environmental disruption, and no air and water pollution.

Tidal Power

Although the use of ocean tides to produce power dates to eleventh-century England, only in recent times has it been seriously considered as an alternative energy resource of significant potential (Steinhart, 1974). Tidal energy has its origin in the kinetic, or mechanical, energy of the orbiting and rotating earth, moon, and sun. As these bodies move relative to one another, the waters of the earth rise and fall in response to changing gravitational forces. The energy of ocean tides can be harnessed simply by damming a partially enclosed bay or inlet and installing an electric generating station. Tidal electric power differs from hydroelectric power in that it depends on alternate filling and emptying of dammed basins rather than on the one-directional flow of a river. The amount of energy available at a site depends on the range of tides and the area of the enclosed basin.

The maximum rate of energy flow at the most promising sites in North America, South America, and Europe has been calculated at 6.4×10^7 Kilowatts (Steinhart, 1974). If 20 percent conversion to electricity is assumed, tidal electric potential is approximately 13×10^6 Kilowatts. Table 8–9 lists characteristics of some of the better sites in the world, ranked according to maximum possible energy. Presently there are only two tidal electric plants—the 240,000-Kilowatt La Rance estuary development in France and the experimental 400-Kilowatt Kislayaguba plant in the U.S.S.R. (Federal Power Commission, 1976).

Figure 8–13 shows tidal ranges on the coasts of the United States. Two areas—Cook Inlet in Alaska, and the Bay of Fundy—have tidal ranges that exceed the 15 feet considered minimal for economical power development. Bay of Fundy tidal ranges, reaching over 35 feet, are the

greatest in the world; and, although Passamaquoddy Bay does not enjoy the highest of these, this site on the Maine-New Brunswick boundary has been studied on and off by Canadian and United States governments for almost fifty years. The most recent plans call for an initial Studies to date have not found the development of this site economically feasible, but further studies are anticipated.

If Cook Inlet is assumed to have the same maximum potential as the Passamaquoddy area, then two large developments may be visualized—each with capacity of 1

Table 8–9 Characteristics of Likely or Producing Tidal Power Sites of the World

SITE	Average Tidal Range (Ft.)	Basin Area (Sq. Mi.)	Maximum Possible Annual Energy (Megawatts)
Minas-Cobequid, Bay of Fundy	35.1	300	19,600
White Sea, U.S.S.R.	18.5	772	14,000
San Jose, Argentina	19.4	289	5,760
Shepody, Bay of Fundy	32.2	45	2,490
Passamaquoddy, U.S.—Canada	18.1	101	1,750
Severn, England	32.2	27	1,490
Mezen Estuary, U.S.S.R.	21.6	54	1,330
La Rance, France	27.6	8.5	343

Source: Healy, Timothy J. *Energy, Electric Power and Man.* (San Francisco: Boyd and Fraser Publishing Company, 1974).

Fig. 8–14 Tidal Energy (La Rance unit).

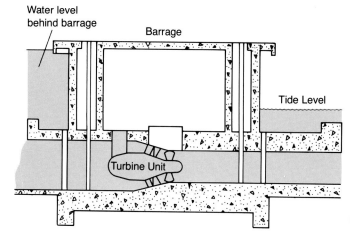

million kilowatts (1,000 Megawatts) and each comparable therefore to a very large conventional steam electric plant. The contribution of tidal power to the Nation's total electrical needs therefore will not be large.

installation of 500,000 Kilowatts capacity and a possible ultimate installation of a million Kilowatts capacity (Federal Power Commission, 1976). Capital costs of the Passamaquoddy project were estimated by the Department of Interior in 1963 at about 750 million dollars with energy costs of about 4 mills per Kilowatt-hour (Healy, 1974).

NOTES

[1] A drainage basin is a large surface area whose waters are drained off into a principal river system, which gives the basin its name.

[2] Planned projects considered here are those included in the reports of April 1, 1976 to the Federal Power Commission by the Regional Electric Reliability Council, for completion by 1985.

[3] The definitions of planned and projected capacities are the same as used in the discussion of conventional hydroelectricity.

[4] The undeveloped hydroelectric power capacity in these river segments was not included in earlier data.

[5] Assumes that a 1000-Megawatt coal-fired electric generating plant uses 3 million tons of coal per year and that a 1000-Megawatt nuclear plant will consume 6,090 tons of U_3O_8 in thirty years.

Ocean Thermal Gradients

- Favorable Areas, and the Resource Potential

- Prospects for Development

- Regional Differences in OTEC Implementation

Each year about 1.5 quadrillion Megawatt-hours of solar energy arrive at the earth's outer atmosphere (Hayes, p. 156). Roughly 47 percent of this reaches the earth's surface—an energy flow that matches the output from 80 million very large electrical generating plants (1,000 Megawatts each) running at full capacity. Because 70 percent of the earth's surface is oceans, and because disproportionate amounts of the solar radiation are received in low latitudes, a great deal of this energy is absorbed by tropical oceans.

One of the results of this energy absorption is that surface waters attain temperatures up to 30 degrees Celsius (85 degrees Fahrenheit). This is definitely low-grade energy: the temperatures are lower than those in a solar collector, and are not adequate for heating a building. At the same time, temperatures in the deep ocean water are about 5 degrees Celsius (40 degrees Fahrenheit) because melting polar icecaps provide a constant supply of cold water that runs under the warm surface water. It is the *contrast in temperatures* (sometimes called ocean thermal gradient) that can be utilized to drive a heat engine in Ocean Thermal Energy Conversion (OTEC).

Efforts to make use of this resource were first made by the French. In 1881, J. D'Arsonval suggested that the temperature difference could be used to run an engine. In the 1920s, Georges Claude persuaded the French government to build several OTEC plants in hopes they would provide inexpensive electrical power to France's tropical colonies. As the overseas empire faded, so did the French interest in the project (Hayes, p. 165). In the United States, OTEC plants were studied seriously in the 1960s. Since 1972, funds for OTEC research and development have been provided by the Solar Energy Program of the Energy Research and Development Administration (ERDA).

Two different types of heat engine are being considered for driving a generator: *open cycle*, and *closed cycle*. In the first type, sea water itself is evaporated at reduced pressure (hence low temperatures) and its vapor drives very large turbine blades. In the closed cycle type, which appears more attractive economically, a fluid with low boiling point, such as ammonia or propane, is evaporated by heat from surface waters, allowed to expand into a turbine, condensed by cool bottom waters, and then returned to the evaporator (Fig. 9–1). The cooling and condensation stage is essential in a closed system to provide lower pressures downstream from the turbine: without lowered pressures there, the vapor would not be

An ocean thermal conversion plant ship; a floating factory which could make 586,000 tons of ammonia a year through the use of inexhaustible solar energy in warm tropical seas. *Courtesy Department of Energy.*

339

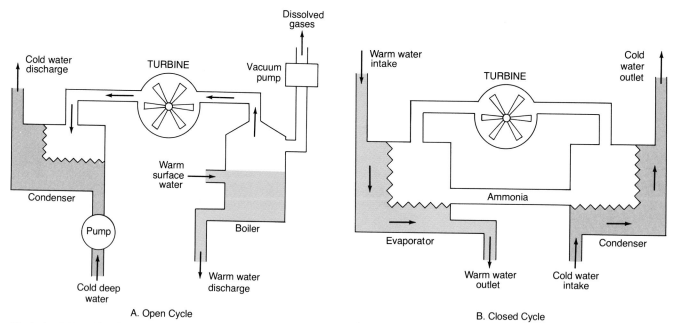

Fig. 9–1 Schematic of closed cycle heat engine using ammonia and open cycle heat engine.

motivated to rush through the turbine. In this kind of system very large volumes of both warm and cool ocean waters must be circulated through heat exchangers. Corrosion of the exchangers and fouling by marine organisms are serious problems to be overcome.

In 1979, a small generating plant of the closed cycle type was established on a barge off the west coast of the island of Hawaii. It uses ammonia as the working fluid, and will generate 50 Kilowatts of electricity. Previous experiments (including the French) have been shore-based, and have used the open cycle. This Hawaiian test, therefore, is important as a first trial of a floating closed cycle OTEC installation. It is funded by the state of Hawaii and by Lockheed Missiles and Space Company.

The volumes of warm and cool ocean waters are very large; and the temperature difference is renewable, since it depends on solar radiation. Nevertheless, it would be possible, by installing too many OTEC plants in one area, to degrade the thermal resource and then have to wait until the original temperature contrasts were re-established.

There are some ecological effects, even in a well-managed program. Disruption of the local environment for marine life is unavoidable. In addition, the cold bottom water is rich in carbon dioxide, and when brought to the surface in large volumes could add substantial amounts of CO_2 to an atmosphere already enriched in that gas as a result of the burning of fossil fuels. One compensating factor is the increased photosynthesis that would occur when the nutrients in the bottom waters are deposited at the surface. The surge in phytoplankton growth could be used, in fact, as the basis for commercial aquaculture systems.

A very attractive feature of the ocean thermal resource is its constancy. The temperature contrast is un-

affected by the night-time interruption of solar energy; and it is only slightly affected by seasonal changes. OTEC electrical generators, if feasible, could produce power round-the-clock and make very effective use of the installation: in the language of the utility industry, they would have a very large *capacity factor*. Electricity produced at sea could be transmitted ashore by cable; or it could be used, instead, to produce various energy-intensive products. Aluminum, for instance, if refined at sea, could free some generating capacity on the continent. As an alternative, energy could be brought to shore in the form of ammonia or hydrogen (both derived from sea-water) for use in fuel cells which produce electricity. Another option would depend on the wide use of new lithium-air batteries. When the batteries are run down they contain lithium hydroxide. This could be shipped to the OTEC platforms where it would be processed for pure lithium which would then be used to replenish the batteries.

Favorable Areas, and the Resource Potential

Temperature contrasts must be 20 degrees Celsius or greater in order to power heat engines; and such contrasts occur only in tropical and sub-tropical areas. Figure 9–2 shows how the general pattern of favorable areas is affected by cold currents: both the Peru and the California currents, which carry cold water from higher latitudes toward the equator, interrupt the tropical band where surface waters are sufficiently warm for a contrast of 20 degrees or more. The map is deliberately simple and does not show that near all shorelines the waters are too shallow for the contrasts to be realized. Areas with bottom depths less than 1,000 meters (3,200 feet) usually are

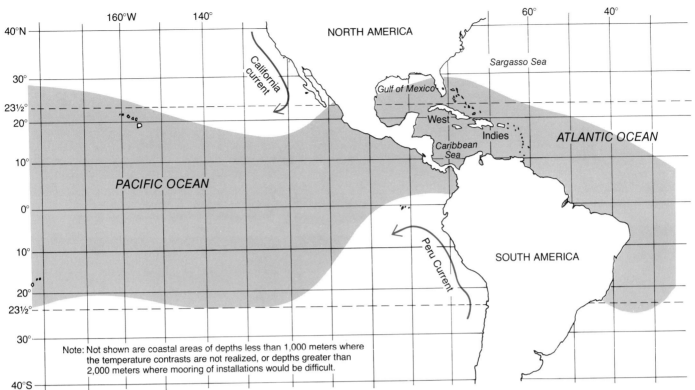

Fig. 9–2 Approximate extent of ocean areas where temperature contrast between surface and 1,000 meters depth is in excess of 20° Centigrade.

Note: Not shown are coastal areas of depths less than 1,000 meters where the temperature contrasts are not realized, or depths greater than 2,000 meters where mooring of installations would be difficult.

unsuitable; and such areas of shallow water overlying continental shelves are quite extensive in the Caribbean.

A number of United States coastal areas have access to OTEC resources. The Gulf of Mexico is within the favorable area, and both Puerto Rico and the Virgin Islands have access to temperature contrasts of the Gulf-Caribbean area. The islands of Hawaii fall within the favored area in the Pacific, and the island of Guam (Lat. 14° North, Long. 143° East) is in a favorable area not mapped on Fig. 9–2 because of the complexity of continental shelf patterns around Indonesia.

THE GULF OF MEXICO

Figure 9–3 shows how bottom depths affect different sites within the generally favorable Gulf areas. Where depths are less than 1,000 meters the bottom waters are not sufficiently cool. Where they exceed 2,000 meters it may be difficult to moor the OTEC installation. The area of potential in the Gulf, therefore, is a relatively narrow realm lying between those two depth lines shown in Fig. 9–3. Highlighted in Fig. 9–3 are the specific areas where temperature soundings have shown that the desired contrasts actually exist.

The electrical power generating potential of the areas shaded in Fig. 9–3 is estimated at 200 to 600 thousand Megawatts, that is, the equivalent of 200 to 600 generating plants of the 1,000-Megawatt size typical of today's large nuclear plants. The estimate takes into account that in continuous operation the projected power plants must

not degrade the temperature resources of Gulf waters (Cohen, p. 19). The potential appears to be very large and attractive. Furthermore, it tends to vary seasonally in a way that coincides with variation in demand in the southern United States. Surface waters are warmer in summer than winter, so the summer contrasts are greater, making possible a greater power output in the season of heavy demand from air conditioners. Figure 9–4 shows how expected output from a 1,500-Megawatt OTEC plant (superimposed on a 2,744-Megawatt baseload) coincides with the actual 1976 power load served by Florida Power and Light Company in Tampa, Florida.

Prospects for Development

Despite the very low conversion efficiencies that inevitably accompany a small temperature difference across the turbine, the possibilities for OTEC development look promising—especially in seas surrounding Hawaii, Guam, Puerto Rico, and the Virgin Islands, which are now totally dependent upon imported fuels for power generation.

The first plant built is expected to cost approximately $2.80 per installed Kilowatt of generating capacity, including transmission cables. With continued production the investment will drop to about $1.50 per installed Kilowatt of capacity, using constant 1975 dollars (Cohen, p. 14). This is higher than the capital costs for coal-fired and nuclear plants which are $0.46 to $0.74 for coal-fired,

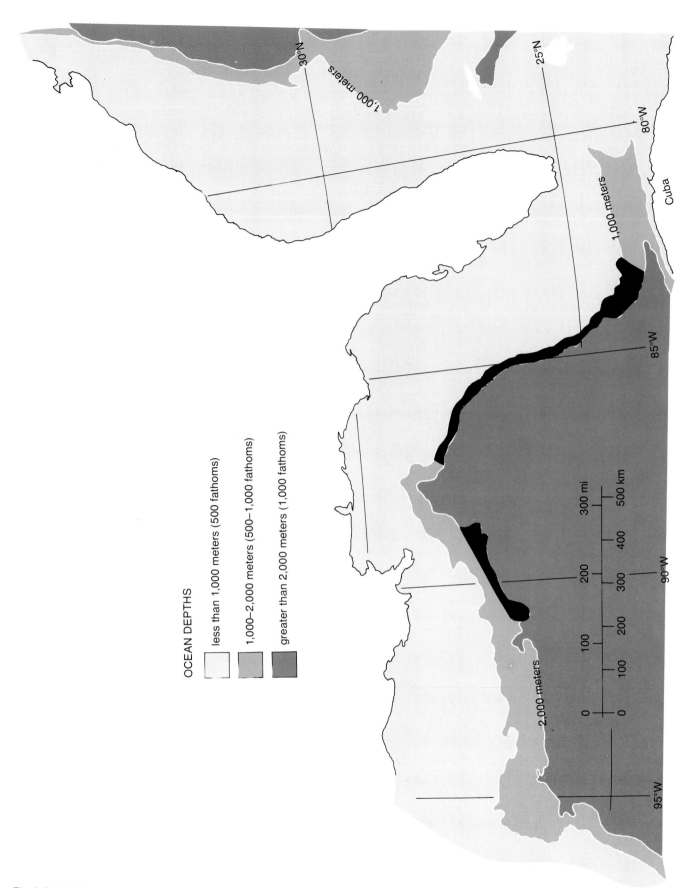

Fig. 9–3 Gulf Coast areas favorable to ocean thermal energy conversion (OTEC).

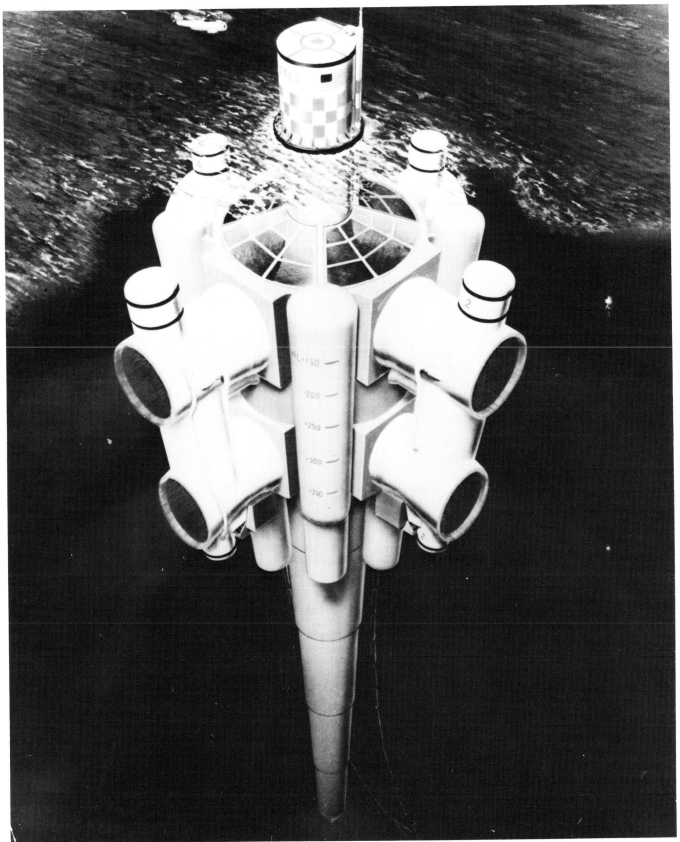

Artist's rendering of a Lockheed Ocean System concept for an OTEC electrical generating platform. The pairs of large ports near the top of the structure move water for the closed cycle turbines mounted around the perimeter. The upper port is an intake for warm surface water; the lower port is an outlet for cool water. Cold bottom water is raised through the tapering vertical pipe that extends downward like a root. *Courtesy Department of Energy.*

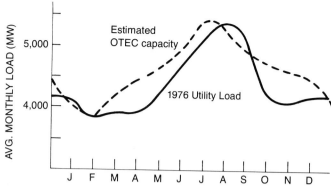

Fig. 9–4 Expected seasonal variations in capacity of a 1,500 Mega-watt OTEC installation off Tampa, Florida (superimposed on a 2,744 MW baseload), compared with variation in the actual load supplied by Florida Light and Power Company in 1976.

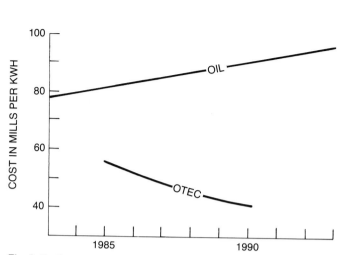

Fig. 9–5 Costs, in 1975 dollars, of baseload electricity derived from fuel oil in Hawaii, and from OTEC power plants of 250 Megawatt size.

Fig. 9–6 Projected OTEC electrical output by economic region (annual) in year 2020, with incentives of the National Energy Plan. Three regions with Gulf shores are considered.

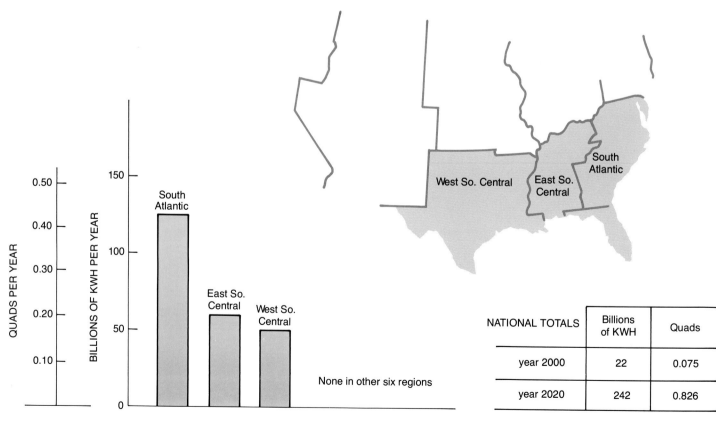

NATIONAL TOTALS	Billions of KWH	Quads
year 2000	22	0.075
year 2020	242	0.826

and $0.67 for nuclear, using 1976 dollars and an assumed 1985 date of installation (Nuclear Energy Policy Study Group, p. 116–122).

Favoring the OTEC plant in competition with fuel-burning plants will be the estimated capacity factor of 85 percent, versus roughly 65 percent for coal or nuclear plants. Freed from fuel costs, the OTEC plants are expected to have operating and maintenance costs that are only 1.5 percent of capital investment (Cohen, p. 12) versus 2.8 and 2.0 percent for coal and nuclear plants respectively (Nuclear Energy Policy Group, p. 126).

In Hawaii, which is now dependent on fuel oil for power generation, the increasing costs of fuel will make the unit costs of output markedly lower for OTEC power (see Fig. 9–5). An experimental OTEC power plant was placed off the Hawaiian coast by ERDA in 1979. OTEC power will be competitive soonest in the American islands partly because of their dependence on oil for electricity, and partly because deep water occurs close to shore, so the plants will not require long transmission lines. As installations become more numerous and the capital costs fall, Gulf markets will be the next to be penetrated, particularly the markets in Tampa, Florida, New Orleans, Louisiana, and Brownsville, Texas. The projected total contribution of OTEC power could be from 6 to 35 thousand Megawatts capacity by the year 2000 (Cohen, p. 20).[1]

Regional Differences in OTEC Implementation

Regional contributions to the national total in 2020 are shown in Fig. 9–6. Only regions bordering the Gulf are involved, and the South Atlantic region (essentially Florida) dominates with 125 billion Kilowatt-hours. The national total of 242 billion Kilowatt-hours of output from OTEC plants compares with 662 from wind machines, 315 from solar radiation, and 47 from biomass for the same year and using the same set of assumptions.

NOTES

[1]Another projection of the future role of OTEC power generation suggests an output of 22 billion Kilowatt-hours in the year 2000, and 242 billion in the year 2020 (see METREK, 1978). The latter output is consistent with the upper end of the Cohen projection for the year 2000, since installations of 35,000 Megawatts with a capacity factor of 85 percent yield 261 billion Kilowatt-hours per year.

Biomass

The term *biomass* is applied to a great variety of plants or plant-derived materials that may be used directly or indirectly for energy supplies. Table 10–1 is an overview of the two major sources of biomass and the two kinds of applications.

The sources of energy are wastes or residues from activities not directly related to energy production, and materials from crops grown specifically for their energy content. In the first category are crop residues such as corn stalks, manures, municipal trash, sewage, and tree tops from forestry operations. These materials are physically available now (if not economically available) for use. Crops grown specifically for energy include trees of various species, sugar cane, sorghum, ocean kelp, and water hyacinth. The economic viability of such crops will depend upon prices for other energy supplies, and could be encouraged through government incentives. Most energy crops will compete for land with food crops and with trees used for lumber and paper.

All plant materials store solar energy in the form of chemical energy created by the process of photosynthesis. In that process, light energy is used to convert water and carbon dioxide into energy-rich carbohydrates plus oxygen. This energy may be released through combustion of materials such as firewood, wood chips from forest wastes, or even urban trash, which typically contains a great deal of paper (Table 10–1). When burned, dry wood will supply roughly the same amount of energy per ton as lignite coal, the lowest of the four ranks of coal discussed in Chapter 1. An alternative to combustion is to use plants or wastes as raw materials for the production of synthetic fuels with high energy content such as alcohols, methane, gas, or fuel oil. As raw material for synthetic fuels, biomass will compete with coal, oil shales, and tar sands. In the production of ammonia for fertilizer use and as feed for fuel cells (producing electricity) biomass will compete with ocean thermal gradient installations (OTEC).

Selected Biomass Materials and Their Conversion to Fuels

Of all the possible biomass materials, five were selected by the Stanford Research Institute (SRI) in its Mission Analysis study conducted for the U.S. Department of Energy (see Ernest, Kent, et al, Jones, Jerry, Scholley, Fred, et al). The types of materials chosen were herbaceous plants, woody plants, aquatic plants, and manure. Herbaceous plants, that is, plants that do not produce persistent woody material, are divided into those with low moisture content and those with high moisture

Euphoribia lathyris, a member of the poinsetta family, which is believed to have the potential to be used to produce petroleum.

Table 10–1 Biomass Sources and Applications

SOURCES

	Wastes from Non-Energy Activities	Energy Crops
Used directly as fuels	Urban trash Tree tops and limbs from forestry operations	Trees used as wood or as charcoal
Used as raw materials for synthetic fuels and products, such as ammonia	Animal manures Urban garbage Sewage Crop residues Tree tops and limbs	Trees for wood Rubber from trees and guayule bush Water hyacinth Sunflower Sugar cane Sea kelp Corn Sorghum

content (Table 10–2). Some of those materials, herbaceous high moisture and woody materials, can be obtained as residues and also from energy crops. Manure and herbaceous low moisture plants are residues only, while aquatic plants are produced only as energy crops. These distinctions are relevant later in this chapter when the assumptions of an optimistic scenario result in projected production of more biomass materials than in the less optimistic base case assumption. Since the optimistic scenario assumes incentives to grow energy crops, it is the herbaceous high moisture, the woody plants, and the aquatic that will be most affected.

Conversion of biomass materials may be considered under two broad headings: thermochemical processes and biochemical processes. Thermochemical processes can be used for all five types of biomass, but the low-moisture herbaceous and the woody materials are most suitable and are the only raw materials shown on the simplified view of the three processes (Fig. 10–1). *Pyrolysis* is a process that converts organic materials into gases or liquids at temperatures of 500–900° Celsius by heating in a closed vessel in the absence of oxygen. The process may be adjusted to favor the production of either liquid fuel or gases. In the version shown, liquid fuels are the major product, while intermediate-BTU gas and char (charcoal, if the raw material is wood) are lesser products. The same two raw materials, low moisture herbaceous plants and woody plants, are favored for *catalytic liquefaction* which yields oil for combustion and chemicals, and for *direct gasification* which yields gases of high and intermediate energy content, along with methanol and ammonia.

Biochemical processes are essentially *anaerobic digestion* and *fermentation* (Fig. 10–2). Both processes are best served by the same raw materials, high moisture herbaceous plants, manure, and marine crops such as giant kelp. Anaerobic digestion yields high-BTU gas, which is methane, and intermediate-BTU gas, which is methane mixed with CO or CO_2. Though not included in the SRI raw materials, sewage and garbage are suitable raw materials for this digestion process: in fact, anaerobic digestion once was widely used at sewage plants and the resulting methane burned to provide heat at the plants. Fermentation is the well-known process that yields alcohols from raw materials such as sugar cane, corn, or algae.

Biomass and Energy Available in the United States

Conversion of solar energy to stored energy in plants is quite inefficient, seldom exceeding 3 percent during the growing season. On a year-round basis, the average efficiency is more like one percent. Nevertheless, if large areas are devoted to the purpose, very substantial amounts of energy can be harvested. For instance, the total biomass grown annually on cultivated land in the United States is roughly 5 billion dry tons (Fowler and Fowler). Assuming that each dry ton will yield 15 million BTU, the energy yield would be 75 Quadrillion BTU (75 Quads) which is roughly the total need for raw energy annually in the United States at present. The 5 billion dry tons for the whole country represents an average yield in the vicinity of 15 tons per acre. Since yields up to 30 tons

Table 10–2 Selected Biomass Types Assumed in the Stanford Research Institute Projections

BIOMASS TYPE	Obtained as residues from agricultural and forestry operations	Grown as energy crops
Herbaceous, Low Moisture	Small grain field residues	
Herbaceous, High Moisture	Wastes from vegetable fields and packing sheds. Residues from sugar cane, sugar beet, corn, sorghum, cotton	Sugar cane, corn, sorghum
Woody Plants	Mill bark, mill wood, logging residues, orchard prunings, standing vegetation not suitable for lumber	Various species
Manure	Only beef manure from feedlots is considered	
Aquatic Plants		Microalgae in fresh water, and giant kelp in the ocean

per acre are possible for certain crops, such as cane sugar, sorghum, or sunflower, the theoretical maximum energy yield from energy crops in the country may be placed higher than 75 Quads.

All of these estimates only provide a perspective. More important are the biomass amounts that will be available *economically* and taking into consideration the use of the land for food crops. Even the question of agricultural and forest residues must be approached re-alistically. One estimate shows that 8.3 Quads of energy could be obtained from wood that is *now physically available* each year as wastes from forestry operations, tree removal and trimming, or land clearing (US Forest Service). How much of this potential could actually be available depends largely on whether the price for fuels in certain regions would justify the collection, processing, and distribution of the wood.

The sections that follow report estimates of biomass

Fig. 10–1 Three thermochemical processes (simplified) for converting biomass into oil, gas, and chemicals.

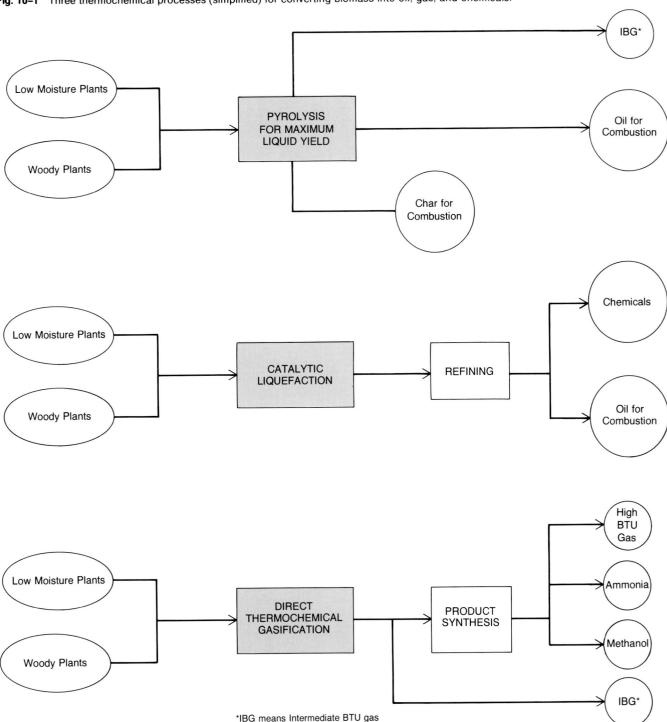

*IBG means Intermediate BTU gas

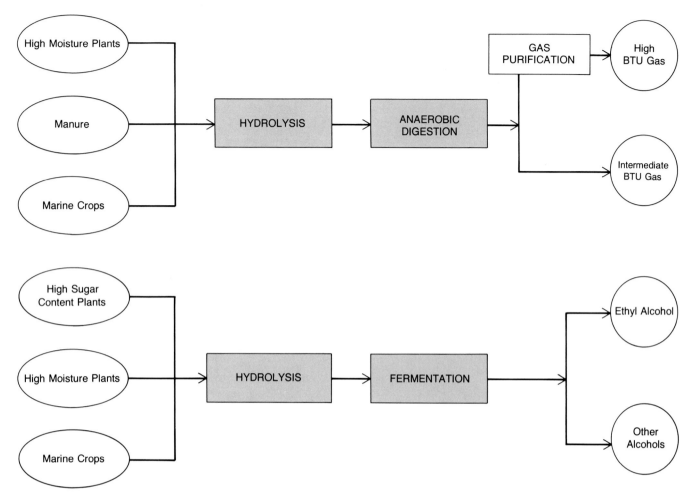

Fig. 10–2 Two biochemical processes (simplified) for converting biomass materials into gas and alcohols.

available for the United States and for its economic regions on the basis of economic models. These models or scenarios specify the prices of competing fuels and also specify whether there are government incentives to promote the use of land for biomass crops.

FOR THE UNITED STATES AS A WHOLE

Most of the following information is taken from a report to the U.S. Department of Energy by the Stanford Research Institute (SRI) which is summarized in Vol. I of a seven-volume series (Schooley, 1979). Estimates of biomass available are dependent upon market price for the biomass materials and upon two separate sets of assumptions called the *base case* and the *optimistic scenario*. The base case assumes crop production (and therefore the production of certain residues) under existing market conditions. The optimistic scenario assumes government actions and incentives designed to increase *energy crop production* and the use of currently-available high and medium potential crop lands.[1] Specifically, the second scenario assumes an investment tax credit designed to increase the net rate of return for both farm owners and the owners of conversion plants. The tax credit would decrease the costs of biomass pro-

duction, and therefore increase demand for it. Regarding forest resources, the optimistic scenario assumes that leases at attractive rates will be made available in public commercial timber lands. In the base case, only a small proportion (less than 2 percent) of total tonnage is expected to be from energy crops; but in the optimistic scenario energy crops constitute 10 percent of the total if the price of biomass is 20 dollars per ton, and constitute 30 percent if the price is 30 dollars per ton.

Figure 10–3 shows a series of estimates of total biomass tonnage (and energy value in Quads) that may be available in the years 1985, 2000, and 2020, assuming a price of 30 dollars per ton for the biomass raw materials. At this price, the amounts of biomass produced are greater in the later years because the prices of competing fuels are assumed to rise sharply (Schooley, p. 23). The difference between base case and optimistic scenarios is evident in each of the three years: more biomass is produced under the optimistic assumptions, and the greatest part of that increment is materials from energy crops, not residues. In the base case for all three years virtually all the materials produced are residues (see Table 10–2 for clarification of the actual materials considered in the analysis). According to the optimistic assumptions, in the

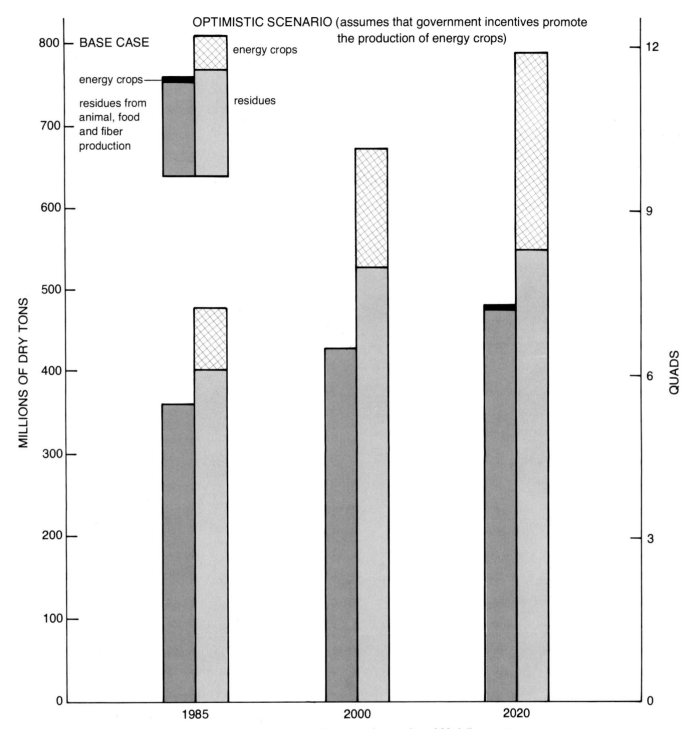

Fig. 10-3 Estimated total biomass available annually, 1985–2000, assuming a price of 30 dollars per ton.

year 1985 enough biomass materials will be produced to supply *over 7 Quads of energy* if the materials were burned. In the year 2000, *over 10 Quads will be available*. These are very substantial amounts, in light of the current total need for around 75 Quads of raw (primary) energy in the United States. A large portion of that energy, however, would be lost in the conversion of raw materials to fuels. In the case of conversion of cattle manure to methane the energy value of the gas is only 30–35 percent of the energy value of the manure if it were used as a fuel (Jones, p. 24).

The effect of more attractive prices for the raw biomass is evident in Figure 10–4. At prices of 40 and 50 dollars per dry ton, the biomass tonnage available is markedly higher than if the price is at 30 dollars per ton. The contrast is especially obvious within the optimistic scenario because the higher prices would encourage the cultivation of energy crops which are assumed to be promoted by the government incentives. In the base case, the higher prices would simply encourage a more thorough gathering of agricultural and forest residues that are physically available regardless of the price.

The importance of the assumed biomass raw material when converted to fuels and chemicals has been estimated by two different studies which both provide estimates for the year 2020. One of these studies (Metrek, 1978) projected gas, oil, methanol, and ammonia production under the assumptions of the National Energy Plan. While the Quad value of those products differs from the

values projected by the SRI (Table 10–3) the two estimates are of the same magnitude in each of the four product categories, and the totals, 2.80 Quads according to Metrek and 1.86 Quads according to SRI, are similar. According to the Metrek study, the gas from biomass would constitute roughly 35 percent of all synthetic gas produced in that year, and the ammonia would be 48 percent of all ammonia produced in that year. The contribution of wood for combustion in the year 2020, estimated only by the SRI study, is a very substantial 8.4 Quads.

FOR NINE ECONOMIC REGIONS OF THE COUNTRY

The relative roles of the nine economic regions (Hawaii and Alaska are excluded) can be seen in Figure 10–5. For simplicity, only the optimistic scenario is considered, and only those biomass amounts that are consistent with a price of 30 dollars per dry ton of the raw materials. Table 10–4 elaborates on the theme somewhat by comparing base case and optimistic scenario for one of the years. On Figure 10–5 the projected biomass production for three years, 1985, 2000, and 2020, shows that the West North Central and the Pacific regions are consistently the largest producers. The other regions, with the exception of the Middle Atlantic, are expected to produce roughly the same amounts of biomass, that is, between 50 and 100 million tons in the year 2000.

The kinds of biomass materials that make up the expected totals in the nine regions are shown on Fig. 10–6, which considers only the year 2000, but shows base case versus optimistic scenario. It is clear that the two leading regions are very different in their materials: the West North Central is expected to supply mostly herbaceous materials, largely of low-moisture type; but the Pacific production includes a great deal of wood and manure. The regions that are involved in production of the five biomass types are best seen by reference to Fig. 10–6 and to Fig. 10–7, which maps the regions producing certain types in the year 2000 according to the optimistic scenario only.

Woody materials are expected to be produced in all regions; but the Pacific, East South Central, and South Atlantic regions show the largest amounts. In the South Atlantic, the incentives assumed in the optimistic scenario are seen to substantially increase the amount of wood produced for energy.

Herbaceous materials of low and high moisture are expected from most regions. New England, Middle Atlantic, and South Atlantic, however, produce little or none (Ernest, R. K., p. 67–70). Low moisture materials (from small grains) account for most of the very large tonnage expected in the West North Central region and most of the tonnage in the West South Central. High moisture materials, which include corn for the production of alcohol, are supplied in largest amounts by the East North Central region which includes much of the nation's corn belt.

Aquatic plants, which will be produced only if the

Fig. 10–4 Estimated total biomass available annually in the Year 2000, showing the effects of five prices and two scenarios.

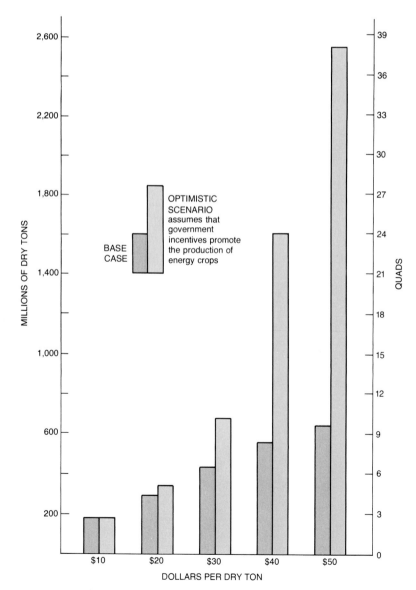

Table 10–3 Biomass Contribution (Annual) to National Totals of Synthetic Fuels and Ammonia in Year 2020

APPLICATIONS	National[1] Production from All Sources Quads/yr.	Biomass Share[1] under Assumptions of National Energy Plan Percent	Quads/yr.	Biomass Share[2] According to SRI Optimistic Scenario Quads/yr.
1. *Synthetic Fuels and Products*				
Natural Gas	3.5	34.8	1.22	0.45
Crude Oil	10.1	6.0	0.61	0.83
Methanol	1.8	16.2	0.29	0.00
Ammonia[3]	1.4	48.5	0.68	0.58
TOTAL	16.8		2.80	1.86
2. *As Fuel for Process Steam or Electrical Generation*				8.39
				10.25

[1]According to METREK, August, 1978, p. 42. In producing 2.8 Quads of fuels and product, 4.4 Quads of biomass material is used.
[2]Stanford Research Institute, 1979, Vol. I, p. 5 (see Schooley, 1979).
[3]Energy value assigned to ammonia is the fuel energy saved by not producing ammonia from natural gas.

Fig. 10–5 Estimated total biomass available 1985, 2000, and 2020, in nine economic regions.

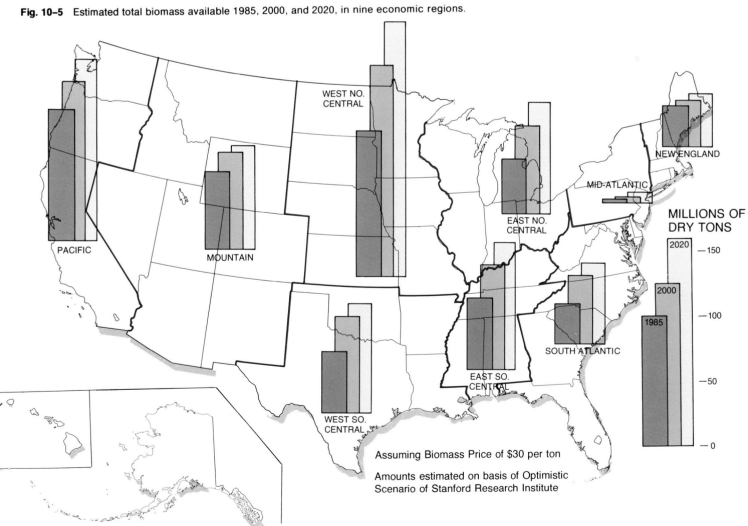

WEST NO. CENTRAL

NEW ENGLAND

MID-ATLANTIC

EAST NO. CENTRAL

PACIFIC

MOUNTAIN

MILLIONS OF DRY TONS

2020 — 150

2000

1985 — 100

SOUTH ATLANTIC

— 50

EAST SO. CENTRAL

WEST SO. CENTRAL

— 0

Assuming Biomass Price of $30 per ton

Amounts estimated on basis of Optimistic Scenario of Stanford Research Institute

price per dry ton rises to 40 dollars (SRI assumption) are expected from three regions that have extensive coast-lines—Pacific, West South Central, and South Atlantic.

Manure, according to the SRI projections, is expected only from the Pacific, Mountain, and West North Central regions. None of the combinations of year, price, or scenario leads to manure production for energy in the

West South Central region. This is surprising in light of the 1976 distribution of cattle feedlots (Fig. 10–8, Table 10–5). Texas held the largest number of feedlots of the required 4,000-head size, while Oklahoma, the other state in that region, held 22 such feedlots. Furthermore, a con-version plant is being planned now near Oklahoma City which will produce over 800 million cubic feet of methane

Fig. 10–6   Estimated biomass available annually in the year 2000 (at 30 dollars per ton) showing four biomass types.

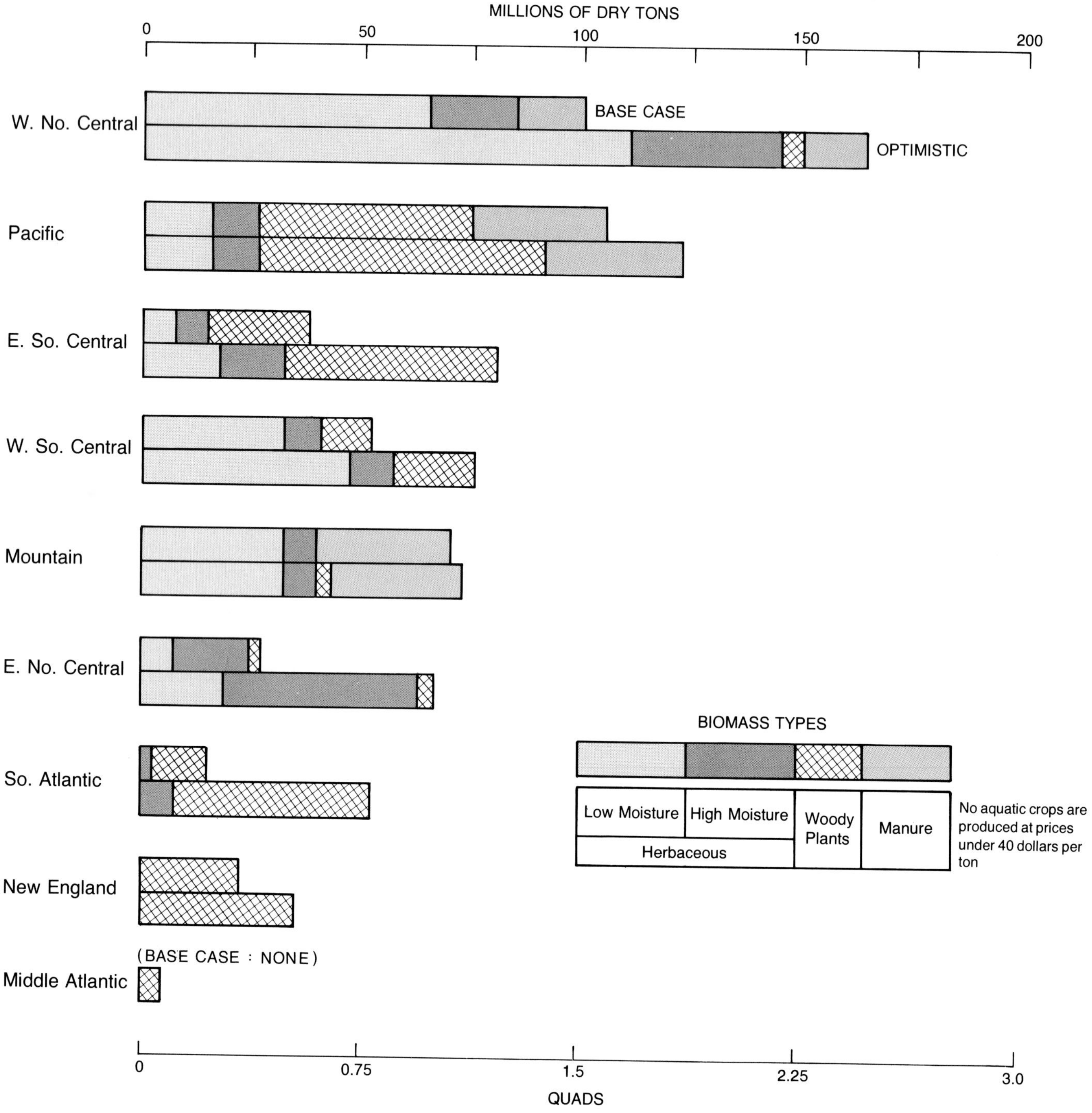

Fig. 10–7 Source regions for five biomass types in the year 2000.

WOODY

HIGH MOISTURE

AQUATIC

Note: Aquatic Biomass is produced only when price is $40 per ton or higher.

New England

Middle Atlantic

So. Atlantic

East No. Central

East So. Central

West No. Central

West So. Central

Mountain

Pacific

LOW MOISTURE

MANURE

Table 10–4 Estimated Biomass Available in Year 2000 in Nine Economic Regions[1]

REGION	Base Case Scenario[2] Millions of Dry tons	Quads	Optimistic Scenario[3] Millions of Dry Tons	Quads
West North Central	99	1.48	163	2.43
Pacific	105	1.57	122	1.82
East South Central	38	0.57	81	1.21
West South Central	51	0.76	76	1.14
Mountain	71	1.06	74	1.11
East North Central	28	0.42	67	1.00
South Atlantic	14	0.21	53	0.79
New England	23	0.34	34	0.51
Middle Atlantic	0	0	5	0.07
TOTAL	429	6.41	675	10.10

[1]At a price of 30 dollars per dry ton, according to two projections by Stanford Research Institute, 1979. Woody crops and residues dominate.
[2]In base case, less than 2 percent of tonnage comes from crops grown specifically for energy, because food and fiber crops compete for land.
[3]In optimistic scenario, at 30 dollars per dry ton, roughly 30 percent of tonnage available comes from energy crops.
[4]Assuming 15 million BTU per ton of dry plant material.
Source: Stanford Research Institute (SRI), March, 1979, p. 22.

Table 10–5 U.S. Feedlots Holding over 4,000 Head of Cattle, 1976

STATE	Number of Lots	Number of head marketed from These Feedlots in 1976 (millions of head)
Texas	125	3.76
Kansas	96	2.48
California	94	1.78
Nebraska	90	1.51
Colorado	67	1.64
Arizona	30	0.78
New Mexico	27	0.30
Oklahoma	22	0.60
TOTAL	551	12.88

Note: Only lots holding over 4,000 head are considered suitable for economical production of manure for methane. In 1976, lots of that size accounted for 58 percent of all U.S. feedlots.
Source: Jones, et al, Dec., 1978, p. 6.

Fig. 10–8 Feedlots with 4,000 head of cattle or more, 1976.

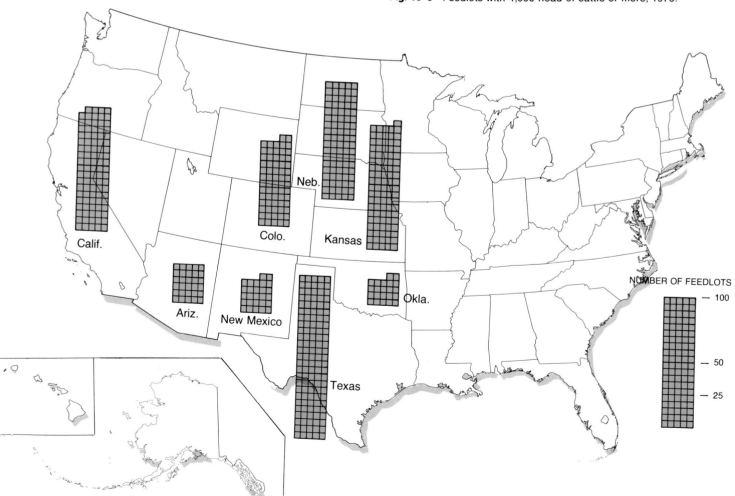

per year from feedlot manure.[2] If there were five hundred methane plants of that scale they could supply roughly 2 percent of the 20 trillion cubic feet of gas used in the country each year.

A group of crops not considered by the SRI study offers a most interesting potential in southwestern dryland areas that include some states in the Mountain region and some parts of the Pacific and West South Central regions. Crops that will thrive in dryland areas are attractive because they use land that is not required for food crops. Of the four dryland crops noted in Table 10–6, two—buffalo gourd and jojoba—yield edible oils. Jojoba oil can be used, in addition, for production of alcohols, lubricants, and waxes. Guayule and the gopher plant can be used for production of rubber, and can, therefore, reduce dependence on crude oil as a raw material for synthetic rubber. The gopher plant (Euphorbia lathyris) is especially interesting because it makes *hydrocarbons* as well as carbohydrates by photosynthesis. Among 200 species of desert plants analysed, only Euphorbia lathyris appears to have the hydrocarbon content and biomass potential that justify development for the sake of hydrocarbons (Johnson and Hinman, 1980).

The areas appropriate for cultivation of the gopher plant are limited by sunlight and moisture as shown on

Figure 10–9. Areas of intense sunlight needed for best growth of the crop are arbitrarily defined as those areas where average daily solar radiation (on an annual basis) is greater than 450 langleys, that is, greater than 450 calories of energy received per square centimeter. The

Table 10–6 Potentially Commercial Dryland Crops for Oils, Rubber, and Crude Oil Substitute

PLANT	Species	Yield and Applications
Buffalo Gourd	Cucurbita	Edible oils, protein, starch
Guayule	Parthenium argentatum	Rubber to reduce dependence on foreign natural rubber and on synthetic rubber made from petroleum
Jojoba	Simmondsia chinensis	Edible oils for use in foods. Source of alcohols, lubricants, waxes
Gopher Plant	Euphorbia lathyris	Hydrocarbons for rubber and as crude oil substitute

Source: Johnson and Hinman, 1980.

Fig. 10–9 Southwestern areas suitable for growing gopher plant (Euphorbia lathyris).

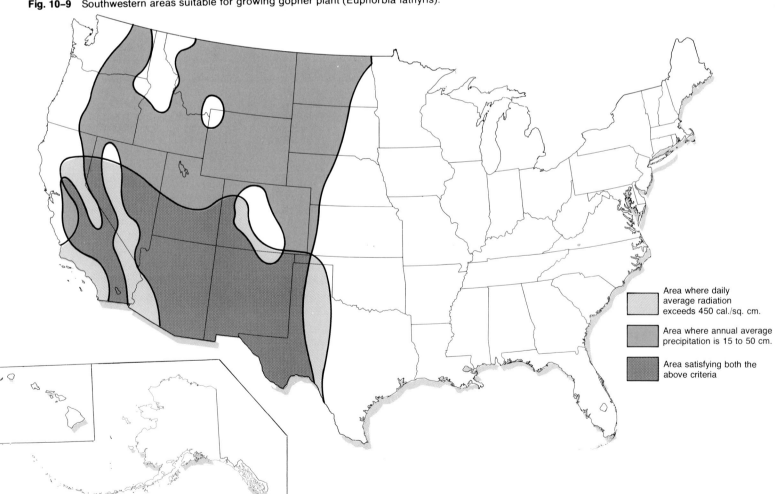

Area where daily average radiation exceeds 450 cal./sq. cm.

Area where annual average precipitation is 15 to 50 cm.

Area satisfying both the above criteria

Steam from Urban trash at Saugus, Massachusetts. The plant in the background accepts residential and commercial refuse from 13 communities on Boston's North Shore and burns 1200 tons per day to produce steam for the nearby General Electric manufacturing plant in the foreground. If it were used to produce electricity, the steam would provide enough energy to light half a million homes.

moisture limitations are as follows. Any area receiving less than 50 centimeters (roughly 20 inches) of precipitation per year is considered too dry for food crops. Areas receiving less than 15 centimeters are too dry for the gopher plant. Areas receiving 15 to 50 centimeters, therefore, are most suitable for growing the plant.

Figure 10–9 shows that suitable moisture coincides with required solar radiation in parts of California, Nevada, southern Utah, western Colorado, Arizona, New Mexico, and western Texas. Studies at the Office of Arid Land Studies at the University of Arizona indicate that roughly 8 to 12 million hectares of this land could be used with little or no irrigation needed.[3] On the basis of present yields of 25 barrels of oil per hectare, 11 million

hectares would supply about 4 percent of the nation's need for crude oil each year. If genetic and agronomic improvements were to increase the yield to 65 barrels per hectare, the production would be 10 percent of the present crude oil need (Johnson and Hinman, 1980).

Alcohol: Areas of Raw Material and Alcohol Production

Alcohol (ethanol and methanol) is the only alternative liquid fuel that is commercially available now, and is the only one likely to be available in quantity before 1985. It can be produced from coal as well as from biomass, using existing technology (see Chapter 1). When added to gasoline (to make *gasohol*) alcohol can serve both to extend the supply of gasoline and to improve the octane rating. The latter is especially attractive now that higher octane lead-free gasolines are in short supply, and nonlead octane enhancers are under restrictions because of their environmental effects.

Present federal incentives (as of 1979) are likely to increase ethanol fuel production from a 1979 level of about 60 million gallons per year to 3 billion gallons per year, or 3 percent of 1979 gasoline consumption, by the year 1982. If fuels containing alcohol are exempted permanently from federal excise tax, investors would be encouraged to build new conversion facilities, and alcohol production from biomass alone would reach 500 to 600 million gallons per year. At this level of production, the alcohol would substitute for 30 to 40,000 barrels per day (up to 14.6 million barrels per year) and could therefore reduce petroleum imports by almost one-half of one percent (Department of Energy, June, 1979).

Such reductions in petroleum needs will be realized *only if large amounts of petroleum are not used in the production of alcohol.* The question of net energy balance is part of this issue: that is, whether the energy value of the alcohol produced will exceed the energy consumed in the growing, transportation, and conversion of the raw materials. Considering energy spent in the growing of corn, its transport, and its conversion, and taking 'energy credit' for the dried grain by-product, the ethanol energy produced from corn is 1.05 times the energy expended. If that energy credit for the by-product is eliminated, then the energy produced is only 0.98 of the energy expended (DOE, June, 1979). Evidently the balance is close. The *savings of petroleum itself* can be greatly enhanced by not burning petroleum fuels in the distilling process.

Coal, wood, or agricultural residues can be burned instead; or in some areas solar energy can be used for heat. Furthermore, waste heat from utilities or from industries could be employed to minimize fuel use in distillation.

The expected biomass production for alcohols is projected with reference to U.S. Department of Agriculture Farm Production Regions in the Department of Energy study referred to above. Raw materials production is summarized by Fig. 10–10 and Table 10–7 which show the corn belt and the Southeast to be the largest produc-

Table 10–8 Projected Maximum Alcohol Production (Ethanol and Methanol) from U.S. Biomass Resources for the Year 2000, by Ten USDA Regions (in billions of gallons per year)

REGION	Ethanol	Methanol	TOTAL
Northeast	3.7	13.3	17.0
Southeast	7.8	18.7	26.5
Appalachian	3.1	11.4	14.5
Lake States	5.7	14.1	15.8
Corn Belt	10.1	29.8	39.9
Delta States	5.3	15.8	21.1
Northern Plains	4.7	9.6	14.3
Southern Plains	5.1	13.3	18.4
Mountain	3.2	11.5	14.7
Pacific	5.3	17.2	22.5
TOTAL	54.0	154.7	208.7

Source: DOE, June, 1979.

Table 10–7 Biomass Materials for Alcohol, in the Year 2000, showing Proportions in Each Region (in percentages)

U.S. DEPARTMENT OF AGRICULTURE FARM PRODUCTION REGIONS	Wood	Agric. Residues	Grains	Sugars	Munic. Solid Waste	Food Wastes	TOTAL
Northeast	10	1			29	6	7
Southeast	17	2		27	14	30	13
Appalachian	12						5
Lake States	9	10	18	16	6	15	10
Corn Belt	8	42	43	10	17	22	21
Delta States	12	8		12	5	5	9
North Plains	1	18	26	16			10
South Plains	9	8	8	15	8		9
Mountain	9	5	5		4	2	6
Pacific	13	6		4	17	20	10
	100%	100%	100%	100%	100%	100%	100%

Source: U.S. Department of Energy, June, 1979, p. 52.

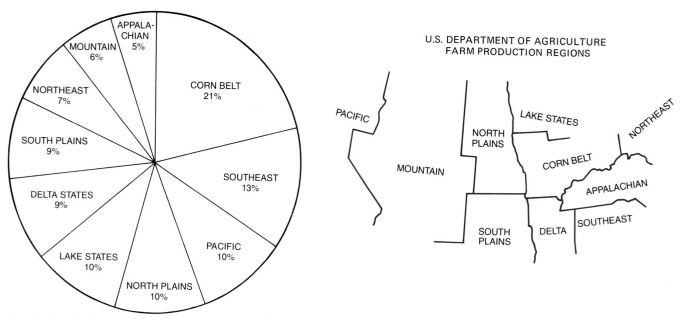

Fig. 10–10 Regional shares of projected biomass available for alcohol in the year 2000.

Fig. 10–11 Maximum ethanol and methanol production from biomass resources in the year 2000 in ten U.S. Department of Agriculture farm production regions.

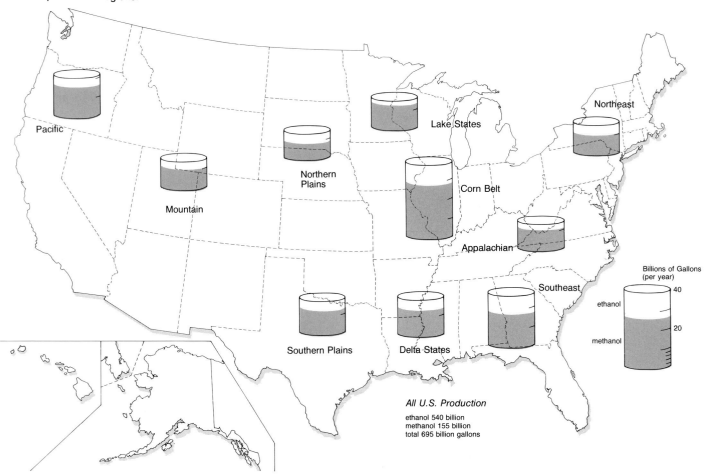

GASES	FREE GAS OR GAS IN SOLUTION IN VARIOUS KINDS OF ROCK	Conventional natural gas in porous and permeable reservoir rocks
		In tight gas sands of western states
		In Devonian shales of Appalachians
		In geopressured reservoirs
		In un-mined coal beds
	SYNTHETIC GASES	From coal gasification at mine-head or in place underground
		From biomass

CRUDE OILS	CRUDE OIL AS SUCH IN VARIOUS RESERVOIR ROCKS	Conventional crude recovered by primary, secondary, or tertiary (enhanced) recovery techniques
	SYNTHETIC CRUDE OILS	From bitumen in tar sands
		From kerogen in oil shales
		From coal liquefaction
		From biomass

Conventional and synthetic crude oils and gases, highlighting the synthetic gas and synthetic crude oil that can be obtained from biomass.

ing regions. The corn belt will supply grains, sorghum, and agricultural residues. The Southeast region will supply wood and cane sugar. The three regions that each are expected to supply roughly 10 percent of the raw materials are varied in their character. The Pacific supply is mostly municipal solid wastes and wood. The North Plains and the Lake states are expected to provide a mixture of materials, with grains being most prominent. In the Delta States and South Plains regions, sugars and wood play a large role. The Northeast region supplies only 7 percent of the national total of raw materials in this projection, but is distinguished by its large contribution to the national total of municipal solid waste. Expected volumes of alcohol produced in the year 2000 are shown on Figure 10–11 and Table 10–8. As expected on the basis of raw materials, the largest production is in the corn belt and the Southeast regions. Methanol is the dominant form in every region.

NOTES

[1] On May 21, 1980, a Senate-House conference committee agreed on a multi-billion dollar program to promote the development of synthetic fuels through loan guarantees, price guarantees, and purchase guarantees. The program encompasses synthetic fuels from coal, oil shales and biomass materials. There is some justification, therefore, for an optimistic scenario that assumes government incentives.

[2] Hayes, *Energy, the Solar Prospect*, p. 42. Apparently for the sake of a snappy acronym, the producing company has contrived to call itself Calorific Recovery Anaerobic Process (Inc.).

[3] Currently in the United States, roughly 34 million hectares are cultivated for corn, and 32 million for wheat. In all, roughly 146 million hectares are cultivated. One hectare equals 2.47 acres.

PART
THREE

Overview

The energy sources reviewed in the foregoing chapters offer supplies of energy that, taken together, have very great potential and wide variety in character. It is reasonable to hope that some mineral fuels can be used in the near term as a bridge to the more remote future in which a combination of renewable energy forms will provide clean and sustainable energy. The fuel options will be narrowed by decisions about environmental safety and about the net energy return of one fuel versus another. The renewable energy forms of the future will depend upon which are proven workable by the current and near-future development work.

Matters of environmental policy or net energy analysis cannot be dealt with here. Neither can this book speculate on the ultimate workability of the various forms of renewable energy. Part Three, however, does offer something that may be helpful in planning: that is, a summary of the energy potentials in each of the energy sources studied earlier. Both national totals and regional characters are compared with a view to a better understanding of the options that exist within the United States.

Conclusion

←

A composite representation of the forms the energy resources discussed in the *Atlas*.

Nonrenewable Sources

The seven nonrenewable energy sources discussed in this atlas—coal, crude oil, natural gas, shale oil, oil from tar sands, nuclear fuels, and geothermal heat—all can be represented by energy amounts thought to be recoverable from identified and from hypothetical or undiscovered deposits. However, the amounts of energy economically recoverable from identified deposits, that is, Reserves, are not firmly definable for shale oil, tar sands, and geothermal heat.

SUMMARY BY STATE

The resource character of states is reviewed in two parts:

1. The *reserves* of fossil fuels and uranium oxide are added together for each state;
2. The energy *possibly recoverable* from identified oil shales and geothermal occurrences is treated in a similar way.

RESOURCES RECOVERABLE ECONOMICALLY AT PRESENT: TRADITIONAL FOSSIL FUELS AND URANIUM OXIDE Reserves of coal, conventional crude oil, conventional natural gas, and uranium oxide are roughly equivalent in their economic feasibility. They may be added, therefore, to show how each one contributes to state totals of recoverable energy.

Coal Reserves are those energy amounts thought to be recoverable from the Demonstrated Reserve Base on the assumption of 90 percent recovery in surface mining and 50 percent recovery in underground mining. Since mineable tonnages are known for the four major ranks separately, the estimates of recoverable coal are translated into energy amounts by using the four appropriate BTU per ton values.

Conventional crude oil and natural gas must be represented by rather conservative numbers. For crude oil, only data for Measured (Proved) plus Indicated reserves are available by state. Missing from this summary are 1. Inferred amounts in known reservoirs that will likely be added to Reserves as young fields are developed by further drilling, 2. future Reserves in undiscovered reservoir rocks, and 3. oil that will be recovered from known and undiscovered reservoirs by advanced recovery techniques. The latter two are rightly excluded as they do not qualify as Reserves. Exclusion of Inferred amounts is less justifiable, but cannot be avoided since data for these amounts are not available by state. For natural gas, Measured (Proved) reserves are used, and again the Inferred and Undiscovered amounts are missing.

Uranium oxide is represented by Reserves that are

365

recoverable at costs of 50 dollars per pound, and therefore embrace the smaller amounts recoverable from richer ores at costs of 15 and 30 dollars per pound. The Department of Energy reports uranium Reserves by state only for New Mexico, Wyoming, and Texas. They lump together the Reserves in Arizona, Colorado, Utah and lesser Reserves in other states, to avoid disclosing the Reserves of mining companies. For this summary by state, the energy in uranium oxide Reserves is assumed to be only that in the fissile U_{235} isotope: the greater energy in the more abundant U_{238} isotope is reflected in tabulations and maps by economic region.

Excluded from this state summary are unconventional or synthetic crude oil from oil shales or tar sands, and unconventional natural gas from tight gas sands of the West or the Devonian shales of Appalachian states. The Reserve numbers selected here are conservative for oil and gas and also for coal. For oil and gas, further drilling will probably locate more new economically recoverable amounts than further exploration for coal. Rising prices and changing technology, however, will open up more

potential for coal than for oil and gas. Improved mining techniques could add substantially to the coal that is mineable; and the advent of in-place gasification could add a similar amount of energy recoverable from deep beds.

Table 11–1 lists energy reserves for coal, petroleum (conventional oil and gas), and uranium oxide for each state. Figure 11–1 shows how the state totals are divided among those three components. On a national basis coal Reserves constitute 90 percent of the total. Furthermore, coal overshadows petroleum in most states where the two coincide. Oklahoma and Alaska, both regarded traditionally as oil states, are dominated by coal, though inclusion of Inferred amounts of crude oil would alter the picture substantially in Alaska where fields are young. Louisiana is clearly an oil and gas state, but the Texas total includes substantial contributions from coal and uranium oxide. In New Mexico and Wyoming the large amounts of uranium oxide contribute heavily to total energy Reserves.

RESOURCES LARGELY NON-ECONOMIC: SHALE OIL AND

Fig. 11–1 State totals of recoverable energy in *reserves* of coal, conventional crude oil, and natural gas, and uranium oxide, showing proportions in each resource.

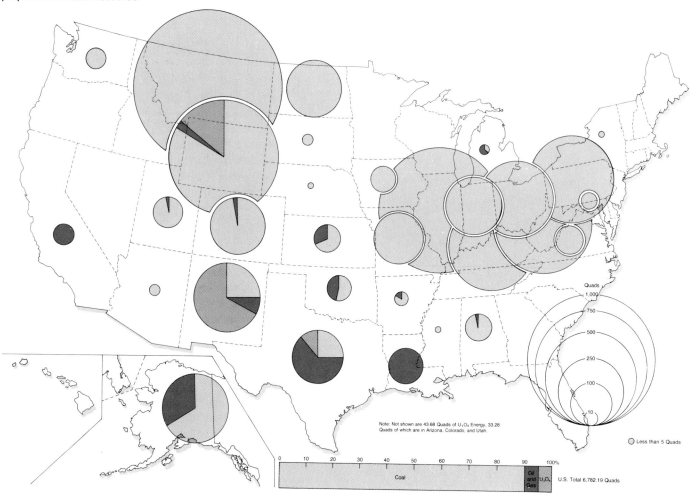

GEOTHERMAL HEAT Energy amounts in oil shales, tar sands, and geothermal occurrences are not necessarily comparable in economic feasibility. Nevertheless, they all require either a rise in fuel prices, a development effort, or both, before the amounts thought recoverable from identified occurrences will be realized. The rather generous amounts defined below are not Reserves, and may not be future Reserves: they are better called *maxi-* *mum potentially recoverable energy from identified occurrences.*

Shale oil amounts recoverable are only those from the Green River formation of Colorado, Utah, and Wyoming. Lower-grade deposits from the Appalachian states and Alaska are omitted completely. Assuming that future fuel prices will justify extracting oil from the leaner as well as the richer beds, this estimate includes oil recoverable

Table 11–1 State Shares of Energy Recoverable from Coal, Natural Gas, Conventional Crude Oil, and Uranium Oxide Reserves (Quads)

STATE	Coal[1]	Conventional[2] Crude Oil	Conventional[3] Natural Gas	Uranium[4] Oxide	TOTAL
Alabama	43.0	0.20	0.77		43.97
Alaska	177.0	51.37	32.69		261.06
Arizona	7.0	0.00	0.00		7.00
Arkansas	11.0	0.59	1.68		13.27
California	0.0	28.53	5.27		33.80
Colorado	187.0	1.31	2.03		190.34
Florida	0.0	0.95	0.16		1.11
Illinois	981.0	0.78	0.43		981.51
Indiana	156.0	0.16	0.05		156.21
Iowa	37.0	0.00	0.00		37.00
Kansas	32.0	1.94	12.71		46.65
Kentucky	408.0	0.18	0.74		408.92
Louisiana	0.0	16.33	51.37		67.70
Maryland	15.0	0.00	0.00		15.00
Michigan	1.5	1.16	1.83		4.49
Mississippi	0.0	1.19	1.46		2.65
Missouri	159.0	0.00	0.00		159.00
Montana	1,400.0	0.92	1.02		1,401.94
Nebraska	0.0	0.16	0.07		0.23
New Mexico	70.0	4.67	13.71	189.6	277.98
New York	0.0	0.06	0.27		0.33
North Dakota	202.0	1.14	0.43		203.57
Ohio	312.0	0.73	1.61		314.34
Oklahoma	21.3	7.23	11.85		40.38
Pennsylvania	415.0	0.39	2.16		417.55
South Dakota	5.4	0.15	0.00		5.55
Tennessee	16.0	0.14	0.00		16.14
Texas	41.0	49.63	56.46	19.8	166.89
Utah	55.0	0.94	0.72		56.66
Virginia	55.0	0.00	0.08		55.08
Washington	24.0	0.00	0.00		24.00
West Virginia	569.0	0.17	2.77		571.94
Wyoming	674.0	5.40	4.46	114.9	798.76
Miscellaneous	0.0	0.00	0.26		0.26
TOTAL	6,074.34	176.25	207.30	324.3	6,782.19

[1]Recoverable from Demonstrated Reserve Base as of December 31, 1974.
[2]Proved plus Indicated reserves as of December 31, 1978.
[3]Proved reserves as of December 31, 1978.
[4]Reserves as 50 dollars per pound as of January, 1979. Other states hold 43.68 Quads, 33.28 of which is in Arizona, Colorado and Utah.

from all identified beds averaging 15 gallons of oil per ton of rock. Potential oil in-place in such beds is about 1800 billion barrels, of which roughly 610 billion (one-third of oil in-place) may be recoverable.

Small deposits of oil sands and other bitumen-bearing rocks occur in a number of states. If deposits holding less than 1 million barrels of bitumen in-place are excluded only the occurrences in Utah remain. They hold about 20 billion barrels of bitumen in-place, of which roughly a third may be recoverable as crude oil.

Geothermal heat recoverable may be estimated for only two of the three major types of exploitable occurrences, because extraction of heat from the volcanic type has not yet been demonstrated. For *hydrothermal* occurrences heat recoverable at well-head is roughly 25 percent of identified heat in-place to a depth of 3 kilometers. For geopressured occurrences in deep sand bodies along the Gulf Coast, both thermal and methane energy are included to a depth of 6.86 kilometers (22,500 feet). For this summary the most ambitious Gulf Coast development plan is assumed, which implies recovery of

roughly 3.3 percent of the heat in-place. Geothermal heat recoverable at well-head without regard to application is loosely analogous to energy in crude oil or natural gas at well-head. The difference is that oil and gas provide very high temperatures and much higher conversion efficiencies than geothermal heat. The comparison with energy in crude oil from shales, therefore, is not entirely appropriate.

Table 11–2 and Fig. 11–2 note the energy amounts recoverable. For the country as a whole, energy in Green River oil shales and in geopressured reservoirs is far greater than that in hydrothermal occurrences and tar sands. Western and southwestern states monopolize the unconventional crude oil and geothermal resources. It is apparent that the two broad types of resource rarely combine in any one state. There are the "shale oil states," Colorado, Utah, and Wyoming; and there are "geothermal states," among which California and Idaho are outstanding for hydrothermal occurrences, and Texas and Louisiana for geopressured reservoirs. Wyoming would be a dual "shale oil and geothermal state" on the

Fig. 11–2 State totals of maximum recoverable energy in identified Green River oil shales and identified geothermal occurrences of hydrothermal and geopressured types, showing proportions in each resource.

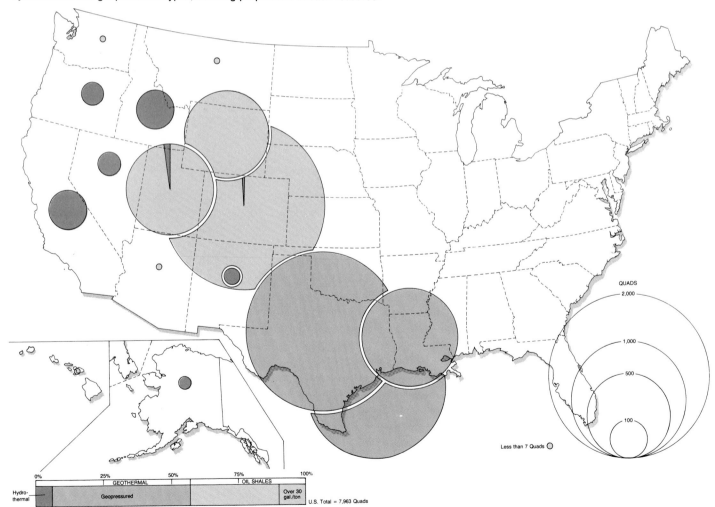

basis of *energy in-place,* but the geothermal energy occurs in Yellowstone National Park and cannot be exploited.

This summary would be more complete if unconventional natural gas from the tight sands of western states and from Devonian shales of the Appalachian states were included. These energy amounts are only vague estimates now, and cannot be clearly assigned to states or to economic regions. An estimate does appear later in the national overview of energy recoverable from identified and hypothetical sources.

SUMMARY BY REGION

In this summary recoverable energy amounts thought to be *economic or near-economic* are brought together in eleven economic regions. As in the foregoing summary by state, the energy amounts from various sources are added for each region using Quads as the energy unit.

FOSSIL FUELS AND GEOTHERMAL HEAT Reserves of coal, conventional crude oil, and conventional natural gas may be compared with Reserves of nuclear fuels (defined below) because they are all roughly equivalent in economic feasibility. Oil from tar sands is ignored, in part because amounts are small, but mainly because extraction is not presently taking place, and seems unlikely in the face of competition from oil shales. For oil shales and geothermal heat, no Reserve amounts are recognized by industry or government; but the current activity in both fields suggests there are some economically obtainable energy amounts. These important resources must be estimated here for inclusion in the summary by region.

For oil shales, only the Green River formation in western states is considered, and only the beds whose crude oil yield is expected to meet or exceed 30 gallons per ton of rock are included. Table 11–2 shows that all three of the western shale oil states have some beds of this richness, but Colorado has the largest share.

For geothermal heat, volcanic occurrences are excluded because no estimate of recoverable heat can be made. Also excluded is heat recoverable from geopressured reservoirs because the proposed application, electric power generation, has yet to be demonstrated. Only hydrothermal systems are considered to hold economic or near-economic energy that is comparable to that in the rich beds of Green River oil shales. For purposes of comparison some portion of the well-head heat estimated recoverable by the U.S. Geological Survey must be selected, but the portion is uncertain. While an earlier

Table 11–2 All Energy Recoverable from Identified Green River Oil Shales, Tar Sands of Utah, and Exploitable Geothermal Occurrences (Quads)

STATE	RECOVERABLE WELL-HEAD GEOTHERMAL ENERGY				RECOVERABLE CRUDE FROM GREEN RIVER OIL SHALE		RECOVERABLE CRUDE FROM UTAH OIL SANDS	STATE TOTALS
	Hydrothermal	Hydrothermal	Geopressured	Total	Over 15 Gal./Ton	Over 30 Gal./Ton		
	150°C	90–150°C						
Alaska	1.78	5.52		7.30				7.30
Arizona	0.27	0.87		1.14				1.14
California	116.54	8.06		124.60				124.60
Colorado	0.27	2.30		2.57	2,222.2	655.6		2,880.37
Hawaii	0.00	1.94		1.94				1.94
Idaho	4.21	108.30		112.51				112.51
Louisiana	0.00	0.00	720.63	720.63				720.63
Montana	0.00	2.58		2.58				2.58
Nevada	27.02	8.45		35.47				35.47
New Mexico	20.87	1.03		21.90				21.90
Oregon	21.43	12.94		34.37				34.37
Texas			1,905.56	1,905.56				1,905.56
Utah	11.39	1.90		13.29	583.3	94.4	43.8	734.79
Washington	0.04	0.00		0.04				0.04
Wyoming	0.00	0.60		0.60	583.3	22.2		606.40
Federal Outer Continental Shelf			1,576.00	1,576.00				1,576.00
TOTAL	203.82	154.49	4,202.19	4,559.90	3,388.8	772.2	43.8	8,764.70

Survey report suggested that *half* the hydrothermal energy recoverable may be called Reserves for electrical applications, the 1978 Survey report declined to make an estimate of Reserves. The experimental power plants being built in California and space-heating applications in Idaho and Oregon justify the inclusion of some of the hydrothermal well-head heat. For the purposes of this summary and for want of a better number, 50 percent of this heat is called Reserves. The 50 percent ratio gives only a very rough estimate of possible Reserves in each region because: 1. the systems with highest temperatures and shallow depths would be most attractive for electrical power generation, 2. only the systems with nearby settlements would be used for space heating, and 3. these characteristics do not occur equally in all areas that have hydrothermal resources. To say, for instance, that in each state half the recoverable heat is economically attractive could be most misleading. The summary is most meaningful when state amounts are agglomerated into regions. Regions which contain only one state, however, such as Alaska or Hawaii, suffer the most distortion in this comparison.

NUCLEAR FUELS   To represent economically recoverable uranium energy the Reserves of uranium oxide producible at forward costs of 50 dollars per pound are assigned to economic regions from the physiographic regions in which data is collected. Thorium energy is represented by Reserves of thorium oxide.

Because transmutation of uranium to plutonium in breeder reactors makes use of the abundant isotope, U_{238}, while non-breeders use the naturally fissile but scarce U_{235}, the energy content of uranium Reserves used for breeders is roughly 140 times greater than if non-breeders are assumed. National totals of 311.5 Quads versus 45,118 Quads demonstrate the difference (Table 11-3). This necessitates *two* regional summaries, one without breeder fuels and one with, the latter to include thorium transmuted to fissile U_{233}.

REGIONAL PATTERNS   Table 11-3 summarizes the energy in all the amounts defined above, for eleven regions, and shows regional totals both with and without breeder fuels. Fig. 11-3 expresses those regional totals in a pair of unconventional maps in which the sizes of regions reveal their total recoverable energy. Since one scaling system

Table 11-3   Regional Shares of Energy Recoverable from Estimated Reserves of Nonrenewable Resources, Including Geothermal

ECONOMIC REGION	Coal	Gas	Oil	Oil Shale*	Oil Sands**	Uranium 235	Uranium 238	Thorium	Geothermal***	REGION TOTALS	REGION TOTALS WITHOUT U-238 OR THORIUM
New England	0.0	0.0	0.0	0.0	0.0	0.0	0.0	0.0	0.0	0.0	0.0
Middle Atlantic	415.0	2.4	0.5	0.0	0.0	0.0	0.0	0.0	0.0	417.9	417.9
East North Central	1450.5	3.9	2.8	0.0	0.0	0.0	0.0	2,899.0	0.0	4,356.0	1,457.2
West North Central	435.4	13.2	3.2	0.0	0.0	0.0	0.0	0.0	0.0	452.0	452.0
South Atlantic	639.0	3.0	1.1	0.0	0.0	0.0	0.0	4,437.0	0.0	5,080.0	643.0
East South Central	467.0	3.0	1.7	0.0	0.0	0.0	0.0	0.0	0.0	472.0	472.0
West South Central	73.3	121.4	73.8	0.0	0.0	19.8	415.7	0.0	0.0	704.0	288.3
Mountain	2,393.0	22.2	13.8	772.0	0.0	314.8	44,207.9	11,122.0	101.55	58,947.0	3,617.0
Pacific	24.0	5.3	28.5	0.0	0.0	0.1	12.0	0.0	79.48	149.4	137.4
Alaska	177.0	32.7	51.3	0.0	0.0	0.0	0.0	0.0	3.73	264.8	264.8
Hawaii	0.0	0.0	0.0	0.0	0.0	0.0	0.0	0.0	0.95	0.9	0.9
Federal Outer Continental Shelf	0.0	0.0	0.0	0.0	0.0	0.0	0.0	0.0	0.0	0.0	0.0
Unspecified Locales[1]	0.0	0.0	0.0	0.0	0.0	81.4	11,432.0	0.0	0.0	11,513.4	81.4
TOTAL	6,074.0	207.3	176.7	772.0	0.0	416.1	56,067.0	18,558.0	185.7	82,357.4	7,831.9

[1]33.4 Quads, mostly in Mountain region, with small amounts in Pacific and West North Central regions. Also 48.0 Quads as by-product of phosphate and copper mining.

*Using only oil recoverable from beds averaging 30 gallons per ton or more. If all beds averaging 15 gallons per ton or more were considered, the recoverable energy would be 3,389 Quads.

**None of the 43.8 Quads potentially recoverable from Utah tar sands is expected to be economic in the near future.

***Near future geothermal reserves exclude all geopressured reservoirs but include one-half of recoverable (well-head) energy from regional totals of heat in identified hydrothermal systems.

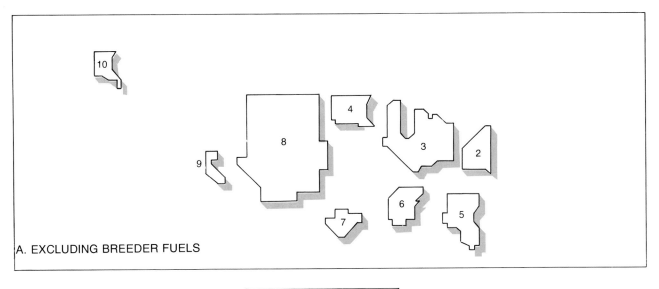

A. EXCLUDING BREEDER FUELS

1 New England   6 E. So. Central
2 Middle Atlantic 7 W. So. Central
3 E. No. Central   8 Mountain
4 W. No. Central   9 Pacific
5 South Atlantic   10 Alaska

REGIONS

100 Quads

New England Region has no recoverable energy; Hawaii has
0.95 quadrillion BTUs from geothermal sources

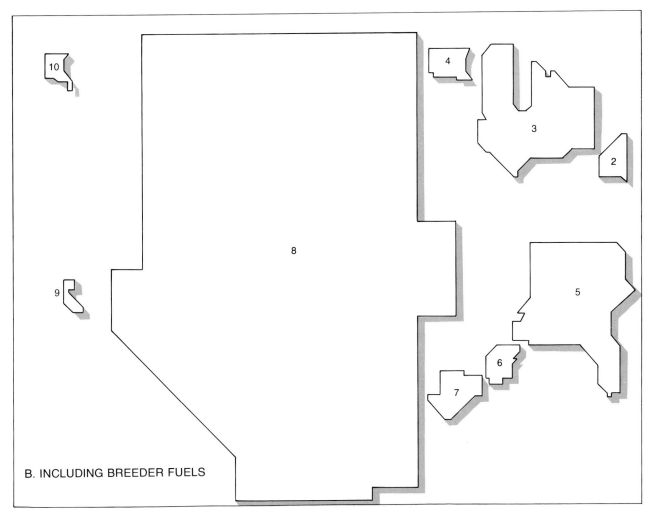

B. INCLUDING BREEDER FUELS

Fig. 11-3 Total energy recoverable from estimated reserves of nonrenewable resources. (A includes breeder reactors and B does not).
This area of each economic region is made proportional to its total recoverable energy in Quads (quadrillion BTUs).

applies to both maps, the impact of including or excluding breeder fuels is readily seen.

Figure 11–4 maps the regional totals and their various resource components assuming that non-breeders are used for conversion. On this basis uranium oxide does not make a conspicuous contribution. In the Mountain region, clearly the richest in energy, uranium contributes only 9 percent of the 3,555 Quads, while coal Reserves account for 66 percent. The East North Central region, with a total of 1,457.2 Quads, is next largest and owes 99.5 percent of its energy to coal Reserves, mostly those found in Illinois. Most other regions are dominated by coal energy, the exceptions being West South Central where oil and gas in Texas and Louisiana overshadow the coal in Texas; Pacific, where geothermal accounts for 58 percent; and Hawaii, where the 0.9 Quads of energy are entirely geothermal.

Figure 11–5 shows that with breeders assumed, uranium and thorium oxide account for 90 percent of the Nation's energy Reserves. In those regions where ura-

nium or thorium occur they play the leading role, constituting 95 percent of the Mountain region's immense total (largely because of uranium) as well as 87 percent and 67 percent in South Atlantic and East North Central regions. Uranium in the Colorado Plateau and Wyoming Basis and thorium in Idaho and Montana are responsible for the Mountain region's character. The South Atlantic and East North Central regions emerge as leading energy suppliers because of thorium Reserves in placer deposits along the Atlantic coast in Georgia and Florida, and in ancient conglomerate rocks in Michigan. Coal is still the dominant resource in the Middle Atlantic, West North Central, and East South Central regions, but those regional totals are eclipsed by breeder fuel energy in the three regions affected.

Renewable Sources of Energy

In this section, the energy from solar radiation itself, and its manifestations in wind, ocean thermal gradients,

Fig. 11–4 Regional totals of recoverable energy in reserves of coal, conventional crude oil, and natural gas, and uranium oxide (U_{235}) combined with energy in possible reserves of Green River oil shales and hydrothermal-type geothermal occurrences. Breeder fuels are excluded.

and biomass, are summarized. Electrical applications are dealt with first, then the electrical and non-electrical together.

ELECTRICAL GENERATION

Wind, hydroelectric, and ocean thermal gradients (OTEC) are the only renewable energy flows that lend themselves to estimates of *maximum* generating potential, that is, the installed generating capacity consistent with saturation of the areas suitable. Tables 11–4 and 11–5 summarize those maximum installed capacities for comparison. It is not realistic, of course, to assume any one of the sources will be used to its fullest extent.

Maximum installed capacity for wind machines is greater than for hydroelectric and OTEC, despite the exclusion of very substantial wind potential in Atlantic and Pacific offshore areas. Maximum annual output, however, is slightly greater from hydroelectric, and is greatest from OTEC. This is because of widely differing

capacity factors: wind machines are expected to run at a rated capacity of only 0.22 to 0.38 of the time, whereas OTEC installations are expected to run steadily at or near their capacity, as expressed by the factor 0.85. If the OTEC assessment is realistic and the technology proves workable, this source of electrical power could be extremely important. Figure 11–6 compares the maximum generating capacities, and provides some estimates of what proportion of those potentials may be installed by the year 2000.

A more complete review of projected electrical contributions, expressed as output, not installed capacity, is in Table 11–6 and Fig. 11–7. For the sake of uniformity, the projection by Metrek, assuming incentives of the National Energy Plan, is the source for all this information—the table shows national totals, while the figure shows regional totals and the solar share of all electrical output in each region. According to this view of the future, wind makes the largest contribution to electri-

Fig. 11–5 Regional totals of recoverable energy in reserves of coal, conventional crude oil and natural gas, uranium oxide (U_{238}) and thorium oxide, combined with energy in possible reserves of Green River oil shales and hydrothermal-type geopressured occurrences. The inclusion of breeder fuels drastically changes the amount and the distribution of recoverable energy.

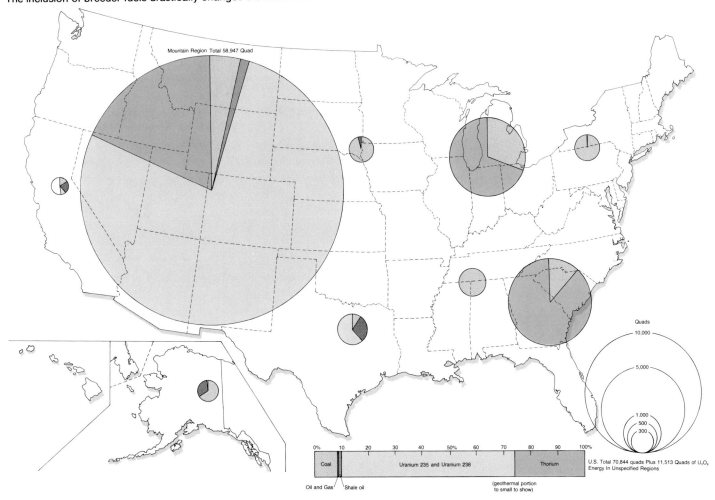

Table 11-4 Total Potential Hydroelectric Capacity and Output for Economic Regions, 1979

REGION	Total Potential Capacity[1] Megawatts	Capacity Factor Percent	Output Quads
New England	5,311.7	46	.073
Middle Atlantic	22,825.3	69	.469
East North Central	5,486.9	54	.090
South Atlantic	19,990.4	31	.187
East South Central	25,576.9	45	.344
West North Central	9,560.3	48	.137
West South Central	10,274.2	31	.095
Mountain	35,565.8	49	.522
Pacific	96,640.2	58	1.676
Alaska	33,825.6	46	.465
Hawaii	84.2	66	.001
UNITED STATES TOTAL	265,141.5	52	4.059

[1]Total potential is the total potential installed capacity reported by the Federal Energy Regulatory Commission for 1979 plus the additional capacity at existing dams reported by the Army Corps of Engineers, July 1979.

Note: Output in Quads =

$$\frac{(\text{Total Potential in Kw})(8760 \text{ hours})(\text{Capacity Factor})}{2.93 \times 10^{11}}$$

cal output in the Nation (Table 11-6) and it is most conspicuous in the South Atlantic region (Fig. 11-7). Wind's contribution is surprisingly large in light of the maximum potential noted above: the projected installed capacity is roughly 10 percent of the maximum, yet the estimated output is roughly 62 percent of the maximum output which assumes capacity factors of 0.22 to 0.38. Two regions, South Atlantic and West South Central show the largest solar-electric outputs, and they, along with Pacific and East South Central show the largest solar shares of all electrical generation within the region.

FUEL SAVINGS FROM ELECTRICAL AND NON-ELECTRICAL APPLICATIONS

The energy value of fuels saved is considerably greater than the energy value of electrical output or heat delivered from a renewable source. If, for instance, the renewable energy displaces coal-fired boilers for electrical generation, the coal energy saved is roughly 2.9 times the output energy because only 38 percent of coal energy in the boiler is converted to electrical energy.

Table 11-7 and Fig. 11-8 show estimates of fuel saved in the year 2020 according to assumptions of the National Energy Plan (Metrek, August, 1978), and according to three scenarios based on three different sets of assumptions (Stanford Research Institute (S.R.I.), March, 1977). For electrical applications the fuel savings projected by

Fig. 11-6 Maximum or saturation generating capacity estimated for wind, ocean thermal gradients (OTEC) and hydroelectric, showing projected installed capacities by the year 2000.

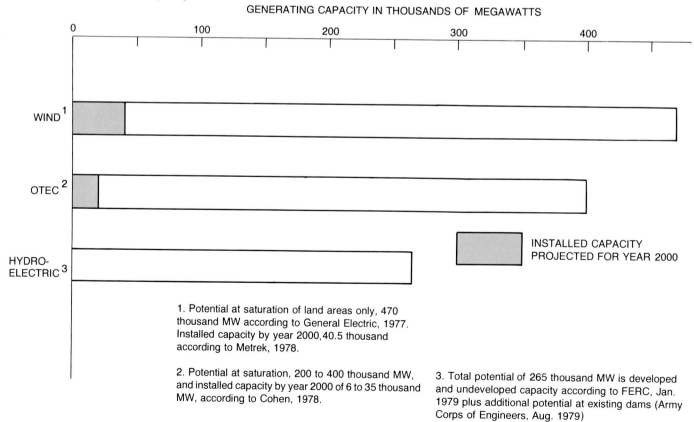

GENERATING CAPACITY IN THOUSANDS OF MEGAWATTS

WIND[1]

OTEC[2]

HYDRO-ELECTRIC[3]

INSTALLED CAPACITY PROJECTED FOR YEAR 2000

1. Potential at saturation of land areas only, 470 thousand MW according to General Electric, 1977. Installed capacity by year 2000, 40.5 thousand according to Metrek, 1978.

2. Potential at saturation, 200 to 400 thousand MW, and installed capacity by year 2000 of 6 to 35 thousand MW, according to Cohen, 1978.

3. Total potential of 265 thousand MW is developed and undeveloped capacity according to FERC, Jan. 1979 plus additional potential at existing dams (Army Corps of Engineers, Aug. 1979)

Table 11–5 National Summary of Maximum Annual Electrical Output (and Fuel Saved) from Hydroelectric, Wind, and Ocean Thermal Gradient Installations—Assuming *Saturation*

SOURCE	INSTALLED CAPACITY Thousands of Megawatts	ANNUAL OUTPUT Billions of Kilowatt-Hours	Quads[4]	ANNUAL FUEL SAVED Quads[5]
Wind	470.3	1,069.5[1]	3.65	9.59
Hydroelectric	265.1	1,440.0[2]	4.06	10.67
OTEC	200–600	1,489–4,468[3]	5.08–15.24	13.36–40.08
Mean	400	2,998	10.16	26.7

[1]Assuming capacity factors of 0.22, 0.33, and 0.38 in Low, Medium, and High regimes respectively (General Electric, 1977).
[2]Assuming capacity factor of 0.62 (FERC, Jan., 1979 and Army Corps, Aug., 1979).
[3]Assuming capacity factor of 0.85 (Cohen, 1978).
[4]Using equivalence, 1 Quad = 2.93×10^{11} Kilowatt-hours output.
[5]Displacing coal-fired generators with efficiency of 0.38. Fuel saved is $2.63 \times$ (output energy).

Fig. 11–7 Projected annual electrical output in the year 2020 from five expressions of solar energy in nine economic regions showing share solar takes in output in each region. Incentives of the National Energy Plan are assumed.

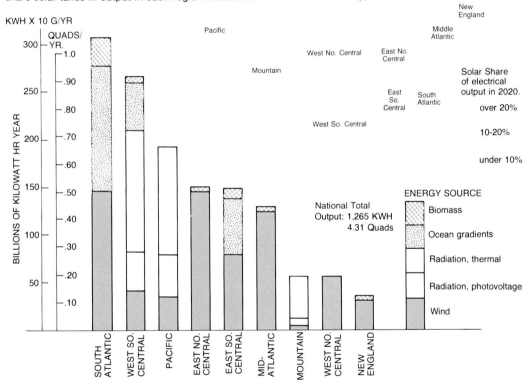

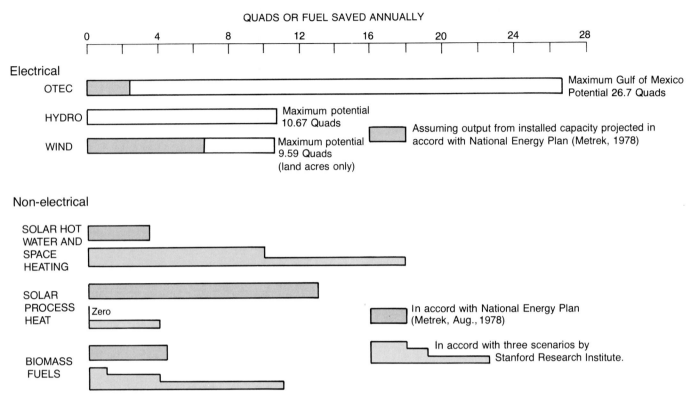

Fig. 11-8 National summary of projected fuel savings from electrical and non-electrical applications of renewable energy flows. Amounts are quads of fuel saved (annually) as of the year 2020.

Table 11-6 Projected Electrical Output in the Year 2020 from Five Expressions of Solar Energy

ENERGY SOURCE	ANNUAL OUTPUT		FUELS[2] SAVED
	Billions of Kilowatt-hours	Quads[1]	
Wind	662	2.26	6.6
Solar radiation, thermal	293	1.00	2.9
Solar radiation, photovoltaic	21	0.07	0.2
Ocean thermal gradients	242	0.82	2.4
Biomass	47	0.16	0.6
TOTAL	1,265	4.31	12.7

Note: Figures assume incentives of the National Energy Plan.
Source: Output in Kilowatt hrs. and Fuels Saved in Quads are taken from Metrek, Aug., 1978.
[1]Using equivalence: 1 Quad = 2.93×10^{11} Kilowatt-hours output.
[2]Except for the biomass entry, Metrek's fuel saved is roughly 2.9 times the output energy. The 0.6 Quads of fuel saved by biomass is 3.75 times the output energy, and may be erroneous.

Metrek appear as a portion of the maximum or saturation savings tabulated earlier. Savings due to wind appear very large in relation to maximum potential, and may assume offshore installations not considered in the maximum potential or may assume much higher capacity factors than those assumed in estimating potential savings. For non-electric applications there is considerable variety among the projections. A striking example is that the projection of fuel saved by solar process heat according to Metrek exceeds the most optimistic projection by S.R.I., but for space heating and biomass fuels the Metrek projections are more modest. Table 11–7 shows that the estimate of the total fuel energy saved according to Metrek lies somewhere between the Low Demand and the Solar Emphasis scenarios of S.R.I. Metrek's ambitious 6.6 Quads saved by wind dynamos has no counterpart in the S.R.I. scenarios because wind power was excluded in the latter as non-competitive.

Energy saved through using renewable sources can be a very substantial part of the total need. In the Reference case, the 15 Quads saved through all renewable sources is only 7.5 percent of the national need; in the Low Demand case, though, the 18 Quads saved is 18 percent of national need; and in the Solar Emphasis case, the 48 Quads saved is about 23 percent of the national need.

Summary of Nonrenewable Energy Amounts and Projected Energy Demand

Figure 11–9 and Table 11–8 summarize energy recoverable from identified and undiscovered (or Hypothetical) occurrences of nonrenewable resources. It is obvious that nuclear fuels predominate overwhelmingly if breeders are assumed. Assuming no breeders, nuclear energy shrinks to an amount slightly greater than the energy recoverable from conventional oil or gas. With this assumption (no breeders), coal energy is the largest amount, followed by geothermal heat (assuming ambitious development in the Gulf Coast area). Shale oil en-energy recoverable, from all beds averaging 15 gallons per ton or better, is more than twice the energy recoverable from conventional crude oil and natural gas together. If energy amount were the only criterion, the cumulative energy need for the period 1975 to 2020, as shown on Fig. 11–9, apparently could be met easily by relying on coal, geothermal or shale oil energy.

DEMANDS UPON NONRENEWABLE ENERGY SOURCES ACCORDING TO THREE SCENARIOS

Figure 11–10 arranges recoverable energy amounts from Table 11–8 in bar graph form, although linear scaling is not the ideal medium for such a large range of energy amounts: energy recoverable from breeder fuels is well off the scale, as is the energy from coal. The arrangement does allow, however, a comparison of energy amounts with possible demands on certain fuels between the present and the year 2020.

Energy demands are based on projections to the year 2020 made by the Stanford Research Institute in a report to E.R.D.A.; their character is shown in Fig. 11–10, and is elaborated in Tables 11–9, 11–10, and 11–11. There is great variety in the three futures: total energy needs for the country in 2020 range from 84 Quads in the Low Demand case to 183 Quads in the Reference scenario; and renewable energy flows (including hydro and geothermal in the S.R.I. analysis) account for only 7.5 percent of the total in the Reference case but 23 percent in the Solar Emphasis case.

Amounts of nonrenewable energy demanded from the various sources in each of the three scenarios are deduced by estimating annual contributions for each source, then accumulating those annual contributions to obtain the total demanded in the period 1975 to 2020. The estimates are listed in Table 11–11 and are represented by the flags attached to the bars in Fig. 11–10. Apparently the demands will consume very large portions of the conventional crude oil and natural gas resources and of the uranium oxide resources, though the Low Demand scenario requires considerably less from uranium oxide than the other two scenarios. The most demanding case consumes almost half of the coal Reserves (coal recoverable from Demonstrated Reserve Base) while the Low Demand case requires less than one-sixth of those Reserves. Little demand is made upon shale oil energy by the Reference or the Solar scenarios, and none is made by the Low Demand case. No flags appear on the geothermal bar because the projections are for conventional fuels *saved* by use of geothermal energy, and the amounts of geothermal energy expended to accomplish those savings

Fig. 11–9 National summary of recoverable and potentially recoverable energy from nonrenewable sources, compared with total cumulative demand for the nation to the year 2020.

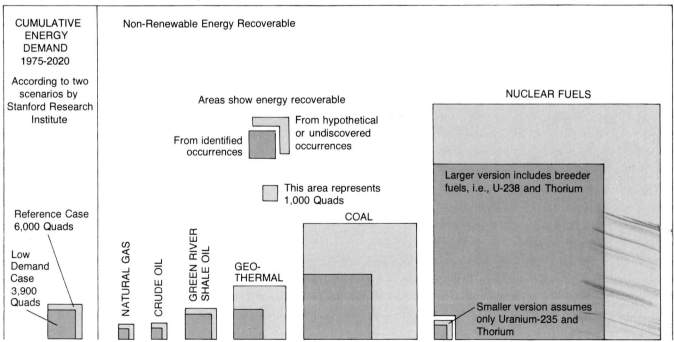

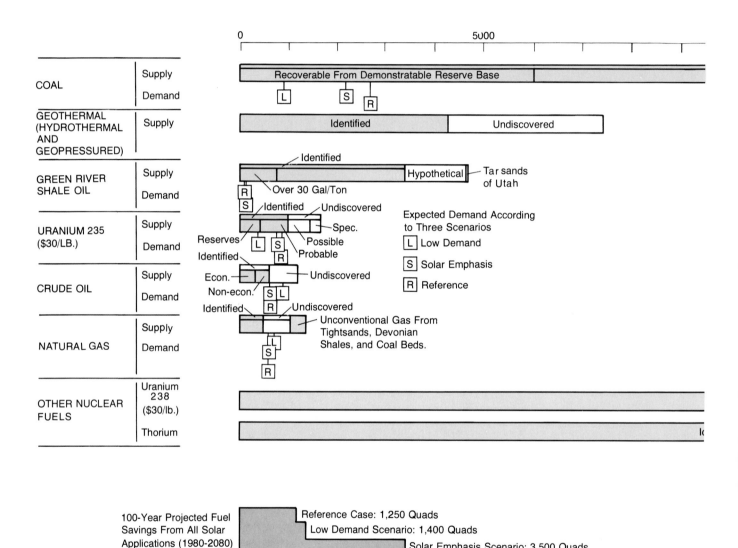

Fig. 11–10 National summary of recoverable and potentially recoverable energy from nonrenewable sources—combined with cumulative demand on each fuel to the year 2020. Also shown are amounts of fuel energy that could be saved during 100 years (1980–2080) on the basis of solar applications.

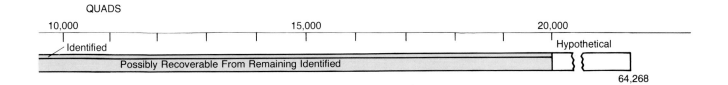

QUADS

10,000   15,000   20,000

Identified

Possibly Recoverable From Remaining Identified   Hypothetical

64,268

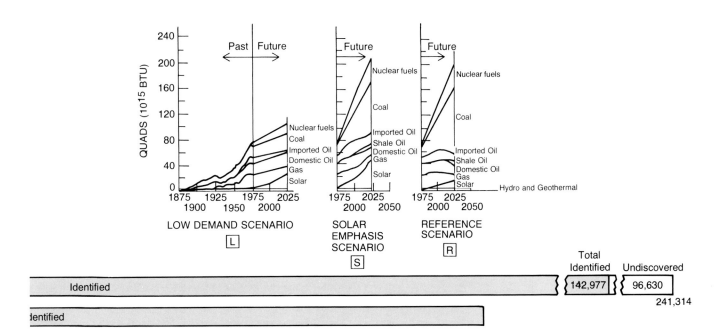

LOW DEMAND SCENARIO
L

SOLAR EMPHASIS SCENARIO
S

REFERENCE SCENARIO
R

Identified   Total Identified   Undiscovered

                                               142,977   96,630

                                                                  241,314

Identified

Table 11-7 Solar Energy Supplied in the Year 2020 According to Stanford Research Institute and Metrek Projections (expressed as Quads of fuels saved)

AMOUNTS PROJECTED BY METREK, WITH INCENTIVES OF NATIONAL ENERGY PLAN[2]	SOLAR APPLICATION	AMOUNTS PROJECTED IN THREE SCENARIOS BY S.R.I.[1]		
		Reference Case	Low Demand	Solar Emphasis
3–5	Hot Water and Space Heating	10	10	18
13	Process Heat in Agriculture and Industry	0	0	4
6.6	Wind Electric	0	0	0
2.9	Solar Thermal Electric			
0.2	Photovoltaic	0	0	11
2.4	OTEC Electric			
0.6	Biomass— Wood for Electric			
4.4	Biomass—Fuels	1	4	11
34.1				
TOTAL SOLAR		11	14	44
Hydro and Geothermal		4	4	4
TOTAL FROM "RENEWABLE" SOURCES		15	18	48
TOTAL NATIONAL NEED FOR PRIMARY ENERGY (Quads)		198	102	204

[1]Stanford Research Institute, 1977. [2]Metrek, Aug., 1978.

cannot be specified here. Tabulated amounts of fuels saved through a combination of hydroelectric and geothermal energy are not very large, and suggest that relatively small demands for geothermal energy are anticipated.

Figure 11–10 includes estimates of all fuels saved through 100 years of renewable energy flows (including geothermal, in the S.R.I. approach). These amounts were taken from the S.R.I. graphs and accumulated in the same way as cumulative demands upon nonrenewable sources, except that the contributions were assumed to *level off* after the year 2020, and were projected at that level to the year 2080. Even the two more modest projections result in fuel savings that, by the year 2080, exceed all the energy now in conventional crude oil resources.

OTHER ENERGY FUTURES

The three S.R.I. scenarios provide yardsticks by which to assess energy recoverable from various sources; but they do not represent ideal energy futures, either in total national need or in the mix of sources that may provide for that need.

The Solar Emphasis scenario displaces coal, and to a lesser extent nuclear fuels and imported petroleum, by means of growth in output from solar energy; but it makes no reduction in demand for domestic crude oil or natural gas. This is because the total national need for energy is expected to increase at 4.5 percent per year. Total cumulative demand through the year 2020 is about the same as for the Reference case (Table 11–11).

The Low Demand scenario greatly reduces nuclear and coal contributions as well as petroleum imports; but it, too, relies on domestic oil and gas production that may be unsustainable. This is in spite of the fact that the total demand for primary energy grows only at the rate of population growth (0.6 percent) and the total cumulative need for energy by 2020 is just two-thirds of that in the Reference case.

It is clear that ever-increasing total energy needs are extremely difficult to meet, especially if the use of environmentally undesirable sources is restricted. A prospect does exist, though, for a future in which total demand peaks at about 100 Quads, then diminishes to 60 Quads by the year 2025 by virtue of greatly reduced conversion and distribution losses (Lovins, 1978). A more extreme future that assumes some alteration of settlement and transportation patterns in the country shows the possibility of the national energy need dropping to 35 Quads by the year 2050 (Steinhart, *et al.*, 1978). Such a level of need could be supplied largely by renewable sources.

Areas of Resource Occurrence and Coincidence

NONRENEWABLE SOURCES

Figure 11–11 maps the literal areas of occurrence of coal of all ranks, conventional oil and gas fields, Green River oil shales, uranium oxide, and identified geothermal occurrences. Excluded for the sake of simplicity are the oil sands of Utah, tight gas sands of the western states, and Devonian (gas) shales of the Appalachians.

Most coincidences of resource location occur in the West, where current activity in subbituminous coal mining, Green River shale processing, and geothermal exploitation foreshadow a future of rapid development. The region generally is one of shortage of both surface and ground water, and at the same time is susceptible to air pollution because of upper-air inversions that accompany the semi-arid and arid conditions. Some of the anticipated development will result in fuels that are transportable and hence will be burned beyond the region;

nevertheless, surface mining damage and water pollution in the region must be dealt with. In the case of geothermal energy, especially for space heating, the energy consumption must be near the source, so some migration to geothermal sites may occur.

Another interesting area is the Gulf Coast, where conventional oil and gas fields and future prospects coincide with uranium oxide deposits and with geothermal energy in geopressured reservoirs.

RENEWABLE SOURCES

Figure 11–12 combines areas of potential for ocean thermal gradients, wind, and solar radiation. Ocean gradients are suitable for electrical generation only in the Gulf of Mexico (and in Hawaii, Virgin Islands, Puerto Rico, and Guam). Favorable areas of wind (Fig. 11–12) are a combination of High, Moderate, and Low wind regimes. Areas favorable for use of solar radiation are represented by two measures. First, the average total daily radiation, direct and diffuse, is used to define areas where various applications, including photovoltaic, would enjoy an abundant resource. Second, the more restricted areas of abundant direct (focussable) radiation, suitable for high-temperature applications, are mapped. Areas where the two coincide are recognized by a darker tone on the

Fig. 11–11 Nonrenewable energy sources: locations superimposed to show areas that are energy-rich and areas of coincidence with shortage of surface water.

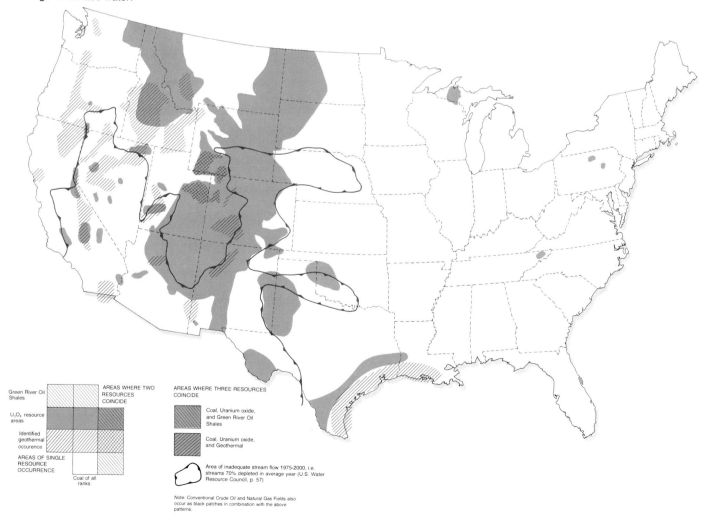

map. Not represented here are the areas in Wyoming, Colorado, and Nevada where exceptional net savings are possible through the use of solar heating systems for buildings: the definition of such areas depends upon specific assumptions about the particular application rather than upon the raw resource.

Especially interesting in Fig. 11–12 is the area embracing southern California, Nevada, Utah, southern Wyoming, Colorado, New Mexico, and west Texas, where favorable wind regimes coincide not only with exceptional total solar radiation but also with greatest direct radiation. As shown in Fig. 11–11, it is apparent that many southwestern areas favored by the sun and wind are short of surface water. Not shown is the fact that south-

Table 11–8 National Summary of Energy Thought to Be Recoverable from Identified and Undiscovered (or Hypothetical) Nonrenewable Sources

RESOURCE	Identified	Hypothetical or Undiscovered	Total	PERCENT OF TOTAL	PERCENT OF TOTAL WITHOUT BREEDER FUELS[1]
Geothermal	4,565	9,079	13,662	3.9	16.0
Geopress.	4,202	1,139	5,341		
Hydro. 150°+ C	204	3,821	4,025		
Hydro. 90–150° C	159	4,119	4,278		
Nuclear Fuels[2]			259,902	75.4	2.0
Uranium-235	1,018	688	1,706		
Reserves	416		416		
Probable	602		602		
Possible		468	468		
Speculative		220	220		
Uranium-238	142,978	96,630	239,608		
Reserves	58,427		58,427		
Probable	84,551		84,551		
Possible		65,731	65,731		
Speculative		30,899	30,899		
Thorium	18,588	0	18,588		
Crude Oil	587	656	1,243	0.4	1.5
Reserves	154	459	613		
Indicated	22		22		
Inferred	128		128		
Sub-economic	283	197	480		
Natural Gas	505	556	1,061	0.3	1.2
Reserves	200	494	694		
Inferred	201		201		
Sub-economic	104	62	166		
Shale Oil[3]	3,388	1,275	4,663	1.3	4.0
Coal	20,000	44,268	64,268	18.7	75.3
From DRBase	6,074		6,074		
Other	13,926	44,268	58,194		
Total	191,629	151,877	344,781		
TOTAL WITHOUT BREEDER FUELS	30,063	55,247	86,585		

[1]Breeder fuels are U-238 and thorium. Total energy recoverable for these is 258,196 Quads.

[2]Includes uranium oxide resources up to cost of 50 dollars per pound. Reserves here include 120,000 tons of uranium oxide recovered as by-product in other refining operations.

[3]Using all beds that together average over 15 gallons per ton of rock. In beds yielding over 30 gallons per ton, the recoverable energy is estimated at 772 Quads.

Table 11–9 Features of the Three Energy Futures to the Year 2020

	REFERENCE CASE	SOLAR EMPHASIS	LOW DEMAND CASE
Basis, or Assumptions	Energy supply shares are based on baseline cost assumptions (those in 1975).	Solar costs 50 to 70 percent lower than in Reference Case. Cost reductions and solar share of market are those of the ERDA-49 study (see ERDA-49). Various solar contributions together would, in 2020, supply as much energy as today's oil and gas industries.	Commitment to conservation. Energy prices forced to double by 1985 and triple by 2000, remaining level thereafter. Economic growth of 2 percent per year after 2000.
Total Need in Year 2020	198 Quads	204 Quads	102 Quads
Coal	Western production rises to 4 billion tons by 2020, and production in East and Midwest also rises rapidly.	Coal shows main impact of growing solar use. Western coal output declines by 14 Quads, eastern by 8 Quads compared to Reference Case.	Western production is reduced from over 3 billion tons annually (Reference Case) to less than 1 billion. Eastern production reduced by more than 50 percent.
Nuclear	Increases from 34 plants of 1,000-Megawatt size in 1975 to over 600 plants of that size.	Energy contribution drops slightly, from 37 Quads in Reference Case to 34 Quads.	Reduced to about the number of plants under construction or planned in 1977.
Solar	Only 11 Quads by 2020.	All solar contributions (biomass, electricity, residential and commercial space heating, and industrial and agricultural heat) total 44 Quads.	20 Quads by 2020.
Oil and Gas	Amounts used decline as coal contribution grows. Shale oil enters market after 1990.	Total crude only slightly less than in Reference Case. Role of imports roughly same as in Reference Case.	Total crude used annually roughly same as in 1975. No development of shale oil.

Table 11–10 Contributions of Various Sources to Total Energy Supply in Year 2020 According to Three Scenarios by Stanford Research Institute

	Low Demand Scenario		Solar Emphasis		Reference Scenario	
	Quads	Percent of Total	Quads	Percent of Total	Quads	Percent of Total
Renewable[1]						
Biomass	4	4	11	5	1	0.5
Space Heating by Solar	10	10	22	11	10	5
Solar-electric	0	0	11	5	0	0
Hydro and Geothermal	4	4	4	2	4	2
NonRenewable						
Nuclear	15	15	34	17	37	18
Domestic Oil and Shale oil	21	21	17	8	18	9
Oil and Gas Imports	3	3	14	7	16	8
Domestic Gas and LPG	13	13	8	4	7	3
Coal	32	32	83	41	105	53
TOTAL	84		156		183	

[1]For renewable sources the energy amounts are fuel saved.   Source: ERDA, March, 1977, p. viii

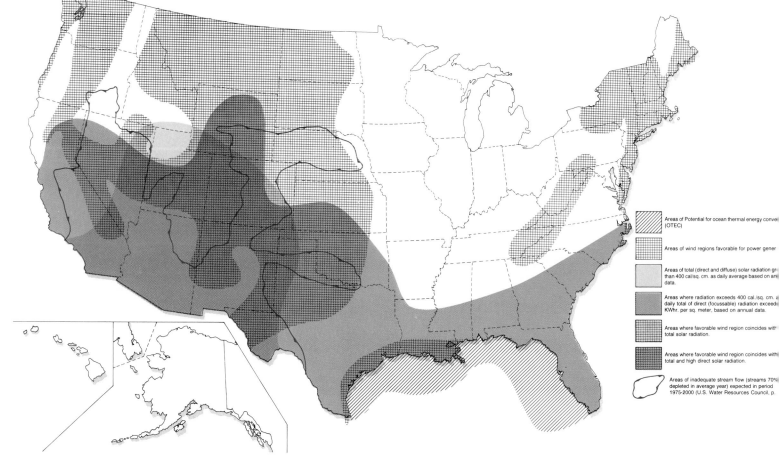

Fig. 11-12 Three renewable energy sources: areas favorable to solar radiation, wind, and ocean thermal gradients superimposed to show areas that are energy-rich and areas of coincidence with shortage of surface water.

Legend (as printed on map):

- Areas of Potential for ocean thermal energy conve[rsion] (OTEC)
- Areas of wind regions favorable for power gener[ation]
- Areas of total (direct and diffuse) solar radiation gr[eater] than 400 cal/sq. cm. as daily average based on ann[ual] data.
- Areas where radiation exceeds 400 cal./sq. cm. [as] daily total of direct (focussable) radiation exceed[s] KWhr. per sq. meter, based on annual data.
- Areas where favorable wind region coincides with total solar radiation.
- Areas where favorable wind region coincides with total and high direct solar radiation.
- Areas of inadequate stream flow (streams 70% depleted in average year) expected in period 1975-2000 (U.S. Water Resources Council, p.

ern plains areas through Kansas, Oklahoma, and west Texas, which are rich in wind and sun, are also areas of overdrawn ground water.

Biomass resource areas and areas of hydroelectric potential are excluded from the summary because of difficulty in defining literal areas of occurrence. In the case of hydroelectric, the areas of greatest undeveloped potential are in the states of Alaska, Washington, California, Idaho, and Oregon.

THE ELEVEN ECONOMIC REGIONS AND THEIR ENERGY RESOURCES

Figure 11–13 summarizes the resources occurring in the eleven economic regions of the country according to two viewpoints. First, a *region's role or contribution with regard to particular resources* is noted: the letter U indicates that the region is unique in holding the resource, while the letter O indicates the region is outstanding. Outstanding status is assigned for mineral fuels on the basis of a region's holding more than 10 percent of the Nation's energy Reserves as defined earlier (Table 11-3). For renewable energy, the assignment of outstanding status is more subjective but still considers the region's share of the resource potential. The second viewpoint

recognizes the *importance of certain resources to the region:* a block of color in a space shows that, in light of other alternatives, the resource indicated may play a large role in the future of the region, and may, on the basis of economic projections, deliver more energy in that region than in regions favored with a greater potential but having smaller demand.

A useful distinction may be made between *national* resources which are transportable fuels that can be shipped throughout the Nation, and *regional* resources, such as solar radiation, wind, hydroelectric, OTEC power, and geothermal heat, which are likely to be used locally or within short distances of their extraction. It is recognized, however, that electrical power produced from such renewable or semi-renewable (geothermal) energy flows can be converted into products such as hydrogen or ammonia and shipped from remote sites to the marketplace for consumption in fuel cells or engines.

MOUNTAIN REGION Because of its size and geological variety this region embraces a remarkable collection of energy sources. It is unique in holding rich oil shale deposits; it is outstanding in its share of the Nation's coal energy, nuclear fuels, and geothermal Reserves; and it is also outstanding with regard to solar radiation, wind

	ENERGY SOURCES													
	NON-RENEWABLE [1]									RENEWABLE				
	Fossil fuels					Nuclear fuels		Geothermal		Hydro-electric (3)	Solar Radiation		Wind (6)	OTEC
	Coal on the basis of			Conv. oil and gas	Green River Oil Shales	U3O8	ThO2	Hydro-Thermal & Volcanic	Geo-pressured		High Temp	Low Temp (5)		
REGIONS	Low Sulfur	Gasification	Total energy											
Mountain	75%	51%	39%		▓	75%	60%	55%		13%				
West So. Central				51%					▓					
East So. Central			24%				16%						▨	
Alaska		36% (2)		22%				▨		13%			▨	
South Atlantic					▨		24%				░		▨	▨
Mid-Atlantic					▨					▨			▨	
West No. Central					▨							▨		
East So. Central					▨						░			
Pacific								43%		36% ▨	(4)		▨	
New England										▨		▨		
Hawaii											▨	▨	▨	▨

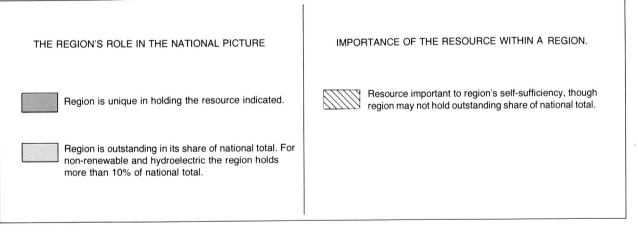

THE REGION'S ROLE IN THE NATIONAL PICTURE

▓ Region is unique in holding the resource indicated.

░ Region is outstanding in its share of national total. For non-renewable and hydroelectric the region holds more than 10% of national total.

IMPORTANCE OF THE RESOURCE WITHIN A REGION.

▨ Resource important to region's self-sufficiency, though region may not hold outstanding share of national total.

1. In most cases, a region's character is based on its share of energy *reserves* as shown by percentages entered in resource columns.

2. Considering hypothetical coal as well as reserves.

3. Percentages show region's proportion of total hydroelectric potential, developed and undeveloped.

4. Mostly in southern California.

5. Regions noted as outstanding are those spanning the southern part of the country where total radiation is abundant. For solar *space heating* the West North Central region and northern part of the mountain region are outstandingly attractive on the basis of net savings estimated by simulation studies.

6. Very attractive wind power available exists off the Pacific, mid-Atlantic, and South Atlantic coasts, though it was not included in the wind chapter assessment of total potential generating capacity.

Fig. 11–13 Eleven economic regions and their energy resource highlights.

Table 11–11 Cumulative Energy Demand from 1975 to 2020, According to Three Energy Scenarios (in Quads)

CUMULATIVE DEMAND
TO YEAR 2020[1]

RESOURCE	Low Demand Scenario	Solar Scenario	Reference Scenario
Solar	473	1,013	270
Hydro and Geothermal	158	158	158
Nuclear Fuels	383	810	878
Domestic Oil	900	653	653
Shale Oil		158	113
Domestic Gas	765	653	630
Coal	923	2,205	2,700
Oil and Gas Imports	360	608	653
TOTAL	3,962	6,258	6,055

[1]Cumulative demand was estimated from the three energy scenario curves, and is only approximate.

Source of Scenarios: U.S. Energy Research and Development Administration. *Solar Energy in America's Future.* Division of Solar Energy. Washington: Government Printing Office, March 1977.

Notes:

1. In addition to the demands on domestic resources, the *Low Demand, Solar,* and *Reference* scenarios call for oil and gas imports of 360, 608, and 653 Quads respectively.

2. The three scenarios, *Low Demand, Solar,* and *Reference* call for 473, 1,013, and 270 Quads of solar energy respectively.

3. In the Stanford Research Institute model used in generating the three scenarios, interaction of price and demand as a resource is depleted depends on certain assumptions about available amounts of resources. The amounts used by S.R.I. may differ considerably from those shown in Fig. 11–11, since their study and this summary of recoverable energy amounts were done independently.

regimes, and hydroelectric potential. Some of the resources, such as oil shales and intense direct radiation, are concentrated in small parts of the region; but some, such as coal Reserves and wind potential are accessible in many parts of the region. Because of its mineral energy which can be processed and shipped, the Mountain region appears as a storehouse of national energy, but at the same time is well-endowed with geothermal, solar, wind and hydroelectric energy which are better regarded as resources for local or regional use.

WEST SOUTH CENTRAL REGION   Because the region includes Texas, Louisiana, and Oklahoma, it holds 51 percent of the Nation's oil and gas Reserves; and it also holds a large portion of the oil and gas that is not yet proven. Substantial, though not outstanding, amounts of uranium oxide occur in the Texas coastal plain along with large amounts of energy in geopressured reservoirs. Although geopressured reservoirs probably exist elsewhere, the large and well-studied geopressured resource is unique to the Gulf Coast and, therefore, to the West South Central

region. Energy in these reservoirs was not included in the summary of energy Reserves (Table 11–3) because of some uncertainty about the economics of extraction; but it probably will prove large and economically feasible. As a southern region, this one has abundant total radiation as a resource, and as a Gulf region it will have access to electrical energy from OTEC installations if they prove workable. Like the Mountain region, therefore, this one is important for national fuel resources and has outstanding regional resources as well.

EAST NORTH CENTRAL REGION   Because of deposits in Illinois and Ohio, this region holds 24 percent of the Nation's coal Reserve energy (Table 11–3). In addition, 16 percent of thorium oxide Reserves are in the Palmer district of upper Michigan. The region is outstanding in no other national resource, and is not well-endowed with solar radiation, wind, or hydroelectric energy; nevertheless economic projections (Metrek, 1978) suggest that electrical output from wind generators may be very important to the region—presumably at sites on the shores of the Great Lakes where local winds are more attractive than those for the region as a whole.

ALASKA   Because of its coal and petroleum, this state (and region) holds a large part of the Nation's transportable energy. For oil and gas, 22 percent of the Nation's Reserves are in Alaska, as well as large parts of anticipated Reserves. While Reserves of coal are not outstanding, Hypothetical amounts of coal are extremely large according to a recent Department of Energy assessment which considers these coals part of the resource for coal gasification. This technology, and/or coal liquefaction, could facilitate long-distance shipment of Alaska's coal energy. In geothermal energy, Alaska holds moderate amounts of hydrothermal heat in exploitable occurrences, and very large amounts of heat in Aleutian volcanic occurrences, many of which have not been measured. Ostensibly, this is a regional resource, but may in the future be shipped southward in the form of products such as hydrogen. The same may be said for the very attractive coastal wind power not included in the wind chapter assessment, and for the hydroelectric potential which is 13 percent of the national total and is now only 0.4 percent developed.

SOUTH ATLANTIC REGION   In the scheme of Fig. 11–13 this is the last ranked region with any outstanding portion of the Nation's mineral energy reserves. Thorium oxide deposits in beach placers and offshore deposits give this region a resource that could be very important if future nuclear power makes use of thorium breeders. The region includes the state of West Virginia, whose coal Reserves are very large, though not enough to make the region outstanding: in a ranking by state shares of coal Reserve energy, West Virginia would be number four (Table 11–1). Offshore wind regimes could be important to the region in the future, as could the potential for OTEC power which is shared by two other regions bordering the Gulf of Mexico.

MID-ATLANTIC, WEST NORTH CENTRAL, AND EAST SOUTH CENTRAL REGIONS These three regions hold similar totals of mineral energy Reserves in which coal is the dominant resource (Table 11–3). In the Mid-Atlantic, offshore wind potential and New York and Pennsylvania's hydroelectric potential are distinguishing features. In West North Central the oil and gas energy of North Dakota is significant; and in East South Central, hydroelectric potential is slightly greater than that of Mid-Atlantic and earns it a marginally outstanding rating in Fig. 11–13.

PACIFIC REGION Because of small coal deposits, the region does not have a large portion of the Nation's mineral energy Reserves; but on a state by state basis, California holds 16 percent of the Nation's crude oil Reserves, while Alaska and Texas each hold 28 percent, and Louisiana only 9 percent. The distinctive character of the Pacific region, though, is due to semi-renewable and renewable resources. In hydrothermal energy Reserves, the region is clearly outstanding with 43 percent, mostly in California and Oregon: these occurrences are largely of the high-temperature type suitable for electrical power generation. Southern California has areas of high total and direct solar radiation suited, therefore, to both low and high-temperature application. The attractive wind potential is mostly off the Washington coast, while outstanding hydroelectric potential is concentrated in Washington state, but is also very large in Oregon and California.

NEW ENGLAND New England stands out as the only region completely without mineral fuels or geothermal occurrences. This is largely because it is small in area and lacks the sedimentary rocks necessary for fossil fuels and also the recent volcanic activity that confers western-style geothermal potential. Some low-grade uranium ores exist here, but they exist in other regions as well. Because of high fuel prices, the region can profitably use solar radiation for space heating and other low-temperature applications. It has outstanding wind potential offshore, and very substantial hydroelectric potential in Maine and New Hampshire.

HAWAII Because of its simple volcanic nature, there are no mineral fuels in Hawaii, but fortunately it has a variety of semi-renewable and renewable energy sources upon which to draw. While its geothermal resources appear small when compared to the national total they are very substantial in relation to the region's needs. It receives large amounts of total and direct solar radiation, as well as predictably steady trade winds. As with many tropical volcanic islands, the steeply sloping ocean floors mean that floating OTEC platforms can find suitable temperature contrasts quite near to the shore. If the technology is workable, OTEC-generated power may be very important to Hawaii.

Glossary

abundant metals Those metals that constitute a relatively large portion of the earth's crust. Aluminum, for instance, is present as roughly 8 percent of ordinary rock. Distinguished from scarce metals whose crustal abundance is less than 0.1 percent.

active solar heating Using a solar collector, a storage device, and pumps or fans to transfer the heat obtained from the sun. Such systems can be added (retrofitted) to existing buildings whether or not the buildings have been designed with using solar energy in mind.

ammonia (NH_3) A colorless pungent gas composed of nitrogen and hydrogen and appearing in a number of energy contexts. It is manufactured for fertilizers because of its nitrogen content. OTEC installations and some biomass operations may produce it and thus save the natural gas (NH_4) now consumed in ammonia manufacture. Ammonia can also be used as a feedstock for fuel cells which produce electricity chemically. It may be used as the working fluid in OTEC generating plants.

aquifer A rock, usually of the sedimentary type, that holds and transmits water.

associated-dissolved gas Natural gas (dominantly methane) which occurs either as a "gas cap" overlying crude oil in the pores of a reservoir rock, or as gas dissolved in crude oil and released when the oil reaches atmospheric pressure.

basin, or geologic basin A present-day area of thick sedimentary rocks which, in the past, accumulated in low areas. Surrounding areas at that time were high and subject to erosion. Examples are the Powder River Basin or the Western Gulf Basin.

breeding ratio In a breeder reactor, the ratio of Plutonium-239 produced to the Uranium-238 consumed. If the value is not greater than 1.0, the reactor is not truly a breeder.

British Thermal Unit (BTU) The energy required to raise the temperature of one pound of water 1 degree Fahrenheit. Equivalent to 252 calories, or 0.252 Calories.

calorie The energy required to raise the temperature of 1 gram of water 1 degree Celsius. Often called the gram calorie to distinguish it from the nutritionists' Kilocalorie or Calorie which is 1,000 calories.

capacity factor As applied to electrical generators, the proportion of generating capacity (in Watts) that is produced on average. Or, the proportion of time during which the generator produces at capacity. Expressed as a fraction or percentage, the factor is multiplied by the generating capacity and by 8,760 hours in a year in order to express the annual output of energy in Kilowatt hours.

conventional crude oil Crude oil that exists as liquid oil in the pore spaces of a reservoir rock and will flow, with varying degrees of recovery effort, through the reservoir rock into a borehole. Distinguished from unconventional (synthetic) crude oils.

conventional hydroelectric power Electrical power produced from water that by natural stream flow accumulates behind a dam of considerable height and thus holds potential energy.

conventional natural gas Gas (dominantly methane) existing *as gas* in a reservoir rock, and able to flow into a borehole with little or no need for special techniques such as fracturing the reservoir rock. Distinguished from unconventional synthetic natural gas which is producible only with special efforts at recovery (from western tight gas sands and eastern shales) and from synthetic gases produced from various raw materials. Also distinguished from the methane which originates in coals and which is now trapped in fractures in coal beds.

conversion of energy, or energy conversion The change of one energy form to another. 1. Chemical energy of coal may be changed into heat energy. 2. That heat energy may be converted into electrical, or 3. The kinetic energy of falling water may be converted into electrical. 4. Electrical energy may be converted into heat. In some conversions, especially conversion of heat into electrical, there is a very substantial loss of energy in the conversion process. See **efficiency**.

critical fuel cost In the analysis of active solar space-heating systems, the cost of fuel (to the consumer) which is just high enough to make a solar installation economically attractive by virtue of the fuel saved.

crystalline rocks A generalization applied to igneous and/or metamorphic rocks to distinguish them from sedimentary.

Cumulative Production For some resources, the accumulation or total of all amounts that have been produced to date.

degree-days The sum of degree differences between mean daily temperature and a desired indoor temperature. The total for each day may be added for a year or for any part of it. Heating degree-days use a base of 65 degrees Fahrenheit, while cooling degree-days use 80 degrees Fahrenheit.

Demonstrated Reserves (of crude oil or natural gas) The total of Measured (Proved) reserves and Indicated additional amounts thought to be recoverable through secondary recovery techniques.

Demonstrated Reserve Base All the coals, of various ranks, that are economically mineable according to criteria of the U.S. Bureau of Mines.

deuterium (Hydrogen-2) An isotope of the element hydrogen that has a nucleus containing one proton and one neutron, and thus has a mass number of 2.

diffuse radiation Solar radiation received under cloudy or hazy skies. Because it is diffuse it cannot be focused by mirrors or lenses for high temperature applications. It is effective, though, in warming solar collectors for space heating, and can also be utilized by photovoltaic cells. Distinguished from **direct radiation**.

direct-normal radiation The direct (or that which can be focused) radiation falling onto a surface which is tilted to intercept radiation at a 90-degree angle. Thus the radiation is normal to the surface.

direct radiation Solar radiation received under clear skies. Since it is not diffused, it can be focused by mirrors or lenses to cause high temperatures needed for power generation and some industrial tasks. Distinguished from **diffuse radiation**.

doubling time In a breeder reactor, the number of years required to double the initial fuel load.

drainage basin The land area encompassing all the tributary streams of the river or rivers that define the basin. The boundary of one basin comprises all the drainage divides that separate the included rivers from those that are excluded.

ductile Capable of being hammered into thin layers or of being drawn out into wire. Describes certain metals.

economic A short-hand expression describing a resource or a process that is economically attractive or feasible. Distinguished from non-economic or subeconomic.

efficiency As applied to a specific energy conversion device, such as a power plant, this is the ratio of the total output of useful energy from a process to the total energy input. Electrical energy delivered from a power plant is roughly 40 percent of coal energy input.

electrical resistance heating An example is residential baseboard heating, in which heating elements convert electrical energy into heat. Since the goal is to make heat, the heating elements are 100 percent efficient. The overall efficiency from fuel through generator and transmission lines, however, is low. Distinguished from electrical **heat pumps.**

end-use energy The amount of energy delivered by some process. Three examples are: heat produced by an electrical heating element; heat supplied to a building by a furnace; or mechanical energy that drives a vehicle.

energy The ability to do work or to raise the temperature of a substance. Also equivalent in physics to the work accomplished. If the energy is expended more quickly (or work done more quickly) then a device of greater power has been used. Energy is distinguished from **power**.

enhanced recovery techniques Often called *tertiary* recovery techniques. Ambitious methods of recovering crude oil that remains in reservoir rock after other methods have ceased to be productive. Methods include use of steam, underground combustion, or gas injection in order to lower the oil's viscosity, or detergents to flush the oil out of the reservoir rock.

enrichment (of uranium oxide) The process in which the abundance of the fissile isotope, U_{235}, is increased from the original 0.7 percent to 2–3 percent of the uranium in the U_3O_8 compound. The process depends on the slight differences in density between U_{235} and U_{238} which is the isotope that constitutes 99.3 percent of natural uranium.

feathering of blades In aircraft propellers or wind machine blades, the adjustment of blade pitch so that blade will bite the air at more or less effective angles. In high winds a wind machine may feather blades in order to slip-by the additional energy and thus maintain a fixed rpm. In very high winds the blades may be feathered completely and become idle for safety.

fission products Unstable isotopes produced when an atomic nucleus is split. Typically, Uranium-235 plus a neutron splits into a heavier and a lighter nucleus; Strontium-90 and Xenon-138 are examples.

fixed carbon in coal That part of the carbon that remains when coal is heated in a closed vessel until the volatile matter is driven off. It is the non-volatile matter in coal minus the ash.

flaring of natural gas Burning of gas at the well-head when it is produced in association with crude oil, and no market or storage exists for the gas.

formation A term applied to a rock unit that persists and is recognizable over many miles and is therefore a useful reference for rock units above and below it. Named for the location of the outcrop (exposure) where the rock was first thoroughly studied. An example is the Green River Formation.

fossil fuels Coal, crude oil of various types, and natural gas: fuels which are derived from the accumulation of plant or animal remains in ancient sedimentary rocks. *Fossil* refers to the preservation of the products of ancient life (organic materials) which themselves depended on the sun's energy utilized through photosynthesis.

fuel saved The amounts of fuel whose burning is averted by substituting some renewable energy flow such as solar radiation or wind. Amounts of fuel saved depend on what task the fuel would have been used for. For instance, to accomplish the energy output of a solar-electric generator, roughly 2½ times the output energy would have been input as fuel because the conversion of heat to electrical energy is only 40 percent efficient. In space heating, however, as much as 85 percent of fuel's energy is delivered as heat to the building by means of a furnace.

generating capacity The maximum number of Watts of electrical power that a generator can sustain. A rate, therefore, of converting some energy form into electrical energy. Expressed in Kilowatts (1,000 Watts) or Megawatts (1,000,000 Watts) electric. Distinguished from **output**, and also distinguished from Kilowatts or Megawatts thermal (KWe as opposed to KWt) which express the rate of producing heat from the fuel.

geologic province An area that is more or less uniform in its geologic history and characteristics and can be distinguished from other large areas with regard to its mineral prospects.

geophysical methods Methods of mineral exploration that make use of seismic shocks, or that study the magnetic and gravity characteristics of an area to find areas with features favorable to mineral occurrence. For oil and gas the features usually sought are structures which might be a trap.

geopressured reservoirs Deeply buried isolated sand bodies which have high pressure due to their inability to compact under the weight of overlying rock, and which hold saline waters whose high temperatures are due to insulating layers of overlying rock.

Gigawatts Power, or electrical generating capacity equivalent to 1,000,000 Kilowatts, or 1,000,000,000 Watts.

half-life The time period required for one-half the atoms in a given amount of a radioactive substance to decay into another isotope.

heat pump A device like an air conditioner, which can move heat 'uphill' from an environment of lower temperatures (such as 50 degrees Fahrenheit air outdoors) to an environment of higher temperatures (such as 65 degrees air indoors). Efficiency of heat pumps can be surprisingly high because the machine is not converting electrical energy to heat (with an efficiency of 100 percent) but is using the electrical energy to *move* heat from one place to another. The energy delivered is more than 100 percent of the electrical energy expended.

heavy crude oil Crude oil that is viscous, but actually low in specific gravity. Its viscosity leads to lower recovery rates than for 'light' (less viscous) crude oils. Extremely heavy crude oils are much like the bitumen in oil sands and must be extracted by special techniques which entail heating the oil-bearing rock in-place.

high-sulfur coal Coal with sulfur content greater than 3 percent by weight. See **low-sulfur coal**.

hot water and space heating The heating (by solar energy for example) of water for domestic uses and to provide heat for buildings.

hydrologic cycle The processes by which water is evaporated from oceans and lakes, and eventually falls as precipitation, some of which infiltrates the ground, some of which evaporates, and some of which runs off in streams to oceans and lakes.

hydrostatic Describes pressure resulting from a standing column of fluid, or a body of water held at some elevation, as by a dam.

hydrothermal The hot spring type of geothermal occurrence in which waters heated at depth flow to the surface as hot water or steam. The water can be used for space heating, or for electrical power generation if temperatures exceed 150 degrees Celsius.

Hypothetical resources Those mineral resources or resource amounts that are thought to exist on the basis of studied geologic predictions or explorations.

Identified resources Those mineral resources or resource amounts known to exist on the basis of information supplied by closely spaced drill holes and/or surface exposures.

igneous rocks Those rocks whose immediate origin is molten rock, either at the earth's surface (volcanic type) or within the earth (plutonic type).

igneous or volcanic geothermal occurrences Relatively shallow molten or very hot rock which may be used as applied energy in the near future by introducing water through drill holes and then retrieving hot water or steam.

Inferred reserves Those amounts of mineral resources that are less certain than *Measured* (Proved) resources but more certain than Hypothetical or Undiscovered.

in-situ recovery Recovery of minerals from deeply buried rock without mining. The mineral-rich rock is left in place; the minerals are extracted at depth, and then brought to the surface in liquid or gas. For oil shales, this process entails distillation underground.

isobars On weather maps, the lines joining points of equal atmospheric pressure. They define areas of high and low pressure and reveal, by their spacing, whether there is a rapid or gradual change in pressure across a given area.

isotopes Different forms of an element that have the same number of protons in the nucleus of each atom but differ in the number of neutrons.

Joule The amount of energy that would be used by a 1-Watt light bulb in 1 second.

kerogen A waxy, organic material containing the remains of aquatic organisms. The constituent of Green River oil shales that is converted into crude oil by distillation.

Kilowatt Power, or electrical generating capacity, of 1,000 Watts.

kinetic energy The energy of some body that results from the body's motion. It is equivalent to ½ mV^2, where m is the mass and V is its velocity. Distinguished from *potential energy* whereby a body raised to some height has the potential to expend energy and do work as it falls.

lacustrine Produced by or related to lakes, such as lacustrine sedimentary deposits, which are sediments deposited in lake basins. Distinguished from *fluviatile* (in rivers) and *marine* (in salt water).

Langley Amount of solar radiation energy equivalent to 1 calorie received per square centimeter of surface.

leaching The process of washing or draining by percolation of water. During the process material is picked up and carried or dissolved and carried.

linear scale On graphs, the commonplace scale in which a 0–40 range of values will be double the length of 0–20.

liquefied natural gas (LNG) Natural gas liquefied at very low temperatures and high pressure for long-distance transport by tanker to special terminals where it is allowed to warm and expand into pipelines.

lithium A very light metal being used experimentally in the development of a rechargeable lithium-anode battery for automotive and peak-energy storage use. Also the source of tritium, a possible fuel for fusion-type nuclear reactors.

Lithium-6 An isotope of the metal lithium, which, under neutron bombardment, is converted into tritium, a radioactive isotope of hydrogen which is a possible fuel for fusion-type nuclear reactors.

logarithmic scale On graphs, a scale in which the logarithms of values are evenly spaced, so the values themselves are more widely spaced at the low end and more crowded at the high end of the scale. Used for values which have a very large range from lowest to highest.

long ton 2240 pounds or 1.016 short tons.

low-sulfur coal Coal with sulfur content 1 percent or less by weight. See **high-sulfur coal**.

market penetration The extent to which a new technology is expected to displace present ones. For solar energy, the proportion of some energy need that will be supplied by solar energy. For heating and cooling of buildings, for instance, the contribution from solar energy may be 10 percent in one region, but 30 percent in another region where economic or physical factors favor the use of solar energy rather than fuels.

median value In any array of numbers the median is the number midway between the highest and the lowest. Distinguished from the *mean* value which is half the sum of the highest and lowest values, and the *mode*, which is the value that occurs most frequently.

Megawatt Power, or electrical generating capacity equivalent to 1,000 Kilowatts, or 1,000,000 Watts.

metamorphic rocks Those rocks derived through the alteration of some pre-existing rock. Marble, for instance, is altered limestone.

metric ton Same as long ton. See **short ton**.

mineable Terms applied (as with coal) to tonnages that are physically and economically accessible. Distinguished from amounts expected to be *recovered* (recoverable) through the mining.

moderators A substance used in nuclear reactors to slow down neutrons, allowing them to be more readily captured by Uranium-235. Water deuterium (heavy water) and carbon in the form of graphite are commonly used moderators.

monazite A mineral that is a phosphate of the cerium metals, but contains small quantities of thorium oxide. The mineral occurs as large crystals in crystalline rocks, and is also found in placer deposits.

natural gas liquids Also called *condensate* and *liquid petroleum gases* (LPGs). Propane and similar heavy hydrocarbons that separate from some natural gases, either naturally through condensation in the reservoir rock, or in separators at the well-head. Distinguished from **liquefied natural gas (LNG)**.

neutron An elemental particle present in all atomic nuclei. It has no electric charge, and a relative mass of 1.

non-associated gas Natural gas (dominantly methane) which occurs as free gas trapped over salt water in the pores of a reservoir rock. Distinguished from the **associated-dissolved** type.

nonrenewable resources Mineral resources that exist in finite quantities, especially mineral fuels which cannot be used again, but also geothermal occurrences which regenerate slowly. Distinguished from **renewable**.

off-peak period Hours of the day when demand for electricity is low. This occurs principally between midnight and 7 A.M. During this period electrical energy is stored for use during peak hours of electrical demand.

oil shale A sedimentary rock containing solid organic material which can be converted to a crude oil by distillation. In the Green River oil shales of Colorado the organic material is **kerogen** and the host rock is technically a marl or shaley limestone.

ore(s) Rock(s) in which one or more valuable minerals exist in concentrations far greater than is average for crustal rocks. Sometimes applied only to rocks whose richness and accessibility make their mining economically attractive.

organic Derived from plant or animal life. The term applies to a live plant, to the animals that eat the plant, and to the manures from and the carcasses of those animals. See **fossil fuels**.

outcrop A surface exposure of all or part of some rock unit, which in other areas is buried and can be studied only by drill holes.

output of electrical energy The estimated or actual electrical energy produced by a generating plant. It is the power produced (in Watts) multiplied by the time (in hours) during which that power is produced—and is expressed therefore in Kilowatt or Megawatt-hours, usually in a year's period. See **capacity factor**.

overburden Soil, and rock of little or no value, which overlie a buried deposit of economic value, such as coal, oil shale, or the ores of metals. In surface mining, the overburden is removed; in underground mining it is penetrated and supported during mining.

passive solar heating Using structural elements of a building to admit, collect, and store solar energy for heating. In its pristine form, heating of this kind relies on convection, and uses no supplemental energy from fans or pumps to distribute heat through the building.

photosynthesis In living plants, the production of carbohydrates from water and carbon dioxide (CO_2) by the action of solar radiation upon chlorophyll.

photovoltaic effect The flow of electrons (hence potential difference, or voltage) that results when certain semi-conducting substances in solar cells are struck by radiation.

placer deposit A water-borne deposit of heavy minerals that have been eroded from their original bedrock and eventually concentrated in gravels of streams or beaches.

polder In the Netherlands, low-lying coastal areas turned into agricultural land by building dikes and pumping out ocean water.

power A rate of doing work or expending energy. One Watt of power is the expenditure of one Joule of energy in one second.

power available in the wind The power that exists instantaneously in winds of a given velocity. Expressed as Watts in each square unit of cross-section perpendicular to wind flow. Distinguished from **power extractable** and from **output**.

power density plot For wind data at a particular site, a graphic expression of annual Kilowatt-hours of wind energy versus wind speeds. It reveals which winds at the site hold the greatest energy on an annual basis by virtue of their velocity and/or their frequency. May be more accurately called an *energy density plot*.

power duration plot For wind data at a particular site, a graphic expression of the number of hours (during a year) that various wind velocities, and hence various values of *power available*, may be expected on the basis of wind records.

power extractable from wind Theoretical Watts of power that can be gathered from wind of a given velocity, assuming: 1. a certain efficiency of extraction, and 2. extraction from a specified area perpendicular to wind flow. It is very different from actual power extracted because specific wind machines do not employ all the power available in areas of low, medium, high, and very high winds.

primary energy The initial energy used for some process or chain of processes. Generally it is fuel energy, as used to generate power or drive engines. Distinguished from *end use energy* which, as the energy delivered, is usually less than the fuel energy input. For a country, the primary energy is the total of all primary energy entering the country's energy systems, and the end use energy is the total of all energy delivered to the consumers, whether as heat, electrical, or mechanical energy.

pumped storage Electrical generating capacity that is due to the fall of water that has been pumped uphill by using excess electrical power—at a power plant that may not be primarily a hydroelectric facility.

Quad One quadrillion (1×10^{15}) British Thermal Units.

rank of coal A designation that groups coals according to their differing maturity and energy content. Those of higher rank are more mature and richer in energy. The ranks are, in decreasing order: anthracite, bituminous, subbituminous, lignite.

rejected heat Amounts of heat thrown off in an energy conversion process such as the conversion of fuel energy to electrical energy in a power plant, or to mechanical energy in a motor. Although the rejection of heat is unavoidable, the waste is not; some rejected heat can be recovered and used for other tasks.

renewable resources A resource which either flows continuously, (e.g. solar radiation and wind), or is renewed within a short period of time, (e.g. wood).

reprocessing (of nuclear fuel) The recovery of uranium and plutonium from spent fuel rods by dissolving them in an acid solution.

Reserve Base for U.S. coal Equivalent to the Demonstrated Reserve Base, which is comprised of the coals that are economically mineable.

reserves For any mineral resource, the amounts that are *recoverable* (extractable) at current prices and using current technology. Differing degrees of certainty may be expressed by the terms Measured and Inferred reserves; but the amounts are those recoverable from *identified* occurrences.

reservoir (or reservoir rock) A porous and permeable rock, such as sandstone, in which crude oil or natural gas accumulations are found.

resource Anything of value to the society. Usually a raw material, such as forests or minerals, or an energy flow, such as wind. A great variety of certainty and economic worth can be embraced by the term "resource."

retort A large pot or chamber in which oil shale is cooked (heated) in order to separate the oil from the shale rock.

retrofit To install new equipment, such as solar heating apparatus, in an existing building. Distinguished from "new installations," or "installations in new buildings."

secondary recovery techniques Methods of recovering additional crude oil after natural pressures in reservoir rock have been depleted. Entails injection of either water or gas into the reservoir rock to maintain pressures.

sedimentary rocks Those rocks formed by the accumulation of grains or fragments of pre-existing rocks, or the accumulation of chemical precipitates such as $CaCO_3$, salt, or gypsum.

shale oil The crude oil resulting from distillation of kerogen in oil shales.

shield areas Parts of continents in which crystalline rocks of the continent's basement are exposed in a non-mountainous region. Distinguished from mountainous areas and from areas where basement rocks are covered by sedimentary rocks.

short ton Equivalent to 2,000 pounds. Distinguished from the **metric ton** or **long ton** which is equivalent to 1,000 kilograms or 2,240 pounds. The long ton is therefore the weight of 1.023 short tons. The short ton is 0.907 long tons.

slurry A mixture of solid and liquid. An example is coal slurry in which finely ground coal is mixed with water and moved by pipeline.

solar fraction The proportion of a building's heating (or cooling) energy need that is provided by solar energy. Distinguished from **market penetration**.

solar-thermal power generation The focussing of solar rays to achieve the high temperatures used to boil fluids for use in vapor-driven dynamos.

space heating Heating of buildings, whether residential, commercial, or industrial.

strata The plural of *stratum* (rarely used). Layers or beds of rock. Usually applied to sedimentary rock lying relatively flat without severe folding.

strip mining Loosely equivalent to surface mining. **Overburden** material is removed; then the valuable rock is taken away in an operation that is everywhere open to the surface.

stratigraphic trap In oil and gas fields, a reservoir rock which is discontinuous, such as a sand bar or a reef, thus permitting accumulation of oil or gas. Distinguished from **structural trap**.

structural trap In oil and gas fields, a reservoir rock that is folded into structures such as domes, or faulted (broken), thus permitting an accumulation of oil or gas.

subduction The process by which a part of the earth's crust is drawn down into the subcrustal zone. This occurs when two crustal plates collide and one of the two is forced under.

subsidence A sinking, settling or otherwise lowering of parts of the crust of the earth.

sulfur oxides Sulfur dioxide (SO_2) and sulfur trioxide (SO_3). Gases produced when sulfur-rich coal and oil are burned. The gases can react with oxygen and water to produce sulphuric acid (H_2SO_4) and thus are dangerous air pollutants.

syncrude A shorthand expression for synthetic crude oils produced from mineral sources such as oil shales or from organic materials, which may be either wastes or crops especially grown for the purpose.

synthetic crude oils Hydrocarbon fluids that can be produced from various raw materials. Mineral sources are the kerogen in oil shales and bitumen in oil sands. Non-mineral sources include urban trash and certain crops grown for the purpose.

synthetic gases Gases produced from various sources including coal and a number of organic materials such as garbage, manures, and certain energy crops.

temperature gradient The rate at which temperature increases with increasing depth below the earth's surface at a specific location. Steeper gradient refers to more rapid change with depth and denotes an area more favorable for developing geothermal resources.

Thorium-232 A radioactive isotope that can be transmuted under neutron bombardment into Uranium-233, which is fissionable in a chain reaction.

total gas reserves Reserve amounts of both **non-associated** and **associated-dissolved** types of natural gas.

total-horizontal radiation The solar radiation (insolation) that is usually reported and mapped. It is the total of **direct** and **diffuse** radiation falling upon a horizontal surface of some particular location. Equivalent to insolation.

trillion Used here with the United States' and French meaning of 1,000 billion, that is, 1×10^{12}. In Great Britain a trillion means 1,000,000 billion, or 1×10^{15}.

tritium (Hydrogen-3) An isotope of the element hydrogen that has a nucleus containing one proton and two neutrons and thus has a mass number of 3.

Ultimate Recovery For crude oil and natural gas the term is applied to the total of Cumulative Production and Proved reserves. It is "ultimate", therefore, only with regard to identified reservoirs.

underground mining Mining in which the valuable rock to be removed is reached by shafts through the **overburden**.

Undiscovered Recoverable amounts (of crude oil or natural gas) Estimates of amounts that may be recovered from reservoir rocks not yet discovered. Usually the historic recovery rates (proportions of oil in-place) are assumed.

uranium oxide (U_3O_8) The compound that is separated from uranium ores. Uranium resources are stated in tons of uranium oxide thought to be recoverable from ores of varying degrees of certainty and economic feasibility.

viscosity The property of fluids that causes them not to flow easily because of the friction of their molecules. It is decreased by raising the fluid's temperature.

volatile matter in coal The portion of the coal that is driven off as gas or vapor when heated. Much of it is combustible (compounds of hydrogen and carbon). Coals with a high proportion of volatile matter may be preferred for applications where a long flame is desirable.

Watt A unit of power, that is, the rate at which energy is converted or consumed. Equivalent to 1 Joule of energy expended in 1 second.

well-head heat Geothermal heat thought to be recoverable at the surface from exploitable occurrences at depth. Neither the application of the heat nor the efficiency of its use are taken into consideration.

Suggested Readings

Introduction

ALLEN, EDWARD LAWRENCE. *Energy and Economic Growth in the United States.* Cambridge, Mass.: MIT Press, 1979.

BANKS, FERDINAND E. *Scarcity, Energy, and Economic Progress.* Lexington, Mass.: Lexington Books, 1977.

BOHI, DOUGLAS R. *U.S. Energy Policy: Alternatives for Security.* Baltimore: Published for Resources for the Future by the Johns Hopkins University Press, 1975.

BOYLE, GODFREY. *Living on the Sun: Harnessing Renewable Energy for an Equitable Society,* London: Calder and Boyars, 1975.

BRUBAKER, STERLING. *In Command of Tomorrow: Resource and Environmental Strategies for America.* Baltimore: Published for Resources for the Future by Johns Hopkins University Press, 1975.

CARR, DONALD EASTON. *Energy and the Earth Machine.* New York: Norton, 1976.

COMMITTEE FOR ECONOMIC DEVELOPMENT. *Nuclear Energy and National Security: A Statement of National Policy.* Washington: Committee for Economic Development, 1976.

COMMONER, BARRY. *The Poverty of Power: Energy and the Economic Crisis.* New York: Knopf, 1976.

COMMONER, BARRY. *The Politics of Energy.* New York: Knopf, 1979.

COOK, EARL. *Man, Energy, and Society.* New York: W.H. Freeman, 1976.

COUNCIL ON ENVIRONMENTAL QUALITY. *The Good News about Energy.* Washington: 1979.

COUNCIL ON ENVIRONMENTAL QUALITY. *Solar Energy: Progress and Promise.* Washington, D.C.: April 1978.

CRABBE, DAVID AND RICHARD MCBRIDE. *The World Energy Book: An A-Z Atlas, and Statistical Source Book.* Cambridge, Mass.: MIT Press, 1979.

CULP, ARCHIE W. *Principles of Energy Conversion.* New York: McGraw-Hill, 1979.

DARMSTADTER, JOEL. *How Industrial Societies Use Energy: A Comparative Analysis.* Baltimore: Published for Resources for the Future by Johns Hopkins University Press, 1977.

DORAN, CHARLES F. *Myth, Oil, and Politics: Introduction to the Political Economy of Petroleum.* New York: Free Press, 1977.

DRYSDALE, FRANK AND CHARLES E. CALEF. *The Energetics of the United States of America: An Atlas.* Upton, New York: Brookhaven National Laboratory, 1977. Prepared for the U.S. Department of Energy, Contract No. EY-76-C-02-0016.

ECKHOLM, ERIK P. *The Other Energy Crisis.* Washington: Worldwatch Institute, 1975.

FORD FOUNDATION ENERGY POLICY PROJECT. *A Time to Choose: America's Energy Future, Final Report.* Cambridge, Mass.: Ballinger Pub. Co., 1974.

FORD FOUNDATION ENERGY POLICY PROJECT. *Exploring Energy Choices: A Preliminary Report.* Washington: Ford Foundation, 1974.

FOWLER, JOHN M. *Energy-Environment Source Book.* Washington: National Science Teacher's Association, 1975.

GABEL, MEDARD. *Energy, Earth and Everyone: A Global Strategy for Spaceship Earth.* San Francisco: Straight Arrows Books, 1975.

GRIFFIN, JAMES M. *Energy, Economics and Policy.* New York: Academic Press, 1980.

HARDWOOD, CORBIN CREWS. *Using Land to Save Energy.* Cambridge, Mass.: Ballinger Pub. Co., 1977.

HAYES, DENIS. *Energy: The Case for Conservation.* Washington: Worldwatch Institute, 1976.

HAYES, DENIS. *Rays of Hope: The Transition to a Post-Petroleum World.* New York: W.W. Norton, 1977.

HEALY, TIMOTHY J. *Energy and Society.* San Francisco: Boyd and Fraser Pub. Co., 1976.

HOYLE, FRED SIR. *Energy or Extinction?: The Case for Nuclear Energy.* London: Heinemann, 1977.

INSTITUTE FOR CONTEMPORARY STUDIES. *No Time to Confuse* (a critique of *A Time to Choose*). San Francisco, 1975.

ION, DAN C. *Availability of World Energy Resources.* London: Graham and Trotman, 1975.

KRUTILLA, JOHN V. *Economic and Fiscal Impacts Coal Development: Northern Great Plains.* Baltimore: Published for Resources for the Future by Johns Hopkins University Press, 1979.

LAPEDES, DANIEL N. *Encyclopedia of Energy.* New York: McGraw-Hill, 1976.

LENIHAN, JOHN AND WILLIAM W. FLETCHER. *Energy Resources and the Environment.* New York: Academic Press, 1976.

LOFTNESS, ROBERT L. *Energy Handbook.* New York: Van Nostrand Reinhold Co., 1978.

LOVINS, AMORY. *Non-Nuclear Energy: The Case for Ethical Strategy.* Cambridge, Mass.: Ballinger Pub. Co., 1975.

LOVINS, AMORY. *Soft Energy Paths: Toward a Durable Peace.* San Francisco: Friends of the Earth International, 1977.

LOVINS, AMORY. *World Energy Strategies: Fact, Issues, and Options.* San Francisco: Friends of the Earth International, 1975.

MANCKE, RICHARD B. *Squeaking By: U.S. Energy Policy Since the Embargo.* New York: Columbia University Press, 1976.

MCMULLEN, JOHN T. *Energy Resources.* New York: Wiley, 1977.

MCMULLEN, JOHN T. *Energy Resources and Supply.* New York: Wiley, 1975.

MILLER, GEORGE TYLER. *Energy and Environment: The Four Energy Crises.* Belmont, Calif.: Wadsworth Pub. Co., 1975.

MONTBRIAL, THIERRY DE. *Energy, the Countdown: A Report to the Club of Rome.* New York: Pergamon Press, 1979.

NADER, RALPH. *The Menace of Atomic Energy.* New York: Norton, 1979.

NASH, HUGH (EDITOR). *The Energy Controversy: Soft Path Questions and Answers by Amory Lovins and His Critics.* San Francisco: Friends of the Earth, 1979.

NATIONAL ENERGY STRATEGIES PROJECT. *Energy in America's Future: The Choices Before Us.* Baltimore: Published for Resources for the Future by Johns Hopkins University Press, 1979.

NORDHAUS, WILLIAM D. *The Efficient Use of Energy Resources.* New Haven: Yale University Press, 1979.

NUCLEAR ENERGY POLICY STUDY GROUP. *Nuclear Power Issues and Choices.* Cambridge, Mass.: Ballinger Pub. Co., 1977.

PETROLEUM PUBLISHING COMPANY. *International Petroleum Encyclopedia.* Tulsa, Oklahoma, 1980 (Published annually).

PHILLIPS, OWEN M. *The Last Chance Energy Book.* Baltimore: Johns Hopkins University Press, 1979.

PIMENTAL, DAVID. *Food, Energy and Society.* New York: Wiley, 1979.

REISCHE, DIANA (EDITOR). *Energy Demand and Supply.* New York: H.W. Wilson Co., 1975.

RIDGEWAY, JAMES. New Energy: *Understanding the Crisis and a Guide to an Alternative Energy System.* Boston: Beacon Press, 1975.

RUEDISILI, LON C. AND MORRIS W. FIREBAUGH. *Perspectives on Energy: Issues, Ideas, and Environmental Dilemmas.* New York: Oxford University Press, 1975.

SONENBLUM, SIDNEY. *The Energy Connections: Between Energy and the Economy.* Cambridge, Mass.: Ballinger Pub. Co., 1978.

STARR, CHAUNCEY (EDITOR). *Current Issues in Energy: A Selection of Papers.* New York: Pergamon Press, 1979.

STEINHART, JOHN S., ET. AL. *Pathway to Energy Sufficiency: The 2050 Study.* San Francisco: Friends of the Earth, 1979.

STOBAUGH, ROBERT AND DANIEL YERGIN. *Energy Future: Report of the Energy Project of the Harvard Business School.* New York: Random House, 1979.

UDALL, STEWART L. *The Energy Balloon.* New York: McGraw-Hill, 1974.

UNITED STATES EXECUTIVE OFFICE OF THE PRESIDENT. *The National Energy Plan.* Cambridge, Mass.: Ballinger Pub. Co., 1977.

UNITED STATES GEOLOGICAL SURVEY. *Energy Perspectives.* Washington: Government Printing Office, 1975.

UNITED STATES GEOLOGICAL SURVEY. *Energy Perspectives II.* Washington. Government Printing Office, 1977.

UNITED STATES LIBRARY OF CONGRESS. *Alternative Energy Conservation Strategies: An Economic Appraisal.* Washington: Government Printing Office.

VAN TASSEL, ALFRED J. (EDITOR). *The Environmental Price of Energy.* Lexington, Mass.: Lexington Books, 1975.

WARKOV, SEYMOUR (EDITOR). *Energy Policy in the United States: Social and Behavioral Dimensions.* New York: Praeger, 1978.

WORKSHOP ON ALTERNATIVE ENERGY STRATEGIES. *Energy Global Prospects, 1985–2000.* New York: McGraw-Hill, 1977.

YANNACONE, VICTOR J. *Energy Crisis: Danger and Opportunity.* St. Paul: West Pub. Co., 1974.

1 Coal

AVERITT, PAUL. *Coal Resources of the United States, January 1, 1974.* U.S. Geological Survey *Bulletin 1412.* Washington: Government Printing Office, 1975.

CAUDILL, HARRY M. *Night Comes to the Cumberlands.* Boston: Atlantic, Little, Brown, 1962.

"Clearing the Way for the New Age of Coal." *Fortune,* (May 1974).

COREY, RICHARD. *Elements of Coal Science and Utilization Technology, and Some Energy Facts and Figures.* Office of Fossil Energy. Washington: Government Printing Office, 1976.

"The Great Black Hope: Coal." *Colorado Business,* (Sept./Oct. 1976).

"A Look at Global Coal Resources." *World Coal,* (Nov., 1975).

"New Carter Plan Places Emphasis on Synfuels." *Oil and Gas Journal*, (July 23, 1979).

"New Fears Surround Shift to Coal." *Fortune*, (Nov. 28, 1978).

NIELSEN, GEORGE F. "Coal Mine Development and Expansion Survey—617.3 Tons of New Capacity through 1985." *Coal Age*, (Feb. 1977).

STOBAUGH, ROBERT AND DANIEL YERGIN, EDS. *Energy Future*. New York: Random House, 1979, Chapt. 4.

"Synfuels: Uncertain and Costly Fuel Option." *Chemical and Engineering News*, (Aug. 29, 1979).

"Underground Coal Gasification Expands U.S. Energy Options." *UA Journal*, Vol. 41, no. 3 (Mar. 1979).

YOUNG, GORDON. "Will Coal Be Tomorrow's Black Gold." *National Geographic Magazine*, (Aug. 1975).

2   Oil and Natural Gas

AMERICAN ASSOCIATION OF PETROLEUM GEOLOGISTS. *Structure of Typical American Oil Fields*. A symposium. Tulsa, Okla., 1948.

CRAM, IRA, ED. *Future Petroleum Provinces of the United States—Their Geology and Potential*. Memoir 15 of the American Association of Petroleum Geologists. Tulsa, Okla., 1971, 2 vol.

EZZATI, ALI. *World Energy Markets and OPEC Stability*. Lexington, Mass.: Lexington Books, 1978.

FORBES, ROBERT J. *Studies in Early Petroleum History*. Leiden: E. J. Brill, 1958.

Geological Estimates of Undiscovered Recoverable Oil and Gas Resources in the United States. U.S. Geological Survey *Circular 725*. Reston, Va., 1975.

HUBBERT, M. K. "Energy Resources." In Preston Cloud (Chairman), *Resources and Man*. National Academy of Sciences and National Research Council: San Francisco: W. H. Freeman, 1969, 157–242.

———. "The Energy Resources of the Earth." *Scientific American*, Vol. 224, no. 3 (Sept. 1971) 60–70.

JIMISON, JOHN. *National Energy Transportation; Volume 1, Current Systems and Movements*. Prepared by Congressional Research Service, with maps jointly prepared by that service and the U.S. Geological Survey. Washington: Government Printing Office, Publication no. 95-115, May, 1977.

KERR, RICHARD A. "Petroleum Exploration: Discouragement about the Atlantic Outer Continental Shelf Deepens." *Science*, Vol. 204, no. 8 (June, 1979) 1069–1072.

LANDES, KENNETH K. *Petroleum Geology of the United States*. New York: Wiley-Interscience, 1970.

LEVERSON, A. I. *Geology of Petroleum*, 2nd ed. San Francisco: W. H. Freeman, 1967.

LOVEJOY, WALLACE F., AND PAUL T. HOMAN. *Methods of Estimating Reserves of Crude Oil, Natural Gas, and Natural Gas Liquids*. Baltimore: Johns Hopkins University Press, 1965.

MCCULLOH, T. H. "Oil and Gas." Chapter in *United States Mineral Resources*, U.S. Geological Survey *Professional Paper 820*, 1973, 477–496.

METZ, W. D. "Mexico: The Premier Oil Discovery in the Western Hemisphere." *Science*, Vol. 202 (1978) 1261–1265.

MOSTERT, NOEL. *Supership*. New York: Knopf, 1974.

ODELL, PETER R. *Oil and World Power: Background to the Oil Crisis*, 4th ed. Baltimore: Penguin Books, 1975.

———. *The Pressures of Oil: A Strategy for Economic Revival*. New York: Harper and Row, 1978.

PETROLEUM PUBLISHING COMPANY. *International Petroleum Encyclopedia*. Tulsa, Okla. Published annually.

———. *Petroleum 2000*. Tulsa, Okla. Special issue of the *Oil and Gas Journal* (Aug., 1977).

SKINNER, BRIAN J. *Earth Resources*. Englewood Cliffs, N.J.: Prentice-Hall, 1969.

STOBAUGH, ROBERT. "After the Peak: The Threat of Imported Oil." Chapter 2 in Stobaugh and Yergin, *Energy Future*, New York: Random House, 1979.

TAIT, SAMUEL W. *The Wildcatters: Informal History of Oil-Hunting in America*. Princeton: Princeton University Press, 1946.

TIRATSOO, E. N. *Natural Gas*. New York: Plenum, 1967.

———. *Oilfields of the World*. Beaconsfield, Eng.: Scientific Press, 1973.

TISSOT, B. *Petroleum Formation and Occurrence: A New Approach to Oil and Gas Exploration*. New York: Springer-Verlag, 1978.

WEAVER, ROBERT B. *Men and Oil*. Chicago: University of Chicago Press, 1938.

WENDLAND, RAY T. *Petrochemicals: The New World of Synthetics*. Garden City, N.J.: Doubleday, 1969.

3   Oil Shales and Tar Sands

BURWELL, E. L., T. E. STERNER, AND H. C. CARPENTER. "Shale Oil Recovery by In-Situ Retorting, A Pilot Study." *Journal of Petroleum Technology*, (Dec. 22, 1970).

CULBERTSON, WILLIAM C. AND JANET K. PITMAN. "Oil Shale." U.S. Geological *Professional Paper, 820*. Washington: Government Printing Office, 1973.

DONNELL, JOHN. "Oil Shale Resources: How Much?" *Shale Country*, (Feb., 1976).

FEDERAL ENERGY ADMINISTRATION. *Project Independence: Potential Future Role of Oil Shale*. U.S. Department of Interior. Washington: Government Printing Office, 1974.

MITCHELL, GUY ELLIOT. "Billions of Barrels of Oil Locked Up in Rocks." *National Geographic Magazine*, (Feb. 1918).

"Oil Shale: Prospects on the Upswing." *Science*, Vol. 190, no. 4321, (Dec. 9, 1977).

PETZRIC, P. A. "Oil Shale—An Ace in the Hole for National Security." *Shale Country*, (Oct. 1975).

"Shale Closest among Liquid Fuels." *Oil and Gas Journal*, (Jan. 17, 1977).

"Shale Oil Finally Rocking off Dead Center." *Oil and Gas Journal*, (June 18, 1979).

VAN WEST, FRANK P. "Green River Oil Shale." *Geologic Atlas of the Rocky Mountain Region, U.S.A.* Denver: Rocky Mountain Association of Geologists, 1972.

WELLES, CHRISTOPHER. *The Elusive Bonanza—The Story of Oil Shale—America's Richest and Most Neglected National Resource*. New York: E. P. Dutton, 1970.

4   Nuclear Fuels

AMERICAN NUCLEAR SOCIETY. *Nuclear Power and Society*. Hinsdale, Ill.: American Nuclear Society, 1976.

BERGER, JOHN J. *Nuclear Power: The Unviable Option*. New York: Dell, 1977.

BOGART, S. LOCKE. "Fusion Power and the Potential Lithium

Requirement." In *Lithium Resources and Requirements by the Year 2000*. U.S. Geologic Survey *Professional Paper 1005*. Washington: Government Printing Office, 1976.

BUPP, IRVIN C. AND JEAN-CLAUDE DERIAN. *Light Water: How the Nuclear Dream Dissolved*. New York: Basic Books, 1978.

FLOWERS, BRIAN J. "Nuclear Power: A Perspective on the Risks, Benefits, and Options." *Bulletin of the Atomic Scientists*, Vol. 34, no. 3, (Mar. 1977).

FREEMAN, S. DAVID, *et al*. *A Time to Choose: America's Energy Future*. Report of the Energy Policy Project of the Ford Foundation. Cambridge, Mass.: Ballinger, 1974.

HAYES, DENNIS. *Nuclear Power: The Fifth Horseman*. Washington: Worldwatch Institute, 1976.

HEWLETT, RICHARD G. *Federal Policy for Disposal of Radioactive Wastes from Commercial Nuclear Power Plants*. Washington: U.S. Department of Energy, Mar. 1979.

KEENEY, SPURGEON, *et al*. *Nuclear Power Issues and Choice*. Cambridge, Mass.: Ballinger, 1977.

LAPP, RALPH E. "We May Find Ourselves Short of Uranium." *Fortune*, (Oct., 1975).

LOVINS, AMORY. "The Case against the Fast Breeder Reactor." *Bulletin of Atomic Scientists*, (Mar. 1973).

———. *Soft Energy Paths*. San Francisco: Friends of the Earth, 1977.

NAJARIAN, THOMAS. "The Controversy over the Health Effects of Radiation." *Technology Review*, (Nov. 1978).

OLSON, MCKINLEY C. *Unacceptable Risk*. New York: Bantam, 1976.

ROSSEN, A. D. AND T. A. RIECK. "Economics of Nuclear Power." *Science* Vol. 201, no. 18, (Aug. 1978).

STAATZ, MORTIMER, *et al*. "Thorium." *United States Mineral Resources*, U.S. Geological *Professional Paper 820*. Washington: Government Printing Office, 1973.

STOBAUGH, ROBERT AND DANIEL YERGIN, EDS. *Energy Future*. Particularly Chapter 5, "The Nuclear Stalemate" by I. C. Bupp. New York: Random House, 1979.

TAYLOR, THEODORE B. AND WILLIRICH MASON. *Nuclear Thefts: Risks and Safeguards*. Cambridge, Mass.: Ballinger, 1975.

WEINBERG, A. M. "Social Institutions and Nuclear Energy." *Science*, Vol. 177, (1972).

YERGIN, DANIEL. "The Terrifying Prospect: Atomic Bombs Everywhere." *Atlantic Monthly*, (Apr. 1977).

5   Geothermal Heat

AMERICAN ASSOCIATION OF PETROLEUM GEOLOGISTS. *Geothermal Gradient Map of North America*. Tulsa, Okla., 1976.

ARMSTEAD, H., ED. *Geothermal Energy: Its Past, Present, and Future Contributions to the Energy Needs of Man*. New York: Wiley, 1978.

———. *Geothermal Energy: Review of Research and Development*. Paris: UNESCO, 1973.

AXTMAN, R. C. "Environmental Impact of a Geothermal Power Plant." *Science*, Vol. 187, no. 4179 (1975) 785–803.

BACON, FORREST, *et al*. *Engineering Geology of the Geysers Geothermal Resource Area*. Special Report No. 122, Sacramento, Cal.: California Division of Mines and Geology, 1976.

BIERMAN, SHELDON L. *Geothermal Energy in the United States: Innovation versus Monopoly*. New York: Praeger, 1978.

CUMMINGS, RONALD G. *et al*. "Mining Earth's Heat: Hot Dry Rock Geothermal Energy." *Technology Review*, (Feb., 1979).

DICKINSON, GEORGE. "Geological Aspects of Abnormal Reservoir Pressure in Gulf Coast Louisiana." *Bulletin of American Association of Petroleum Geologists*, Vol. 37, no. 2 (1953) 410–432.

ELLIS, ALBERT J. *Chemistry and Geothermal Systems*. New York: Academic Press, 1977.

FINNEY, J. P. "Design and Operation of the Geysers Power Plant." A chapter in Kruger and Otte, 1972. (see below)

HEACOCK, J. G., ED. *The Earth's Crust: Its Nature and Physical Properties*. American Physical Union, *Geophysical Monograph No. 20*, 1977.

KRUGER, PAUL, AND CAREL OTTE, EDS. *Geothermal Energy: Resources, Production, Stimulation*. Papers presented at session on geothermal energy at American Nuclear Society annual meeting, June, 1972. Stanford, Cal. Stanford University Press, 1973.

MILORA, S. L., AND J. W. TESTER. *Geothermal Energy as a Source of Electric Power*. Cambridge, Mass.: M.I.T. Press, 1976.

MUFFLER, L. J. P. "Geothermal Resources." In *United States Mineral Resources*, U.S. Geological Survey *Professional Paper No. 820*. Washington, D.C.: U.S. Government Printing Office, 1973, 251–261.

———, AND R. CATALDI. "Methods for Regional Assessment of Geothermal Resources." *Geothermics*, Vol. 7, no. 2–4, (1978).

MURRAY, G. E. *Geology of the Atlantic and Gulf Coastal Province of North America*. New York: Harper, 1961.

NATIONAL RESEARCH COUNCIL. *Natural Gas from Unconventional Sources*. National Academy of Sciences, 1976.

POLLACK, HENRY N., AND D. S. CHAPMAN. "The Flow of Heat from the Earth's Interior." *Scientific American*, Vol. 237, no. 2 (Aug., 1977) 60–76.

UNITED NATIONS. *Proceedings of UN Conference on New Sources of Energy: Solar, Wind, and Geothermal*. Rome, Aug., 1961, New York: U.N., 1963–64.

U.S. LIBRARY OF CONGRESS. *Energy from Geothermal Resources*. Prepared for Committee on Science and Technology of U.S. House of Representatives by Science and Policy Research Division of the Library of Congress. Washington: U.S. Government Printing Office, 1978.

WAHL, EDWARD F. *Geothermal Energy Utilization*. New York: Wiley, 1977.

WHITE, D. E. *Geothermal Energy*. U.S. Geological Survey *Circular 519*. Washington, D.C., 1965.

6   Solar Radiation

ANDERSON, BRUCE. "Low Impact Solutions." *Solar Age*, (Sept., 1976).

BECKMAN, W. A., *et al*. *Solar Heating Design by the F-Chart Method*. New York: Wiley, 1977, $14.95.

BLISS, RAYMOND. "Why Not Build the House Right in the First Place?" *Bulletin of the Atomic Scientists*, (Mar., 1976).

CHALMERS, BRUCE. "The Photovoltaic Generation of Electricity." *Scientific American*, (Oct., 1976).

DANIELS, FARRINGTON. *Direct Use of the Sun's Energy*. New York: Ballantine, 1964.

DAVIS, C., AND M. DAVIS. *People Who Live in Solar Houses, and What They Say about Them*. New York: Popular Library, 1979, $2.25.

DAVIS, NORAH D., AND L. LINDSEY. *At Home in the Sun.* Charlotte, Vt.: Garden Way, 1979.

DUFFIE, JOHN AND W. A. BECKMAN. "Solar Heating and Cooling." *Science,* Vol. 191 (1976) 143–149.

GILMORE, C. P. "Solar Assisted Heat Pumps." *Popular Science,* (May, 1978) 86–90.

GOODENOUGH, JOHN. "The Options for Using the Sun." *Technology Review,* (Oct.–Nov., 1976).

HOWELL, D. *Your Solar Energy Home.* Elmsford, N.Y.: Pergamon Press, 1979, $8.80.

HOWELL, Y., AND J. A. BERENY. *Engineer's Guide to Solar Energy.* Solar Information Services, PO Box 204, San Mateo, California: 94401, 1979, $27.50.

MC DANIELS, D. K. *The Sun: Our Future Energy Source.* New York: Wiley 1979, $12.95.

MC PHILLIPS, M., ED. *The Solar Age Resource Book.* New York: Everest House, 1979, $9.95.

MAIDIQUE, MODESTO A. "Solar America." Chapter 7 in Stobaugh and Yergin, *Energy Future.* New York: Random House, 1979.

MEINEL, A. B., AND M. P. MEINEL. *Applied Solar Energy.* Addison-Wesley, 1976.

———. "Is It Time for a New Look at Solar Energy?" *Bulletin of the Atomic Scientists,* (Oct., 1971) 32–37.

MERRIGAN, JOSEPH. *Sunlight to Electricity: Prospects for Solar Energy Conversion by Photovoltaics.* Cambridge, Mass., MIT Press, 1975.

METZ, W. D., AND A. L. HAMMOND, ED. *Solar Energy in America.* Washington, D.C.: American Association for the Advancement of Sciences, 1978, $8.50.

MORROW, WALTER E., JR. "Solar Energy: Its Time Is Near." *Technology Review,* (Dec. 1973).

Popular Science; Solar Energy Handbook 1979. New York Times Mirror Magazines, 1979.

SHURCLIFF, W. A. *Solar Heated Buildings: A Brief Survey,* 13th ed. San Diego, Cal.: Solar Energy Digest, 1977.

———. *Solar-Heated Homes of North America.* Harrisville, N.H.: Brick House, 1978, $8.95.

STEIN, RICHARD G. *Architecture and Energy.* Garden City, N.Y.: Anchor Books, 1977.

U.S. DEPARTMENT OF HOUSING AND URBAN DEVELOPMENT (HUD). *Solar Dwelling Design Concepts.* Washington, D.C.: AIA Research Corporation, June, 1976.

WELLS, M., AND I. SPETGANG. *How To Buy Solar Heating without Getting Burnt.* Emmaus, Pa.: Rodale Press, 1978, $6.95.

WILLIAMS, R. H., ED. *Toward a Solar Civilization.* Cambridge, Mass.: MIT Press, 1978, $12.50.

The First Passive Solar Home Awards. Franklin Research Service, GPO Stock No. 023-000-00571-4, 1978, $4.40.

Solar Energy in America's Future: A Preliminary Assessment. ERDA, GPO Stock No. 060-000-00051-4, Mar., 1977, $2.00.

Solar Heating and Cooling Demonstration Program: A Descriptive Summary of HUD Cycle 4 and 4A Solar Residential Projects. AIA Research Corp., GPO No. 023-000-00531-0, 1979, $4.75.

A Survey of Passive Solar Buildings. AIA Research Corp., GPO Stock No. 023-000-00437-2, 1978, $3.75.

Note: Current lists of books, periodicals, and educator's materials can be obtained from the National Heating and Cooling Information Center, PO Box 1607, Rockville, Maryland 20850.

7 Wind Power

ELDRIDGE, FRANK R. *Wind Machines.* Report prepared for the National Science Foundation and Research Applied to National Needs (RANN). Washington: Government Printing Office, 1975.

GOLDING, EDWARD W. *The Generation of Electricity by Wind Power.* N.Y.: Halsted Press, 1976.

HACKLEMAN, MICHAEL A. *Wind and Windspinners: A Nuts and Bolts Approach to Wind-Electric Systems.* Acton, Cal.: Earthmind, 1974.

HERONEMUS, W. E. "Power from Offshore Winds." Proceedings of the 8th Annual Marine Technology Conference, Washington, D.C., 1972.

———. "Wind Power: A Near-Term Partial Solution to the Energy Crisis." In Ruedelisi, L. C. and M. W. Firebaugh, eds. *Perspectives on Energy.* N.Y.: Oxford University Press, 1975.

HEWSON, E. W. "Generation of Power from the Wind." *Bulletin of the American Meteorological Society,* Vol. 36, no. 7 (July, 1975) 660–675.

HICKOCK, FLOYD. *Handbook of Solar and Wind Energy.* Boston: Cahners Books, 1975.

INGLIS, DAVID R. *Wind Power and Other Energy Options.* Ann Arbor, Mich.: University of Michigan Press, 1978.

———. "Wind Power Now." *Bulletin of the Atomic Scientists,* (Oct. 1975) 20–25.

JUSTUS, CARL J. *Winds and Wind System Performance.* Philadelphia, Pa.: Franklin Institute Press, 1978.

LINDSLEY, E. F. "Wind Power: How New Technology Is Harnessing an Age-Old Energy Source." *Popular Science,* (July, 1974) 54–59.

MC GUIGAN, DERMOT. *Harnessing the Wind for Home Energy.* Charlotte, Vt.: Garden Way, 1978.

MERRIAL, MARSHALL. "Wind Energy for Human Needs." *Technology Review,* (Jan., 1977) 29–39.

METZ, W. D. "Wind Energy: Large and Small Systems Competing." *Science,* Vol. 197, (1977) 971–973.

PIERSON, RICHARD E. *Technician's and Experimenter's Guide to Using Sun, Wind, and Water Power.* West Nyack, N.Y.: Parker, 1978.

PUTNAM, PALMER C. *Power from the Wind.* New York: Van Nostrand Reinhold, 1948 (1974).

REED, JACK W. "Wind Power Climatology." *Weatherwise,* (Dec., 1974) 236–242.

REYNOLDS, JOHN. *Windmills and Watermills.* New York: Praeger, 1970.

SIMMONS, DANIEL M. *Wind Power; Energy Technology Review No. 6.* Park Ridge, N.J.: Noyes Data Corp., 1975.

SORENSEN, BENT. "Energy and Resources." *Science,* Vol. 189. no. 4199 (July 25, 1975) 255–260.

UNITED NATIONS, *Proceedings of UN Conference on New Sources of Energy: Solar, Wind, and Geothermal.* Rome, Aug., 1961, New York: United Nations, 1963–64.

ZELBY, LEON W. "Don't Get Swept away by Wind Power Hopes." *Bulletin of the Atomic Scientists,* (Mar. 1976) 59.

8 Hydroelectric Power

FEDERAL POWER COMMISSION. *Hydroelectric Power Resources of the United States, 1976.* Washington: Government Printing Office, 1976.

FISHER, JOHN. *Energy Crises in Perspective.* New York: Wiley, 1974.

HAYES, DENNIS. "The Solar Prospect." *Worldwatch Paper 11.* Washington: Worldwatch Institute, Mar. 1977.

HEALY, TIMOTHY. *Energy, Electric Power and Man.* San Francisco: Boyd Fraser, 1974.

LILIENTHAL, DAVID F. "Lost Megawatts Flow over Nation's Myriad Spillways." *Smithsonian Magazine,* Vol. 8, no. 6 (Sept. 1977).

MCGUIGAN, DERMOT. *Harnessing Water Power for Home Energy.* Charlotte, Vt.: Garden Way, 1978.

METZ, WILLIAM D. "Solar Thermal Energy: Bringing the Pieces Together." *Science,* Vol. 197 (Aug. 12, 1977).

U.S. ARMY CORPS OF ENGINEERS. *National Hydroelectric Power Resources Study—Preliminary Inventory of Hydropower Resources.* Washington: U.S. Army Corps of Engineers, July 1979, 6 vol.

9 Ocean Thermal Gradients

"Energy from the Sea." *Popular Science,* (June, 1975).

HABER, GEORGE. "Solar Power from the Oceans." *New Scientist,* (Mar. 10, 1977).

HAGEN, ARTHUR W. *Thermal Energy from the Sea; Energy Technology Review No. 8.* Park Ridge, N.J.: Noyes Data Corp., 1975.

METZ, W. D. "Ocean Temperature Gradients: Solar Power from the Sea." *Science,* Vol. 180 (1973) 1266–1267.

————. "Ocean Thermal Energy: the Biggest Gamble in Solar Power." *Science,* Vol. 198 (1977) 178–180.

PICKARD, GEORGE L. *Descriptive Physical Oceanography.* New York: Pergamon Press, 1975.

U.S. LIBRARY OF CONGRESS. *Energy from the Ocean.* Report prepared for House of Representatives by Congressional Research Service, Library of Congress. Washington, D.C.: Government Printing Office, 1978.

WILLIAMS, R. H. "The Greenhouse Effect for Ocean Based Solar Energy Systems." *Working Paper No. 21.* Princeton, N.J.: Center for Environmental Studies, Princeton University, Oct., 1975.

ZENER, CLARENCE. "Solar Sea Power." *Bulletin of the Atomic Scientists,* (Jan., 1976).

10 Biomass

ANDERSON, RUSSEL, E. *Biological Paths to Self-Reliance: A Guide to Biological Solar Energy Conversion.* New York: Van Nostrand Reinhold, 1979.

BENEMANN, J. R., et al. "Energy Production by Microbial Photosynthesis." *Nature,* Vol. 268 (1977) 19–23.

BIO-ENERGY COUNCIL. *Capturing the Sun through Bioconversions.* Proceedings of a Conference, March, 1976. Washington, D.C.: Bio-Energy Council, 1976.

CALVIN, MELVIN. "Photosynthesis as a Resource for Energy and Materials." *American Scientist,* Vol. 64, no. 3 (1976) 270–278.

ECKHOLM, ERIK P. *The Other Energy Crisis: Firewood.* Worldwatch Paper No. 1, Washington, D.C.: Worldwatch Institute, 1976.

GRANTHAM, J. B. "Potentials of Wood for Producing Energy." *Journal of Forestry,* Vol. 72, no. 9 (1974).

HAMMOND, A. L. "Alcohol: A Brazilian Answer to the Energy Crisis." *Science,* Vol. 195 (1977) 564–66.

LIETH, H. AND R. H. WITTAKER, EDS. *Primary Productivity of the Biosphere.* New York: Springer-Verlag, 1975.

PIMENTAL, DAVID, et al. "Biological Solar Energy Conversion and U.S. Energy Policy." *BioScience,* Vol. 28, no. 6 (1978) 376–381.

POOLE, ALAN AND ROBERT H. WILLIAMS. "Flower Power: Prospects for Photosynthetic Energy." *Bulletin of the Atomic Scientists,* (May, 1976).

SLESSER, MALCOLM. *Biological Energy Resources.* New York: Wiley, 1979.

WASHINGTON CENTER FOR METROPOLITAN STUDIES. *Capturing the Sun through Bioconversion.* Washington, D.C., 1976.

References

Introduction

LOBECK, ARMIN K., *Physiographic Diagram of the United States*, The Geographical Press, a division of Hammond Inc., 1957.

———, *World Energy Strategies: Facts, Issues, and Options*, Friends of the Earth, Ballinger, 1975.

MC KELVEY, VINCENT E., AND F. H. WANG, *World Subsea Mineral Resources*, U.S. Geological Survey Miscellaneous Geologic Investigation I-632, U.S. Department of the Interior, 1969.

SKINNER, BRIAN, *Earth Resources*, (second edition), Prentice-Hall, 1976, p. 330.

NATIONAL ACADEMY OF SCIENCES, *Energy in Transition, 1985–2010*, the CONAES report to Capitol Hill, San Francisco, W.H. Freeman, March, 1980.

1 Coal

AVERITT, PAUL. "Coal." U.S. Geological Survey Professional Paper, 820. Washington: Government Printing Office, 1973.

———. *Coal Resources of the United States, January 1, 1974, U.S. Geological Survey Bulletin 1412*. Washington: Government Printing Office, 1975.

BATEMAN, ALAN M. *The Formation of Mineral Deposits*. New York: John Wiley and Sons, 1951.

BAYLEY, W. S. *A Guide to the Study of Non-Metallic Mineral Products*. New York: Henry Holt, 1930.

COMMITTEE ON INTERIOR AND INSULAR AFFAIRS AND PUBLIC WORKS. "Greater Coal Utilization." *Joint Hearings United States, Ninety-Fourth Congress, First Session*. Washington: Government Printing Office, 1975.

COREY, RICHARD. *Elements of Coal Science and Utilization Technology, and Some Energy Facts and Figures*. Washington: Office of Fossil Energy, 1976.

EXECUTIVE OFFICE OF THE PRESIDENT. *The National Energy Plan*. Washington: Government Printing Office, 1977.

FEDERAL ENERGY ADMINISTRATION. "U.S. Coal Resources and Reserves." *Report No. FEA/B—76/210*. Washington: F.E.A., 1976.

"A Look at Global Resources." *World Coal*, Nov., 1975.

MOORE, ELWOOD S. *Coal*. New York: John Wiley and Sons, 1922.

NATIONAL COAL ASSOCIATION. *Bituminous Coal Facts, 1972*. Washington: N.C.A., 1972.

"New Carter Plan Places Emphasis on Synfuels." *Oil and Gas Journal*, July 23, 1979.

NIELSEN, GEORGE F. "Coal Mine Development and Expansion Survey—617.3 Tons of New Capacity through 1985." *Coal Age*, Feb., 1977.

OFFICE OF COAL RESEARCH. *Coal Technology—Key to Clean Energy*. Washington: Government Printing Office, 1973.

"Synfuels: Uncertain and Costly Fuel Option." *Chemical and Engineering News*, Aug. 29, 1979.

"Underground Coal Gasification Expands U.S. Energy Options." *UA Journal*, Vol. 41, no. 3. March 1979.

U.S. BUREAU OF MINES. "Bituminous Coal and Lignite Distribution, 1976." *Mineral Industry Surveys*. Washington: Division of Fuels Data and Division of Coal, 1976.

―――. "Coal—Bituminous and Lignite in 1975." *Mineral Industry Surveys*. Washington: Division of Fuels Data and Division of Coal, 1977.

―――. "Demonstrated Coal Reserve Base of the United States, by Sulfur Category on January 1, 1974." *Mineral Industry Surveys*. Washington: Division of Fuels Data and Division of Coal, 1975.

―――. "Long-Distance Coal Transport: Unit Trains or Slurry Pipelines." *Information Circular, 8690*. Washington: Government Printing Office, 1975.

―――. *Mining Technology Research*. Washington: Bureau of Mines, 1976.

―――. "Projects to Expand Fuel Sources in Eastern States." *Information Circular, 8726*. Washington: Government Printing Office, 1976.

―――. "Projects to Expand Fuel Sources in Western States." *Information Circular, 8719*. Washington: Government Printing Office, 1976.

―――. "The Reserve Base of Coal for Underground Mining in the Western United States." *Information Circular, 8678*. Washington: Government Printing Office, 1975.

―――. "The Reserve Base of Bituminous Coal and Anthracite for Underground Mining in the Eastern United States." *Information Circular, 8655*. Washington: Government Printing Office, 1974.

―――. "The Reserve Base of U.S. Coals by Sulfur Content—The Eastern States." *Information Circular, 8680*. Washington: Government Printing Office, 1975.

―――. "The Reserve Base of U.S. Coals by Sulfur Content—The Western States." *Information Circular, 8693*. Washington: Government Printing Office, 1975.

―――. "Strippable Reserves of Bituminous and Lignite in the United States." *Information Circular, 8531*. Washington: Government Printing Office, 1971.

―――. "Sulfur Content of United States Coals." *Information Circular, 8312*. Washington: Government Printing Office, 1966.

U.S. DEPARTMENT OF ENERGY. *Energy Data Reports, Production of Coal, Bituminous, and Lignite*. Washington, Apr. 20, Mar. 30, 1979.

―――. *Fossil Energy Research and Development Program of the U.S. Department of Energy*. DOE/ET 0013 (78). Washington: D.O.E., Mar. 1978.

―――. *Fossil Energy Research and Development Program of the U.S. Department of Energy*. DOE/ET-0014 (78). Washington: D.O.E., March, 1978.

―――. *Underground Coal Conversion Program*, Volume III, *Resources*. Washington: Government Printing Office, Feb. 1978.

U.S. ENERGY RESEARCH AND DEVELOPMENT ADMINISTRATION. *Energy from Coal—A State of the Art Review*. Washington: Government Printing Office, 1976.

U.S. DEPARTMENT OF INTERIOR. *Energy Perspectives 2*. Washington: Government Printing Office, 1976.

ZINDER-NERIS INC. *Coal International: A Monthly Inventory of Information*. Washington, D.C.; New York, N.Y., Vol. 1, no. 1, May, 1979.

2   Oil and Natural Gas

AMERICAN GAS ASSOCIATION. *The Underground Storage of Gas in the United States and Canada*. Washington, D.C., 1974.

―――. Data tabulations in *Reserves of Crude Oil, Natural Gas Liquids, and Natural Gas in the United States and Canada as of Dec. 31, 1978*, Vol. 33. Published by the American Petroleum Institute, Washington, D.C., May, 1979.

AMERICAN PETROLEUM INSTITUTE. *Reserves of Crude Oil, Natural Gas Liquids, and Natural Gas in the United States and Canada as of Dec. 31, 1978*, Vol. 33. Published by American Petroleum Institute, Washington, D.C., May, 1979.

―――. *Annual Reserve Reports*. As in above entry, published annually.

―――. *Basic Petroleum Data*. Washington, D.C., 1977.

Canada Today/D'aujourd'hui, Vol. 8, no. 5, 1977. Published by Canadian Embassy, Washington, D.C.

COOK, EARL. *Man, Energy, Society*. San Francisco: W. H. Freeman and Co., 1976.

CRAM, IRA, ed., *Future Petroleum Provinces of the United States—Their Geology and Potential*. American Association of Petroleum Geologists, *Memoir 15*, 1971.

DEUL, MAURICE. Quoted in "Natural Gas: United States Has It If Price Is Right." *Science*, Feb. 13, 1976.

DOSCHER, T. M., AND F. A. WISE. "Enhanced Crude Oil Recovery Potential—an Estimate." *Journal of Petroleum Technology*, May, 1976, 575–577.

EXXON CORPORATION. *Energy Outlook, 1977–1990* (Jan., 1977) and *Energy Outlook 1978–1990* (May, 1978). Houston, Tex.: Exxon Public Affairs Department.

GRANVILLE, MAURICE F. "Petroleum's Role from Now to End of the Century." In *Petroleum 2000*, special issue of *Oil and Gas Journal*, Aug. 1977, p. 61.

HENDRICKS, T. A. *Resources of Oil, Gas, and Natural Gas Liquids in the United States and the World, U.S.G.S. Circular 522*, 1965.

HUBBERT, KING. "Energy Resources." In *Resources and Man*. Committee on Resources and Man, National Academy of Sciences and National Research Council. W. H. Freeman and Co., 1969.

JIMISON, JOHN W. *National Energy Transportation; Volume 1, Current Systems and Movements*. Prepared by Congressional Research Service, with maps jointly prepared by that service and the U.S. Geological Survey, U.S. Gov't. Printing Office, Publication No. 95–115, May 1977.

KERR, RICHARD A. "Petroleum Exploration: Discouragement about Atlantic O.C.S. Deepens." *Science*, Vol. 204, no. 8, (June 1979).

KLEMME, H. D. "What Giants and Their Basins Have in Common." *Oil and Gas Journal*, Vol. 69, no. 9 (Mar. 1, 1971), 85–99.

KUUSKRAA, V. E., *et al.* "Vast Potential Held by Four Unconventional Gas Sources." *Oil and Gas Journal*, June 12, 1978, 48–54.

LAWRENCE, GEO. H. "Gas Industry Outlook Good for 2000." In *Petroleum 2000*, special issue of *Oil and Gas Journal*, Aug. 1977.

MASTERS, CHARLES D. "Recent Estimates of U.S. Oil and Gas." *Oil and Gas Journal*, Mar. 19, 1979.

MEYER, RICHARD F. "A Look at Natural Gas Resources." *Oil and Gas Journal*, May 8, 1978.

MOODY, J. D., AND R. W. ESSER. Address to 9th World Petroleum Congress, Tokyo, May, 1975, reported in *World Oil,* Sept., 1975, 47–56.

NATIONAL ACADEMY OF SCIENCES. *Mineral Resources and the Environment.* A report by Committee on Mineral Resources and the Environment, National Research Council, Washington, D.C., 1975.

NATIONAL PETROLEUM COUNCIL. *U.S. Energy Outlook—Oil and Gas Availability.* Washington, D.C., 1973.

Oil and Gas Journal. Published weekly by Petroleum Publishing Company, 1421-South Sheridan Rd., Tulsa, Oklahoma.

SCHOLLE, P. A., ed. *Geological Studies in the East B-2 Well, U.S. Mid-Atlantic O.C.S. Area; U.S. Geological Survey Circular, 750.*

TIRATSOO, E. N. *Oilfields of the World.* Beaconsfield, England: Scientific Press, 1973.

U.S. CONGRESS, OFFICE OF TECHNOLOGY ASSESSMENT. *Transportation of Liquefied Natural Gas.* Washington, D.C., Sept., 1977.

U.S. DEPARTMENT OF ENERGY. *Fossil Energy Research and Development Program of the U.S. Department of Energy, FY 1979.* Assistant Secretary for Energy Technology, Washington, D.C., Mar., 1978. DOE/ET-0013(78).

————. *Natural Gas Annual; Natural Gas Production and Consumption: 1976.* Prepared Feb. 7, 1978 in Office of Energy Data and Interpretation, U.S. Department of Energy, Washington, D.C.

————. *Petroleum Statement, Annual; Crude Petroleum, Petroleum Products, and Natural Gas Liquids: 1976.* Prepared Jan. 31, 1978 in Office of Energy Data and Interpretation, Department of Energy, Washington, D.C.

————. *Supply, Demand, and Stocks of All Oils by P.A.D. Districts, and Imports into the U.S.A. by Country: Final, 1977.* Washington, D.C. Feb. 8, 1979.

————. *United States Imports and Exports of Natural Gas, 1978.* Washington, D.C., June 8, 1979.

U.S. DEPARTMENT OF THE INTERIOR. *Crude Oil Transportation Valdez, Alaska to Midland, Texas.* Draft Environmental Impact Statement prepared by California State Office of Bureau of Land Management, 1977(?).

————. *Final Environmental Statement, Outer Continental Shelf Oil and Gas Lease Sale No. 40 (Baltimore Canyon Area).* Bureau of Land Management, May 25, 1976.

————. *Final Environmental Statement, Outer Continental Shelf Oil and Gas Lease Sale No. 42 (George's Bank Area).* Released 1976 (?) by Bureau of Land Management.

————. *Final Environmental Statement, Outer Continental Shelf Oil and Gas Lease Sale No. 43 (Southeast Georgia Embayment Area).* Released by Bureau of Land Management, Feb. 16, 1977.

U.S. ENERGY RESEARCH AND DEVELOPMENT ADMINISTRATION. *Fossil Energy Research Meeting.* Washington, D.C., December, 1977. CONF 7706100.

U.S. GEOLOGICAL SURVEY. *Circular 725, Geological Estimates of Undiscovered Recoverable Oil and Gas Resources in the United States.* 1975.

————. *Circular 726; Assessment of Geothermal Resources of the United States—1975.* Arlington, Va.: U.S.G.S., 1975.

3 Oil Shales and Tar Sands

BALL ASSOCIATES. *Surface and Shallow Oil-Impregnated Rocks and Shallow Oil Fields in the United States. U.S. Bureau of Mines Monograph No. 12,* 1965.

CASHION, W. B. "Bituminous-Bearing Rocks." In *United States Mineral Resources,* U.S. Geological Survey Professional Paper No. 820, Washington: Government Printing Office, 1973.

COLORADO CONSERVATION BOARD. Report on Economic Potential of Western Colorado. Boulder: Bureau of Research, School of Business, University of Colorado, 1957.

CULBERTSON, WILLIAM C. AND JANET K. PITMAN. "Oil Shale." *U.S. Geological Professional Paper, 820.* Washington: Government Printing Office, 1973.

DONNELL, J. "Oil Shale Resources: How Much?" *Shale Country,* Feb., 1976.

DUNCAN, DONALD C. AND VERNON E. SWANSON. "Organic-Rich Shale of the United States and World Land Areas." *U.S. Geological Survey Circular, 523.* Washington: Government Printing Office, 1965.

EAST, J. H. AND E. D. GARDNER. "Oil Shale Mining, Rifle, Colorado, 1944–56." *U.S. Bureau of Mines Bulletin, 611.* Washington: Government Printing Office, 1964.

GOVIER, G. W. "Alberta's Bitumen Sands in the Energy Supply Picture." Presented at symposium: *Oil Sands—Fuel of the Future, Calgary, Alberta,* Sept., 1973. Reported in *Daily Oil Bulletin,* Calgary, Alberta, Sept. 10, 1973.

NATIONAL PETROLEUM COUNCIL. *U.S. Energy Outlook—Oil Shale Availability.* Washington: National Petroleum Council, 1973.

"Oil Shale: Prospects on the Upswing." *Science,* Vol. 190, no. 4321, Dec. 9, 1977.

"Oil Shale 1976/1977." *Shale Country,* Dec. 1976.

PETZRIC, P. A. "Oil Shale—An Ace in the Hole for National Security." *Shale Country,* Oct., 1975.

PHIZACKERLY, P. H., AND L. O. SCOTT. "Major Tar Sand Deposits of the World." In *Proceedings of the 7th World Petroleum Congress.* Mexico, 1967, Vol. 3, 551–571.

SCHRAMM, L. W. "Shale Oil." *Mineral Facts and Problems, 1975; Bureau of Mines Bulletin, 667.* Washington: Government Printing Office, 1975.

"Shale Closest among Liquid Fuels." *Oil and Gas Journal,* Jan. 17, 1977.

"Shale Oil Finally Rocking off Dead Center." *Oil and Gas Journal,* June 18, 1979.

U.S. DEPARTMENT OF THE INTERIOR. *Final Environmental Statement for the Oil Shale Leasing Program;* (Volume I of VI) *Regional Impacts of Oil Shale Development.* Washington: Government Printing Office, 1973.

————. *Final Environmental Statement for the Oil Shale Leasing Program;* (Volume II of VI) *Energy Alternatives.* Washington: Government Printing Office, 1973.

————. *Final Environmental Statement for the Oil Shale Leasing Program;* (Volume III of VI) *Specific Impacts of Prototype Oil Shale Development.* Washington: Government Printing Office, 1973.

WELLES, CHRISTOPHER. *The Elusive Bonanza—The Story of Oil Shale—America's Richest and Most Neglected National Resource.* New York: E. P. Dutton, 1970.

WHITECOMBE, JOHN A. *et al. Shale Oil Production Costs and the Need for Incentives for Pioneer Plant Construction.* Colorado Springs: Symposium on the Commercialization of Synthetic Fuels, February 1–3, 1976.

WORLD ENERGY CONFERENCE. *Survey of Energy Resources.* New York: U.S. National Committee—World Energy Conference, 1974.

4 Nuclear Fuels

BOGART, S. LOCKE. "Fusion Power and the Potential Lithium Requirement." *Lithium Resources and Requirements by the Year 2000; U.S. Geological Survey Professional Paper 1005.* Washington: Government Printing Office, 1976.

CHILENSKAS, A. A. *et al.* "Lithium Requirements for High-Energy Lithium-Aluminum/Iron Sulfide Batteries for Load-Leveling and Electric Vehicle Applications." *Lithium Resources and Requirements by the Year 2000. U.S. Geological Survey Professional Paper 1005.* Washington: Government Printing Office, 1976.

FINCH, WARREN I. *et al.* "Uranium." *United States Mineral Resources. U.S. Geological Professional Paper 820.* Washington: Government Printing Office, 1973.

FOWLER, JOHN M. *Energy and the Environment.* New York: McGraw-Hill, 1975.

EXXON CORPORATION. *Energy Outlook, 1977–1990.* Exxon Corporation, U.S.A., 1977.

HUBBERT, M. KING. "The Energy Resources of the Earth." *Scientific American,* Vol. 224, no. 3, Sept., 1971.

HULME, H. R. *Nuclear Fusion.* London and Winchester: Wykeham Publications, 1969.

LAPP, RALPH E. "We May Find Ourselves Short of Uranium." *Fortune,* Oct. 1975.

MCKELVEY, V. E. "Mineral Potential of the Submerged Part of the U.S." *Ocean Industry,* Vol. 3, no. 9, 1968.

MILLER, G. TYLER, JR. *Living in the Environment—Concepts, Problems and Alternatives.* Belmont: Wadsworth, 1975.

NUCLEAR ENERGY POLICY STUDY GROUP. *Nuclear Power Issues and Choices.* Cambridge: Ballinger, 1977.

SINGLETON, RICHARD H. AND HIRAM WOOD. "Lithium." *Mineral Facts and Problems. Bureau of Mines Bulletin 667.* Washington: Government Printing Office, 1975.

SKINNER, BRIAN J. *Earth Resources.* Englewood Cliffs, N.J.: Prentice-Hall, 1969.

SONDERMAYER, ROMAN V. "Thorium." *Mineral Facts and Problems, 1975. Bureau of Mines Bulletin 667.* Washington: Government Printing Office, 1975.

SORENSON, BENT. "Wind Energy." *Bulletin of the Atomic Scientists,* Sept., 1976.

STAATZ, MORTIMER, *et al.* "Thorium." *United States Mineral Resources. U.S. Geological Professional Paper 820.* Washington: Government Printing Office, 1973.

TURK, AMOS, *et al. Environmental Science.* Philadelphia: W. B. Saunders, 1974.

U.S. ATOMIC ENERGY COMMISSION. *Nuclear Fuel Reserves and Requirements,* WASH-1243. Washington: Government Printing Office, 1973.

————. *Nuclear Fuel Supply,* WASH-1242. Washington: Government Printing Office, 1974

U.S. BUREAU OF MINES. *Mineral Commodity Summaries, 1978.* Washington: Government Printing Office, 1978.

U.S. DEPARTMENT OF ENERGY. *Statistical Data of the Uranium Industry, January 1, 1979.* Grand Junction, Colorado, 1979.

U.S. DEPARTMENT OF INTERIOR. *Energy through the Year 2000.* Washington: Government Printing Office, 1975.

U.S. ENERGY RESEARCH AND DEVELOPMENT ADMINISTRATION. *National Uranium Resource Evaluation Preliminary Report.* Grand Junction: E.R.D.A., 1976.

————. *Statistical Data of the Uranium Industry.* Grand Junction: E.R.D.A., 1976.

————. *Nuclear Power Reactors in the United States.* Oak Ridge: E.R.D.A. Technical Information Center, 1976.

U.S. NUCLEAR REGULATORY COMMISSION. *Program Summary Report.* NUREG-0380, Vol. 3, no. 2, Feb. 16, 1979.

VINE, JAMES D. "The Lithium Resource Enigma." *Lithium Resources and Requirements by the Year 2000. U.S. Geological Survey Professional Paper 1005.* Washington: Government Printing Office, 1975.

WOODMANSE, WALTER C. "Uranium." *Mineral Facts and Problems, 1975. Bureau of Mines Bulletin 667.* Washington: Government Printing Office, 1975.

5 Geothermal Heat

COOK, EARL. *Man, Energy, Society.* San Francisco: W. H. Freeman, 1976.

CUMMINGS, RONALD G., *et al.* "Mining Earth's Heat: Hot Dry Rock Geothermal Energy." *Technology Review,* Vol. 81, no. 4, Feb. 1979, 58–78.

DICKINSON, GEORGE. "Geological Aspects of Abnormal Reservoir Pressures in Gulf Coast Louisiana." In *Bulletin of American Association of Petroleum Geologists,* Vol. 37, no. 2, 1953, 410–432.

DIMENT, W. H., *et al.* "Temperatures and Heat Contents Based on Conductive Transport of Heat." In White, D. E., and D. L. Williams, *U.S. Geological Survey Circular 726,* 1975.

LOMBARD, DAVID (Physicist with U.S. Department of Energy). Quoted in *Philadelphia Inquirer,* Sept. 3, 1978.

NATHENSON, M., AND L. J. P. MUFFLER. "Geothermal Resources in Hydrothermal Convection Systems and Conduction-Dominated Areas." In White, D. E., and D. L. Williams. *U.S. Geological Survey Circular 726,* 1975.

NUCLEAR ENERGY POLICY STUDY GROUP. *Nuclear Power, Issues and Choices.* Cambridge: Ballinger, 1977. (The Mitre Study).

PAPADOPULOS, S. S., *et al.* "Assessment of Onshore Geopressured-Geothermal Resources in the Northern Gulf of Mexico Basin." In White, D. E., and D. L. Williams. *U.S. Geological Survey Circular 726,* 1975.

POLLACK, HENRY N., AND D. S. CHAPMAN. "The Flow of Heat from the Earth's Interior." *Scientific American,* Vol. 237, no. 2, Aug., 1977, 60–76.

RENNER, J. L., D. E. WHITE, AND D. L. WILLIAMS. "Hydrothermal Convection Systems." In White, D. E., and D. L. Williams. *U.S. Geological Survey Circular 726,* 1975.

SMITH, R. L., AND H. R. SHAW. "Igneous-Related Geothermal Systems." In White, D. E., and D. L. Williams. *U.S. Geological Survey Circular 726,* 1975.

SONDERMAYER, ROMAN V. "Thorium." In *Mineral Facts and Problems. Bureau of Mines Bulletin, 667,* Washington, Government Printing Office, 1975.

U.S. DEPARTMENT OF ENERGY. Interagency Geothermal Coordinating Council. *Geothermal Energy Research Development and Demonstration Program.* Third Annual Report. DOE/ET-0090, Mar., 1979.

U.S. DEPARTMENT OF THE INTERIOR. *Energy through the Year 2000.* (Revised) Bureau of Mines, Dec. 1975.

U.S. GEOLOGICAL SURVEY. *Circular 726; Geothermal Resources of the United States—1975.* Arlington, Va.: U.S.G.S. Branch of Distribution, 1976.

————. *Circular 790; Assessment of Geothermal Resources of the United States—1978,* Arlington, Va.: U.S.G.S. Branch of Distribution, 1979.

WHITE, DONALD E. "Characteristics of Geothermal Resources and Problems of Utilization." In Kruger, P., and C. Ott, eds. *Geothermal Energy: Resources, Production, Stimulation.* Stanford, Cal.: Stanford University Press, 1973.

———, AND D. L. WILLIAMS. *Assessment of Geothermal Resources of the United States—1975. U.S. Geological Survey Circular 726,* 1975.

———. Personal communication Jan. 5, 1978.

WILSON, HOWARD M. "Pace of Action Quickens in U.S. Geothermal Areas." *Oil and Gas Journal,* Dec. 18, 1978, 15–19.

6 Solar Radiation

BECKMAN, W. A., of University of Wisconsin Solar Energy Lab. Personal communication, July, 1978.

BENNET, IVEN. "Monthly Maps of Mean Daily Insolation for the United States." In *Solar Energy,* Vol. 9, no. 3, July–Sept., 1965, 145–152.

BEZDEK, ROGER H. *An Analysis of the Current Economic Feasibility of Solar Water and Space Heating,* Washington D.C.: U.S. Dept. of Energy, Division of Solar Application, Jan. 1978.

———, ALAN S. HIRSHBERG, AND WILLIAM H. BABCOCK. "Economic Feasibility of Solar Water and Space Heating." *Science,* Vol. 203, Mar. 23, 1979, 1214–1220.

BOES, ELDON C., et al. *Distribution of Direct and Total Solar Radiation Availabilities for the USA.* Albuquerque, N.M.: Sandia Laboratories, Aug. 1976.

BOOZ, ALLEN, AND HAMILTON INC. *HUD Residential Solar Economic Performance Model.* Bethesda, Md., 1977.

CLARKE, JOHN, of the Phila. Office National Solar Heating and Cooling Information Center. Personal Communication, June, 1978.

COUNCIL ON ENVIRONMENTAL QUALITY (CEQ). *Solar Energy: Progress and Promise.* Washington, Government Printing Office, 1978.

CURTO, PAUL A., AND Z. D. NIKODEM. *Solar Thermal Repowering.* McLean, Va.: Metrek Division of the Mitre Corp., May, 1978.

DUFFIE, J. A., W. A. BECKMAN, AND M. J. BRANDEMUEHL. "A Parametric Study of Critical Fuel Costs for Solar Heating in North America." Paper obtained from authors, Solar Energy Lab, University of Wisconsin, 1978.

DUFFIE, J. A., W. A. BECKMAN, AND J. R. DEKKER. "Solar Heating in the United States." Paper presented at winter Annual Meeting Amer. Soc. Mechanical Engineers, New York City, Dec. 5, 1976.

DURRENBERGER, ROBERT W., AND ANTHONY J. BRAZEL. "Need for a Better Solar Radiation Data Base." *Science,* Vol. 193, Sept. 17, 1976, 1154–1155.

GILMORE, C. P. "Solar Assisted Heat Pumps." *Popular Science,* May, 1978, 86–90.

KLEIN, S. A., W. A. BECKMAN, AND J. A. DUFFIE. "A Design Procedure for Solar Heating Systems." Presented at meeting of International Solar Energy Society, July, 1975.

LANDSBERG, HELMUT E. "Solar Radiation at the Earth's Surface." In *Solar Energy,* Vol. 5, no. 3, 1961, 95–98.

LANTZ, LOREN, of Environmental Engineering Co., Ft. Collins, Colorado, Personal communication, July, 1978.

LIU, Y. B. H., AND R. C. JORDAN. "Availability of Solar Energy for Flat Plate Collectors." *Low Temperature Engineering Application of Solar Energy.* New York: ASHRAE, 1967, 1–18.

LÖF, G. O. G., AND R. A. TYBOUT. "Solar Energy Heating." *Natural Resources Journal,* Vol. 10, 1970.

———. "The Design and Cost of Optimized Systems for Residential Heating and Cooling by Solar Energy." *Solar Energy,* Vol. 16, no. 9, 1974.

MEINEL, ADEN B. AND MARJORIE P. MEINEL. "Is It Time for a New Look at Solar Energy?" *Bulletin of Atomic Scientists,* Oct., 1971.

———. "Physics Looks at Solar Energy." *Physics Today,* Vol. 25, no. 2, 1972.

METREK. *Solar Energy: A Comparative Analysis to the Year 2000.* Prepared for E.R.D.A. under contract No. E-(4818)-2322, MITRE Technical Report MTR-7579, The Mitre Corporation, Metrek Division, McLean, Va., 1978.

NATIONAL OCEANOGRAPHIC AND ATMOSPHERIC ADMINISTRATION, ENVIRONMENTAL DATA SERVICE. *Annual Summary of Local Climatological Data.* National Climatic Center, Asheville, N. Carolina.

RIDENOUR, STEVE. "Solar Water Heaters." In Eccli, Eugene, ed. *Low-Cost Energy-Efficient Shelter.* Emmaus, Pa.: Rodale Press, 1976.

SCHULZE, WILLIAM D., B. SHAUL, J. D. BALCOMB, et al. *The Economics of Solar Home Heating.* Report prepared for the Joint Economic Committee of the U.S. Congress, Jan., 1977.

"Solar-Heating Your House—Would It Pay?" *Changing Times: The Kiplinger Magazine,* April, 1978.

"Solarville, Ariz.: Down to $7 a Watt." *New York Times,* June 24, 1979.

TREWARTHA, GLENN. *An Introduction to Climate.* N.Y.: McGraw-Hill, 1968, 23.

TRW SYSTEMS GROUP. *Solar Heating and Cooling of Buildings (Phase Zero).* Vol. 1. Executive summary, prepared for NSF-RANN, Washington, D.C., May 31, 1974.

U.S. DEPARTMENT OF COMMERCE. *Climates of the United States.* Washington, D.C., 1973.

U.S. DEPARTMENT OF ENERGY. *All-Electric Homes: Annual Bills—Jan. 1, 1977.* Energy Data Report.

———. *Availability of Solar Energy Reports from the National Solar Data Program.* SOLAR/0020-79/37, Washington, D.C., May, 1979.

———. *An Economic Assessment of Solar Energy Systems in the Commercial Demonstration Program.* SOLAR/0823-79/01. Prepared by P.R.C. Energy Analysis Company, 1979.

———. *Results of Thermal Performance Analysis of Passive Solar Space Heating Systems in the National Solar Data Network.* SOLAR/0022-79/39, Washington, D.C., July, 1979.

———. *Solar Energy for Agricultural and Industrial Process Heat. Program Summary.* DOE/CS-0053, Sept., 1978.

———. *Thermal Performance of Space Cooling Energy Systems in the National Solar Data Network.* SOLAR/0023-79/40, Washington, D.C., July, 1979.

———. *Thermal Performance of Hot Water Systems in the National Solar Data Network.* SOLAR/2024-79/41, Washington, D.C., July, 1979.

———. *Thermal Performance Analysis of Space Heating Systems in the National Solar Data Network.* SOLAR/0025-79/42.

U.S. DEPARTMENT OF HOUSING AND URBAN DEVELOPMENT. *Regional Guidelines for Building Passive Energy Conserving Homes.* HUD-PDR-355, Washington, D.C., Nov., 1978.

———. *Solar Heating and Cooling Demonstration Program: A Descriptive Summary of HUD Solar Residential Demonstrations, Cycle 1.* Washington, D.C., July, 1976.

———. *Solar Heating and Cooling Demonstration Program: A*

Descriptive Summary of HUD Cycle 2 Solar Residential Projects. Washington, D.C., Apr., 1977.

————. *Solar Heating and Cooling Demonstration Program: A Descriptive Summary of Cycle 3 Solar Residential Projects.* Washington, D.C., Nov., 1977.

————. *Solar Heating and Cooling Demonstration Program: A Descriptive Summary of Cycle 4 and 4A Solar Residential Projects.* Washington, D.C., July, 1979.

U.S. DEPARTMENT OF LABOR. BUREAU OF LABOR STATISTICS. *Retail Prices and Indexes of Fuels and Utilities; Residential Usage.* Mar., 1978.

U.S. ENERGY RESEARCH AND DEVELOPMENT ADMINISTRATION. *Central Receiver Solar Thermal Power System.* U.S. Government Printing Office, 1976. 1976-0-220-067.

————. Division of Solar Energy. *An Economic Analysis of Solar Water and Space Heating.* Nov., 1976.

————. Division of Solar Energy. *An Introduction to SOL-COST* (2nd ed.), June, 1977. And: *SOLCOST Solar Hot Water Handbook.*

U.S. GEOLOGICAL SURVEY. *National Atlas of the United States of America.* Washington, D.C., 1970.

7 Wind Power

BETZ, A. "Windmills in the Light of Modern Research." *Die Naturwissenschaften*, Vol. 15, no. 46, Nov. 18, 1927.

CRUTCHER, H. L. *Upper Wind Statistics Charts of the Northern Hemisphere.* NAVAER 50 1C-535, Vol. 1, Chief of Naval Operations, Washington, D.C., Aug. 1959.

ELDRIDGE, FRANK R. *Wind Machines.* Report prepared for National Science Foundation and RANN, with cooperation of Energy Research and Development Administration, Oct., 1975.

ELLIOT, DENNIS. *Synthesis of National Wind Energy Assessments.* Batelle, Pacific Northwest Laboratories, Richland, Washington, July, 1977.

GENERAL ELECTRIC COMPANY. *Wind Energy Mission Analysis, Executive Summary.* Contract No. EY-76-C-02-2578. Philadelphia, Pa.: G.E. Space Division, Feb. 18, 1977.

————. *Wind Energy Mission Analysis, Final Report.* Contract No. EY-76-C-02-2578. Philadelphia, Pa.: G.E. Space Division, Feb. 18, 1977.

HERONEMUS, W. E. "Power from Offshore Winds." Proceedings of the 8th Annual Marine Technology Society Conference. Washington, D.C., 1972.

————. "The United States Energy Crisis: Some Proposed Gentle Solutions." Paper at joint meeting of American Society of Mechanical Engineers, and Institute of Electrical and Electronics Engineers, W. Springfield, Mass., Jan. 12, 1972.

LOCKHEED-CALIFORNIA CO. *Wind Energy Mission Analysis, Final Report.* Burbank, Calif.: Lockheed-California Co., 1976.

METREK. *Solar Energy: A Comparative Analysis to the Year 2020.* Prepared for E.R.D.A. under Contract No. E-(4918)-2322, Technical Report MTR-7579. McLean, Va.: The Mitre Corporation, Metrek Division, Aug., 1978.

REED, J. W. *Wind Power Climatology of the United States.* SAND 74-0348. Albuquerque, N. Mex.: SANDIA Laboratories, June, 1975.

————, R. C. MAYDEW, AND B. F. BLACKWELL. *Wind Energy Potential in New Mexico.* SAND 74-0071. Albuquerque, N. Mex.: SANDIA Laboratories, July 1974.

TREWARTHA, GLENN T. *An Introduction to Climate*, 4th Ed. N.Y.: McGraw-Hill, 1968.

U.S. DEPARTMENT OF THE INTERIOR. BUREAU OF MINES. *United States Energy Through the Year 2000.* 1975.

U.S. ENERGY RESEARCH AND DEVELOPMENT ADMINISTRATION (E.R.D.A.). Wind Systems Branch Division of Solar Energy. *Federal Wind Energy Program.* Washington, D.C., Jan. 1, 1977.

————. *Federal Wind Energy Program: Summary Report.* Washington, D.C.: Jan. 1, 1977. *Federal Wind Energy Program: Summary Report,* Jan. 1, 1978.

U.S. GEOLOGICAL SURVEY. *The National Atlas of the United States of America.* Washington, D.C., 1970.

8 Hydroelectric Power

COOK, EARL. *Energy: The Ultimate Resource. Resource Paper No. 77-4.* Washington: Association of American Geographers, 1977.

FEDERAL ENERGY REGULATORY COMMISSION. Data on conventional and reversible hydroelectric potential (Tables III and IV) supplied by Neal C. Jennings, Director, Division of River Basins, 1978.

FEDERAL POWER COMMISSION. *Hydroelectric Power Resources of the United States, 1972.* Washington: Government Printing Office, 1972.

————. *Hydroelectric Power Resources of the United States, 1976.* Washington: Government Printing Office, 1976.

HEALY, TIMOTHY J. *Energy, Electric Power and Man.* San Francisco: Boyd and Fraser, 1974.

STEINHART, CAROL AND JOHN STEINHART. *Energy Sources, Use and Role in Human Affairs.* North Scituate: Duxbury Press, 1974.

U.S. ARMY CORPS OF ENGINEERS. *Estimate of National Hydroelectric Power Potential at Existing Dams.* Washington: U.S. Army Corps of Engineers, July 20, 1977.

————. *National Hydroelectric Power Resources Study— Preliminary Inventory of Hydropower Resources,* 6 Volumes. Washington: U.S. Army Corps of Engineers, July, 1979.

U.S. DEPARTMENT OF INTERIOR. *Energy Perspectives 2.* Washington: Government Printing Office, 1976.

————. *United States Energy Through the Year 2000.* Rev. ed. Washington: Government Printing Office, 1975.

9 Ocean Thermal Gradients

COHEN, ROBERT. "An Overview of the U.S. OTEC Development Program." Paper delivered to American Society of Mechanical Engineers, Engineers Energy Technology Conference, Houston Tex., Nov. 6–9, 1978.

GENERAL ELECTRIC TEMPO. *OTEC Mission Analysis Study. Phase I* and *Phase II.* Washington, D.C., Nov., 1976 and Mar., 1978.

HAYES, DENNIS. *Rays of Hope: the Transition to a Post-Petroleum World.* New York: Norton and Co., 1977.

METREK. *Solar Energy: A Comparative Analysis to the Year 2020.* Prepared under E.R.D.A. Contract No. E-(4918)-2322. Technical Report MTR-7579. McLean, Va.: The Mitre Corporation, Metrek Division, Aug., 1978.

OCEAN DATA SYSTEMS INC. "Large Scale Distribution of OTEC Thermal Resource." Maps prepared for Division of Solar

Technology, U.S. Department of Energy under Contract No. ET-78-C-01-2898, 1978.

U.S. GEOLOGICAL SURVEY. *National Atlas of the United States*. Washington, D.C., 1970.

10 Biomass

ERNEST, KENT, ET AL, *Mission Analysis for the Federal Fuels from Biomass Program, Volume III: Feedstock Availability, Final Report*, Menlo Park, Calif., Stanford Research Institute, January, 1979 (Available from National Technical Information Service, Springfield, Virginia).

FOWLER, JOHN M., AND KATHRYN FOWLER, *Fuels From Plants: Bioconversion*, Factsheet Number 1, National Science Teachers Association, Oak Ridge, ERDA Technical Information Center.

HAYES, DENNIS, *Energy: The Solar Prospect*, Worldwatch Paper 11, The Worldwatch Institute, Washington, D.C., March, 1977.

JOHNSON, JACK D., AND C. WILEY HINMAN, "Oils and Rubber from Arid Land Plants", *Science*, Vol. 208, No. 4443 (May 2, 1980), pp. 460–464.

JONES, JERRY L., ET AL, *Mission Analysis for the Federal Fuels from Biomass Program, Vol. V, Biochemical Conversion of Biomass to Fuels and Chemicals*, Menlo Park, Calif., Stanford Research Institute, Dec., 1978 (Available from National Technical Information Service, Springfield, Virginia).

METREK, 1978, *Solar Energy: A Comparative Analysis to the Year 2020*, prepared for ERDA under contract No. E-(4918)-2322, Technical Report MTR-7579, The Mitre Corporation, Metrek Division, McLean, Virginia, Aug., 1978.

SCHOOLEY, FRED A., ET AL, *Mission Analysis for the Federal Fuels from Biomass Program, Volume I: Summary and Conclusions, Final Report*, Menlo Park, Calif., Stanford Research Institute, March, 1979 (Available from National Technical Information Service, Springfield, Virginia).

U.S. DEPARTMENT OF ENERGY, *The Report of the Alcohol Fuels Policy Review*, DOE/PE-0012, Washington, D.C., U.S. Government Printing Office, June, 1979.

U.S. FOREST SERVICE, *Wood for Energy: A Renewable and Expandable Resource*, Washington, D.C., November 23, 1977.

11 Conclusion

BUTLER, ARTHUR P. "Uranium Reserves and Progress in Exploration and Development." Paper presented to Department of Interior Coal Industry Conference, Washington, D.C., Aug. 16, 1967.

LOVINS, AMORY B. "Energy Strategy: The Industry, Small Business, and Public Stakes." In Ruedilisi, L. C., and Morris W. Firebaugh, eds. *Perspectives on Energy*. New York: Oxford University Press, 1978.

METREK. *Solar Energy: A Comparative Analysis to the Year 2020*. Mitre Technical Report MTR-75-79. Prepared for ERDA under Contract No. E-(4918)-2322 by the Mitre Corp., METREK division, McLean, Virginia.

STEINHART, JOHN S., *et al.* "A Low Energy Scenario for the United States: 1975–2050." In Ruedilisi, L. C., and Morris W. Firebaugh, ed. *Perspectives on Energy*. New York: Oxford University Press, 1978.

U.S. ENERGY RESEARCH AND DEVELOPMENT ADMINISTRATION. Division of Solar Energy. *Solar Energy in America's Future: a Preliminary Assessment (Executive Summary)*. Washington, D.C.: Stanford Research Institute, Mar., 1977. Superintendent of Documents Stock No. 060-000-00051-4.

U.S. WATER RESOURCES COUNCIL. *The Nation's Water Resources, 1975–2000. Vol. 1: Summary*. The second national assessment, Government Printing Office 052-045-00051-7, Dec., 1978.

Appendix

Energy Equivalents

ONE QUAD IS EQUIVALENT TO THE FOLLOWING:

Fundamentally
1×10^{15} BTUs
252×10^{15} calories or 252×10^{12} K calories

In Fossil Fuels
180 million barrels of crude oil
0.98 trillion cu. ft. of natural gas
37.88 million tons of anthracite coal
38.46 million tons of bituminous coal
50.00 million tons of subbituminous coal
71.43 million tons of lignite coal

In Nuclear Fuels
2500 tons of U_3O_8 if only U_{235} is used
17.8 tons of U_3O_8 if all U_{238} is transmuted to PU_{239} and U_{235} is used as well
15.87 tons of ThO_2 if all thorium is transmuted to U_{233}

In Electrical Output
2.93×10^{11} Kilowatt-hours electric

In Biomass
58.43×10^6 dry tons of wood

ENERGY VALUES OF SINGLE FUEL UNITS

Uranium Oxide	
U_{235} content only	1 ton equiv. to 4.0×10^{11} BTU
U_{235} and U_{238} content	1 ton equiv. to 5.618×10^{13} BTU
Thorium Oxide (as U_{233})	1 ton equiv. to 6.301×10^{13} BTU
Crude Oil	1 barrel equiv. to 5.56×10^6 BTU
Natural Gas	1 cu. ft. equiv. to 1.035×10^3 BTU
Coal	
Anthracite	1 ton equiv. to 26.4×10^6 BTU
Bituminous	1 ton equiv. to 26.0×10^6 BTU
Subbituminous	1 ton equiv. to 20.0×10^6 BTU
Lignite	1 ton equiv. to 14.0×10^6 BTU
Wood	1 dry ton equiv. to 17.1×10^6 BTU

Conversion Factors

To Convert From:	Into These Units:	Multiply by This Factor	To Convert From:	Into These Units:	Multiply by This Factor
Length			**Weight**		
Feet	Meters	0.305	Metric tons	Short tons	1.1025
Miles	Meters	1,609	Kilograms	Pounds	2.20
Microns	Meters	1×10^{-6}	Kilograms	Tons	0.0011
Area			**Radiation**		
Acres	Square feet	43,560	BTU/square feet	Langleys	0.271
Acres	Square meters	4,047	Langleys	BTU/square feet	3.69
Square centimeters	Square feet	0.00108	Langleys/minute	Watts/square centimeters	0.0698
Square centimeters	Square inches	0.155	**Energy and Power**		
Square feet	Square inches	144	BTUs	Calories	252
Square feet	Square meters	0.0929	Kilogram calorie (food Calorie)	Calories	1,000
Square inches	Square centimeters	6.45			
Square meters	Square feet	10.8	BTU	Joules	1,055
Square meters	Square miles	3.68×10^{-7}	BTU	Kilowatt-hour	2.93×10^{-4}
Square miles	Acres	640	BTU	Megawatt-year	3.34×10^{-11}
Square miles	Square feet	2.79×10^{7}	Calories	BTU	3.97×10^{-3}
Square miles	Square meters	2.59×10^{6}	Calories	Foot-pounds	3.09
Volume			Calories	Joules	4.18
Barrels	U.S. gallons	42	Calories/minute	Watts	0.0698
U.S. gallons	Imperial gallons	0.8326	Gigajoules	BTU	0.95×10^{15}
Imperial gallons	U.S. gallons	1.201	Kilowatts	Watts	1×10^{3}
U.S. gallons	Cubic inches	231	Megawatts	Watts	1×10^{6}
Imperial gallons	Cubic inches	277.42	Kilowatt-hour	BTU	3,413
U.S. gallons	Liters	3.79	Kilowatts	Horsepower	1.34
Weight			Kilowatt-hour	Foot-pounds	2.66×10^{6}
Short tons	Pounds	2,000	Watt-hours	Joules	3,600
Short tons	Kilograms	907	Therm	BTU	1×10^{5}
Short tons	Metric tons	0.907	Joule	Therms	9.4782×10^{-7}

Geologic Time Scale

Era	Period	Epoch	Millions of Years Before Present
Cenozoic	Quarternary	Recent Pleistocene	2.5–Present
	Tertiary	Pliocene Miocene Oligocene Eocene Paleocene	63–2.5
Mesozoic	Cretaceous		135–63
	Jurassic		181–135
	Triassic		230–181
Paleozoic	Permian		280–230
	Carboniferous	Pennsylvanian Mississippian	345–280
	Devonian		405–345
	Silurian		425–405
	Ordovician		500–425
	Cambrian		600–500
Precambrian			3,500–600

Index

Source Notes

I-1: Schurr, Sam H., *et al,* June 1979; I-3: Areas of crystalline rocks generalized from USGS *National Atlas of the U.S.A.,* p. 75; cross-section taken from Lobeck, Armin; 1-1: Federal Energy Administration, 1976; 1-2: Federal Energy Administration, 1976, and D.O.E., Mar. 30 and April 20, 1979; 1-5: Bureau of Mines, I.C., 8693, 1975; 1-6: American Petroleum Institute; 1-8: Bureau of Mines, I.C. 8678, 1975, p. 3; 1-10: *World Coal,* Nov. 1975, p. 37; 1-11: Federal Energy Administration, 1976, and *Coal International,* Vol. 1, No. 1, May 1979; 1-13 Averitt, 1973, p. 137; 1-15: Bureau of Mines, I.C. 8655, 8678, and Averitt, 1975; 1-16: Bureau of Mines, I.C. 8655, 8678; 1-17: Bureau of Mines I.C., 8655, 8678; 1-18: Bureau of Mines, I.C. 8655, 8678; 1-19: Bureau of Mines, I.C. 8655, 8678; 1-20: Bureau of Mines, I.C. 8655, 8678; 1-21: Bureau of Mines, I.C. 8655, 8678; 1-23: Bureau of Mines I.C. 8680, 8693; 1-30: Bureau of Mines, I.C. 8678, 8655, and Averitt, U.S.G.S. Bulletin 1412, 1975; 1-33: U.S. D.O.E. *Energy Data Reports,* April 20, 1979, March 30, 1979; 1-37: Sabotko & Co., 1973; 1-38: A.B.C.: USGS, 1976; 1-39: Bureau of Mines, Mineral Industry Survey, *Bituminous Coal and Lignite Distribution,* Calendar Year 1976; 1-40: ERDA, *Energy from Coal,* 1976; 1-41: D.O.E., *Fossil Energy Research & Development Program,* DOE/ET-0013, March 1976; 1-42: D.O.E., Mar. 1978; 1-43 Coal areas from U.S.G.S. *The National Atlas of the U.S.A.,* 1970; Resource estimates from D.O.E., Mar. 1978; 2-1: Taken largely from Cook, Earl, p. 93; 2-2: American Petroleum Institute; 2-5: 1890 to 1970, Tiratsoo, 1973; 1970 to 1977 *Oil & Gas Journal,* Feb. 27, 1978, p. 179; 2-6: *Oil & Gas Journal,* Dec. 26, 1977, Dec. 25, 1978; 1890 to 1970 Tiratsoo, 1973; 1970 to 1977, *Oil & Gas Journal,* Feb. 27, 1978, p. 179; 2-8: History to 1970, Tiratsoo, 1973, p. 5; Revision to 1977, *Oil & Gas Journal,* Feb. 27, 1978, p. 179; 2-9: American Petroleum Institute; 2-11—2-15: USGS Circ. 725; 2-16: USGS I.C. 725, p. 44; 2-18: For 15 regions, USGS 725; For 3 locations on Atlantic Shelf, U.S. Department of Interiors, Bureau of Land Management Final Environmental Statements; 2-11: U.S. Department of Interior, Sale No. 40; 2-21—2-24: American Petroleum Institute, 1979; 2-25: 1969-1978, American Petroleum Institute Annual Reports; 1900-1968, Hubbert, King, 1969; 2-27: Derived from American Petroleum Institute, Vol. 33, June 1979; 2-31: American Petroleum Institute, Vol. 33, June 1979; 2-32: Reported in *Oil & Gas Journal,* July 25, 1977; 2-33: ERDA, Dec., 1977, p. 221; 2-35: 1875 to 1968: Hubbert, K., 1969, p. 164; 1969-1978, American Petroleum Institute Annual Reports; 2-36: American Petroleum Institute, June 1979; 2-37: USGS, 1976; 2-38: U.S. D.O.E., Petroleum Statement Annual 1978; 2-40: Derived from tabulations by U.S.B.M. reported in A.P.I. Basic Petroleum Data, 1977, and D.O.E., Feb. 8, 1979, for year 1977; 2-41: *Oil & Gas Journal,* Dec. 25, 1978; 2-42—2-51: USGS Circ. 725; 2-52—2-55: American Gas Association, June, 1979; 2-60: American Gas Association, June, 1979; 2-61: U.S. D.O.E., *National Gas Annual,* Feb. 7, 1978; 2-62: *National Gas Annual,* Feb., 1978; 2-63: For pattern of occurrence, American Gas Association, 1974; 2-66: Office of Technology Assessment, Sept. 1, 1977, and *New York Times,* Dec. 2, 1977; 2-68: Compiled from information in *Canada Today/D'Audjoud'hui,* Vol. 8, No. 5, 1977; 3-1: Culbertson and Pitman, 1973; 3-3: Source for pattern of occurrence USGS Professional Paper, 820; Sources for resource magnitudes USGS Circular 523; 3-4: Donnelly, 1976; 4-4: U.S. Department of Energy, *Statistical Data of the Uranium Industry,* 1979, Grand Junction, Colorado; 4-5—4-15: U.S. Department of Energy, January 1, 1979; 4-18: Atomic Energy Commission, 1973; 4-19: Bureau of Mines, 1975, and ERDA, 1976;

4-20: ERDA, 1976; 4-21: D.O.E., 1977; 4-22: D.O.E., 1979; 4-23: *Energy in Focus* (Washington, D.C., Federal Energy Administration, 1977; 4-24: U.S. Nuclear Regulatory Commission, 1/31/79; 4-25: Nuclear Energy Policy Study Group, 1977; 4-26: Atomic Energy Commission, 1973; 4-27: Data from Atomic Energy Commission, 1973, and Department of Energy, 1979; 4-28: Sondermayer, 1975; 4-29: Staatz, 1973, and Sondermayer, 1975; 4-30: Sondermayer, 1975; 4-31: Adapted from illustration in Miller, 1975; 4-32: Singleton and Wood, 1975, and Vine, 1976; 4-33: Information derived from Vine, 1976; 4-34: Vine, 1976; 5-1: Heat flow areas, Pollack & Chapman, p. 74; Plate edges and geothermal systems from White, Donald F., 1973; 5-2: Diment, *et al,* Fig. 11; 5-3: Derived from Diment, *et al,* 1975, Fig. 12; 5-7: Derived from information in White and Williams (Circ. 726) and Dickinson, 1953, Figs. 9-11; 5-8: Modified from chart in USGS Circ. 790, p. 5; 5-10: USGS Circ. 790; 5-12: USGS Circ. 790; 5-13: USGS, Circ. 790, Map 1; 5-14: USGS Circ. 790; 5-15: USGS Circ. 790; 5-16: USGS Circ. 790, p. 135; 5-18: USGS Circ. 790; 5-19: USGS Circ. 790; 5-20: USGS Circ. 790, p. 146; 5-27: Interagency Coordinating Council, Mar. 1979, pp. 45-58; 6-2: USGS, *National Atlas of the U.S.A.,* 1970, p. 93; 6-3: USGS *National Atlas of the U.S.A.,* 1970, p. 93; 6-5: Deduced from Ridenour, p. 321; 6-6: *Changing Times,* April, 1978; 6-7: Department of Commerce, *Climates of the U.S.,* p. 68, Period of record 1931-1960; 6-8: Department of Commerce, *Climates of the U.S.,* Period of record 1931-60, p. 69; 6-9: TRW Systems Group, p. 2-2. Period of record not stated; 6-10: TRW Systems Group, p. 2-3. Period of record not stated; 6-14: Duffie, Beckman, and Dekker, 1976, Fig. 2; 6-15: Duffie, Beckman, and Dekker, 1976, Fig. 3; 6-19: Duffie, Beckman, and Dekker, 1976, Fig. 5; 6-21: Duffie, Beckman, and Dekker, Figs. 9, 10; 6-23: A. HUD, July, 1976; B. HUD, Apr. 1977; C. HUD, Nov. 1977; D. HUD, July, 1979; 6-26: U.S. D.O.E. Solar/0025-79/42, July 1979 (for active space heating); Solar/0022-79/39, July 1979 (for passive space heating); Solar/0023-79/40, July 1979 (for active cooling); Solar/0024-79/41, July 1979 (for domestic hot water); For catalog of monthly summary reports, see Solar/0020-79/37, May 1979; 6-26: D.O.E./CS-0053, Sept. 1978, p. 115; 6-27: DOE/CS-0053, Sept. 1978, pp. 113; 6-28: Values calculated fr. monthly means in Boes, *et al,* based on data 1958 to 1962 for 26 stations; 6-29: DOE/CS-0053, Sept. 1978, pp. 113; 6-30: METREK, Aug. 1978, p. 32; 6-31: METREK, Aug. 1978, p. 34-5; 6-32: ERDA, Central Receiver Solar Thermal System, 1976, p. 3; 7-1: Eldrige, p. 5; 7-2: Trewartha, G. *An Introduction to Climate;* 7-5: General Electric Final Report, p. 3-23; 7-6: General Electric Final Report, p. 3-22; 7-7: General Electric, Executive Summary, 1977, p. 7; 7-8: Reed, 1975, pp. 14-15; 7-9: Elliot, p. vi; 7-10: TRW Systems Group, p. 2-3; 7-13: Derived from data in Reed, 1975; 7-15—7-22: General Electric Final Report, 1977; 7-23: METREK, Aug. 1978, pp. 40-41; 7-24: ERDA, 1977; 8-1: U.S. Department of Interiors, 1976; 8-2—8-3: Federal Power Commission, 1976; 8-4: Data from Federal Energy Regulatory Commission, 1978; 8-9: Federal Power Commission, 1976; 8-10: U.S. Army Corps of Engineers, July, 1979; 8-12: U.S. Army Corps of Engineers, July, 1979; 8-13: USGS *National Atlas of the U.S.A.,* 1970, p. 81; 9-1: Cohen, p. 6; 9-2: Ocean Data Systems, Inc.; 9-3: Cohen, p. 11; 9-4: Federal Energy Regulatory Commission, 1978; 9-6: METREK, Aug. 1978, pp. 40-41; 10-3: Ernest, 1979, Appendix; 10-11: U.S. D.O.E., June 1979, p. 57; 11-10: ERDA, Mar. 1977, p. viii; 11-12: U.S. Water Resources Council, p. 57.